The Human Fossil Record

The Human Fossil Record

Series Editors

Jeffrey H. Schwartz
Department of Anthropology
University of Pittsburgh
Pittsburgh, Pennsylvania

Ian Tattersall
Department of Anthropology
American Museum of Natural History
New York, New York

Forthcoming volumes:

The Human Fossil Record, Volume Two: Craniodental Morphology of Genus *Homo* (Africa and Asia)

by Jeffrey H. Schwartz, Ian Tattersall

The Human Fossil Record, Volume Three: Craniodental Morphology of Early Hominids

by Jeffrey H. Schwartz, Ian Tattersall

The Human Fossil Record, Volume Four: Brain Endocasts

by Ralph L. Holloway, Michael S. Yuan, Douglas C. Broadfield

THE HUMAN FOSSIL RECORD

Volume One

Terminology and Craniodental Morphology of Genus *Homo* (Europe)

Jeffrey H. Schwartz
Department of Anthropology
University of Pittsburgh
Pittsburgh, Pennsylvania

Ian Tattersall
Department of Anthropology
American Museum of Natural History
New York, New York

A John Wiley & Sons, Inc., Publication

This book is printed on acid-free paper. ∞

Published simultaneously in Canada.

For ordering and customer service, call 1-800-CALL WILEY.

Library of Congress Cataloging-in-Publication Data:

The human fossil record.
p. ; cm.
Includes bibliographical references.
Contents: v. 1. Terminology and craniodental morphology of genus Homo (Europe)/
Jeffrey H. Schwartz, Ian Tattersall
ISBN 0-471-31927-9 (v. 1 : cloth : acid-free paper)
1. Fossil hominids. I. Schwartz, Jeffrey H. II. Tattersall, Ian.

GN282 .H83 2002
569.9—dc21

200102664

Printed in the United States of America.

10 9 8 7 6 5 4 3 2 1

For

Lynn Emanuel

and

Jeanne Kelly

CONTENTS

Preface to Volumes 1 and 2

These books began in an attempt to understand systematic diversity among later Pleistocene hominids, specifically the Neanderthals. Almost immediately, however, it became evident that it is impractical to limit systematic study to one specific group of hominid fossils or to one particular period of human evolution; reference to outgroups is invariably necessary. Yet we were unable to find any single source to which we could turn to provide material for the broader comparisons that it proved necessary to make. Of course, the literature of paleoanthropology is replete with descriptions of hominid fossils; but, as we soon realized, most of these descriptions of individual fossils or fossil assemblages are not amenable to direct or at least complete comparison with others. This is, ironically, precisely because the tradition in paleoanthropology has been to describe fossils not as isolated entities, but comparatively. For, although providing a sense of how one fossil differs from another in the group being compared, the comparative descriptive approach often makes it difficult or impossible for the reader to extract the morphological information necessary for making comparisons with fossils not under immediate consideration. There is thus a clear need for a resource in which hominid fossils are described in detail on their own individual terms, using a consistent protocol from one fossil to the next.

We have attempted to supply such a resource here. The volumes in this series present uniform descriptions and illustrations, almost all based on the examination and photography of original specimens, of the most significant among the major fossils comprising the human fossil record. To these descriptions are added ancillary information on dating, archaeological context, and so forth. We also clearly define the anatomical terminology we use, adapting this terminology to make it applicable not simply to the extant species, *Homo sapiens*, but to fossil hominids as a whole. The first two volumes of the series are devoted to fossils that have been allocated to the genus *Homo* (whether or not it is likely that they will ultimately be found to warrant this appellation), and they focus on skull and dental morphology. Volume 1 presents our descriptive protocol and the craniodental fossils from Europe; Volume 2 covers the African and Asian hominid fossil records and concludes with a systematic analysis of the genus *Homo*. Volume 3 will cover the early hominid fossil record, together with a discussion of morphological nomenclature and of the systematics of these hominids; it will conclude with a general overview of hominid systematics. Volume 4 will describe cranial endocasts. A volume on the postcranial skeleton is also envisioned. Each of the volumes in the series will be published separately, but in order, and as close together in time as possible.

Clearly, the definitive account of the human fossil record will never be written. This is partly for the best of reasons: the paleoanthropological record is already very extensive, and it is growing at a rate with which it is hard to keep up. Partly, though, it reflects the fact that certain human fossils, even ones that have been comprehensively published, are in some cases surrounded by a wall of curatorial protectionism that amounts almost to paranoia. Interestingly enough, this attitude is largely foreign to other areas of vertebrate paleontology, but in human paleoanthropology it constitutes a real stumbling block to progress in what is after all an essentially comparative science. We have, however, done our best to make these volumes as comprehensive as possible. These remarks having reluctantly been made, it is important to acknowledge that we have been enormously gratified by the help and hospitality extended to us by the great majority of those responsible for the fossils that make up the human historical record, without whose active assistance this project would never have been possible. Warm welcomes and extraordinary helpfulness all over the world have transformed a potentially Herculean labor into a pleasure.

In preparing Volumes 1 and 2 of this series we have thus been hugely impressed by the generosity, helpfulness and hospitality of very many curators and other colleagues. In particular, we would like to extend our warmest thanks and appreciation to the following, who gave us access to fossils or provided other valuable help: Susan Antón, Aomar Akerraz, Juan-Luis Arsuaga, the late Antonio Ascenzi, Graham Avery, Fachroel Aziz, Cecilio Barroso-Ruiz, Peter Beaumont, Abdelwahed Ben-Ncer, Directrice Joudia Benslimane, Amilcare Bietti, Luca Bondioli, Günter Bräuer, James Brink, Tim Bromage, Ralf Busch, Jean-Francois Bussière, Miguel Caparros, Chang Mee-Mann, Mario Chech, Ron Clarke, Jean-Jacques Cleyet-Merle, Silvana Condemi, Alfred Czarnetzki, Hilary Deacon, Janette Deacon, Miluse Dobisikova, Viola Dobosi, Sigrid Dusek, Christophe Falguères, Noor Farsan, Larry Flynn, Rob Foley, Heidi Fourie, Jens Franzen, the late Leo Gabunia, Lena Godina, Dominique Grimaud-Hervé, Almut Hoffmann, Hou Yamei, F. Clark Howell, Huang Yunping, Jean-Jacques Hublin, Ato Jara Haile Mariam, Etty Indriati, Teuku Jacob, Jan and Kveta Jelinek, Hans-Eckert Joachim, Don Johanson, Kebede Worke, Vitaly Kharitonov, Bill Kimbel, László Kordos, George Koufos, Beverley Kramer, Robert Kruszynsky, Kathy Kuman, Viteslav Kuzelka, Marta Mirazón Lahr, André Langaney, Henry and Marie-Antoinette de Lumley, Meave Leakey, Li Tianyuan, David Lordkipanidze, Lu Zune, Angiolo del Lucchese, Roberto Macchiarelli, Nasser Richard Malit, Mamitu Yilma, Dietrich and Ursula Mania, Giorgio Manzi, Marie-Hélène Marino-Thiault, Emma Mbua, Veronique Merlin-Anglade, Janet Monge, Rosine Orban, Marcel Otte, Ildiko Pap, Maja Paunovic, David Pilbeam, Eddy Poty, Jakov Radovcic, Yoel Rak, Antonio Rosas, Alain Roussot, Friedemann Schrenk, Betsy Schumann, Aldo and Eugenia Segre, Horst Seidler, Patrick Semal, Suzanna Simone, Petr Skrdla, Malcolm Smale, Giuseppina Spadea, Gabriella Spedini, Chris Stringer, Emmy Suparka, Jiri Svoboda, Maryse Tavoso, Maria Teschler-Nicola, Francis Thackeray, Herbert Thomas, Alan Thorne, Phillip Tobias, Michel Toussaint, Javier Trueba, Sophie Tymula, Bernard Vandermeersch, John de Vos, Alan Walker, Wang Youping, Diethard Walther, Gill Watson, Karin Wiltschke-Schrotta, Wu Xinzhi, Yahdi Zaim, Zhao Linxia, Zhu Min, Joe Zias, and Reinhard Ziegler. The institutions in which the fossils reside, and whose cooperation was obviously essential, are listed individually by site entry. We are grateful to all. The Spy remains reside in the Institut Royal des Sciences Naturelles, Brussels, by courtesy of the estate and family of Max Lohest (1857–1926). Photographs of the Ceprano calvaria are by Antonio Solazzi, courtesy of Giorgio Manzi, Italian Institute of Human Paleontology (Rome). Photographs of the Gran Dolina and Sima de los Huesos specimens were taken and provided by Javier Trueba (Madrid Scientific Films S.L.). With the exception of these photographs and the digital scans of the St. Césaire, Shanidar1 and Biache specimens by Ken Mowbray, all other photographs were taken by Jeffrey H. Schwartz. Thanks to André Langaney for permission to photograph the specimens in the collections of the Musée de l'Homme, which retains copyright on the illustrations. All other photographs are copyright © Jeffrey H. Schwartz. We thank Petica Barry for the black and white drawings, Ken Mowbray and Bridget Thomas for the maps.

Many other friends and colleagues have also been indispensable in making these volumes a reality. Our initial editor at Wiley, Robert Harington, enthusiastically embraced the notion of this series, which could not have come to fruition without the commitment of Luna Han, who steered it through to completion. Also at Wiley, Joe Ingram and numerous individuals at all stages of production and marketing deserve our warmest appreciation, while at the American Museum of Natural History, Ken Mowbray rendered indispensable help, and at the University of Pittsburgh, Jim Burke and Kolleen Mitchell of Photographic Services undertook the painstaking task of scanning the black and white negatives and enhancing each image to bring out as much detail as possible. In addition to our personal financial contributions, additional funding for this project was made through funds administered through the Department of Anthropology, American Museum of Natural History, as well as grants from the L. S. B. Leakey Foundation, John Wiley & Sons, Nevraumont Publishing Co., and the University of Pittsburgh [Central Research Development Fund, University Center for International Study, and Nationality Rooms Programs (J. G. Bowman)].

Finally, it should be noted that no paleoanthropologists embarking on a project such as this one could ever ignore the fact that they are standing on the shoulders of some very illustrious predecessors. Notable among these forerunners are the authors and editors of the *Catalogue des Hommes Fossiles,* edited by H. V. Vallois and H. L. Movius and published in 1953; the three volumes of the *Catalogue of Fossil Hominids*, edited by K. P. Oakley, B. G. Campbell, and T. I. Molleson and published and revised between 1967 and 1977; and M. H. Day's multi-edition *Guide to Fossil Man*, which first appeared in 1965. The *Catalogue* has been very usefully updated in recent years by the several volumes of *Hominid Remains: An Up-Date*, edited by Rosine Orban and Patrick Semal and published by the Royal Belgian Institute of Natural Sciences in Brussels. None of these works had exactly the same intentions as this one; for example, the *Catalogue of Fossil Hominids* aimed at comprehensiveness of sites but ignored morphology and illustration, whereas the *Guide to Fossil Man* did provide some general morphological information and illustration but was highly selective in site choice. Nevertheless, we are conscious that we are following a road that has already been partly trodden, and that our task has thereby been rendered easier.

Jeffrey H. Schwartz
Ian Tattersall

PART ONE

Terminology and Craniodental Morphology of Genus *Homo* (Europe)

DESCRIPTIVE PROTOCOL

In these volumes we describe as many as possible of the major fossils that make up the fossil record of the genus *Homo*. The arrangement is by site: we describe each fossil or fossil assemblage individually and without reference to specimens from other sites. To make this possible we have adopted a single descriptive protocol and a uniform nomenclature for morphological features of the hominid skull. Armed with these descriptions, the reader will be able to make direct comparisons among whatever fossils he or she desires. We should note that we follow the precedent set by F. H. Smith and F. Spencer in *The Origins of Modern Humans* (1984, Alan R. Liss, NY) in not presenting site and person names with attendant diacritical marks.

In the next few pages we present the descriptive format we have developed, and where necessary we discuss details of nomenclature. Each fossil description in these volumes follows as closely as possible the order presented below, even where individual specimens are incomplete. Deviations from this format are, however, sometimes dictated by the completeness and state of preservation of the specimen or specimens. Often where a homogeneous assemblage of fossils is described from the same site, the most complete specimen (or, if appropriate, specimens) is taken as the exemplar, and other individuals are described only to the extent to which they differ. In some cases, however, even when the homogeneity of an assemblage is obvious, some specimens are described individually (for example, for historical reasons, because the specimens may not be widely known, previous descriptions were not detailed or features were described without accompanying illustration, or to illustrate ranges of variability). The views of a specimen are generally organized in the following sequence: for the skull, frontal, side(s), posterior, superior, inferior, and then close-ups (e.g., teeth, basicranium, internal structures such as the petrosal and drainage sinuses, and nasal cavity structures such as the floor, the conchal crest, and medial projection); for the mandible, frontal, side(s), occlusal, inferior, posterior, and close-ups (such as the mandibular foramen and lingula, sigmoid notch crest and condyle, and medial pterygoid tuberosities). In the following outline of our descriptive protocol we highlight the principal bones and structures to which attention is paid, region by region; necessarily, there is some overlap between descriptions of adjacent regions and structures. For nomenclature, refer also to Figures 1–9.

Many will fault us for the fact that we have deliberately refrained from providing measurements for the specimens described. Partly this was to save space in an already very bulky series of books (another reason for the terse morphological descriptions), and partly it was because measurement criteria vary so much among practitioners. Size is, of course, a significant factor in paleoanthropology, and we hope that the fact that the vast majority of photographs are to scale will be an adequate guide to the size of each fossil illustrated.

Our reluctance to emphasize metrical analyses also derives from an ongoing concern with the need to better understand and incorporate morphology in systematics. Often, it seems, especially in the case of hominids, either the landmarks chosen for measurement are not obviously correlated with developmentally relevant features, or the attempt to characterize mathematically the "shape" of an entire bone or structure fails to discriminate between features that are not unique to the taxon under study and those that are. We do not mean to imply that pursuing and refining metrical approaches is irrelevant. We do, however, believe that, especially with studies in regulatory genetics throwing increasing light on the interrelatedness of features (either tooth and face or of the limbs), we are just at the threshold of beginning to comprehend how to think about morphology, systematics, and phylogenetic reconstruction.

Because it is "morphology" that metrical studies are trying to capture, we must first focus on the developmental constraints that produce the features that we use in our various scholarly endeavors. Having said this, we must also acknowledge, especially in the case of craniodental morphology, that we are still very much in the dark as to the ontogenetic interrelatedness of the details of morphology that are potentially relevant to the delineation of morphs, species, and even higher taxonomic groups. As such, while knowing that various regulatory molecules (for example, Sonic hedgehog) can affect the development of teeth (at least increasing or decreasing their field and number) as well as of the skull (even the development of eyes in bony sockets), we are still faced with the task of describing morphology.

In the spirit of anticipating knowing what causes the developmental appearance or disappearance of, for example, hypoconulids on lower molars, or the emergence of a bipartite versus smoothly rolled and bilaterally continuous brow from a fetally unadorned frontal, we have tried to capture in our descriptions as much general and detailed information as possible. Although we have used relative terms in our descriptions of size and shape—for example, slightly, somewhat, moderately, markedly—we also recognize that some readers of this work might not have described specimens as we did. Our attempt to be consistent in the application of these evaluations, however, will allow these individuals to assess the specimens themselves from the accompanying photographs and, if necessary, to apply their own consistent terms of relative size and shape.

DESCRIPTIVE FORMAT

General Comments. General preservation and completeness of the specimen(s).

Cranium—Overview. Overall form and proportions of the cranium; general bone thickness.

Supraorbital Region and Splanchnocranium (Figures 1–3). Supraorbital structures, glabella, frontal sinuses, orbits, infraorbital region and zygomas, nasal bones, nasal aperture and cavity, nasoalveolar region, palate, and pterygoids.

Cranial Roof (Figures 1 and 2). Contours and external details of frontal and parietals.

Cranial Walls (Figure 2). Temporal bone and attendant fossae, posterior part of zygomatic arch, mastoid region and auditory meatus, mandibular fossa.

Cranial Rear (Figures 2 and 4). General contour, occipital plane and associated structures.

Cranial Base (Figure 4). Nuchal plane and possibly some of the contiguous mastoid area, spheno- and basioccipital region, foramen magnum, and occipital condyles.

Cranial Sutures. Configuration of their margins.

Anterior Endocranial Compartment (Figure 5). Anterior cranial fossa and associated structures (may be included with discussion of specific bones in cases of more fragmentary specimens).

Middle Endocranial Compartment (Figure 5). Middle cranial fossa and associated structures, including petrosals (may be included with discussion of specific bones in cases of more fragmentary specimens).

Posterior Endocranial Compartment (Figure 5). Posterior cranial fossa and associated structures, excluding petrosals (may be included with discussion of specific bones in cases of more fragmentary specimens).

Major Endocranial Sinus Impressions (Figure 5).

Mandible (Figures 6 and 7). Overview and detailed morphology.

Dentition—Overview. Condition, general size, and proportions.

Upper Dentition (Figure 8). By tooth, mesial to distal. Major points of variation among the permanent teeth are discussed below; descriptions of deciduous teeth, where known, follow the same protocol.

Lower Dentition (Figure 9). By tooth, mesial to distal, as above.

ABBREVIATIONS

To save space, we have abbreviated certain frequently occurring terms. We use R and L for right and left, respectively, a/p for anteroposteriorly, s/i for superoinferior, and m/l for mediolateral. Specifically for tooth descriptions, m, d, b and l stand for mesial, distal, buccal, and lingual, respectively. We use standard terminology for denoting teeth, for example, I_2 for the lower lateral incisors, M^1 for the first upper molars, and di^2 for the upper deciduous lateral incisors; we also use P1 and P2 for the anterior and posterior premolars, in preference to the alternatives P3 and P4, and dm1 and dm2 for the anterior and posterior deciduous molars, in preference to the alternatives dp3 and dp4. Super- and subscripts are not used to denote upper and lower teeth, respectively, when it is obvious from the specimen under discussion which teeth they are.

CRANIODENTAL MORPHOLOGY AND TERMINOLOGY

In the next few dozen pages we outline the general range of dental and cranial morphologies exhibited within the genus *Homo*, following the order of the descriptive protocol outlined above. Our nomenclature of features is further shown in Figures 1–9. The reader is advised to consult this section in the event that he or she requires any clarification of the terminology used in the descriptions presented in Part 2. It is useless to hope that all readers will approve of all of the terms we have chosen to use for anatomical features. We can only plead that choices have had to be made and that we are consistent throughout the volumes. Most of the features identified in the anatomical Figures 1–9 are noted in *italics* the first time they are mentioned. Additionally, so that readers may check our criteria for themselves, examples are given of individual specimens possessing each of the principal anatomical configurations described; overwhelmingly, of course, these examples are representative rather than unique. No systematic implications are intended by our choice either of specimens or of characters, as the examples show. This is because we have designed our descriptive protocol to be as dispassionate as possible, our aim being to provide compact yet comprehensive descriptions that will be of maximum utility for diverse purposes. We do not wish in any way to bias comparisons among specimens. Our descriptions simply reflect what is preserved in each fossil, without any selection of characters for systematic reasons. We introduce our own evaluation and comparison of the morphologies we have recorded only at the end of Volume 2, in the section on "Systematics and Phylogeny." By following our descriptions, it should be possible for readers to form usable mental images of the fossils discussed, regardless of whether or not our terminologies are fully congenial to them.

Cranium—Overview

Within the genus *Homo*, as conventionally understood, there is a huge variety in general cranial form. Under this rubric we describe the general features of the cranial vault of each specimen, noting relative length, breadth, and height, as well as the general sagittal contour including that of the supraorbital region. In addition, where preservation is adequate we record the relative size and spatial relationships between splanchno- and neurocrania and measured or estimated endocranial volume.

Supraorbital Region and Splanchnocranium (Figures 1–3)

Supraorbital morphology is highly variable within the genus *Homo*. The supraorbital margins may (e.g., Abri Pataud) or may not (e.g., AMNH VL/3229) be distended (into what are generally known as *supraorbital tori*). This has, unfortunately, become a blanket term tending to obscure the differences between prominent supraorbital regions of distinct types, as in the simplistic notion of a gradual decrease in size from an apelike torus to the smooth or weakly developed brow of modern human type. Supraorbital tori are emphatically not a presence/absence character, and we have thus tried to minimize our use of this term except as a general indicator of the morphological area involved. If the superior margins are distended, they may protrude superiorly (Kabwe), anteriorly (Ndutu), or both (KNM-ER 3733), and they also may (Monte Circeo) or may not (Ngandong 12) be confluent across *glabella* (which is technically a single point, the most anterior in the midline between the orbits, but which descriptively has come to refer to the region above the frontonasal suture as a whole—a usage that we will continue here). Above each orbit, the marginal surface may be continuous (Skhul 5) or disjunct (bipartite). In the latter case, this margin may be divided into a smoothly and anteriorly protrusive *medial component* and a distinct, platelike, and posteriorly inclined *lateral part* (Qafzeh 9) or into a vertically flat medial portion and a more swollen and anteriorly protrusive lateral part (Zhoukoudian XII). In cases where the superior margin is continuous, there are several alternative character states. The rim may be thickened superiorly; in this case it may be evenly thickened from side to side (Tabun C1), or it may be thinner (Krapina C) or thicker (Zhoukoudian XII) at one of its extremities or even centrally (Qafzeh 6). Sometimes the presence of a gutter above a supraorbital notch gives the impression of superoinferior thinning of the supraorbital rim (KNM-ER 3733); we do not regard this effect as a discrete trait, although the presence of a notch is noted. The contour of the superior orbital rim is a result of orbital shape, because bone is passively deposited along this margin during development, and is quite variable. In some cases the orbit is squarish, and

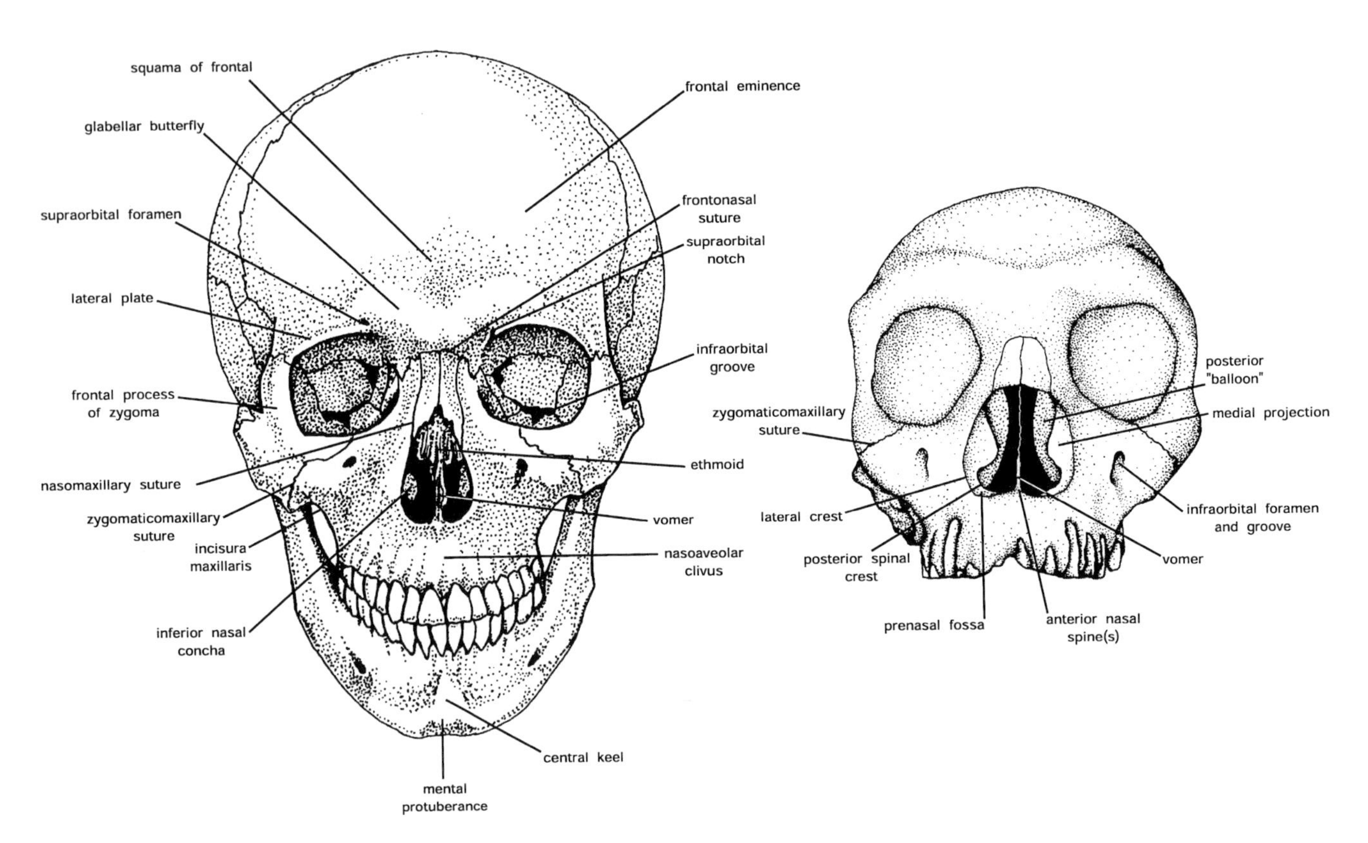

Figure 1. Anterior view of two crania, one with mandible, with identification of major features discussed in this volume.

its superior margin, inflated or not, is straight across (Skhul 5); in other instances it is more ovoid or rounded, and the margin is hence arcuate (Feldhofer Grotto). In some specimens there is a distinct superior margin to the anterior surface of the vertically thickened brow (Kabwe).

In profile, the superior orbital rim displays varied contours. The brows may be smoothly rolled from the orbital roof forward, then up and back into the lower frontal squama (i.e., in a regular curve: La Ferrassie 1), or they may form an angle with the orbital roof, with (KNM-ER 1813) or without (KNM-ER 3733) a posterior slope (rather than a curve) above. Viewed from above, the brows may retreat sharply from glabella (La Quina 5), may curve more gently backward (Kabwe), or may be fairly straight across (Zhoukoudian 12). Postorbital constriction isolates the lateral orbital cones to varying degrees: negligible (Qafzeh 9), small (Kabwe), and moderate (KNM-ER 3883). The superior surface of the orbital margins may flow smoothly up into the frontal squama (KNM-ER 1470); there may be a more or less horizontal posttoral plane from which the frontal squama rises posteriorly (Tabun C1); or there may be a distinct sulcus between the posterior aspect of the elevated supraorbital rim and the rising frontal squama (KNM-ER 3733).

Glabella itself (as defined above) is highly variably developed. Relative to the superior orbital margins flanking it the superior interorbital area may be protrusive anteriorly (Monte Circeo 1), flush (Krapina C), or depressed (Ngandong 12). If protrusive, it may generally swell out over the frontonasal suture in its entirety (Qafzeh 6), or it may be inclined, such that it presents an inferior surface that sharply overhangs this suture (Skhul 5). Glabella always lies below the most superior point along the supraorbital margin, but it varies in the extent to which it does so (contrast KNM-ER 1813 with ER-3883). As noted above, the glabellar eminence, if any, may form a continuous bridge between the prominences on either side (La Chapelle-aux-Saints), or it may display a central depression (KNM-ER 1813) or a vertical crease (Petralona) dividing the lateral supraorbital arcs. In some cases (Zhoukoudian XII) a distinct angulation delineates the superior margin of glabella. We will use the term *glabellar butterfly* to refer to the configuration

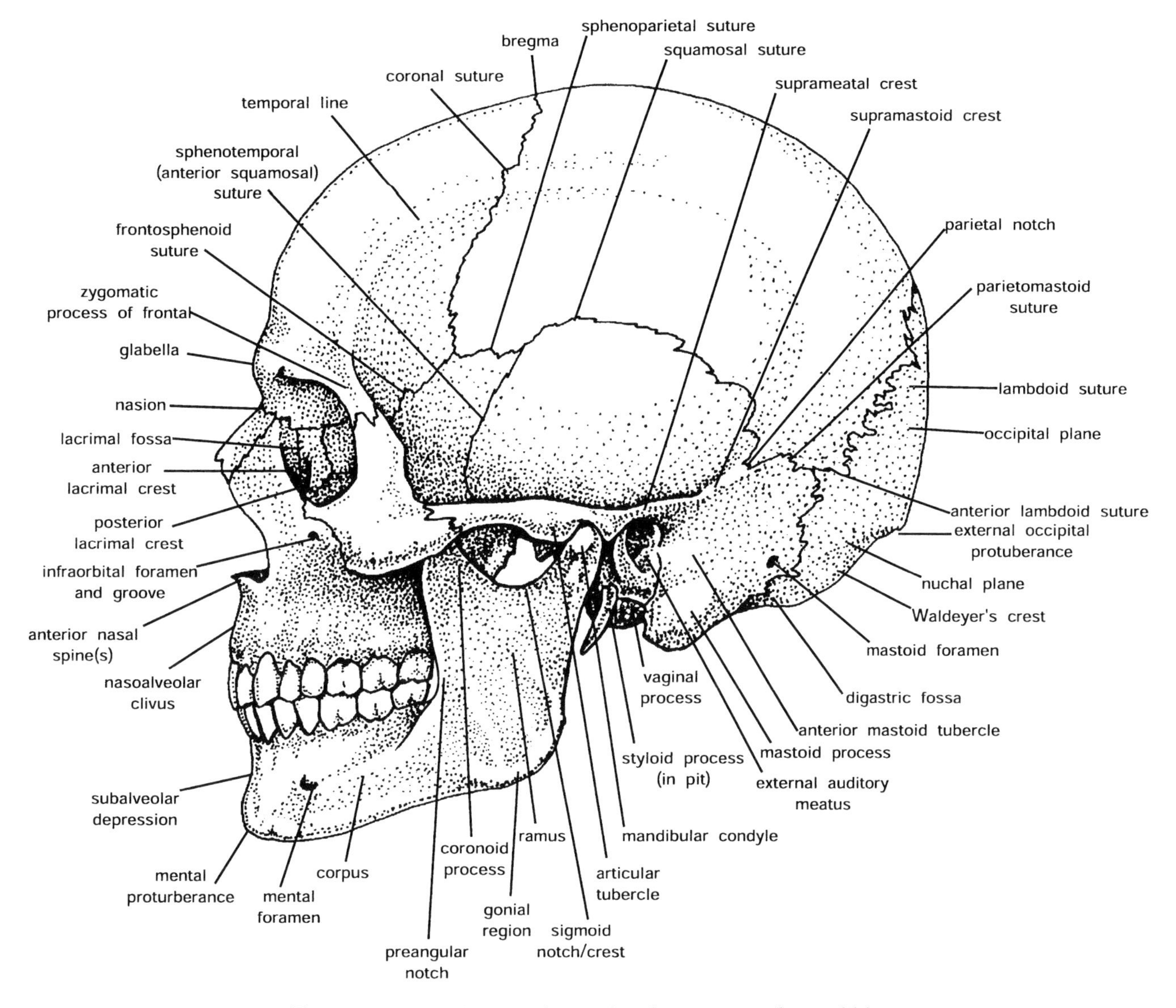

Figure 2. Lateral view of articulated cranium and mandible.

whereby glabella is continuous with the medial component of a bipartite supraorbital configuration in which the lateral part is platelike (Qafzeh 9).

Almost invariably, the frontal squama above and behind the supraorbital margins is pneumaticized to a greater or lesser extent. In many cases it appears that such *frontal sinuses* are derived from the *ethmoidal air cells* below. There are normally at least two frontal sinuses; these lie generally within the area behind glabella, separated by a bony septum. They may be confined to this area (Steinheim), or they may expand laterally as far as the frontozygomatic suture (Monte Circeo) and/or extend far back into the frontal (Petralona). Expansive frontal sinuses are invariably multifocular (Kabwe).

Orbital shapes vary greatly. Orbits may be taller than wide (KNM-WT 15000), wider than tall (Jebel Irhoud 1) and rounded (Tabun C1) or square cornered (Skhul 5) superiorly. Occasionally, the m/l axis slopes down laterally (Qafzeh 6). Inferiorly, the orbital margin may be more or less horizontal (Qafzeh 9), or it may be angled up toward the midline throughout its length (Kabwe) or just in its medial portion (Ndutu), a configuration we describe as "medially constricted." Structures within the orbit are rarely preserved intact but also show some variation. The infraorbital groove/canal may lie centrally (Gibraltar 1) or laterally (Kabwe). Medially, the lacrimal fossa may lie right on the orbital margin, unbounded by an *anterior lacrimal crest* (Kabwe); it may lie at least partly external

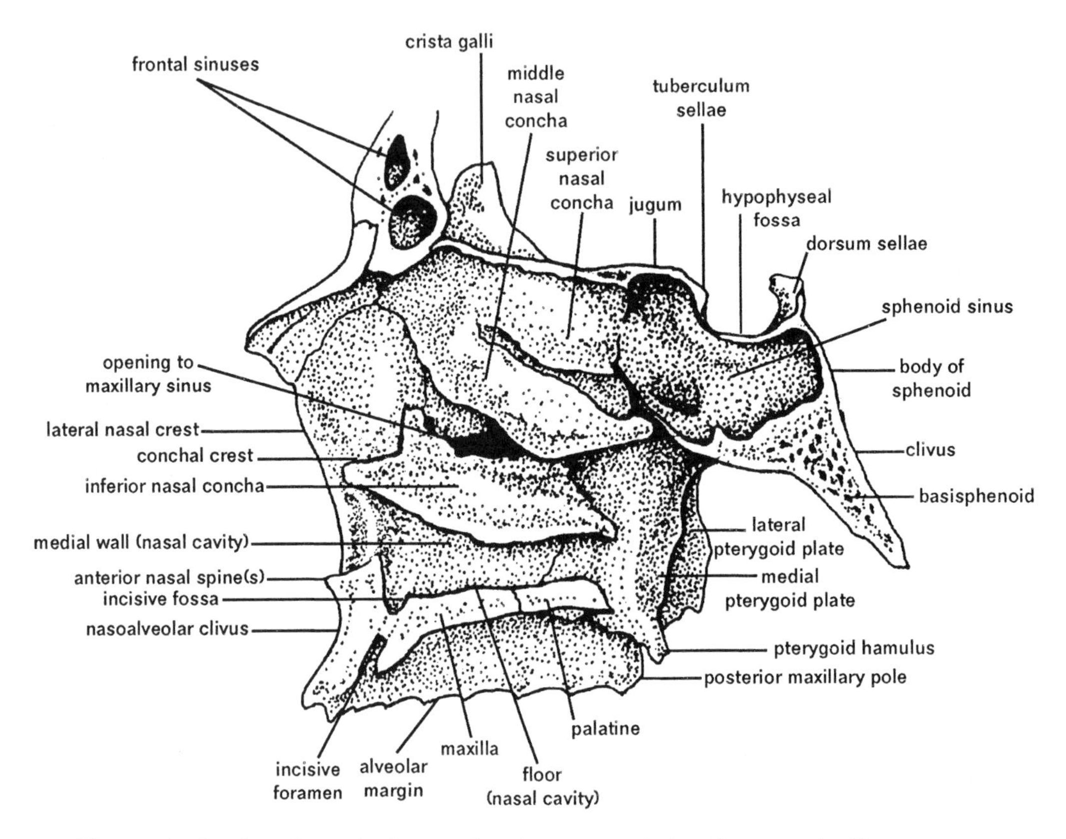

Figure 3. Section through the nasal and pterygopalatine fossae and adjacent areas.

to the orbit (Bodo) or completely within (Sima Hominid 5), in which case the anterior lacrimal crest defines the orbital margin. The anterior and *posterior lacrimal crests* may be confluent superiorly in a smooth curve (Kabwe), or they may remain separated (Ndutu). In some individuals the posterior crest is more prominent than the anterior (Kabwe) and in others less prominent (Gibraltar 1). The floor of the orbit may be delineated anteriorly by a crest-like infraorbital margin (Ndutu) or may flow uninterrupted on to the face (La Ferrassie 1).

The upper part of the infraorbital portion of the *frontal process* of the premaxilla may be more or less vertical (Qafzeh 6) or may be almost horizontal (La Ferrassie). The frontal process may be slightly concave (Ndutu), slightly convex (Kabwe), or markedly puffed outward (La Ferrassie). The zygomatic portion of the *infraorbital plane* may be swept back and more laterally facing (La Chapelle-aux-Saints), fairly sharply angled laterally (KNM-WT 15000), or smoothly curved from front to side (KNM-ER 3733). In the latter case it may be more vertical (KNM-ER 3733) or may flare out inferiorly (KNM-ER 3883). Finally, it may be quite forwardly facing (Jebel Irhoud 1), or it may retreat inferiorly (Qafzeh 6). The *infraorbital foramen* may be single (KNM-WT 15000) or multiple (Kabwe) and relatively close to the infraorbital margin (La Ferrassie) or well below it (KNM-ER 1813). The orifice may open downward (Monte Circeo) or medially (Kabwe) and may continue down the face as a shallow (Gibraltar 1) or deep (Steinheim) groove. The inferior border of the *anterior root of the zygomatic arch* may be more or less horizontal (Jebel Irhoud 1), it may arc concavely (Kabwe) or convexly (KNM-ER 15000) downward, or it may angle steeply down toward the alveolar margin (Gibraltar 1). It may terminate close to the alveolar margin (Monte Circeo) or high above it (Jebel Irhoud 1). At the inferior terminus of the zygomaticomaxillary suture the inferior margin of the bone may be downwardly distended into a *maxillary tubercle*, producing an *incisura maxillaris* medial to the latter (Jebel Irhoud 1).

In the midline, the paired nasal bones show a variety of configurations. They may be long (La Ferrassie) or short (KNM-ER 1813) in the midline and broad (Krapina C) or narrow (KNM-ER 3733) superiorly. They may be smoothly arced across from side to side (KNM-WT 15000), or they may bear a midline keel (KNM-ER 1813). They may broaden continuously away from the frontonasal suture, either gently (KNM-WT 15000) or markedly (Monte Circeo), or they may be concave laterally (Kabwe). In profile the nasals may be fairly gently concave below nasion (KNM-ER 3883), or they may be sharply flexed below nasion (Gibraltar). They may continue quite steadily downward, in concert with the thin frontal process of the premaxilla (Qafzeh 6), or they may run close to horizontally in their distal portion, along a much thicker frontal process (La Ferrassie). Specimens such as Ndutu are intermediate.

As in all large-bodied hominoids, the inferior margin of the nasal aperture is wider than the superior margin. At one end of the range the aperture is rather tall and narrow (KNM-ER 1813), whereas at the other it is relatively short and broad (Qafzeh 6). Most commonly, the lateral margins of the aperture are sharply delineated (Kabwe), although occasionally these edges are rather blunt and less distinct (KNM-WT 15000), particularly inferiorly. The inferior margin of the nasal aperture shows a number of distinctive configurations. The floor of the nasal cavity may flow smoothly outward and down onto the *nasoalveolar clivus*, with no visible discontinuity (KNM-ER 1813), or spinal crests may delineate the floor of the nasal cavity from the clivus below. This latter may involve only the *posterior spinal* (*conchal*) *crests* (Jebel Irhoud 1), or both posterior and *anterior spinal crests* (Kabwe) with a *prenasal fossa* between them (Kulna). In both of these cases, the *lateral crests* of the nasal aperture terminate at or below the floor of the nasal cavity, fading out on reaching the clivus. Alternatively, the floor of the nasal cavity may be delineated from the clivus by the lateral crests (Qafzeh 6) as they swing around to meet the *anterior nasal spine*(*s*). Some individuals lack distinct protruding nasal spines (KNM-WT 15000).

In the nasal cavity itself, most structures are rarely preserved intact, if at all. The cavity floor may be relatively flat (Qafzeh 6), it may slope gently downward posteriorly (Jebel Irhoud 1), or it may be sharply "stepped" downward (Gibraltar 1). The lateral wall may bear a low-lying horizontal *conchal crest*, with which the *inferior nasal concha* may remain associated (Cro-Magnon 1), or it may display a vertically oriented *medial projection* (La Ferrassie), which may be confluent below with the posterior spinal crest (Gibraltar 1). The medial projection may be large (Spy 1), moderate in size (Krapina C), or quite diminutive (Steinheim). It may bear a crease or ridge along the posteroinferior margin (La Chapelle-aux-Saints), along which an inferior conchal crest might have articulated. In individuals with a medial projection the *lacrimal groove*, posterior to the medial projection, may not have been covered by a bony sheet (La Chapelle-aux-Saints); otherwise, there is a covered *lacrimal canal* (Abri Pataud). Behind the lacrimal region the wall of the nasal cavity may be flat or slightly concave (Cro-Magnon 1), strongly concave (some San), or it may protrude into the nasal cavity in a *posterior swelling* (Gibraltar 1).

The nasoalveolar clivus may be long (KNM-WT 15000) or relatively short (Qafzeh 6) superoinferiorly. It may be essentially vertical (KNM-ER 1470), or it may slope markedly forward and down (Jebel Irhoud 1). It can be quite flat across from side to side (KNM-ER 1470) or strongly arced (Kabwe).

Palates vary from long and narrow (KNM-ER 1813) to broad and short (Kabwe). In *Homo* the dental arcade generally arcs smoothly across the front (Skhul 5), although in some cases the arcade is squared off in front (KNM-ER 3733). The palate may be more or less uniformly deep (Qafzeh 6), or slope gradually up and back from behind the anterior teeth (Kabwe). Posteriorly, the sides of the palate may be rather vertical (Steinheim), or they may slope medially (Jebel Irhoud 1). Often the *incisive foramen* lies in the midline immediately behind the incisors (Monte Circeo), but it may lie more posteriorly (KNM-WT 15000). At the rear of the tooth rows, the *maxillary pole* may (Skhul 5) or may not (KNM-ER 1813) extend far behind the M3. Behind, the *pterygoid plates* may unite superiorly in a smooth curve (Steinheim), or they may remain separated (Kabwe). The *foramen ovale* may lie medial to the lateral pterygoid plate (Skhul 5) or lateral to it (Steinheim).

CRANIAL ROOF (Figures 1 and 2)

Sagittal profiles of the cranium differ dramatically. The frontal may rise directly from glabella (KNM-ER 1470), either sharply (Abri Pataud) or in a rapidly retreating curve (Kabwe), or it may begin rising some distance behind glabella (Jebel Irhoud 1). Maximum cranial height may be attained well anterior to (KNM-ER 3733), near (Monte Circeo), or well behind *bregma* (KNM-ER 3883). The sagittal curve itself may be low (Kabwe), strongly curved upward (Qafzeh 6), or in between (Zhoukoudian XII), and it may be relatively symmetrical from front to back (Skhul 5) or strongly asymmetrical, with its peak lying anteriorly (KNM-ER 3733) or posteriorly (KNM-ER 3883). Coronal profiles are equally variable, with maximum width lying low down on the temporal bone (KNM-ER 1813), around the squamous suture (Jebel Irhoud 1), or high on the parietals (Skhul 5). Maximum width always lies posteriorly. On some individuals a low longitudinal eminence, the *frontal keel*, adorns the frontal in the midline, over a short distance (KNM-ER 3733) or a rather longer one (Kabwe). Further back, a midline *sagittal keel* is sometimes present (Sangiran 2). Occasionally, keeling may be present along the coronal suture across bregma (Ngandong 7). Internally, we note the occurrence, if any, of *Pacchionian depressions*, otherwise known as arachnoid pits (Hahnofersand).

CRANIAL WALLS (Figure 2)

In coronal profile, at around the mastoids, the walls of the cranium may slope out and up before curving inward (Abri Pataud), swell symmetrically (Monte Circeo), or tilt inward (KNM-ER 1813). The *temporal lines* always originate from the lateral border of the frontal process of the zygoma but vary widely in their expression and in the degree to which they extend up on to the cranial roof. They may be traced up, back, and around on the parietals almost to the mastoid notch (KNM-ER 3733), or they may become more or less indistinct posterior to the coronal suture (KNM-ER 3883). In some individuals the temporal lines circumscribe swollen *frontoparietal bosses* (Trinil 1), whereas in others (KNM-ER 3883) the anterior temporal lines overhang the bone beneath, thus subtending well-defined *frontal depressions.*

The *temporal fossa* can be subequal in its maximum a/p and m/l dimensions (Qafzeh 6), or it may be longer than wide (La Ferrassie 1) or wider than long (KNM-ER 3883). On its anterior border a *malar tubercle* may be present at the lateral zygomatic margin (Qafzeh 6) or on the posterior wall of the frontal process of the zygoma (KNM-ER 3883) or may not be present at all (Kabwe). The fossa may (Gibraltar 1) or may not (Cro-Magnon 1) be delineated posteriorly by a *medial flexure* along the sphenotemporal suture (the *anterior squamosal suture*). The fossa is always bounded superiorly by some development of the temporal lines; but inferiorly it may (KNM-ER 3883) or may not (La Chapelle-aux-Saints) curve down smoothly toward the lateral pterygoid plate. In the latter case, a *horizontal flexure* runs along the greater wing of the sphenoid below the level of the zygomatic arch, delineating an *infratemporal fossa* inferiorly.

The *posterior root* of the *zygomatic arch* may originate anterior to the *external auditory meatus* (La Ferrassie 1), above its midline (Kabwe), or above its posterior margin (KNM-ER 1813). From the posterior root the arch may flare laterally markedly (Gibraltar 1), moderately (KNM-ER 3733), or hardly at all (Kabwe). The arch itself may be more or less straight as viewed from above (La Ferrassie) or may be laterally bowed (Skhul V). The greater the flare, the larger the area of the shelf between the cranial wall and the lateral margin of the zygomatic arch. The posterior root may be continuous posteriorly with a *suprameatal crest.* This latter may be pronounced (KNM-ER 1813) or weak (Monte Circeo). A *supramastoid crest* may also be present; it may be horizontal (KNM-ER 1470), or slightly (Steinheim) or markedly (La Quina 5) upwardly swept. The suprameatal and supramastoid crests may also be confluent (Ndutu). A *mastoid tubercle* may be present level with the meatus; it may be tiny (La Ferrassie 1), modestly sized (Kabwe), or quite large (KNM-ER 1813). It may be horizontal (KNM-ER 1813) or anteroinferiorly inclined (Kabwe). The *mastoid process* itself varies from very broad-based a/p (KNM-ER 3883) to narrow-based (Steinheim). Its long axis may be directed anteroinferiorly (Kabwe) or directly down (Krapina C). Its projection varies from minimal (KNM-ER 1813) to extensive (Qafzeh 9). It may be pointed (Krapina C) or broad and blunt (Qafzeh 6). Its lateral surface may be smoothly convex (Kabwe), flat and laterally facing (KNM-ER 1813), or flattish and somewhat posteriorly facing (KNM-ER 3883).

The auditory meatus varies both in size [large (Ndutu) to tiny (KNM-WT 15000)], and in shape [round (Monte Circeo) to a/p compressed (Steinheim)]. The meatus may closely contact the anterior face of the mastoid process (Qafzeh 6), or it may be slightly separated from it inferiorly (KNM-ER 1813).

The *squamous portion* of the temporal, between the parietal notch and the sphenotemporal suture, varies from relatively short (Steinheim) to quite long (KNM-ER 3883). It may be tallest anteriorly (KNM-ER 3733) or around its midpoint (La Ferrassie). Its superior border may be relatively smoothly arced (Kra-

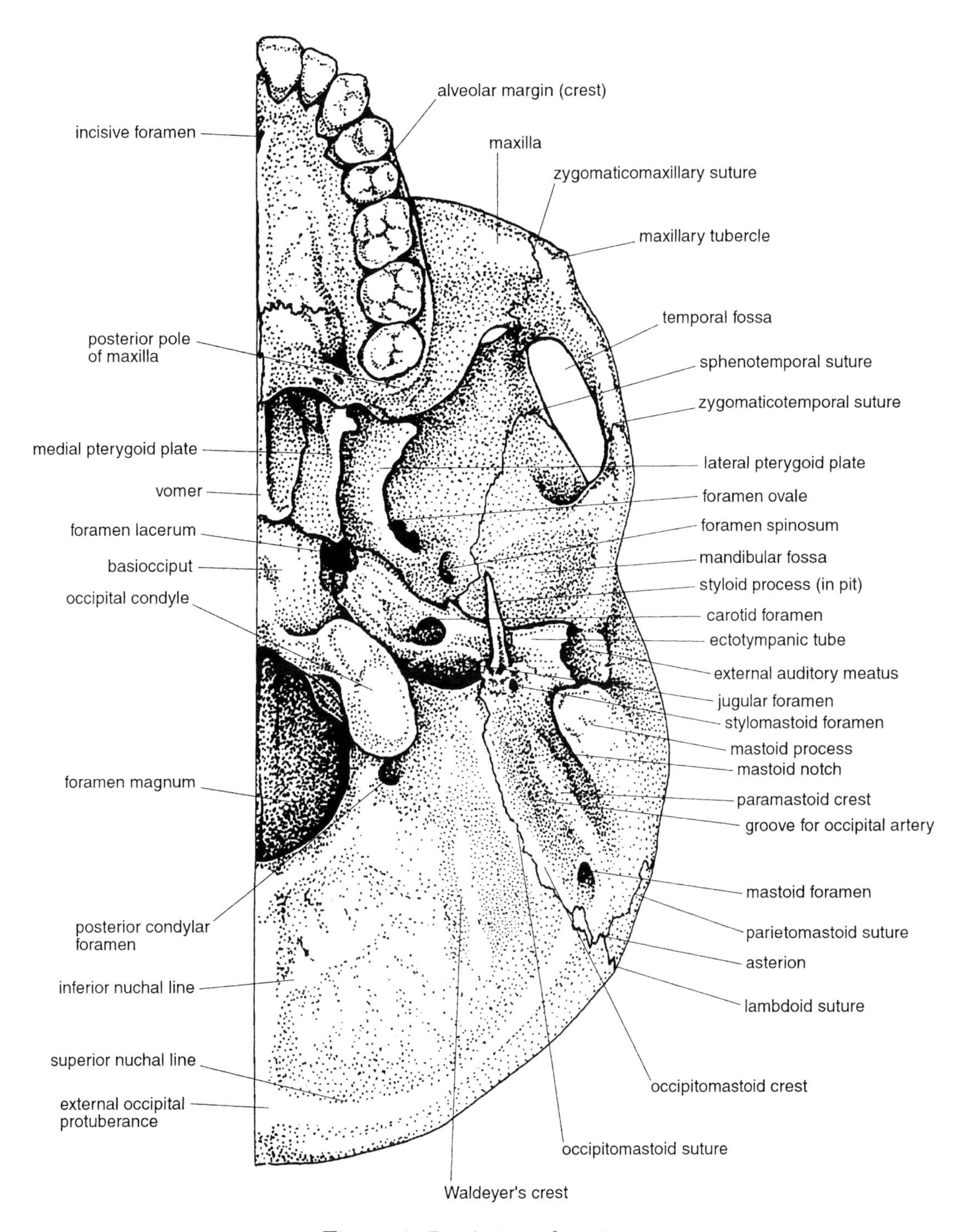

Figure 4. Basal view of cranium.

pina C) or slope posteriorly (KNM-ER 3883). It varies in relative size from tiny (Petralona) to large (KNM-ER 3883). The *parietal notch* may be deeply incised (Steinheim), form a simple angle (Monte Circeo), or be indistinct (KNM-ER 3883). The *parietomastoid suture* may be quite long (Skhul 5), of moderate length (KNM-WT 15000), or shortish (KNM-ER 1470). It may be relatively horizontal (Krapina C) or slightly tilted upward (Steinheim) or downward (Kabwe).

CRANIAL REAR (Figures 2 and 4)

Sagittal profiles of the skull rear vary from a smooth, tall curve that continues virtually uninterruptedly into the cranial base (Qafzeh 9) to various combinations of intersecting planes. In some individuals the *nuchal plane* markedly undercuts the *occipital plane*, the juncture of these surfaces being either a smooth flexure (Monte Circeo) or a more severe angulation (Kabwe). In some cases, only the posteriormost part of the occipital plane is strongly curved (Monte Circeo); in others, this region is flatter (Kabwe). When viewed from behind, the border between the occipital and nuchal planes may be relatively horizontal and straight across (La Chapelle-aux-Saints) or *bow shaped*, with bilateral arches converging to an inferiorly directed external occipital protuberance (KNM-ER 1813). In the bow-shaped configuration, the *highest nuchal lines* may be modestly expressed (Skhul 5) or more markedly developed (Kabwe). In the horizontal condition, the highest nuchal lines invariably traverse straight across, delineating the lower border of the occipital plane. In both configurations, it is the distinct lower border that is the anatomical divider. It is simply the undercutting below this line that gives the impression of a distinct *occipital torus*—a term we prefer to avoid here because it implies a morphology (a raised bar) that does not actually cxist.

Occasionally, the occipital plane bears a shallow depression at the midline. In some individuals (Feldhofer Grotto) this depression lies just above the horizontal angulation between the occipital and nuchal planes and is known as the *suprainiac depression*. This usually ovoid depression varies in size from tiny (Spy 1) to laterally extensive (Swanscombe) and is rugose in surface texture. It is only along the inferior margin of the suprainiac depression, where it is present, that a distinct superior border to the junction of the occipital and nuchal planes can be observed. The other type of depression in this region (the *protuberal depression*) is found only where an enlarged *external occipital protuberance* is present (Predmosti 3). It is restricted to the superior base of the protuberance, and is typically deep although small in circumference.

The occipital plane varies between tall and narrow (Steinheim) and low and broad (KNM-ER 3883). The lambdoid suture may follow a low, broad, continuous arc between the *asterions* (KNM-ER 3733), smoothly curving across *lambda* as it does so; or it may rise steeply biasterionically to peak at lambda (Mladec 5). In some individuals (Steinheim), the lambdoid suture does not descend directly from lambda to asterion. Instead, this suture descends from lambda to behind the region of asterion as the *superior lambdoid suture*; it then flexes sharply to become the *anterior lambdoid suture*, which runs horizontally to this landmark. In individuals with anterior lambdoid sutures, the superior lambdoid suture may peak (Steinheim) or curve smoothly across (La Ferrassie 1) at lambda.

CRANIAL BASE (Figure 4)

The nuchal plane shows great variation in degree of muscle scarring. Below the *superior nuchal lines* are variably impressed scallop-shaped depressions that may (Predmosti 3) or may not (La Ferrassie 1) be separated by a median crest. Posterior to these depressions the nuchal plane may be relatively flat (Ndutu) or may bulge bilaterally (Skhul 5). Laterally, the nuchal plane may (La Chapelle-aux-Saints) or may not (KNM-ER 3733) bear a *Waldeyer's crest* (a crest of variable length and prominence that lies entirely on the occipital, paralleling the occipitomastoid suture). An *occipitomastoid crest* lies by definition only along the occipitomastoid suture and may (Tabun C1) or may not (La Chapelle-aux-Saints) be present. Similarly, a *paramastoid crest*, which is found between the occipitomastoid suture and the mastoid process, may (Ndutu) or may not (Qafzeh 9) be present. There has been considerable confusion over the nomenclature of crests in the region lateral to the mastoid process, and the terminology adopted here, together with the abandonment of equivocal terms such as "juxtamastoid process," should, we hope, help to clarify matters. The *mastoid notch* lies obliquely at the internal base of the mastoid process and may be narrow and relatively short (Ndutu), deep and wide (Monte Circeo), deep, moderately wide and very long (Kabwe), or barely impressed at all (KNM-ER 3883). The notch may open posteriorly onto a *digastric fossa* (Liujang), or there may be no discernible fossa (KNM-ER 1813). Basal area of the mastoid process in ventral view varies from m/l thin (Steinheim) to thick and bulky (KNM-ER 3883). Its apex may come to a point (KNM-ER 1813); it may be a/p ridgelike (La Ferrassie 1), or it may be blunted (Skhul 5).

The basiocciput may be broad laterally and thin superoinferiorly (La Chapelle-aux-Saints), or narrow and thicker (KNM-ER 3733). The *foramen magnum* may be relatively small (KNM-ER 3883), or large (Kabwe); it may be almost round (KNM-ER 3733); it may form a short (Liujang) or a large (Swanscombe) ellipse, or it may approximate a long ovoid (La Chapelle-aux-Saints). The *occipital condyles* may be small and only gently convex (Skhul 5) or long and strongly arced inferiorly (Kabwe); they may be quite anteriorly placed on the rim of the foramen magnum (KNM-ER 3733) or more anterolaterally situated (Sangiran 4).

The *vaginal process*, which runs inferiorly along the midline of the *ectotympanic tube*, may peak around the *styloid process* (Skhul 5), or it may reach its greatest vertical dimension laterally, near the meatus (Sangiran 4). It may remain separate from (Ndutu) or may abut (Cro-Magnon 1) the anterior surface of the mastoid process. The styloid process itself is generally broken in fossils or may only be represented by the *styloid pit* in which it sits, but it may be thin (Abri Pataud) or thick (Barma Grande 3) at its base. The *stylomastoid foramen* is most often situated far medial to the styloid process (Skhul 5), but in some cases (Cro-Magnon 1) it is found near its base. The *foramen spinosum* in all *Homo* lies medial to foramen ovale and is entirely contained within the sphenoid. The *carotid foramen* may face directly downward (Monte Circeo) or posteriorly (Kabwe). The *jugular foramen* may face directly downward (Sangiran 4) or forward (Sangiran 17). The *alae of the vomer* may lie level with the *sphenooccipital synchondrosis* (KNM-ER 3883) or quite anterior to it (Cro-Magnon 1). The superior face of the *medial pterygoid plate* may extend laterally beyond the margin of the basisphenoid (Kabwe) or lie even with it (AMNH VL/4674).

The *mandibular fossa* (referred to in the older literature as the "glenoid" or "articular" fossa) is often remodeled in fossils. This aside, it may be oriented directly laterally (KNM-ER 1813 or angled forward and out (KNM-ER 3883). It may be deep (Sangiran 4) or relatively shallow (KNM-WT 15000). It may (KNM-ER 3883) or may not (Skhul 5) extend laterally beyond the cranial wall. It may be long a/p (Sangiran 17) or rather constricted (Ndutu). It may be bounded anteriorly by a markedly developed *articular eminence*, or it may open almost directly anteriorly (KNM-ER 3733). Medially the fossa may (La Ferrassie 1) or may not (Kabwe) be bounded by a distinct, swollen *medial articular tubercle.* The mandibular fossa is closed off posteriorly by a short (Monte Circeo) or tall (Sangiran 4) vaginal process.

Cranial Sutures and Thickness

Cranial sutures vary in the depth and regularity of their interdigitations. In some specimens interdigitations are insignificant (La Quina 5), whereas in others they are marked (Cro-Magnon 1). In some specimens the sutures are fairly uniform throughout their length (Steinheim), whereas in others they vary at different points along their length (Chancelade). Sometimes the latter condition along the sagittal suture is recognized by naming discrete zones of interdigitation (s1, s2, s3) from front to back. The bone of the parietals and occipital is particularly variable in thickness, both among and within individuals.

Anterior Endocranial Compartment (Figure 5)

Sometimes damage or X ray (and, latterly, CT scanning) permits the examination of endocranial morphology in fossils, although delicate detail is relatively rarely preserved. As far as is presently known, two major configurations of the *anterior cranial fossa* present themselves in genus *Homo*. This fossa may lie substantially above, and forward along, the orbital cones (Arago 21), or it may lie largely behind them (Kabwe). The *cribriform plate* may be situated well below the floor of the fossa (Gibraltar 1) or essentially flush with it (AMNH 99/8176). The *lesser wings of the sphenoid* may be short a/p (Kabwe) or longer (Gibraltar 1). The *anterior clinoid processes* may be thin (Kabwe) or thick (Gibraltar 1).

Middle Endocranial Compartment (Figure 5)

The *optic groove* (=chiasmatic groove) may be faintly impressed (Kabwe) or better defined (Gibraltar 1). The *hypophysial fossa* may be moderately shallow (Kabwe) or deeply excavated (Gibraltar 1). The *middle clinoid processes* may be absent (Kabwe) or tiny (Gibraltar 1). The *tuberculum sellae* may be barely (Kabwe) or more clearly (Gibraltar 1) defined. The *dorsum sellae* may be almost level with (Kabwe) or may lie significantly below (Gibraltar 1) the plane of the *jugum.* The *posterior clinoid processes* may be small and pointed (Kabwe) or large (Gibraltar 1). The *frontal crest* is sometimes long and quite pronounced (Gibraltar 1), or it may be low and indistinct (Bilzingsleben 1). The *crista galli* may be tall and long (Gibraltar 1) or low and short (Dar es Soltane).

The superior surface of the *petrosal portion of the temporal* may be relatively flat (Spy 1), or it may bear a somewhat elevated, domelike *arcuate eminence* (Qafzeh 6). On its superomedial edge, the petrosal

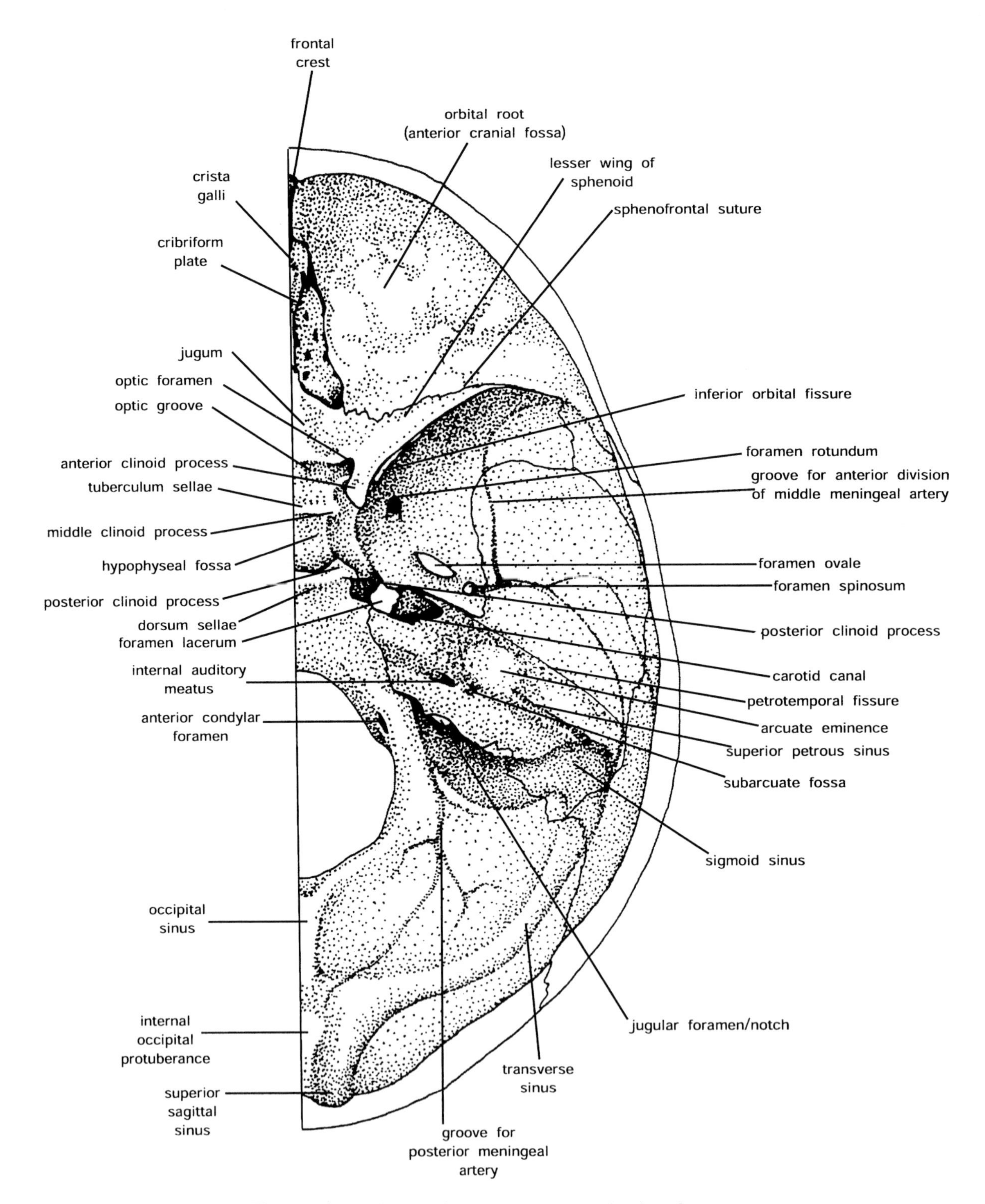

Figure 5. Endocranial compartments and other features.

may exhibit a *superior petrous sinus* (Abri Pataud) or it may not (Tabun C1). The *subarcuate fossa* may (Krapina C) or may not (Cro-Magnon 1) be fully closed over. The superior plane of the petrosal may be very wide (Sangiran 4) or relatively narrow (Cro-Magnon 1).

POSTERIOR ENDOCRANIAL COMPARTMENT (Figure 5)

The *internal occipital protuberance* may be well defined by its surrounding grooves (Swanscombe), low (Bilzingsleben 1), or poorly defined (Sangiran 4). The *internal occipital crest* may similarly be well defined by the impressions of the occipital lobes adjacent to it (Swanscombe) or less well defined (Sangiran 4).

MAJOR ENDOCRANIAL SINUS IMPRESSIONS (Figure 5)

The *superior sagittal sinus* is always identifiable when accessible but presents little notable variation in morphology. However, the *transverse sinuses* may be clearly asymmetrical in their divergence from the superior sagittal sinus (Swanscombe), or they may diverge more symmetrically (Sangiran 4). In addition, they may be quite deeply (Swanscombe) or more faintly (Sangiran 4) impressed. The inferiormost portion of the *sigmoid sinus* may be short (Gibraltar 1) or relatively long (Engis 1).

MANDIBLE (Figures 6 and 7)

Lower dental arcades vary from long and narrow (Qafzeh 9) to wide and short (Mauer). They may be broadly curved around the front (Tabun 2), tightly curved (Montmaurin), or squared off (Krapina 59). The *symphyseal region* may be quite broad from side to side (Mauer) or narrow (Isturitz 1934–65). It may be tall vertically (Tabun 2) or short (Montmaurin). Externally, it may bear a central *keel* that may be thin (Qafzeh 11) or moderately thick (Tighenif 1), and may (Cro-Magnon 1) or may not (Tighenif 2) expand into a distended inferior margin. Where present, the symphyseal keel is bounded by variably deep *mental fossae* (Predmosti 3). The symphyseal region may bear a moundlike swelling externally; this may be either low and broad (Spy 1) or distinctly teardrop shaped (Skhul 6). There may (Krapina 55) or may not (Qafzeh 11) be a *subalveolar depression* running horizontally below the incisor roots. The inferior border of the mandible may (Mauer) or may not (Tabun C1) be thickened on either side of the *mental foramen*. This foramen may lie posteriorly under M1 (Ehringsdorf 1009/69), or more anteriorly, under P2 (Abri Pataud). *Inferior marginal tubercles* may (Krapina 59) or may not (Tabun C1) lie on the inferior margin of the corpus below M1. The lower margin of the front of the jaw between the M1s may (Spy 1) or may not (Montmaurin) be elevated relative to the rest of the *mandibular corpus*. This latter may be taller anteriorly than posteriorly (Tabun 2), or it may be relatively uniform in height throughout its length (Montmaurin).

When viewed from the side, the anterior margin of the *ramus* may lie entirely behind (La Ferrassie 1), or may to some extent cover, the last molar (Tighenif 1); in the former case, a *retromolar space* is present. In some individuals (Skhul 5) the anterior margin of the ramus is indented to exhibit a *preangular notch*. The *gonial region* may be relatively smooth (La Ferrassie) or rugose (Mauer) externally. Viewed from behind, the *gonial angle* may be straight (AMNH 99.1/92), inflected medially (Kebara 1), or everted (Oberkassel). Internally, the gonial region may be weakly muscle scarred (Skhul 5) or generally rugose (Chancelade), or it may bear a distinct *pterygoid tubercle* (La Ferrassie 1), which may be flanked by subsidiary rugosities (Amud 7). The bone of this region may be as thick as the rest of the ramus (Spy 1), or it may thin toward its posterior margin (Skhul 5).

The *mandibular foramen* may be laterally compressed (Skhul 5), or it may be more circular (Amud 1). It may (Chancelade) or may not (La Ferrassie 1) bear a *lingula*. The foramen may be directed quite posteriorly (Cro-Magnon 1) or face upward (Kebara). The internal aspect of the coronoid process may (Mauer) or may not (Predmosti 3) bear a distinct vertical *internal coronoid pillar* that runs from close to the tip of the process down to the *internal alveolar crest*. The *coronoid process* may (Tighenif 1) or may not (Skhul 5) rise noticeably above the level of the *mandibular condyle*. The deepest point of the *sigmoid notch* may lie at its midpoint (Predmosti 3), or close to the condyle (Tabun C1). The coronoid process may be long a/p at its base (Mauer) or more slender (Grimaldi 4). The *sigmoid notch crest* may run to the midpoint of the mandibular condyle (La Quina 5) or to its lateral extremity (Skhul 5).

The *mylohyoid lines* may be faint (Cro-Magnon 1), or relatively rugose (La Ferrassie 1); the more rugose, the more strongly this line emphasizes the *subman-*

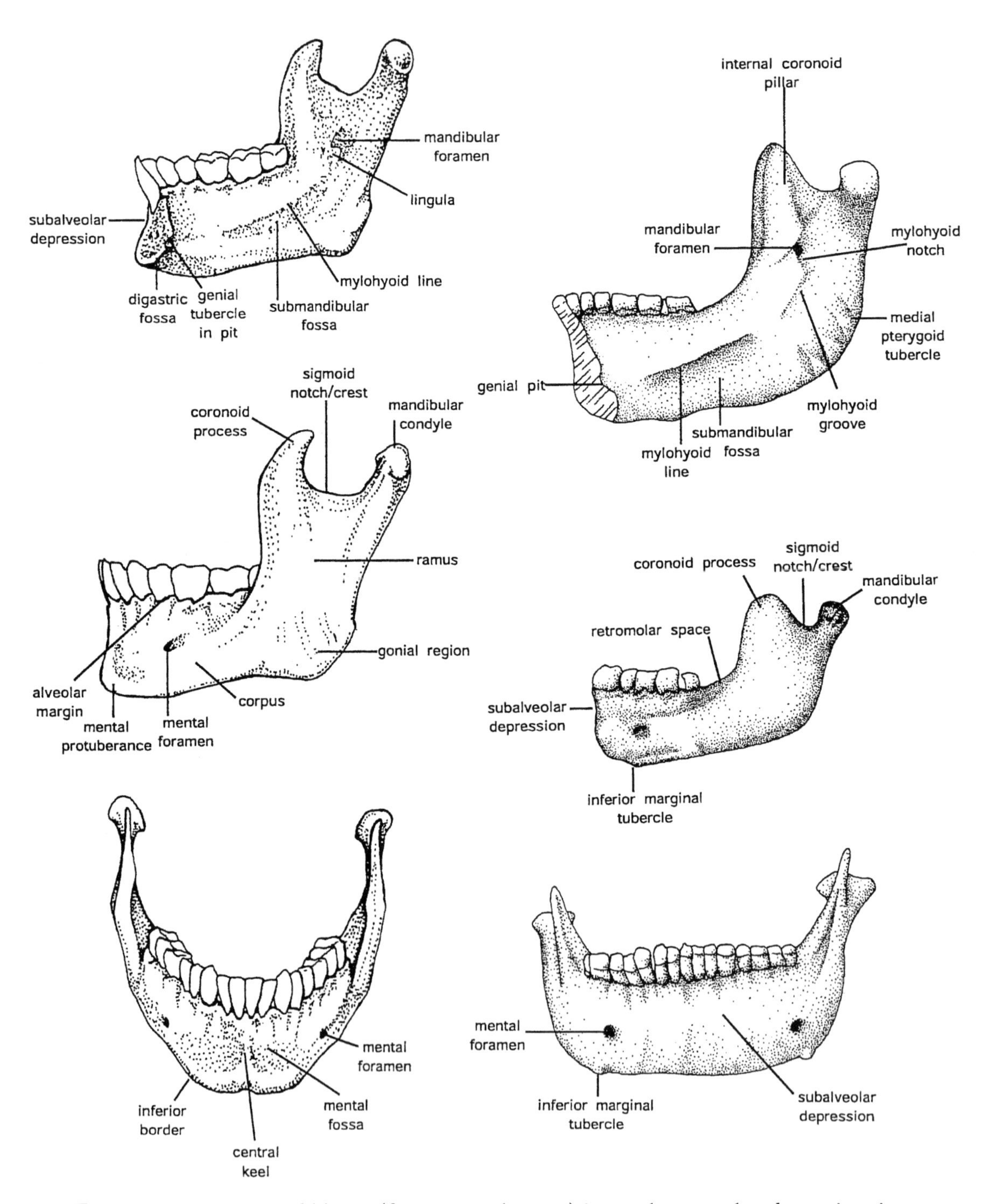

Figure 6. Various mandibles in (from top to bottom) internal, external and anterior views.

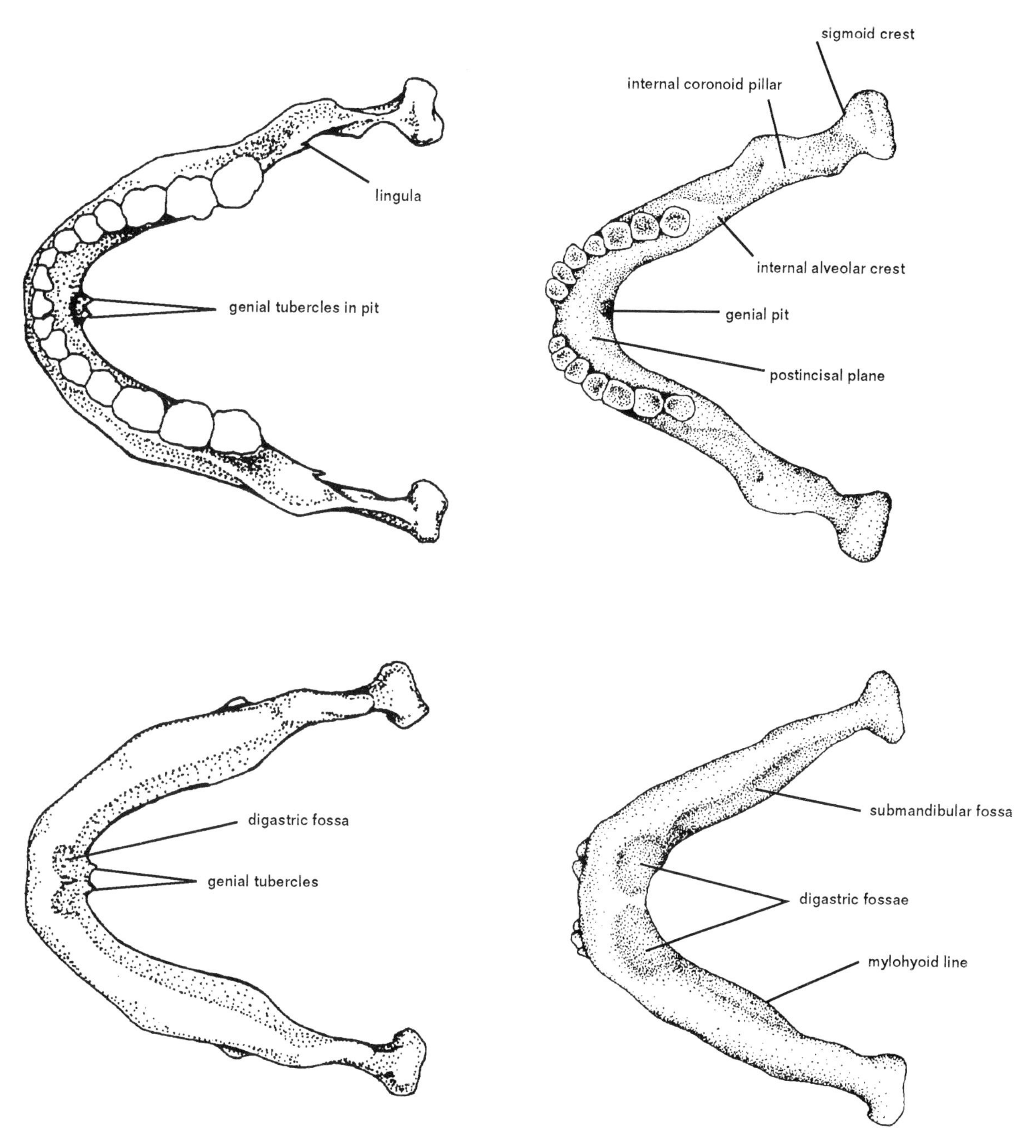

Figure 7. Various mandibles, viewed from below and above.

dibular fossa beneath it. The genial region may be excavated (*genial pits*), either simply (Skhul 5) or with the addition of small *genial tubercles* (Skhul 5), or this region may bear genial tubercles alone. The *postincisal plane* may be relatively vertical (Predmosti 3) or inclined slightly (Tabun C1) or markedly (Tabun 2). Inferiorly, the paired *digastric fossae* may be deeply (Mauer) or shallowly (Kebara) excavated. They may be broad from side to side (La Ferrassie 1) or short from side to side (Cro-Magnon 1). They may face directly downward (Amud 1) or face slightly posteriorly (Chancelade).

From below, the symphyseal region may be relatively uniformly thick from front to back (Zafarraya), or it may be thickest at the midline (Isturitz 1934–65).

DENTITION—OVERVIEW

More often than not, teeth are missing from even quite complete specimens and, when present, are commonly worn to a greater or a lesser extent. We report on the completeness and state of wear of the dentitions described. Whenever possible, we also report on the relative sizes of tooth classes. Individual teeth are described in the following sections. As already noted, we adopt the following dental notation for the upper and lower dentitions: I1, I2 for the central and lateral incisors; C for the canine, P1 and P2 for the anterior and posterior premolars, and M1, M2, and M3 for the anterior through posterior molars.

Upper Dentition (Figure 8)

The central and lateral incisors vary greatly in relative size. The lateral upper incisor is always smaller than the central incisor, but the discrepancy may be great (KNM-WT 15000) or moderate (Qafzeh 9). Both incisors may be shoveled (KNM-WT 15000), shoveled with *lingual pillars* (Krapina 158), or normally concave lingually (Qafzeh 9). The crowns and roots may be more or less in line with each other (Qafzeh 6) or may form an obtuse angle, with the roots flexed back (Krapina 123).

Canines may bear a large (Krapina 49) or small (KNM-ER 1813) lingual tubercle or none at all (Qafzeh 9). Crowns may lie in the same axis as the root (Skhul 5), or the root may be flexed back (Krapina 37).

The upper premolars are typically subequal in size, with well-developed *paracones* and *protocones*, and their occlusal surfaces may be variably wrinkled. The paracone and protocone may be relatively close to one another (Krapina 48) or farther apart (KNM-ER 1813). They may be subtended by mesiodistally short and shallow (Krapina 58), mesiodistally short and deep (KNM-WT 15000), or long and moderately shallow (Qafzeh 9) *anterior* and *posterior foveae*. Their internal surfaces may bear crisply distinct (KNM-WT 15000), blunt and distinct (Krapina 58), or very faint (Qafzeh 9) *lophs*. The buccal sides of these teeth, especially, may be markedly swollen beyond the neck (Krapina 49), or they may be essentially vertical (KNM-ER 1813).

The upper molars may decrease in size, notably m/d length, from M1-M3 (Skhul 5), or they may be of subequal in size (KNM-ER 1813). In the former case, the *hypocone* decreases in size disproportionately and, indeed, may be absent on M3 (Kabwe). The upper molars may also be squarish (KNM-ER 1813), or rectangular, with a long m/d axis (KNM-WT 15000). The occlusal surfaces may be highly (KNM-WT 15000), or moderately (Krapina 48) wrinkled or not wrinkled at all (AMNH 99/8423). The paracone and *metacone*, at least on M1 and M2, may be subequal in size (Qafzeh 9), or the latter cusp may be smaller on M2 (KNM-ER 1813). The *preprotocrista* and *postprotocrista* may be subequal in size (Qafzeh 9), or the postprotocrista may dominate (Krapina 48). The preprotocrista may swing broadly around the paracone, encircling a large anterior fovea (Qafzeh 9), or it may hug the anterior face of the paracone, restricting the anterior fovea (KNM-WT 15000). Similarly, the hypocone may be separated from the postprotocrista by a well-developed *posterior fovea* (Qafzeh 9), or the two structures may abut, minimizing the fovea between them (KNM-WT 15000). The hypocone may (Krapina 48) or may not (KNM-ER 1813) distend the distolingual corner of the tooth. The buccal surfaces of the molars may be somewhat swollen (Krapina 48), or they may be vertical (Skhul 5). The protocone may lie far internally (Krapina 48), restricting the size of the *trigon basin*, or it may lie more lingually (KNM-ER 1813). The molar roots may be well separated (M1 of Kabwe), poorly separated (M2 of Qafzeh 6), or not separated at all (Krapina 160). In the latter case, a huge pulp chamber extends a long way down the almost barrel-shaped root.

Lower Dentition (Figure 9)

The lower lateral incisors are always larger than the lower central incisors, but the size discrepancy varies

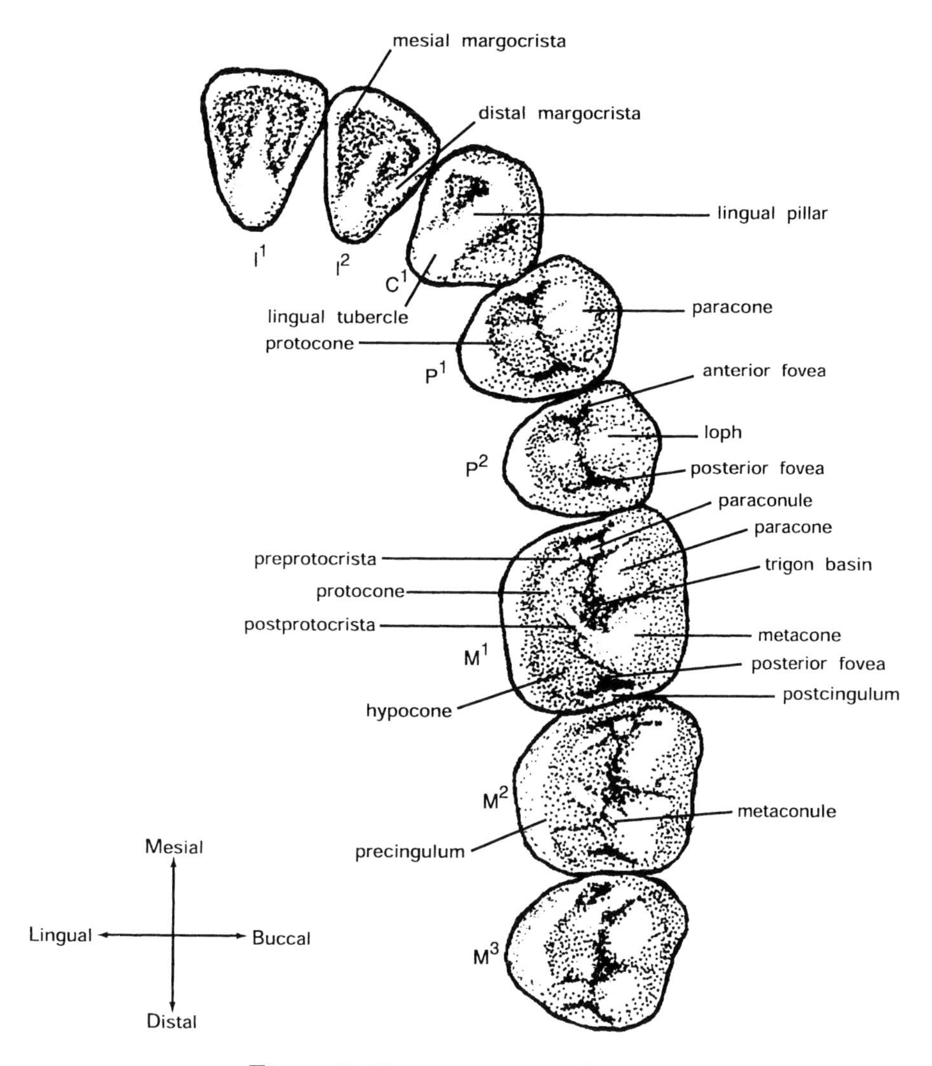

Figure 8. Upper permanent dentition.

from moderate (Rabat) to marked (Skhul 5). Occasionally (Krapina 5) the lingual surfaces of both incisors may be somewhat shoveled and may even bear a *lingual tubercle* (Krapina 69). The lower incisor crowns may lie along the same axis as the roots (Predmosti 3), or the crowns may curve back moderately (Skhul 5) or markedly (Krapina 55).

The lower canines may bear a small lingual tubercle (Skhul 5), a larger tubercle and a lingual pillar (Krapina 120), or only a lingual pillar, which may be modestly (Krapina 119) or strongly (Krapina 138) developed. The crown may lie in the same axis as the root (Predmosti 3) or may be modestly (Tabun C1) or strongly (Krapina 55) flexed back.

In the lower jaw, P1s tend to possess a well-developed *protoconid* and a variably developed lingual tubercle, which may resemble a small *metaconid* (Tabun C1). In the P2s the protoconid and metaconid are usually subequal in size (Skhul 5). The metaconid may be "twinned," either subequally (Krapina 5) or with a larger distolingual cusp (Tabun C1). Anterior and posterior foveae are always present; they may be subequal in size (Tabun C1), or the posterior fovea may be larger (Krapina 58).

Lower molars may decrease in size distally (Tabun C1), or they may be subequal in size (Predmosti 3). All three may bear five cusps (Krapina 58), or the *hypoconulid* may be missing on M3 (AMNH VL/4676) or M2 and M3 (U. Pittsburgh uncatalogued). The *hypoconid* may (Mauer) or may not (Krapina 1) extend lingually beyond the midline of the tooth; the hypoconulid, if present, may be centrally (Mauer) or

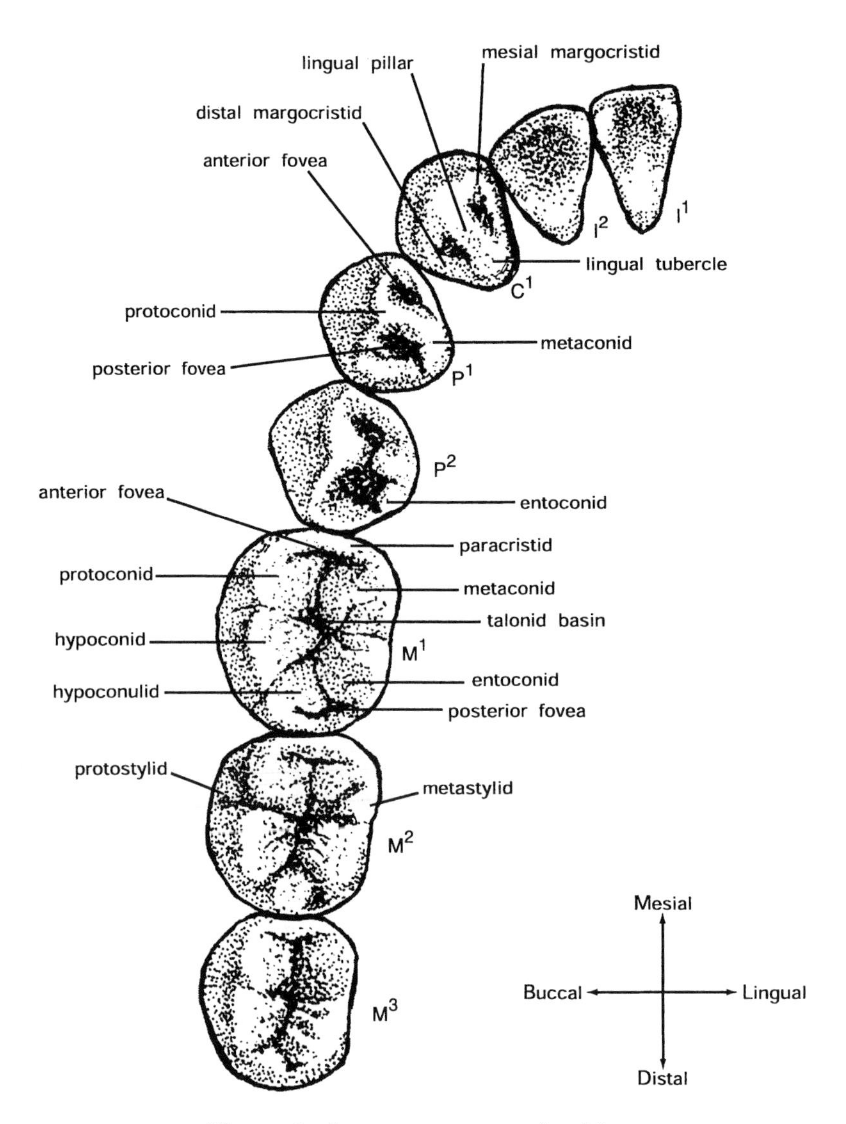

Figure 9. Lower permanent dentition.

slightly buccally (Krapina 80) placed. The anterior fovea may be quite large (Krapina 81), of moderate size (Qafzeh 11), or indistinct (Mauer). Its size depends on how mesially the *paracristid* is placed. The buccal sides of the lower molars may be somewhat swollen (Krapina 59) or more vertical (Skhul 5). The apices of the buccal cusps may be more lingually placed, constricting the *talonid basin* (Krapina 27), or more peripherally positioned (Qafzeh 9). The occlusal enamel may be moderately (Qafzeh 9) or more markedly (Krapina 80) wrinkled. Lower molar roots may be completely separated (Tighenif 3), moderately separated (Skhul 5), or not differentiated at all, with huge pulp cavities (Krapina 8).

MAPS

For the convenience of the reader, all European sites mentioned here are located in the maps in Figures 10–18. These maps are grouped in a single section to make individual sites as easy as possible to find.

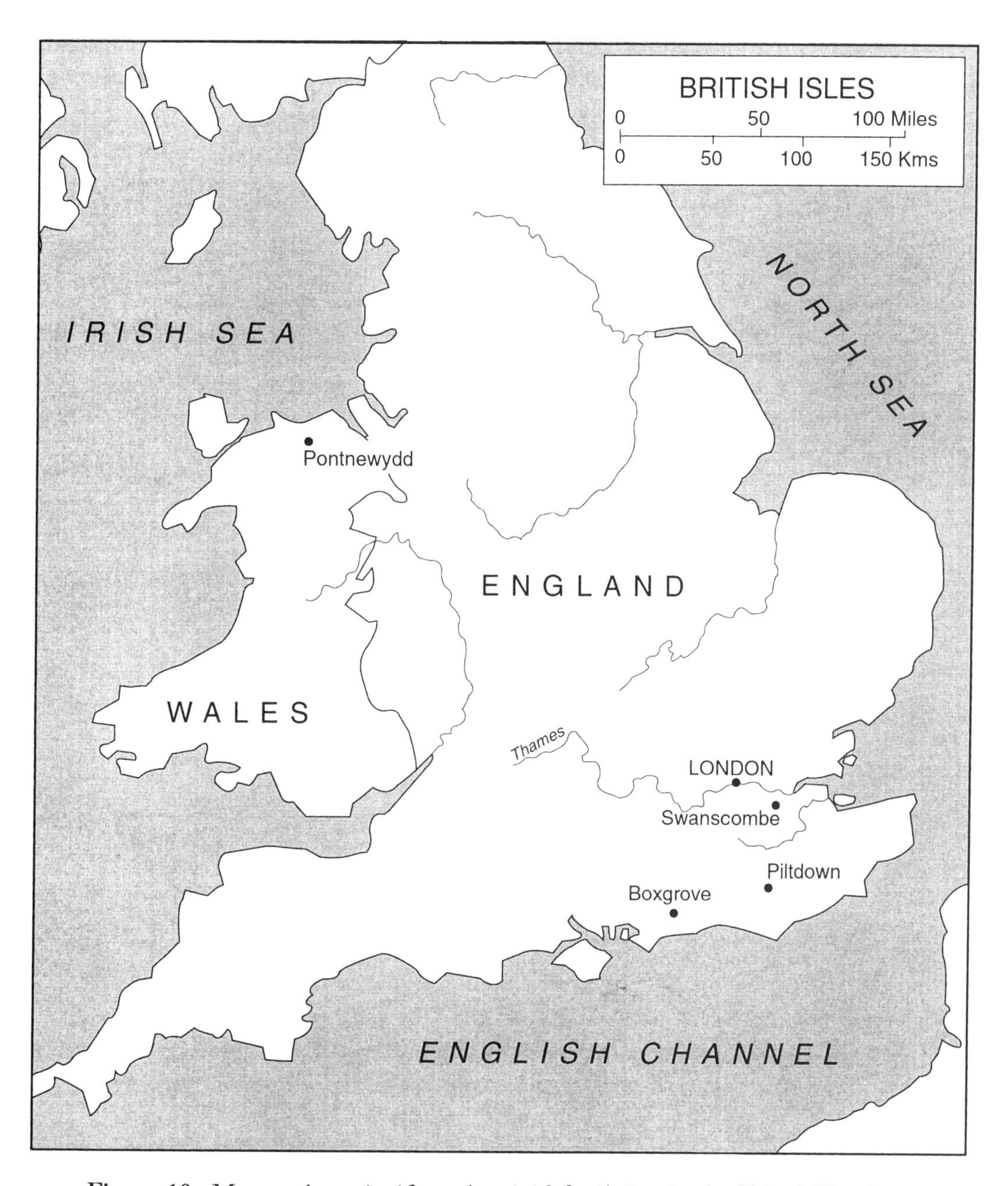

Figure 10. Map to show significant hominid fossil sites in the United Kingdom.

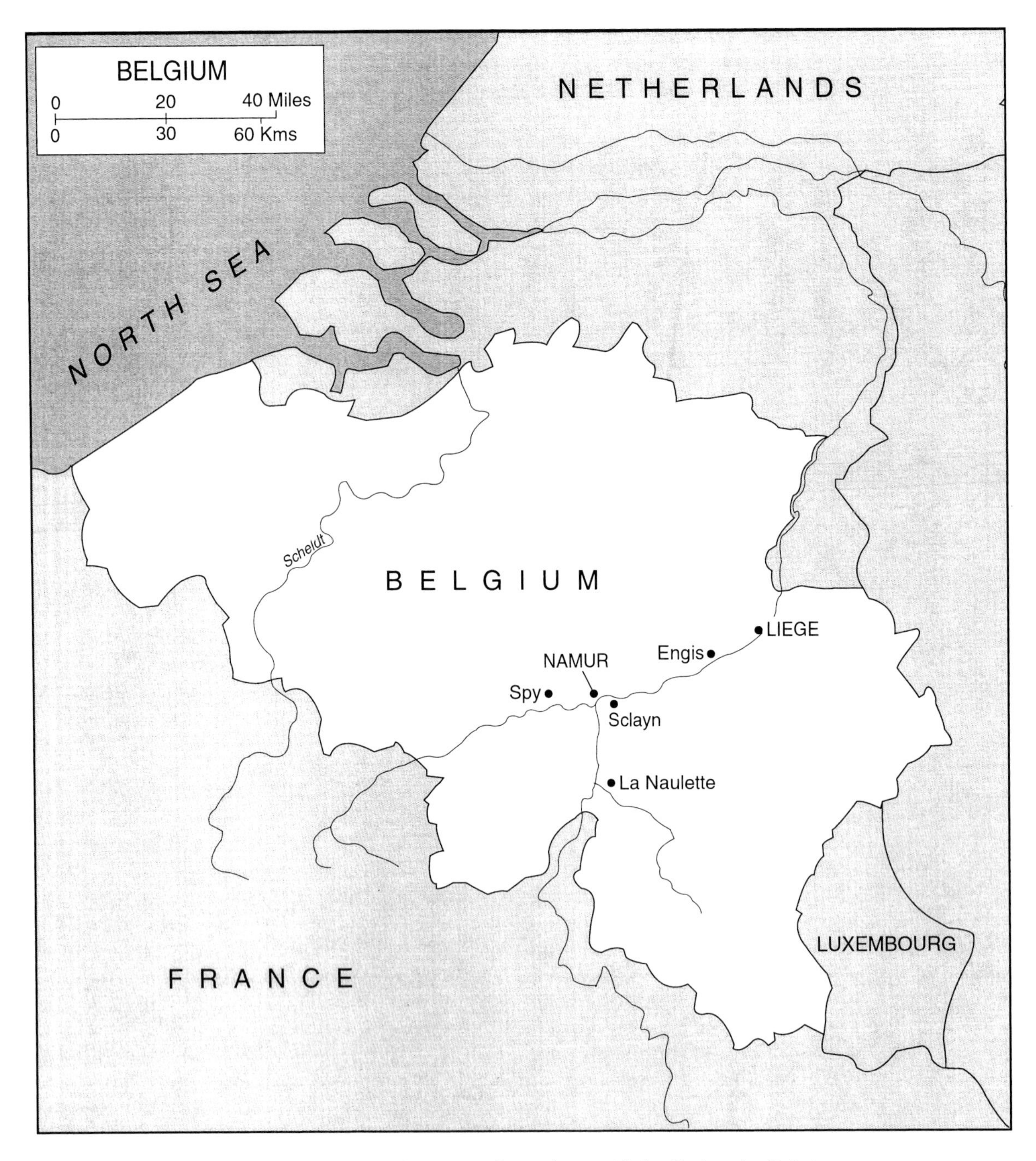

Figure 11. Map to show significant hominid fossil sites in Belgium.

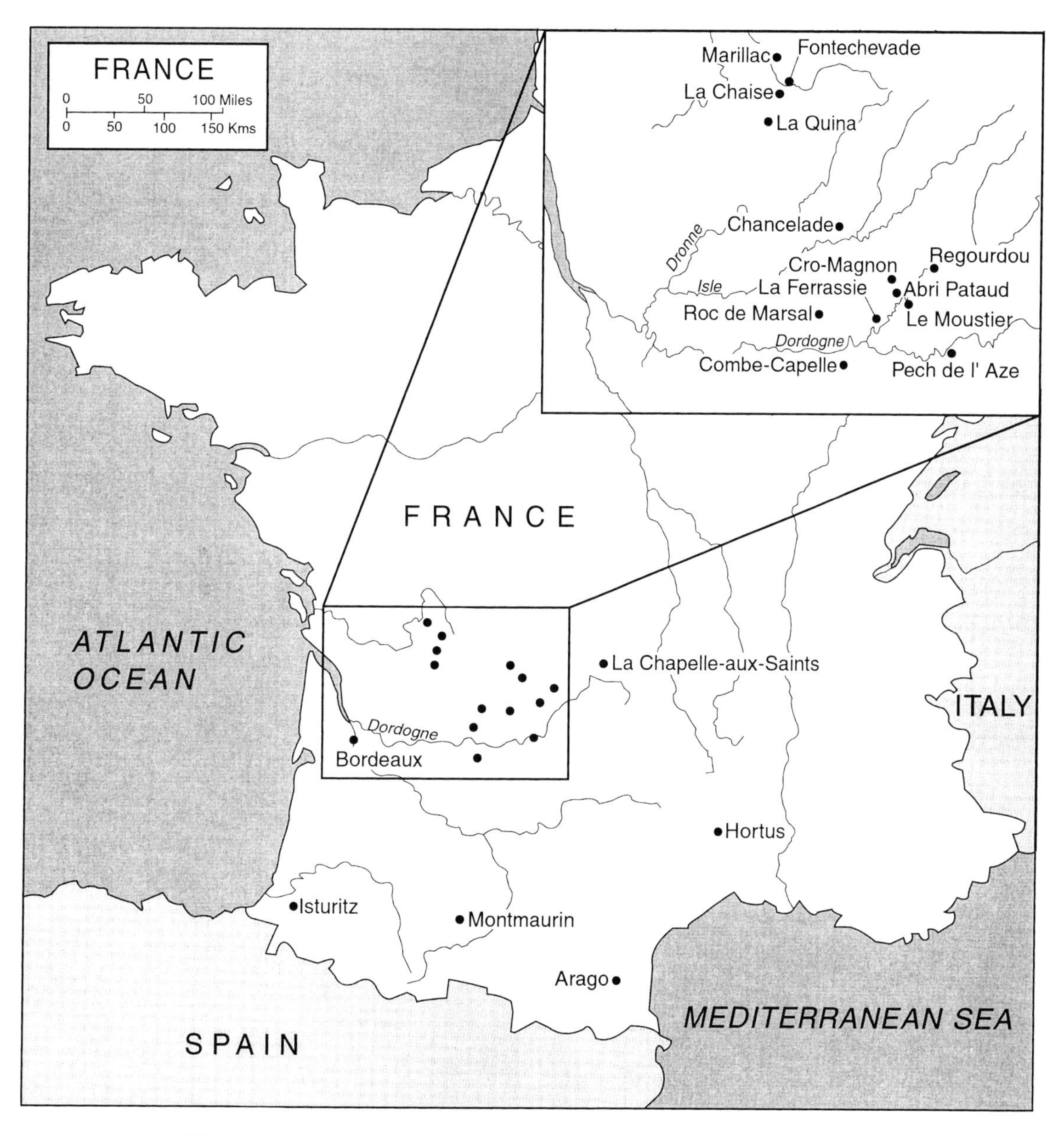

Figure 12. Map to show significant hominid fossil sites in France.

Figure 13. Map to show significant hominid fossil sites in Germany.

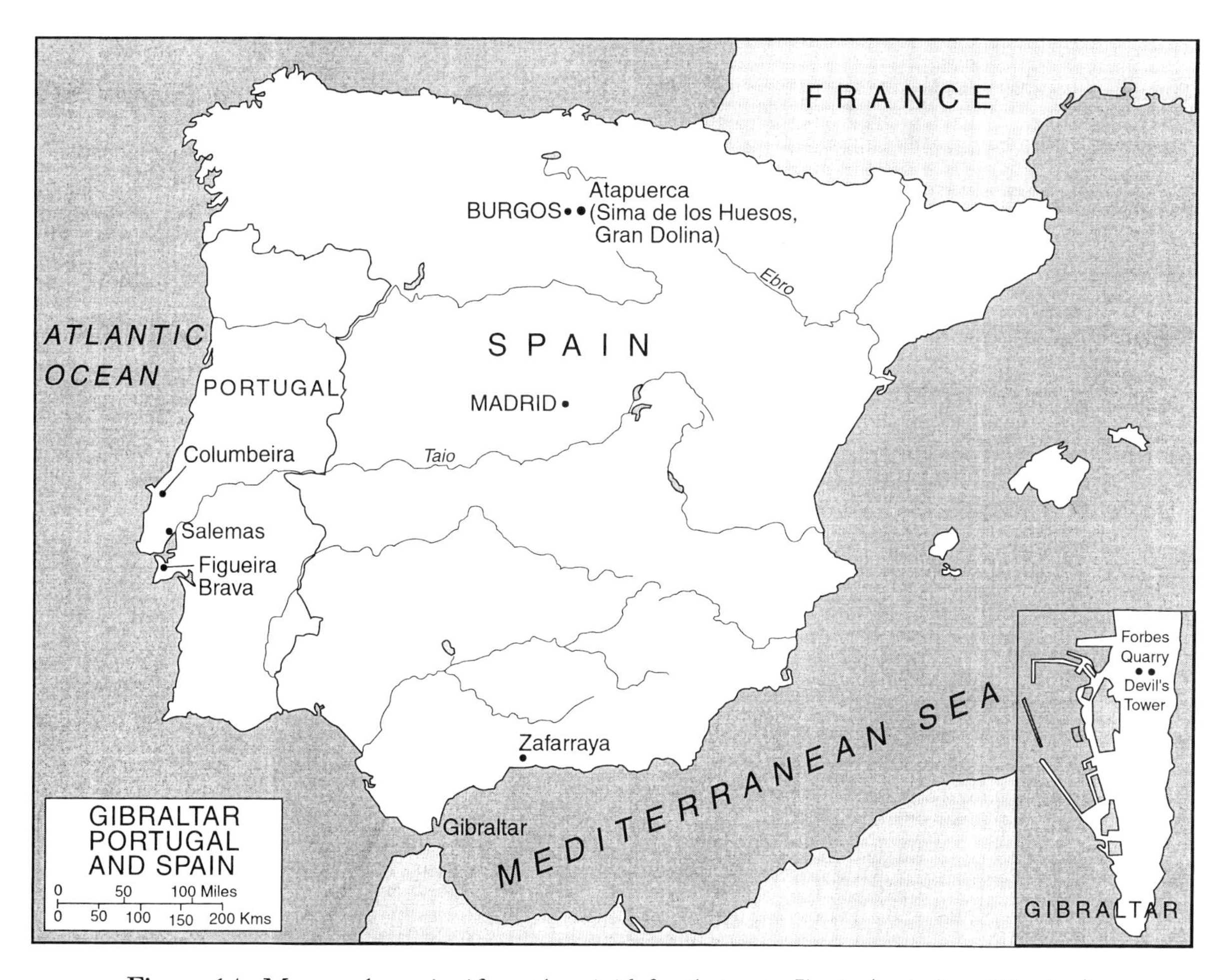

Figure 14. Map to show significant hominid fossil sites in Iberia (including Gibraltar).

Figure 15. Map to show significant hominid fossil sites in Italy.

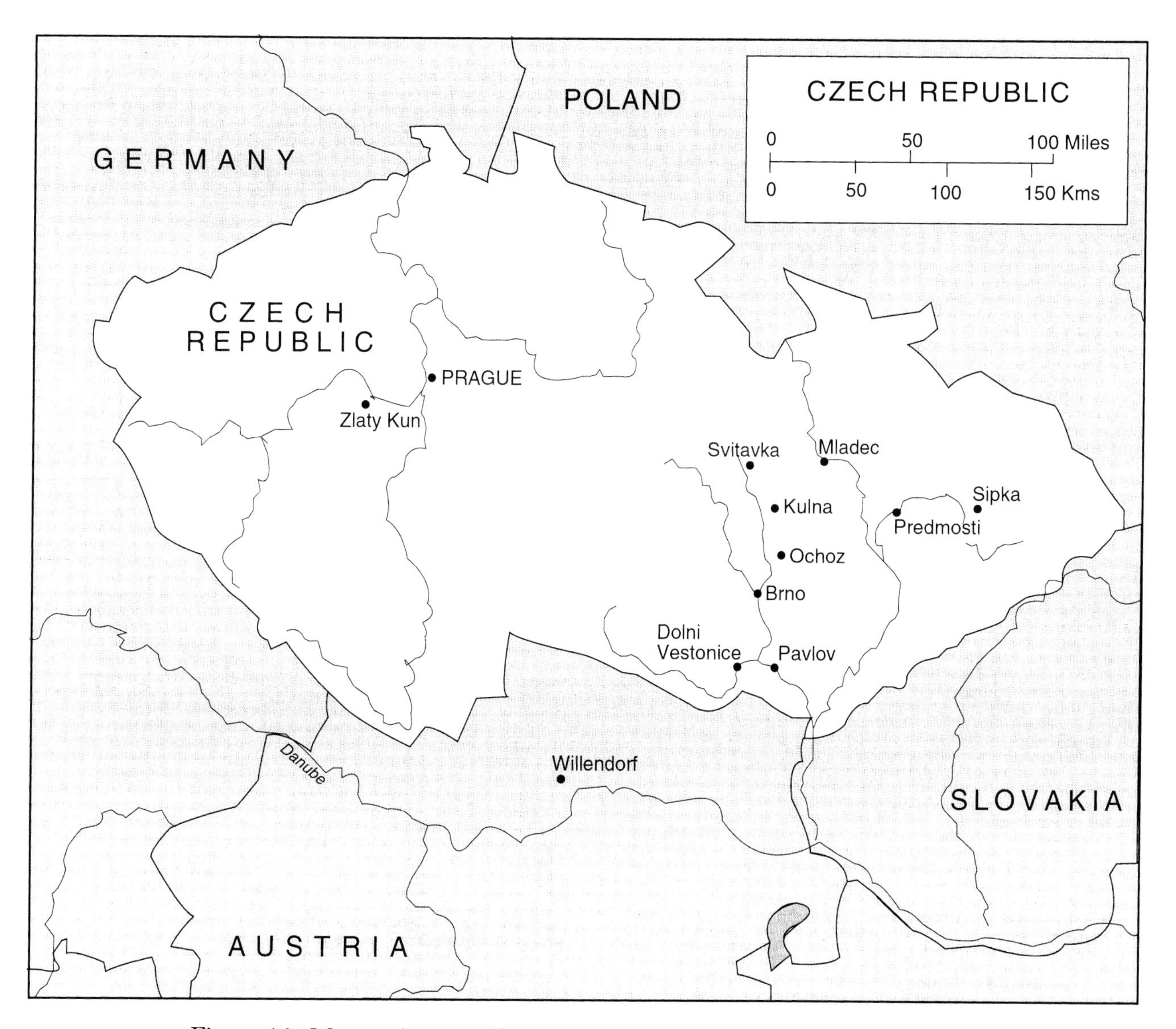

Figure 16. Map to show significant hominid fossil sites in the Czech Republic.

Figure 17. Map to show significant hominid fossil sites in Croatia and adjacent countries.

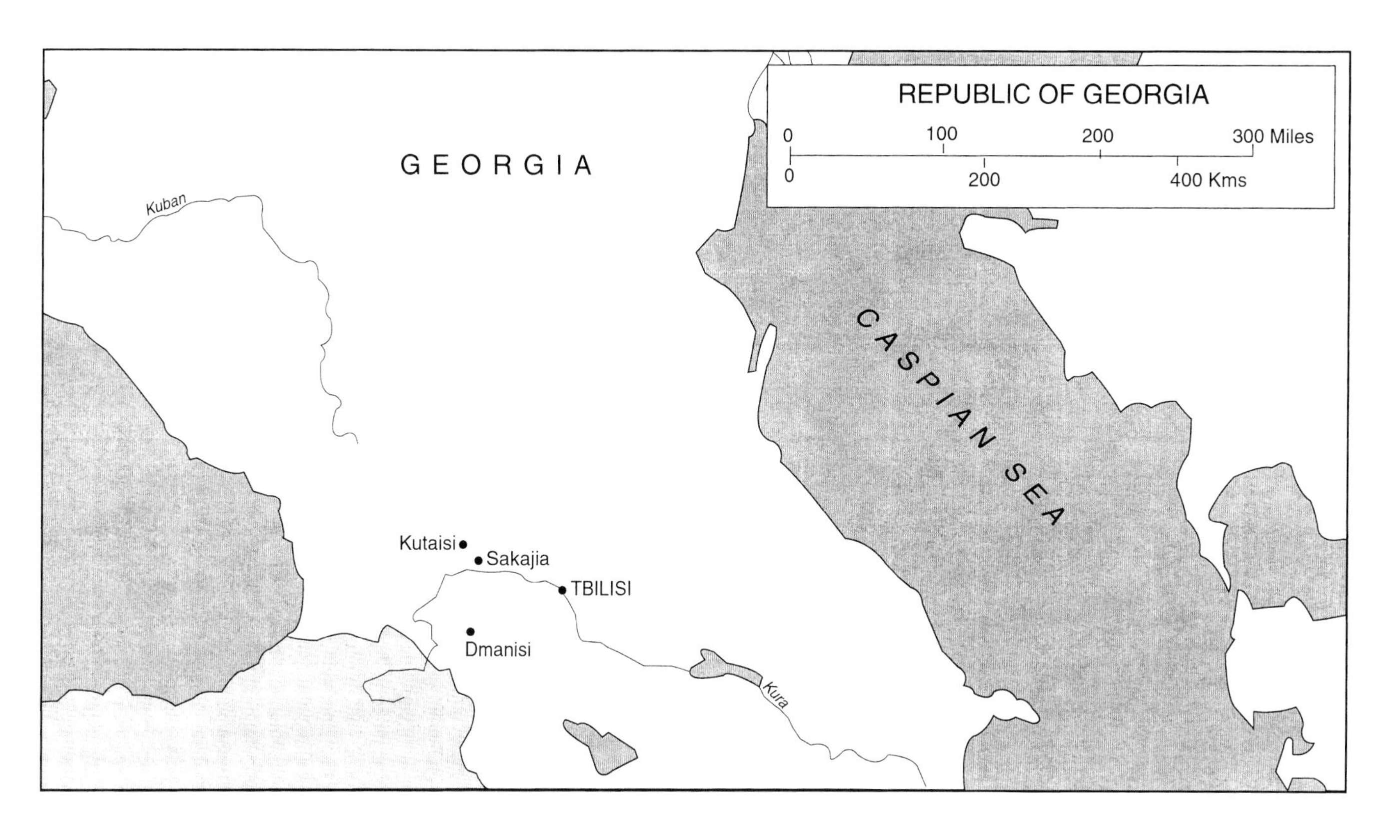

Figure 18. Map to show significant hominid fossil sites in Republic of Georgia.

PART TWO

Site-by-Site Atlas of European Hominid Fossils

INTRODUCTION

In this section we provide documentation of the principal fossils that constitute the European record of the genus *Homo*, from the first Europeans until close to the end of the Pleistocene. It is, of course, impossible to be totally comprehensive: literally thousands of fragmentary European fossils have been attributed to various species of *Homo*, from a very large number of sites. We have had to be selective, and every reader will doubtless find shortcomings in our selection, with favorite fossils and sites left out in favor of possibly less deserving rivals. We can only plead that this was inevitable whatever choices we made and that we have done our best to provide the most generally useful overview we could of the human fossil record. In some cases choice was constrained by accessibility, for we have striven to provide our own first-hand descriptions in as many cases as possible and to minimize our reliance on casts and the literature in our morphological characterizations of the fossils. This approach has been essential for maintaining the consistency of our descriptive protocol; but one unfortunate result, already noted, is that certain fossils that we would have liked to include, but to which we were for one reason or another unable to obtain access, remain lamentably unrepresented in these pages. Such cases are, however, gratifyingly rare, and we have in certain instances resorted to secondary sources where an unacceptably large gap would have been left by excluding materials to which direct access was impossible. This is the case, for example, with the Moravian fossils tragically lost in the fire at Mikulov Castle in 1945. In a couple of instances, too, kind colleagues have contributed direct descriptions or photographs that we were for one reason or another unable to make ourselves. Accessibility also had to enter into decisions concerning the less complete or more obscure fossil fragments that ideally would have qualified for inclusion. This said, the only major group of fossils among which truly arbitrary decisions about inclusion had regularly to be made is that comprised of modern *Homo sapiens* with Upper Paleolithic associations. This group is simply too numerous to be represented comprehensively here, even where access has been no problem; and in making our selection we have generally favored very early representatives of our species, or fossils that have figured prominently in arguments over potential "continuity" between more "archaic" and "modern" types.

In the following accounts, the fossil record of genus *Homo* is presented in the form of a Europe-wide alphabetical listing of sites. The name of the site first given is that by which it is most commonly known; any alternative or complementary names follow in parentheses. Within each site entry, information is presented in the following categories:

1. **Location**. Where the site is: country, region, and in most cases distance and direction to nearest village and/or major town. See also Maps section.
2. **Discovery**. Date(s) of discovery of the fossil(s) plus the name of the individual(s) who made the discovery(ies) or the name of the excavation director(s). Note that the date is not that of the discovery of the site itself, and the names are not necessarily those of the discoverers of the site.

3. **Material.** A short note on what the human fossil(s) from the site consist(s) of. Further details are found under "Morphology."
4. **Dating and Stratigraphic Context.** A brief review of absolute dates (if any) obtained for the site and/or hominids and of the stratigraphic context(s), geological or archaeological, of the locality(ies) of fossil recovery. Where dating is by archaeological association there is some overlap with the next section.
5. **Archaeological Context.** A brief résumé of the cultural association(s) of the hominid fossil(s).
6. **Previous Descriptions and Analyses.** An overview of the history of description and analysis of the fossil(s). This is not intended to be comprehensive or discursive but is simply a very general summary and pointer toward the literature.
7. **Morphology.** Here we present a brief but comprehensive account of the morphology of the cranial, mandibular, and dental *Homo* fossil(s) known from the site. Each account is based on the approach to terminology and description presented in Part 1 of this volume and is made according to a consistent protocol that makes descriptions directly and conveniently comparable from one site to the next. Where the hominid remains consist of a series of fragments, we describe as complete a composite as representation allows. Where there is one particularly well-preserved specimen, our description is based on it, with references added to any less well-preserved fossils from the site that depart from it in one or more morphological features. Where there is more than one well-preserved specimen representative of a morph, we describe more than one if this provides additionally useful information or if the sample is not particularly well represented in the literature. If more than one distinctive morph (as opposed to simple character variation) is represented at a site, we describe each morph separately, using a similar protocol. Where both adults and juvenile specimens are known, these are described separately. Although we note, in the case of multiple morphs, which fossils belong to which morph, we make no attempt at systematic analysis here, and we do not attempt to compare any of the fossil(s) under description with any others beyond those from the same site and belonging to the same morph. The focus is exclusively on individual morphology described by way of the characters discussed in Part 1. The reader will thus be able to make objective comparisons among fossils from every site in the book, unimpeded by confusing comparatives.
8. **References.** For the reader's convenience, all literature citations made in previous sections of the site entry are quoted in full here. Note that this is not, and is not intended to be, a comprehensive bibliography on each site and the fossils found there.
9. **Repository.** The location where the fossils are held.

Abri Pataud

Location

Rock shelter in the town of Les Eyzies de Tayac, valley of the Vézère River, Dordogne, France. Now incorporated into a site museum.

Discovery

Excavations of H. Movius, 1958 (Pataud 1–5), 1959 (Pataud 8–10), 1960 (Pataud 13), 1963 (Pataud 6, 7).

Material

Remains of 13 individuals of varying ages have been recovered at this site, but most are very incomplete and 5 are simply isolated teeth. The best preserved specimens are Pataud 1, the skull of a young female, and Pataud 6, much of an adult skeleton lacking the skull.

Dating and Stratigraphic Context

The thick section at the Abri Pataud includes levels running from the earliest Aurignacian to the Protomagdalenian and Solutrean. Because these levels were dated in the very early days of radiocarbon (e.g., Movius, 1963; Vogel and Waterbolk, 1963), this site became a major yardstick for the chronology of the Upper Paleolithic in western France (e.g., Movius, 1963). Ages of the strata were found to range from 32 to 20 ka. Human remains were concentrated in the upper levels of the site. The complete skull Pataud 1 came from Bed 2, most recently dated to 22 ka (in good agreement with the earlier dates) by AMS radiocarbon (Bricker and Mellars, 1987).

Archaeological Context

Bed 2 contains a Protomagdalenian industry, a distinctive Upper Paleolithic assemblage that underlies the Solutrean (e.g., Movius, 1958).

Previous Descriptions and Analyses

The Pataud 1 skull was described by Movius and Vallois (1959) as that of a fully modern young adult female. This conclusion has never been contested.

Morphology

Pataud 1. Virtually complete cranium and mandible lacking ethmoturbinals, vomer, inferior nasal conchae, entire R side of ethmoid; nasal margin damaged, especially inferior halves of nasal bones; RP2–LC missing. Complete mandible retains R and LP1s and M1–3.

Cranium moderately long, somewhat domed; forehead relatively vertical; frontal and parietal eminences low. Trace of metopic suture on glabella. Glabellar "butterfly" moderately swollen and well defined; its lateral extremities extend beyond large supraorbital notches, to about midpoint of superior orbital margin. More laterally, supraorbital region flat, platelike, inclined posteriorly. Temporal ridges emerge from high up on zygomatic processes of frontal and fade out by coronal suture, posterior to which temporal lines take form of broad, smooth surface that posteriorly recurves smoothly toward parietal notch.

Nasal bones narrow superiorly, angle (not flex) outward below nasion, are quite strongly arced from side to side (nasomaxillary suture appears to lie in trough). Apparently egg-shaped nasal aperture was moderately sized in height and width at superior and inferior portions. Crisp lateral crests terminate lateral to and below thick, not very projecting double anterior nasal spines. Anteriorly positioned groove between spines for cartilaginous nasal septum very thick. On both sides of nasal cavity, low conchal crests slope somewhat up and back; bone below them concave and, posteriorly, not expanded. Anterior lacrimal crest within nasal cavity extends partly over lacrimal groove. Essentially flat nasal cavity floor slopes gently downward posterior to aperture. Short nasoalveolar clivus angles slightly forward. Interorbital region relatively broad. Anterior lacrimal crests poorly defined; wall of lacrimal fossa slightly convex superiorly. Orbits large, rectangular, with thick, slightly everted inferior margin. Orbital floor slightly sunken; infraorbital canal and groove of approximately equal lengths. Large, single infraorbital foramina lie close to inferior orbital margin, within upper part of distinctly excavated canine fossae. Viewed from below, infraorbital region faces anteriorly; from side, slopes back gently. Anterior root of zygomatic arch originates well above P^2/M^1, projects directly laterally; inferior margin concave and bounded laterally by fairly robust maxillary tubercles. From this point, body of zygoma (and arch as well) runs somewhat outward and back to reach fairly wide posterior root.

Squamosals fairly long with quite tall, arced sutural edges. Anterior squamosal descends steeply, smoothly into deeply concave alisphenoid (temporal fossa deepest anteriorly); inferiorly, curves strongly, smoothly toward basicranium (no distinguishable infratemporal fossa). Posterior root of zygomatic arch originates above auditory meatus. Confluent with posterior root of zygomatic arch is moderately developed suprameatal crest that flows posteriorly into taller s/i, lower suprameatal crest that arcs gently upward and is delineated below by shallow groove.

Mandibular fossae very wide laterally, deep, quite long a/p; bounded anteriorly by prominent articular eminences; fossae not completely closed off medially; tubular ectotympanic forms posterior wall. Small, circular auditory meati with rather thick bony walls. Ectotympanic tubes fused to bases of mastoid processes; bear thick, consistently low vaginal processes that run along midlines, do not contact mastoid processes. Styloid pits, with bases of very thin processes, lie very far laterally along vaginal processes; posterolaterally very close to them lie very small stylomastoid foramina. Slightly posteriorly facing carotid foramina lie well back on petrosals. Rather small jugular foramina face forward. Petrosals rather thick externally, not fully ossified along their entire lengths. R foramen lacerum larger than L. R carotid canal incompletely closed over internally.

Mastoid processes very thick, quite long a/p, and muscle scarred on lateral surfaces with blunt, stubby tips that project moderately downward and forward. Mastoid notches (deeper on L) narrow and lie over posterior part of process; from notch, a relatively long parietomastoid suture gently descends to asterion. Notches bounded medially by thick, quite well-developed paramastoid crests, medial to which occipitomastoid sutures lie sunken in narrow creases. Broad, low, quite long, obliquely oriented Waldeyer's crests lie further medially.

Basiocciput rather thick, stubby, quite long a/p. Foramen magnum elliptical, fairly small; posterior margins of relatively long, arced occipital condyles lie at about its midpoint. Postcondylar foramina closed bilaterally; both condyloid canals patent. Bow-shaped superior nuchal line originates just below asterion, becomes more prominent toward midline, where it forms low, crescentic, downwardly pointing crest (not true occipital protuberance). Lambdoid suture rises steeply from asterion, peaks at lambda (to the R of large ossicle). Below lambda, broadly triangular occipital plane protrudes very slightly.

Sagittal and lambdoid sutures only slightly more deeply denticulated than coronal suture; none is strongly segmented.

Upper dental arcade forms moderately narrow parabola. Palate shallow at front, deepens posteriorly with vertical walls above Ms. Low, narrow, triangular palatal torus terminates in blunt, projecting posterior nasal spine. Incisive foramen just behind I1 alveolae. Quite closely approximated medial and lateral pterygoid plates confluent only at inferior ends; foramen ovale lies just at base of lateral pterygoid plate; superior roots of medial pterygoid plates extend laterally beyond margins of basiocciput. Posterior superior root of vomer quite inflated; its alar region lies in front of incompletely fused sphenooccipital synchondrosis.

Internally, petrosals moderately wide, with peaked arcuate eminences (especially on L); superior petrous sinuses present on both sides; region of subarcuate fossa indicated by small pit.

Complete mandible retains R and LP_1s and M_{1-3}. Corpus moderately deep, ramus rather wide. Front of jaw rather narrow but quite strongly arced; corpora diverge posteriorly from it. Low, moderately wide central keel along symphysis fans out, becomes more protrusive toward inferior margin. Wide, very shallow mental fossae lie on either side of the keel; from below, anterior surface of symphysis comes to peak; lower part of symphysis thick from front to back. Inferior margin lowest under Ms, producing "rocker" configuration. Large mental foramina, with an accessory on the L, lie under P_2s. Internally, no postincisal plane; symphyseal surface rather vertical; genial region adorned with tall, thin crest (no tubercles or pits). Well-defined, a/p elongate digastric fossae face somewhat backward. Mylohyoid lines represented by low muscle scarring; moderately excavated submandibular fossae lie well below. Rather vertical rami bear shallow preangular sulci on their anterior margins; most of M_3s remain obscured in side view. Gonial angle tightly curved with modest muscle scars internally and externally on inferior margin; rest of rami muscle scarred externally. Small, ovoid mandibular foramina, incomplete inferiorly, oriented up and back; superior to them lie short, shelflike lingulae. Short, stubby, pointed coronoid processes lie lower than condyles. Broad sigmoid notches deepest at midpoints. Sigmoid notch crests course to lateral extremities of condyles, which are somewhat convex in coronal plane. Internal surfaces of coronoid processes bear broad, thickened bony pillars that terminate behind above and behind M3s.

I^1 roots were thick, conical (judged from alveoli); subtriangular C^1 roots were not very large. RP^1 was almost completely double rooted; RP^2 bore single root with distal groove. LP^{1-2} quite similar in morphology; P^2 only slightly smaller; protocones slightly lower than paracones, but two cusps essentially subequal in area (are slightly closer together on P^1 than on P^2). Molar sizes and morphologies (especially M^{2-3}s) compromised by pathological development. On both sides, but particularly on the L, molars bear accessory cuspulelike structures on distal margins. Quite heavily worn, apparently unpathological M^1 typified by large, distolingually swollen hypocone; fissurelike trigon basins essentially confined to crease between subequal paracone and metacone.

I_{1-2} alveolae not very compressed laterally: C_1 alveoli ovoid, moderate size. Rather small P_1s bear tiny anterior and posterior foveae; on RP_1, anterior fovea continues down side of tooth as a thin crease. Region of metaconid low on both P_{1-2}. Single P_2 alveolae large, ovoid. M_2 and especially M_3 bear accessory cusps. Rather small, somewhat elongate M_{1-2} have large, slightly buccally shifted hypoconulids. Preserved on M_2, transversely fissurelike anterior fovea, some wrinkling of enamel. On M_{1-2}s buccal and lingual cusps of each pair lie opposite one another; hypoconids do not extend lingually beyond midline of teeth.

REFERENCES

Bricker, H. and P. Mellars. 1987. Datations ^{14}C de l'Abri Pataud (Les Eyzies, Dordogne) par le procédé "accélérateur–spectomètre de masse." *L'Anthropologie* 91: 227–234.

Movius, H. 1958. The Proto-Magdalenian of the Abri Pataud, Les Eyzies, Dordogne. *Ber. Vth. Intern. Kongr. Frühgesch.* Hamburg: 561–566.

Movius, H. 1963. L'âge du Périgordien, de l'Aurignacien et du Proto-Magdalénien en France sur la base des datations au carbone 14. *Bull. Soc. Mérid. Spel. Préhist.* 6–9: 138.

Movius, H. and H. Vallois. 1959. Crâne Proto-Magdalénien et Vénus du Périgordien final trouvés dans l'Abri Pataud. *L'Anthropologie* 63: 213–232.

Vogel, J. and H. Waterbolk. 1963. Gröningen radiocarbon dates IV. *Radiocarbon* 14: 163–202.

Repository

Laboratoire d'Anthropologie Biologique, Musée de l'Homme, Place du Trocadéro, 75113 Paris, France.

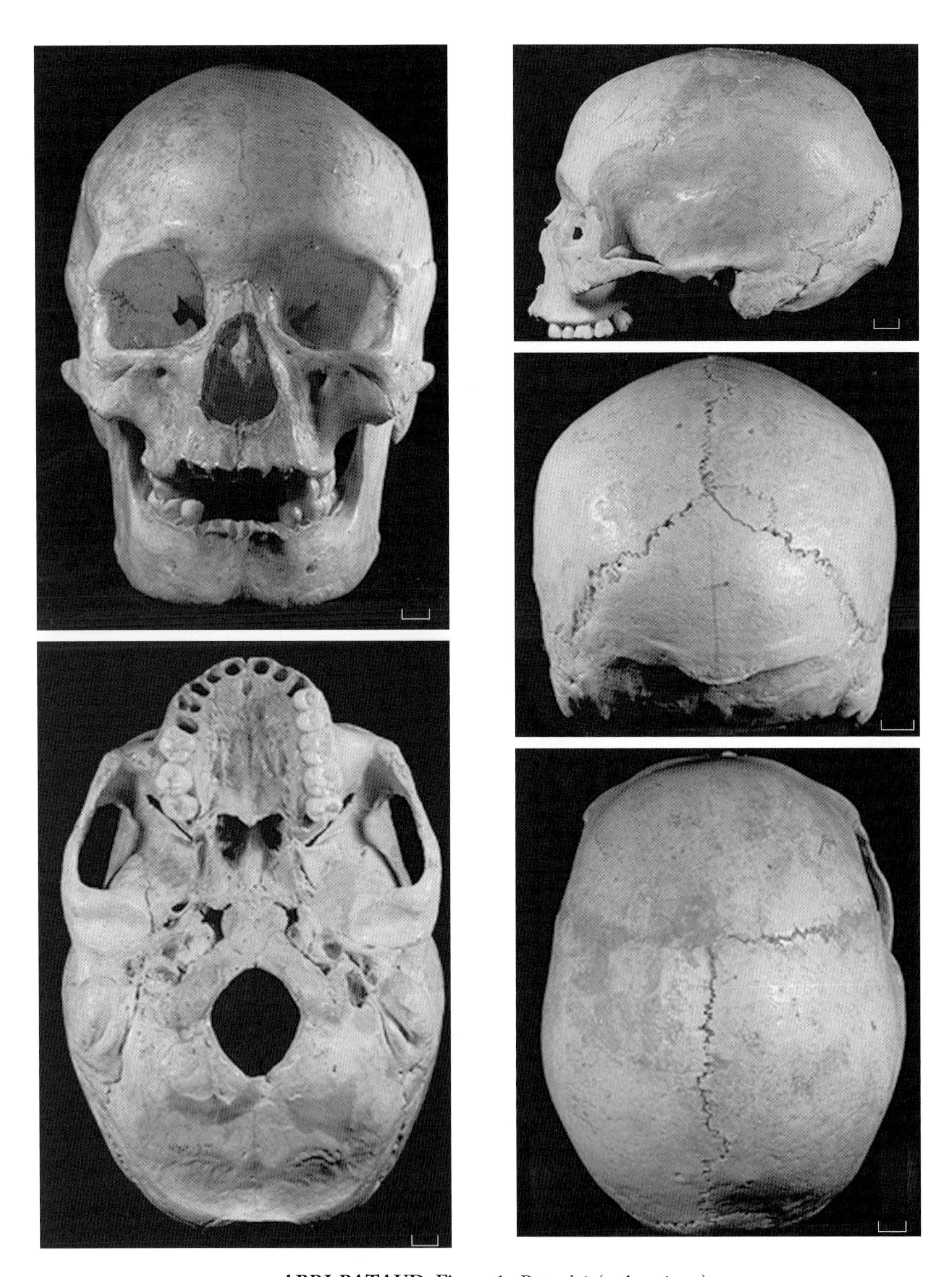

ABRI PATAUD Figure 1. Pataud 1 (scale = 1 cm).

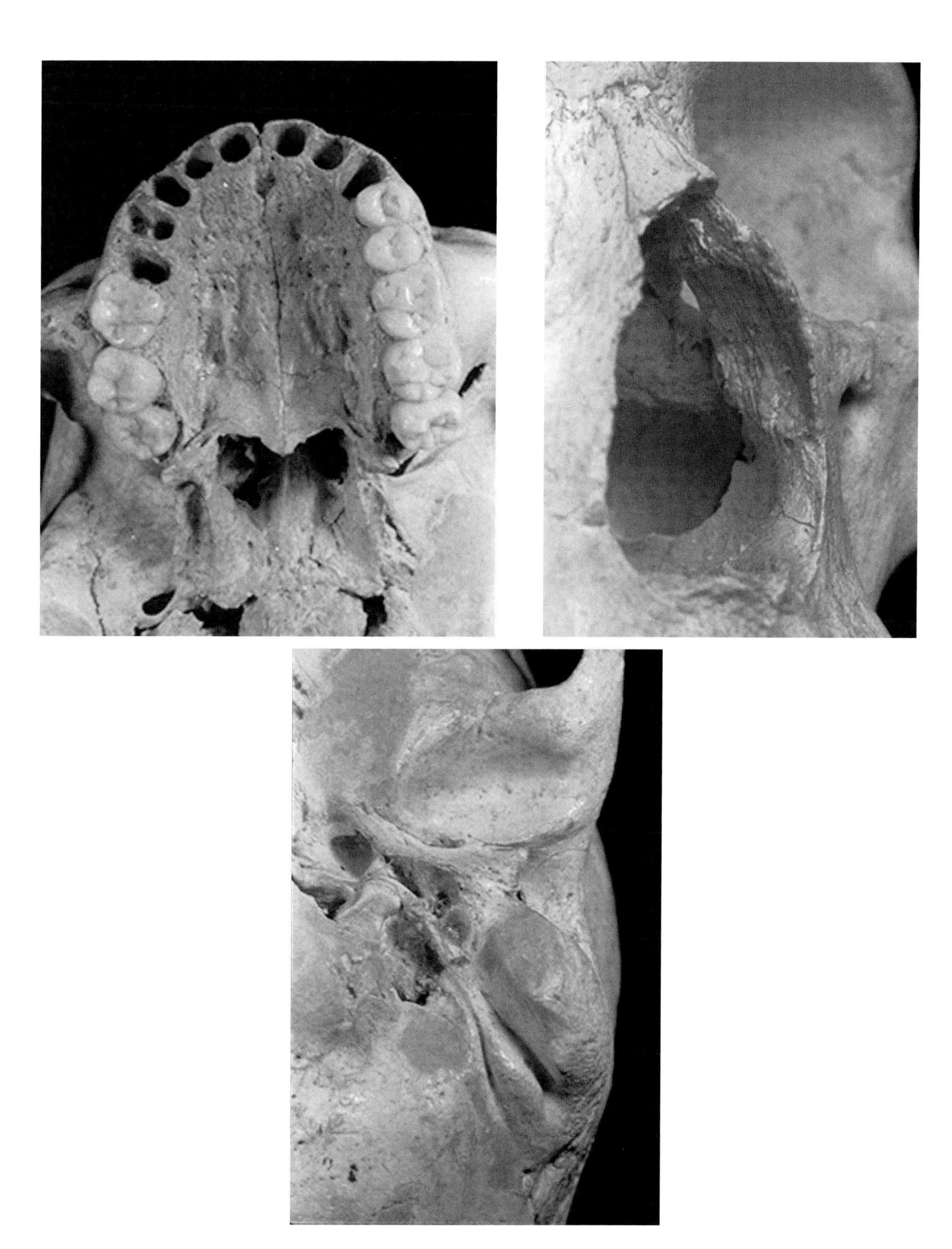

ABRI PATAUD Figure 2. Pataud 1 (not to scale).

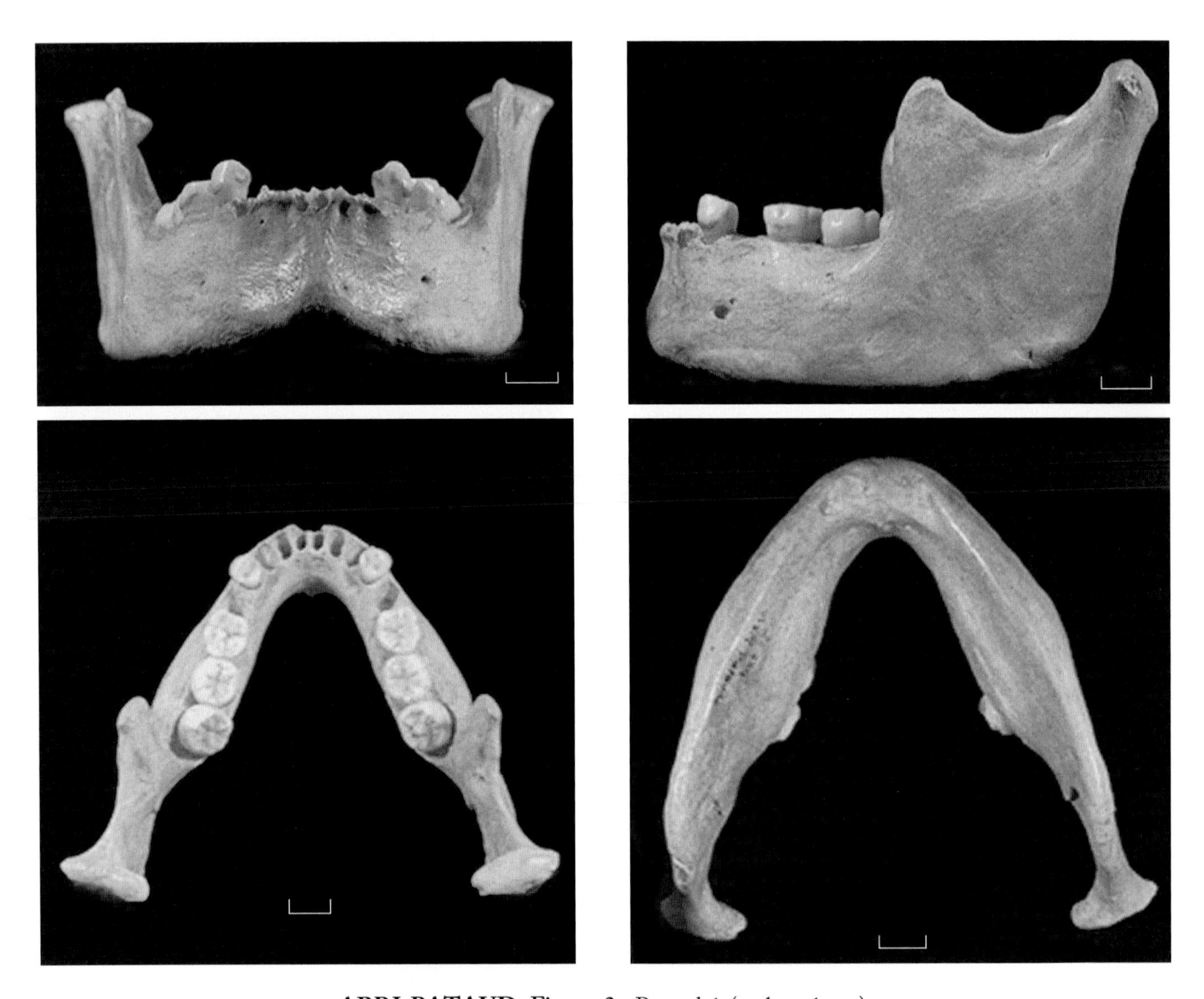

ABRI PATAUD Figure 3. Pataud 1 (scale = 1 cm).

Arago (Tautavel)

Location

Karst cave in the valley of the Verdouble river, between the towns of Tautavel and Vingrau, some 19 km NW of Perpignan, southern France.

Discovery

Ongoing excavations of H. de Lumley. First hominid found April 1964, with a steady trickle of discoveries since. Arago 21 was discovered in 1970.

Material

Some 70 hominid fossil fragments (including around 50 isolated teeth) had been found by the time of writing. The most significant of the cranial remains so far discovered are Arago 21, a face and frontal bone, with which a parietal bone, Arago 47, may be associated. Important mandibular finds include the Arago 2/2bis lower jaw and the Arago 13 hemimandible. Postcranial finds are mainly fragmentary but include most of a left innominate bone (Arago 44).

Dating and Stratigraphic Context

A thick section of Pleistocene deposits has been divided into a basal and an upper unit, the latter divided into five major layers, each with multiple occupation levels. Various approaches to dating have been tried, including fission track, ESR, TL, and paleomagnetism (summarized in de Lumley and Labeyrie, 1981; see also de Lumley et al., 1984), not always with consistent results. A recent synthesis by Iacumin et al. (1996) suggests that the upper unit was laid down between about 650 and <92 ka, and that the part of Layer III from which most of the hominid fossils come (including those mentioned above) is about 450 ka old. A direct U-series date on the Arago 21 specimen by nondestructive gamma spectrometry gave almost exactly this age, albeit with a very wide margin of error (Yokoyama and Nguyen, 1981). Nevertheless, the possibility of a somewhat later date, as suggested by certain faunal analyses (e.g., Chaline, 1971), has yet to be completely eliminated.

Archaeological Context

The lower levels at Arago contain a Tayacian (mode 1) stone tool assemblage, with "choppers" and numerous flake tools, mostly in quartzite. Flakes tend to be small and are often retouched (e.g., de Lumley, 1976). Above the hominid levels Acheulean assemblages appear, most often using schist as a raw material (de Lumley, 1976). The various occupation levels that have been identified do appear to represent living floors of some kind, but to what extent the numerous broken faunal (notably reindeer) remains found in them represent hunted rather than scavenged carcasses remains debated.

Previous Descriptions and Analyses

The authors charged with the initial descriptions of the Arago hominids at first gave them the noncommittal epithet of "anteneanderthals" (e.g., de Lumley and de Lumley, 1971) based on time rather than morphology (and on an implicitly linear view of human phylogeny). Shortly afterwards a divided viewpoint developed whereby one school (mainly French) preferred to call these fossils "advanced *Homo erectus*" (e.g., M.-A. de Lumley and Spitéry and other contributors to de Lumley, 1982), while another group of (mostly anglophone) authors elected to dub them "archaic *Homo sapiens*" (e.g., Wolpoff and others in de Lumley, 1982). See Tattersall (1995a) for a discussion of this bizarre dichotomy and Day (1986) for a compromise proposal that recognizes the Arago fossils as a "*Homo erectus/Homo sapiens* transitional form" (p. 53). Stringer et al. (1984) pointed out certain apomorphies that could implicate the Arago hominids in the ancestry of the Neanderthals, and some authors have grouped these fossils, along with such forms as Petralona, Bodo, and Kabwe, into the species *Homo heidelbergensis* (e.g., Tattersall, 1986, 1995b; Rightmire, 1990; Stringer and Gamble, 1993). Those who accept this species have generally felt that *Homo heidelbergensis,* thus defined, could stand in the common ancestry of both *Homo sapiens* and *Homo neanderthalensis*, but Bermudez de Castro et al. (1997) have proposed that it stands uniquely in the lineage leading to the Neanderthals. The cranial capacity of Arago 21 was roughly 1150 cc (Holloway, 1985).

MORPHOLOGY

Arago 2 (and 2bis)

Adult. Partial mandible missing all of L ramus, R coronoid process, small pieces of R ramus and corpus. Only teeth present: LP2, broken LM1, RM1, R and LM2, and LM3, all very worn and chipped. Alveoli for remaining teeth complete or virtually so.

Very long, lightly built mandible; corpora shallow, only moderately thick. Symphyseal region broad, broadly arced from side to side; viewed in profile, anterior tooth alveolae angle forward, overhanging bone below. Inferior border of mandible significantly raised in front of LM1 and RM2, i.e., shortest s/i at symphysis. Below roots of anterior teeth, symphyseal profile curves smoothly down and back to inferior margin. Internally, oblique postincisal plane absolutely short, quite long relative to modest corpus height. At bottom of postincisal plane, slight swelling continues posteriorly and quite horizontally, becoming confluent with mylohyoid line, which runs obliquely up to meet internal alveolar crest just behind M3. Long, well-excavated fossa lies below mylohyoid/alveolar crest ridge; continues forward almost to midline, overlying very m/l wide, well-excavated, posteriorly placed, downwardly and somewhat posteriorly directed digastric fossae. R mandibular foramen (crushed) upwardly and posteriorly directed; may have been compressed. Large R and L single mental foramina lie below M1. Anterior root of ramus lies behind alveolus for M3 on both sides (would have been very long retromolar space).

Sigmoid notch was not very deep, at least near condyle where preserved. Sigmoid notch crest runs lateral to midline of very m/l wide, moderately long a/p, coronally curved mandibular condyle. Gonial region very damaged; bone quite thin; margin of angle may have been slightly everted.

Relative to jaw size, tooth roots appear average in size, but molar crowns seem small. P_{1-2} alveoli wide b/l and m/d compressed, not conical.

Arago 13

Adult. R hemimandible, broken at symphysis, missing tip of coronoid process and bone internal to M3; RP1–M3 preserved, also RI1–C alveolae.

Extremely long jaw, corpus very shallow s/i in relation to length. Corpus very wide because of bulge (mandibular torus) below teeth, thinning inferiorly below cheek teeth. Only small portion of symphyseal region to L of midline preserved. In profile, alveolar region smoothly confluent with fairly vertical symphysis, which, inferiorly, curves gently back into inferior border. On the preserved R, below and posterior to fully exposed (possible abscessed) canine alveolus, inferior margin thickens into torus of sorts that extends to level of front of M2. Large mental foramen lies below anterior root of M1. Viewed from below, preserved symphyseal surface relatively flat across (may have been opened as result of abscessing).

As seen on the R, digastric fossa shallow, very long a/p, wide m/l, down and slightly forwardly pointing, very anteriorly placed; its anterior margin extends onto front of jaw, producing low ridge where corpus angles back. Postincisal plane short, somewhat steeply inclined; more vertical surface below is slightly concave; possible genial pit, not tubercle, present. From below, symphyseal region was probably narrow from side to side; corpus diverged strongly posteriorly. Large mental foramen lies below region between P2 and M1; directly below it, inferior margin of mandible raises slightly, reaching greatest superior extent in region of digastric fossa. Inferior border apparently descended again at midline, beyond digastric fossa.

Internally, just lateral to midline, upper half of corpus very bulky, thick. Corpus thins posteriorly, becomes ridgelike internal alveolar crest behind M3, then curves up internal face of ramus below coronoid process; thick inferiorly, this ridge fades out well below lowest level of sigmoid notch. Mandibular foramen tall, compressed, points posteriorly and slightly up; from its notchlike inferior border, deep mylohyoid groove descends rather steeply and forward. Superior margin of foramen thickened; not distended into lingula.

Ramus quite long a/p; not very tall at level of condyle (tip of coronoid process missing). Anterior root of ramus originates just below distal part of M3, which is exposed in front of the backwardly sloping anterior ramal margin, which becomes almost vertical higher up. Inferior margin of gonial region relatively straight, thickened along its rim; this thickening more pronounced laterally in anterior part and more pronounced medially in more posterior part. This thickening aside, ramal surface quite smooth externally; internal surface bears series of muscle scars (most inferior is largest). Base of broken coronoid process is long a/p; its tip probably stood well above level of moderately m/l broad, a/p long, gently coronally arcuate mandibular condyle. A/p short sigmoid notch smoothly and tightly curved; deepest point lies close to condyle. Sigmoid notch crest runs to midline of condyle.

I1–2 alveolae wide b/l, very deep; C alveolus very large in surface and also extremely deep. Preserved crowns quite massive, not very heavily worn. P1 appears longer buccally m/d than P2. Strong lingual tubercle on P1 lies well below protoconid, to which it is connected by a stout crest, beside which lie distinct, deep anterior and posterior foveae (latter is the larger). From either side of somewhat centrally placed protoconid, stout crests descend to either side of lingual tubercle, from which they remain separated by fairly deep grooves. Apex of protoconid somewhat internally placed, giving crown quite pronounced buccal slope. P2 slightly longer lingually than buccally; moderately developed metaconid lies even more mesially than mesially placed, somewhat taller metaconid (posterior part of tooth much larger than anterior part). Stout, short crest runs between protoconid and metaconid, enclosing tiny, deep fovea. Even stouter crest runs distally from protoconid around much more expansive talonid basin and around lingually to the metaconid. Somewhat centrally on its distal margin, this crest bears slightly raised area; at base of protoconid on its internal side, this crest bears an even smaller cuspulid-like swelling. Somewhat internally placed protoconid apex produces long buccal slope.

All Ms elongate, ovoid in occlusal outline. M2 larger in all dimensions than M1; M1 only slightly wider, longer than M3. All three Ms have internally placed buccal cusps (rounded buccal surfaces), well-developed trigonid and talonid basins; their four major cusps all converge at midline of crown. Enamel of slightly worn M2–3 deeply but sparsely crenulated; M1 was probably similarly configured. On M1–3, stout paracristid encloses trigonid basin. On M1, large hypoconulid lies just buccal to midline of crown; on M2–3, is much wider b/l and straddles midline. On M2, buccal side of protoconid also noticeably swollen.

Arago 21

Adult. Very crushed face including frontal, slightly reconstructed in upper part of frontal. Externally and internally much of R side of sphenoid present, plus much of ethmoid (albeit crushed). Missing part of palate posteriorly. Only upper teeth present: LM1–2, RM1 (partial), RM2–3. Bone moderately thick.

Glabellar region somewhat "spindle shaped," its midline less anteriorly projecting when viewed from above, and less superiorly projecting when viewed from in front, than its lateral edges (seen on undistorted R side). This lateral glabellar extremity lies above nasomaxillary suture, forming blunt edge with torus lateral to it. On the L, torus displaced upward at this point. Each torus bears (quite far medially) supraorbital notch (wider and shallower on the L, deeper on the R). Shallower L notch reveals how gently convex anterior surface of torus twists laterally. Medially, this surface tall, faces almost completely forward. As torus continues laterally, this surface twists gradually, so that by midline of orbit it, still long and flattish, faces obliquely upward. By its lateralmost extent this surface becomes almost rolled, more compressed from top to bottom; superior surface faces almost directly upward. At least until midline, flatter medial component of torus delineated above by blunt margin that fades laterally. Lateral portion, although still relatively thick, much thinner when viewed from front than medial portion. As seen on the R, superior orbital border long, straight (not arcing); orbital roof descends anteriorly to recurve fairly acutely onto anterior border of torus. Viewed from above, undistorted R supraorbital margin retreats gently from projecting lateral border of glabellar region.

On the R, lower part of frontal, from just lateral to glabella to zygomaticofrontal suture, is undistorted. Small domed region lies above extremely broad glabella; for most of its width, plane above orbital region fairly straight in side view as it proceeds toward bregma. In front view, this plane arcs gently above each orbit. Slope up to coronal suture short, relatively straight. Viewed from above, unsegmented coronal suture appears to be finely denticulate; viewed straight on, it evidently received strong fingerlike interdigitations from parietals; appears to have swept backward as well as upward toward midline from each pterion.

Interorbital region only slightly less broad than glabella region. Nasal bones apparently relatively broad superiorly, slightly convex from side to side; too damaged inferiorly to judge width. Clearly some forward flexion below nasion. Lacrimal fossae, better preserved on the R, were apparently relatively shallow, probably not very tall. Anterior lacrimal crest was probably absent. Lacrimal bone confluent with inferomedial corner of orbit. Crushing distorts orbital shape; was probably something between the now very wide and rectangular R orbit and the tall, squarish L orbit.

On the R, the infraorbital groove, which lies just lateral to midline of orbital floor, is relatively short, being shorter than roofed-over portion of orbital floor in front. As seen in exposed L maxillary sinus, infraorbital bony canal runs from posterior portion of orbital floor down and forward, to emerge on face through relatively large, downwardly directed infraorbital foramen that lies quite far below infraorbital margin. Although this region somewhat crushed on the R, is clear that the R infraorbital foramen was similar in size and position to the L and was probably the aperture for a downwardly descending bony canal that ran through maxillary sinus. Shallow groove descends vertically from infraorbital foramen; it indents an otherwise very slightly puffy infraorbital region.

Especially on the L, appears that orbital floor flowed smoothly out and down onto facial surface, being interrupted by little or no defined inferior orbital border. As seen on the L, frontal process projected outward (anteriorly) slightly; in vertical view, was moderately convex, i.e., appearing slightly puffy. Frontal process was probably not long a/p (indicating only modest midfacial projection). Posterior surface of zygoma extremely broad m/l; faces directly back upon temporal fossa.

As seen on the L, lateral crest runs continuously around nasal aperture, thickening toward midline, where it dips and proceeds forward to meet its right-hand counterpart. All evidence lacking as to form or even presence of anterior nasal spines. Lateral nasal margin, as seen especially well on the L, thick, despite lateral crest sharpness. Nasal cavity floor flat, not depressed below level of aperture. Original shape of nasal aperture impossible to determine with confidence; lateral internal walls preserved (still covered with matrix on R). On the L, nasal cavity wall immediately behind lateral crest slightly puffy, with slight roughening about two-thirds of the way up it. No sign of medial projection. Low on L lateral wall, faint crest parallels lateral crest, runs toward midline, and joins anterior part of vomer. Prenasal fossa between crests absent. Farther back, preserved lower wall of L maxillary sinus very strongly inflected into nasal cavity, essentially dividing nasal passage into funnel-like inferior portion and constricting superior portion where ethmoturbinals would have been. Far back inside nasal cavity, three vertical plates of bone descend from roof of nasal cavity. Thickest and most medial appears to correspond to perpendicular plate that meets vomer; lateral to this is a thinner plate, and lateral to that a yet thinner one. These may be equivalents of superior and middle ethmoturbinals of other terrestrial mammals, although they are platelike and parallel to one another (rather than scrolled) and one lies superior and lateral to the other. Except for these apparent ethmoid derivatives, the ethmoid itself seems to have no presence within the nasal cavity.

On the L, exposed maxillary sinus partly filled with matrix (obscuring full extent both medially and laterally); did not extend very far inferiorly relative to floor of flat nasal cavity. Nasoalveolar clivus moderately long, slopes obliquely forward somewhat; from below, it probably arced smoothly and gently from side to side.

Temporal ridges (in form of quite pronounced muscle scars) arise quite abruptly from just behind superiormost extent of zygomatic process of frontal. Almost as abruptly, and still on frontal, these scars transform into broad temporal bands that run quite straight up rather than curving back, paralleling oblique profile of frontal. R and L malar tubercles, almost ridgelike, distended posteriorly. Anterior root of zygomatic arch, on less distorted L side, originates relatively close to level of M^1. In front view, its inferior border runs quite steeply upward before turning more aggressively outward. From below, the inferior border turns very quickly posteriorly on both sides. Zygomatic temporal processes on both sides run relatively straight back, with no pronounced flare.

Thick-boned alisphenoid preserved on the R; its contribution to anterior squamosal suture oriented laterally (suggesting cornering down from suture into quite deeply excavated anterior temporal fossa). Inferiorly, alisphenoid flexes medially and posteriorly quite strongly, clearly delineating infratemporal fossa.

Palate was long and probably (see R side) diverged somewhat posteriorly; its maximum depth lay quite far posteriorly (around M2–3), where side walls most vertical. From behind the Is, plane of palate slopes gently backward toward its deepest point. Probable incisive foramen lies level with Cs. Lateral pterygoid plate almost completely preserved; medial plate represented only by vestige superiorly. Clearly, lateral plate flared laterally; medial plate was oriented directly a/p. For at least the middle of their lengths, these plates would have been parallel.

Internally in braincase, cribriform plate very long a/p, narrow m/l, sunken quite low between orbital cones. Anteriorly bulky, posteriorly tapering, moderately tall crista galli extends for most of length of plate, almost entirely occuping its anterior half. Frontal lobes lie well over orbital cones, extending about as far as anterior border of zygomaticofrontal suture. Frontal crest moderately tall, becoming even more so as it runs down to large foramen caecum. As seen on the R, lesser wing of sphenoid short but projection of anterior clinoid process quite pronounced posteriorly; this clinoid process overhangs medially the fairly deep anterior portion of the middle cranial fossa. Jugum appears to have been relatively long a/p, flat; R optic canal rather long. Superior orbital fissure restricted medially; inferior orbital fissure restricted inferiorly (posterior opening of orbital cone restricted to small C-shaped fissure at apex). On the R, groove for optic chiasm was apparently not developed. Large sinus anterior to broken tuberculum sellae underlies jugal region. L sphenoid sinus penetrated up into jugal region; R sphenoid sinus, although expanding out into base of alisphenoid, was superiorly confined to region just below optic canal.

Alveoli of all missing upper teeth except RP2 (lost antemortem) present. Alveolar crest destruction on both sides buccally along molar rows. Anterior tooth roots (including double-rooted P1s) were huge in cross section but not very tall. I1 roots were thick, conical; the not much smaller alveolae for I2s are thickly trapezoidal. C alveolae medially flattened but b/l wide. LP2 represented by single, lingually placed, subcircular alveolus. On broken RM1, lingual root was m/d very long, b/l very wide. LM1 and both M2s somewhat worn; RM3 only minimally worn. Judging from the L, M1 was probably longer mesiodistally than M2; M2 is b/l wider. M1–2 bear large hypocones, more swollen distolingually on M1. Small trigon basins invaded by expansive base of protocone (restricted to depression at juncture of the three trigon cusps). On M1–3, metacone mesiodistally compressed; much shorter than paracone. Short preprotocrista runs from expansive protocone on all Ms to mesial side of base of paracone. On M2s and apparently also M1, extremely short postprotocrista separated from metacone base by horizontal crease. Preserved on M2s and the M3, postcingulum, which runs from distal side of metacone to hypocone region, is beaded. Buccal molar sides fairly straight; lingual sides may have been more rounded (given apparent central displacement of now-worn protocone apices).

Arago 45

Adult. L maxillary fragment with L zygoma, somewhat crushed, but containing LM2 (very worn, somewhat damaged).

Comparison with Arago 21 suggests that curvature of lateral border of orbit, and medially as far as the zygomaticomaxillary suture, similar, although inferior margin on 45 not as thickened as it is, if slightly, in 21. Malar tubercle taller, blunter in 45 but more posteriorly projecting in 21. Posterior surface of zygoma also very wide m/l; would also have faced directly back on temporal fossa. Depth of lower face appears similar in 21 and 45; in 45, relatively small infraorbital foramen lay somewhat below inferior orbital margin. Internally, bony tube runs from well back along orbital floor and down to infraorbital foramen (suggesting that the foramen was well below the margin and that the infraorbital groove was relatively short). Anterior root of zygomatic arch took origin probably well above M2. About level with inferior orbital margin, frontal process of maxilla appears to have been long a/p (suggesting a little more midface projection than in 21); tables of bone that converge to form lateral crest converge in both specimens at similarly wide angle. Maxillary sinus extended into this area of convergence; a similar conformation in 21 would account for slight puffiness of this area.

What is preserved of palate includes uncrushed alveolae for LM1 and P2 and crushed alveolae for C–P1; palate deep along its side wall; would have been pretty steep postincisal slope. Preserved M2 had moderately large hypocone, centrally placed protocone, and fissure between very short postprotocrista and metacone base. As in 21, evidence of alveolar crest destruction both lingually and buccally in molar region. P2 root buccolingually wide, somewhat compressed m/d; in cross section, its alveolus quite large; the root may not have been very long. As in 21, lingual root of M1 (judged by alveolus) was quite large m/d as well as b/l, especially at distal extremity. In 45, alveolus for M1 mesiobuccal root much wider b/l than distobuccal root.

Arago 47

Adult. R partial parietal missing its more medial anterior portion. Suture contacting alisphenoid is present. Coronal suture is broken off. More than half of sagittal suture present (posterior portion), all of R lambdoid suture (with at least three ossicles), posterior two-thirds of squamosal suture to parietal notch, plus all of parietomastoid suture.

This bone extremely thick, dense (little diploe), short a/p; would have been even shorter along sagittal suture because of rearward sweep of frontal. Lambdoid and sagittal (had begun to fuse) sutures massively denticulate throughout their thickness. With contribution to parietal notch and squamosal suture rays oriented vertically, bone flexes into two planes. A very short, vertical plane above squamosal suture contributes to lateral vault system (lateral vault wall); its superior border defined by thick, raised, arced, smooth-surfaced bone of temporal band (= superior + inferior lines) that evidently lay very low on cranial vault. Larger portion of parietal lies superior to this temporal band; it angles very sharply with cranial wall, suggesting very low vault height, with almost transversely flat skull top and great cranial breadth, especially relative to length. At peak of low temporal band lies a low but not typical parietal eminence (where, topographically, the various planes of the bone converge). Temporal band curves to asterion in gentle arc. Just above asterion, bone reaches its maximum thickness (reflected externally in large, broad mound that would have projected laterally over mastoid process). Medial to temporal band and above bulge, is shallow, relatively long and wide depression. Squamosal suture apparently did not arc very strongly (seen in cross section); bone in this region remains thick close to sutural margin, after which it thins very rapidly. Region of "parietal eminence" reflected internally as most concave region of inner surface. Parietomastoid suture long, horizontal. Lambdoid suture rises very steeply directly from asterion. Grooves for branches of middle meningeal artery broad, not deeply impressed in bone, show minimal branching. Trace of superior sagittal sinus indicates it had been very broad, very shallow. With sagittal suture oriented horizontally, profile of side of skull angles inward strongly, suggesting deep infratemporal fossa and marked postorbital constriction.

A fragment, also Arago 47, is hard to associate with main parietal piece; is probably from another individual. This small fragment of cranial bone was two sutures meeting at right angle. Bone thinner than Arago 47.

Arago 3

Another cranial fragment, looks very similar to smaller Arago 47. Is thinner than primary Arago 47.

Arago Isolated Teeth

Some 50 isolated teeth, allegedly representing 21 individuals.

Deciduous upper molars. dm1 protocone large with centrally displaced apex. Protocone fold courses to small hypocone. Paracone large, b/l compressed, with small metacone at its base distally. Distinct postprotocrista courses to metacone. Preprotocrista swings around paracone. Parastyle small.

Upper premolars. P1 large crowned. Apices of subequal protocone and paracone slightly centrally shifted. Postprotocrista not markedly thicker than preprotocrista. Distal slope of protocone descends farther down crown than mesial slope. P1 somewhat truncated on distal side. Pre- and postprotocristae each cut in the middle by a crease. Buccal side slopes; base bulbous. P2 fairly large crowned. Peripherally placed protocone and paracone subequal in size and slightly mesially shifted; postprotocrista much thicker than preprotocrista (distal part of tooth more swollen than mesial). Enamel slightly wrinkled; protocone distal slope descends farther down crown than mesial slope.

Upper molars. Typically deeply wrinkled or creased. M1 cusps somewhat internally placed; protocone and metacone apices approximated; pit(s) on sides of protocone; postcingulum is short; hypocone is swollen; talon basin restricted, buccally placed. M2 similar to M1 with metacone smaller than protocone; preprotocrista stouter than postprotocrista.

Deciduous lower molars. dm1 has compressed protoconid and metaconid. Protoconid continuous with cristid that runs back and incorporates basally swollen hypoconid; deep notch between small entoconid and base of hypoconid; thin paracristid courses from apex of protoconid steeply down and around to base of metaconid, delineating tall, somewhat lingually open, large trigonid basin. Buccal side of crown somewhat swollen, with enamel a bit exodaenodont over anterior root; hypoconid swelling goes to center of tooth; trigonid basin apparently narrower than in dm2.

dm2 somewhat elongate, high cusped, with hypoconid swelling into center of large, deep talonid basin. Hypoconulid is large, buccally placed; notch between hypoconulid and entoconid. Protoconid and metaconid closely connected at their bases with stout paracristid between apices that enclose large deep trigonid basin.

Lower premolars. P1 very tall crowned. Protoconid compressed and centrally shifted with a long buccal slope; protoconid apex somewhat mesially shifted (mesial edge shorter than very long distal edge). Crestlike margins course up sides of medium-sized, very pointed metaconid that lies opposite protoconid, to which it is connected by low crest. Anterior fovea very deep, medium sized; talonid basin very large, deep.

Lower molars. M2 very long, narrow, with rounded occlusal outline and deep, thick wrinkling; lingual cusps more centrally placed than the buccal. Hypoconulid large, centrally placed; trigonid basin quite large, deep; lingual side of crown bulbous. The four major talonid cusps meet at midline of tooth; roots long, very closely appressed. M1 was apparently similar.

REFERENCES

Bermudez de Castro, J. et al. 1997. A hominid from the Lower Pleistocene of Atapuerca, Spain: Possible ancestor to Neandertals and modern humans. *Science* 276: 1392–1395.

Chaline, J. 1971. L'âge des Hominiens de la Caune d'Arago à Tautavel (Pyrenées-Orientales). *C. R. Acad. Sci. Paris* 272: 1743–1746.

Day, M. 1986. *Guide to Fossil Man*, 4th ed. Chicago, University of Chicago Press.

de Lumley, H. 1976. Les civilisations du Paléolithique inférieur en Languedoc mediterranéen et en Roussillon. In: *La Préhisoire Francaise*, vol. 1–2. Paris, CNRS, pp. 852–874.

de Lumley, H. (ed.). 1982. L'*Homo erectus* et la place de l'Homme de Tautavel parmi les hominidés fossiles. Prétirage, Coll. Cong. Int'l Pal. Hum., Nice.

de Lumley, H. and M.-A. de Lumley. 1971. Découverte de restes humaines anténéandertaliens datés au début de Riss à la Caune d'Arago (Tautavel, Pyrénées-Orientales). *C. R. Acad. Sci. Paris* 272: 1729–1742.

de Lumley, H. and J. Labeyrie, eds. 1981. *Datations Absolues et Analyses Isotopiques en Préhistoire.* Tautavel, CNRS.

de Lumley, H. et al. 1984. Stratigraphie du remplissage Pléistocène moyen de la Caune d'Arago à Tautavel: Etude de huit carottages effectuées de 1981 à 1983. *L'Anthropologie* 88: 5–18.

Holloway, R. L. 1985. The poor brain of *Homo sapiens neanderthalensis*; see what you please. In: E. Delson (ed), *Ancestors; The Hard Evidence.* New York, Alan R. Liss, pp. 319–324.

Iacumin, P. et al. 1996. A stable isotope study of mammal skeletal remains of mid-Pleistocene age, Arago Cave, eastern Pyrenees, France. Evidence of taphonomic and diagenetic effects. *Palaeogeog. Palaeoclimat. Palaeoecol.* 126: 151–160.

Rightmire, P. 1990. *The Evolution of* Homo erectus. Cambridge, Cambridge University Press.

Stringer, C. and C. Gamble, 1993. *In Search of the Neanderthals.* London, Thames and Hudson.

Stringer, C. et al. 1984. The origin of anatomically modern humans in Western Europe. In: F. Smith and F. Spencer (eds), *The Origins of Modern Humans.* New York, Alan R. Liss, pp. 65–175.

Tattersall, I. 1995a. *The Fossil Trail: How We Know What We Think We Know About Human Evolution.* New York, Oxford University Press.

Tattersall, I. 1995b. *The Last Neanderthal: The Rise, Success and Mysterious Extinction of Our Closest Human Relatives.* New York, Macmillan.

Tattersall, I. 1986. Species recognition in human paleontology. *J. Hum. Evol.* 15: 165–175.

Yokoyama, Y. and H.-V. Nguyen. 1981. Datation directe de l'homme de Tautavel par la spectrométrie gamma, non-destructive, du crâne humain fossile Arago XXI. *C. R. Acad. Sci. Paris*, II, 292: 927–930.

Repository

Laboratoire d'Anthropologie, Université de la Méditerranée, Faculté de Médécine, 13916 Marseille, France.

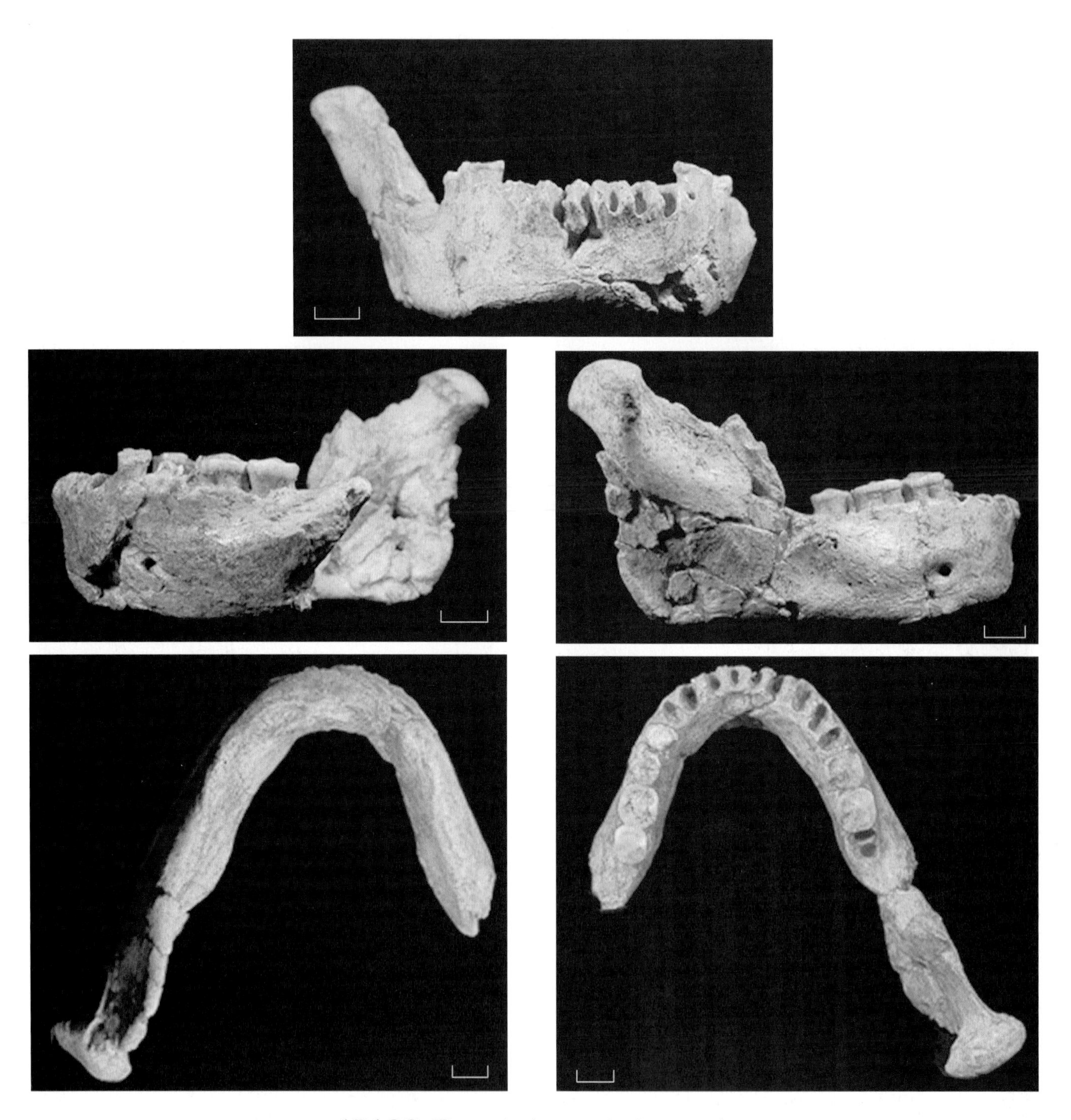

ARAGO Figure 1. Arago 2 (scale = 1 cm).

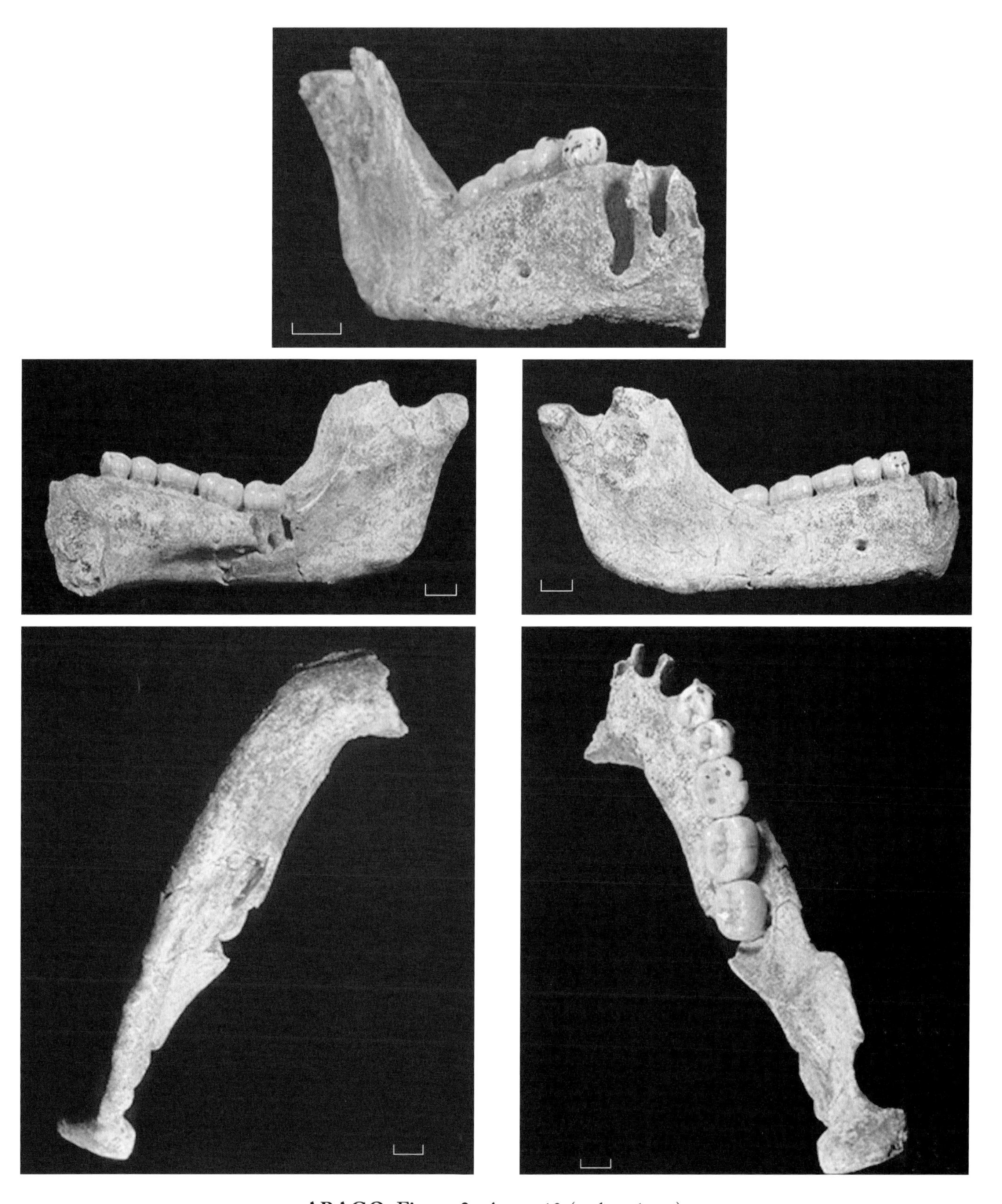

ARAGO Figure 2. Arago 13 (scale = 1 cm).

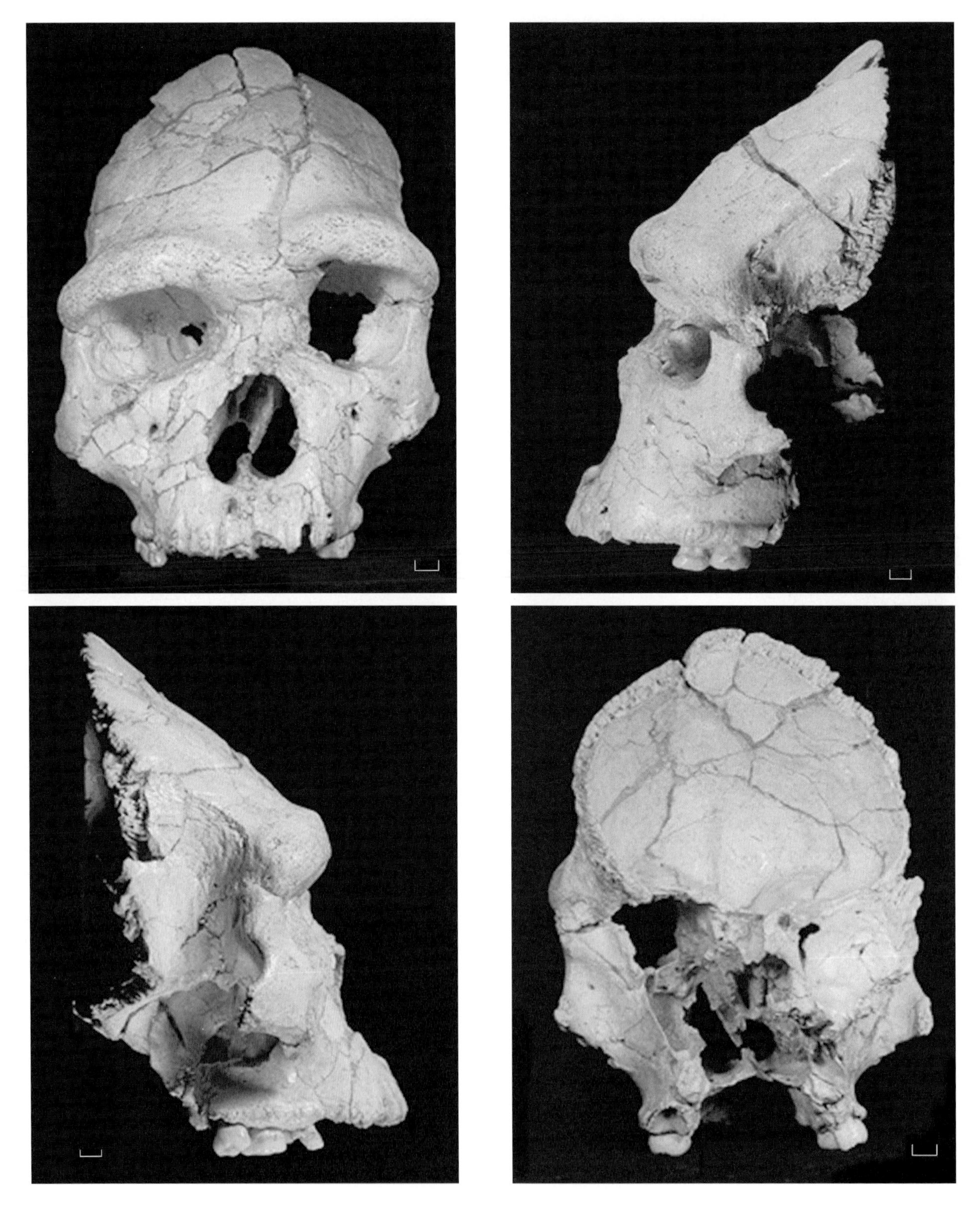

ARAGO Figure 3. Arago 21 (scale = 1 cm).

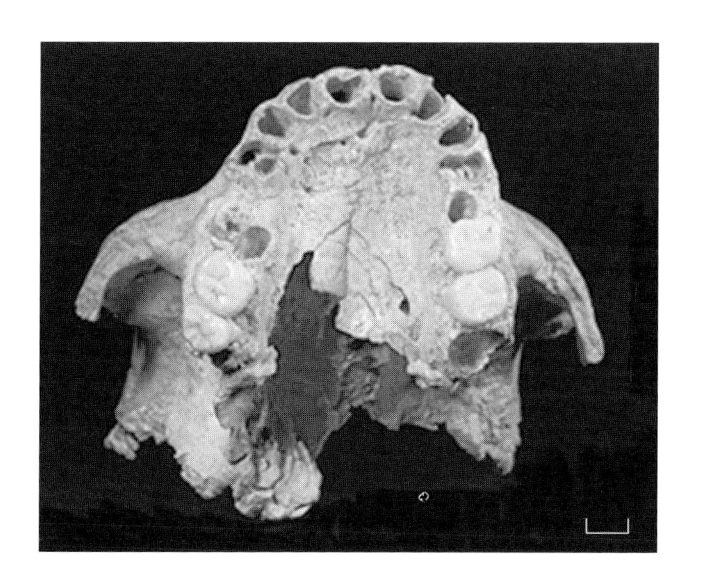

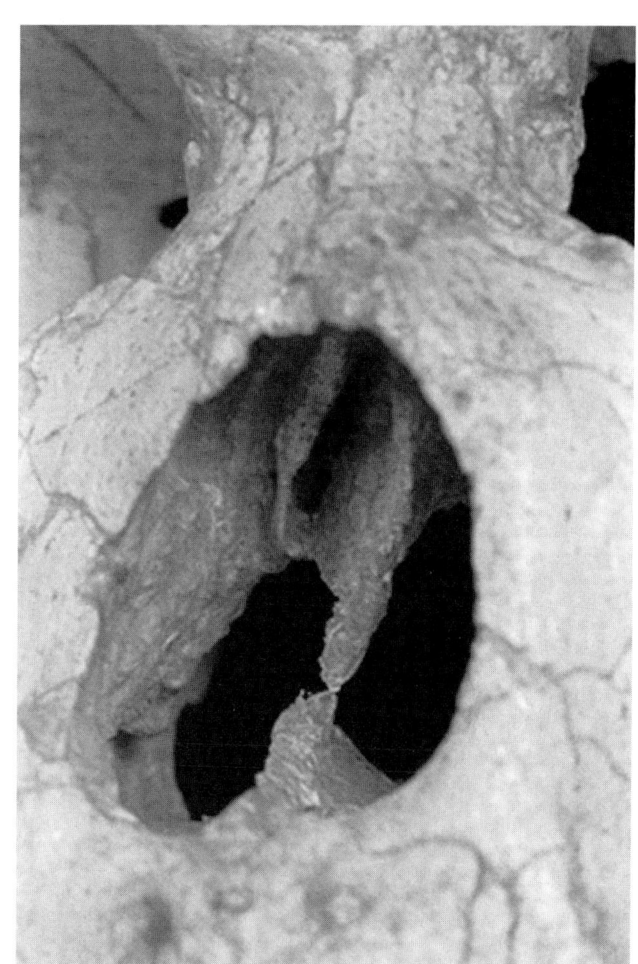

ARAGO Figure 4. Arago 21 (scale = 1 cm).

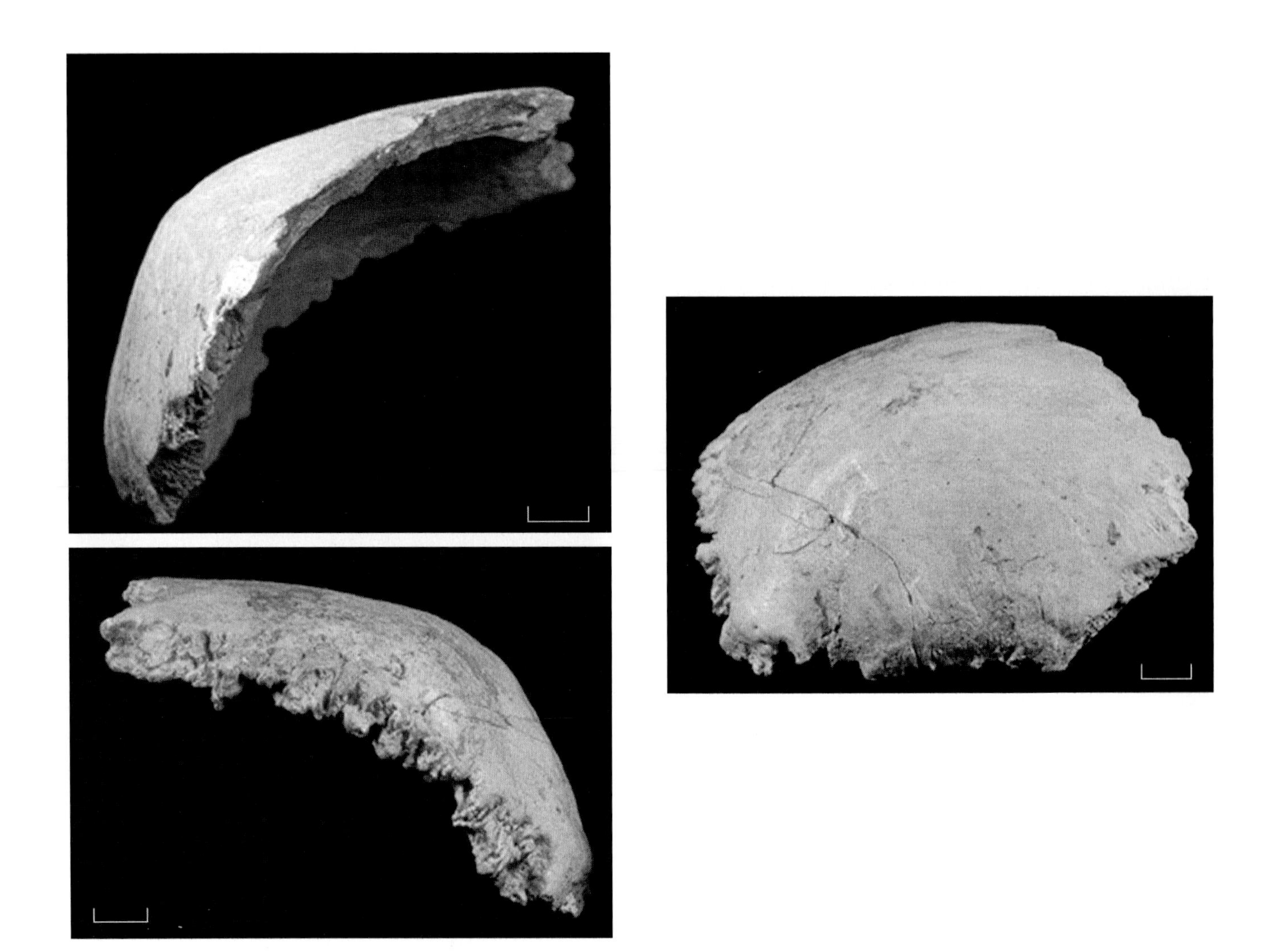

ARAGO Figure 5. Arago 47 (scale = 1 cm).

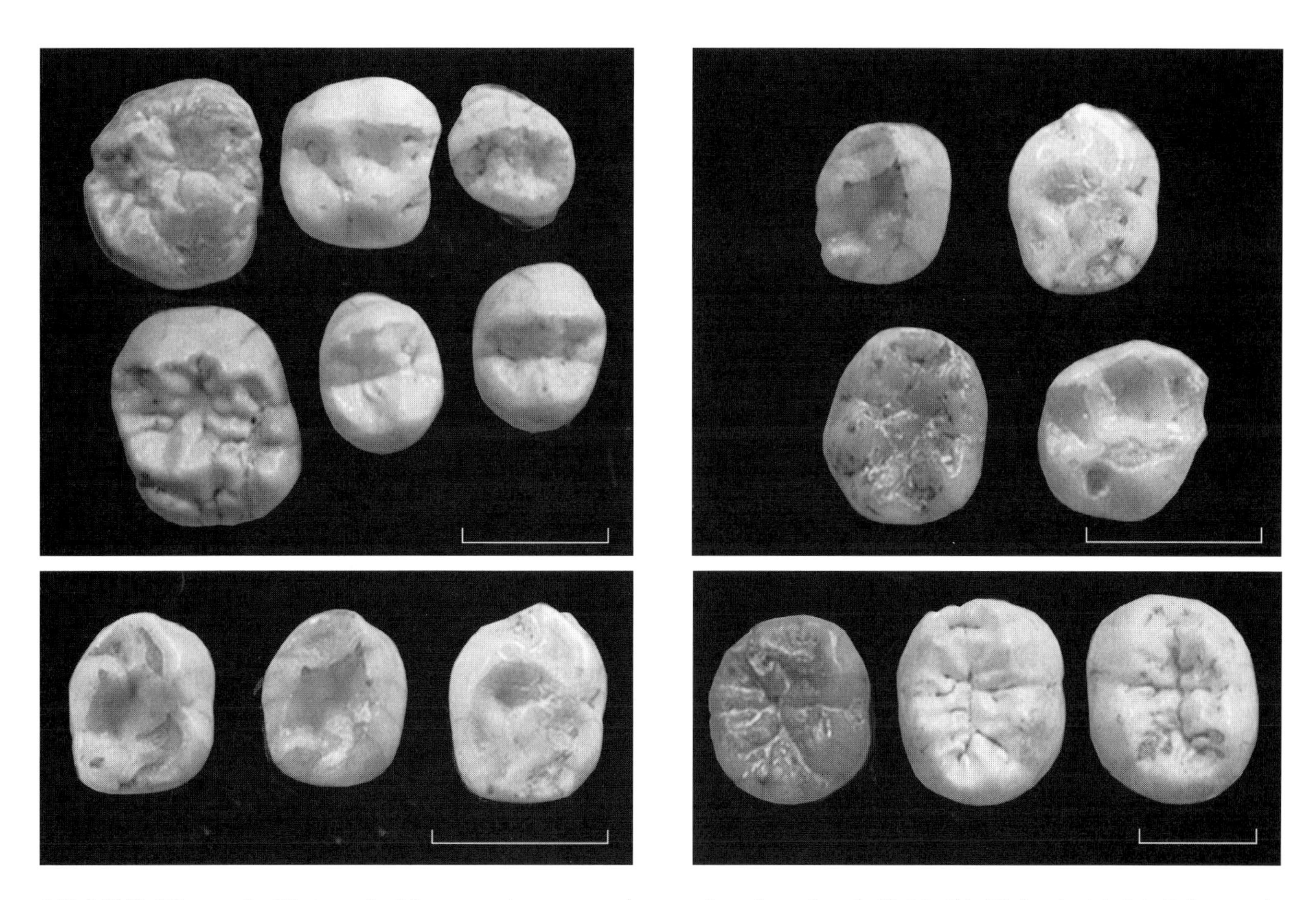

ARAGO Figure 6. Various deciduous molars, premolars, and molars: (top left) 31, 54, 71 (top), 14, 26, 7 (bottom); (top right) 55, 66 (top), 5, 12 (bottom); (bottom left) 34, 55, 66; (bottom right) 69, 10, 32, (scale = 1 cm).

Archi

Location

Road cut in the western foothills of the Aspromonte range just S of the town of Archi, near Reggio Calabria, Italy.

Discovery

A. Berdar, September 1970.

Material

Juvenile mandible lacking rami and some teeth.

Dating and Stratigraphic Context

Sandy gravel deposits cut through by the Torbido River and faunally dated to some time in the Upper Pleistocene (Ascenzi and Segre, 1971).

Archaeological Context

None.

Previous Descriptions and Analyses

Ascenzi and Segre (1971) described the mandible as Neanderthal, probably male because of its massiveness. As Ascenzi and Segre had done, Stringer et al. (1984) remarked on particular similarities to the Gibraltar Devil's Tower child's mandible. No one has contested the claim that the specimen is Neanderthal.

Morphology

Partial juvenile mandible lacking both rami and retaining Ldc–dm2 and Rdm1–2. Certainly <5, probably 2–3 years old.

For its age, mandible thick boned from side to side and all around; corpora deep in front of dm2s. In profile, symphyseal region slopes gently backwards, with slight convexity toward inferior border; inferior border of symphyseal region also elevated between regions of lateral deciduous incisors. Inferior border begins to descend quite dramatically behind dm2. In external view, midline between roots of di1s is apex of very low triangular mound, which inferiorly forks out laterally just above inferior margin and continues laterally along anterior border of inferior margin on both sides as low ridge that continues as far back as level of dm2. These ridges thicken somewhat, becoming slightly outwardly reflected below large mental foramen, which lies below dm1. Shallow gutter above ridges throws them into relief from region of di2 to region of dm2. In inferior view, front of jaw arcs broadly from side to side between regions of Cs. In midline, low mound appears as slight rounded elevation in contour; small cleft marks midline of anterior symphyseal surface at base of this swelling. Narrow digastric fossae anteriorly placed; point down, backward.

Internally, postincisal plane quite long, rather rounded; low down pair of thin, vertical, genial ridges; scar of symphysial fusion runs down midline. On both sides, steeply descending mylohyoid are prominent, ridgelike; overhang well-excavated fossae below.

Ldc crown relatively small; swells out buccally toward neck. Cusp's apex situated well forward; short, convex medial edge crest descends from apex to curve around lingually (enclosing narrow fovea) to fade into slightly swollen basal lingual region. Longer distal edge at first slightly concave; then becomes convex toward base, above which it curves in slightly lingually, enclosing shallow, somewhat larger posterior fovea.

dm1s markedly exodaenodont over anterior root; this buccal enamel swelling continues up and back to region of hypoconid. Slightly posteriorly, centrally placed protoconid met by three stout, compressed crests: a paracristid extends a short distance directly mesially before turning back sharply and descending to become confluent with the lingual side of tooth (enclosing deep, wide trigonid basin); a second crest runs up and forward from base of well-developed entoconid and then curves buccally to meet the apex of the protoconid (no distinct metaconid in this area); the third crest is a stout cristid obliqua that runs directly forward from the compressed hypoconid to the protoconid apex. Distally, a shallow, broad notch separates entoconid and hypoconid; centrally, a long, wide, deep talonid basin opens distally. There may be a "metastylid" in the cristid obliqua.

dm2s distinctive with columnar cusps delineated from one another by very long vertical creases (particularly on buccal side). Major cusps quite compressed b/l at their apices and swollen at bases; protostylid distinct, tall. Stout paracristid encloses wide, deep trigonid basin; runs medially around between closely appressed bases of protoconid and metaconid. Talonid basin very large, deep, with coarsely crenulated enamel (also some areas of internal cusp surfaces); surrounding tall cusps quite peripherally placed. Large hypoconulids placed slightly buccal to midline, sharply delineated by clefts from their neighbors. In occlusal outline, teeth quite rounded.

References

Ascenzi, A. and A. Segre. 1971. A new Neanderthal child mandible from an Upper Pleistocene site in southern Italy. *Nature* 233: 280–283.

Stringer, C., J.-J. Hublin and B. Vandermeersch. 1984. *The Origin of Anatomically Modern Humans in Western Europe*. New York, Alan R. Liss, 51–135.

Repository

Istituto Italiano di Paleontologia Umana, Piazza Mincio 2, 00198 Roma, Italy.

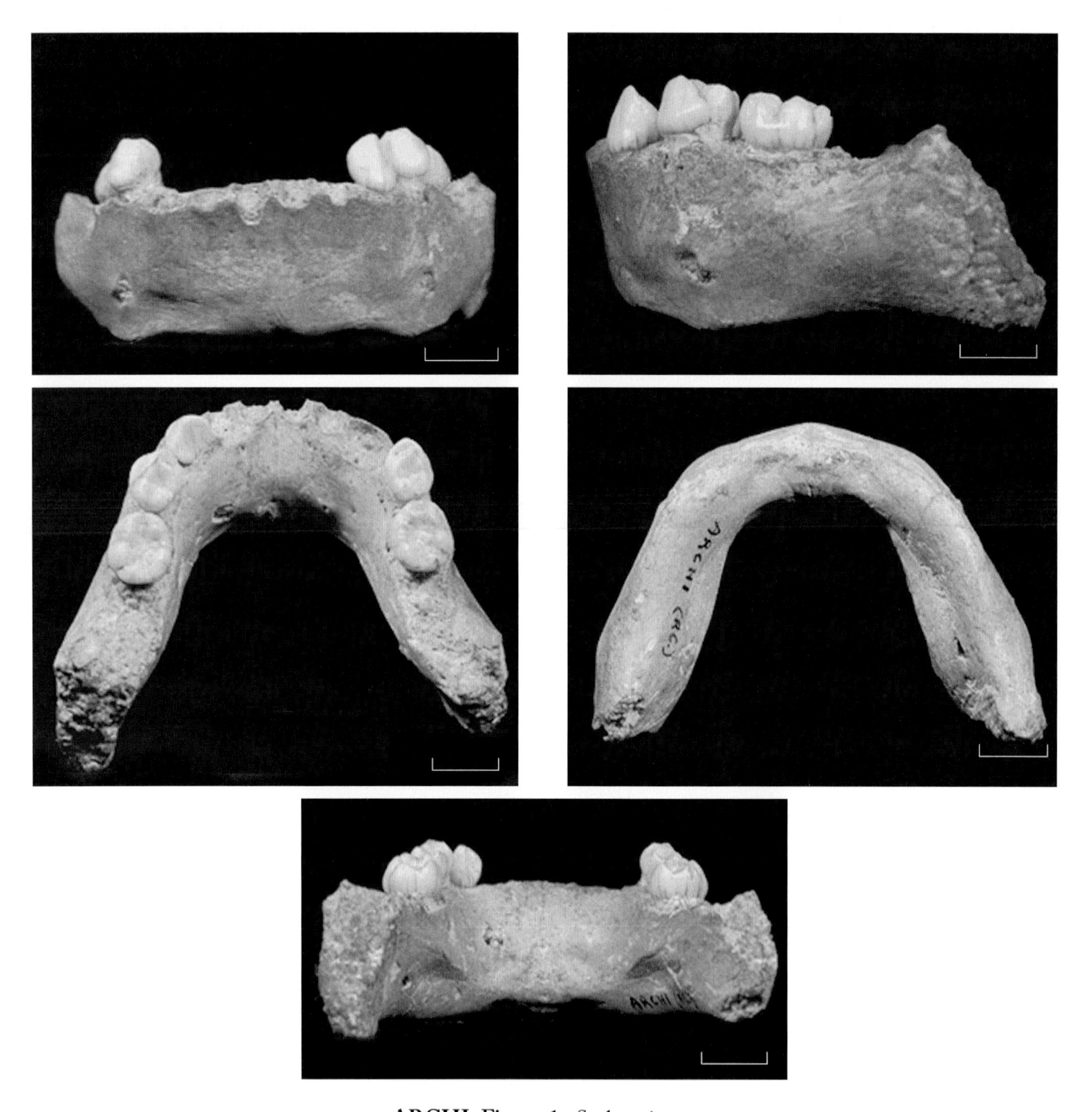

ARCHI Figure 1. Scale = 1 cm.

Atapuerca: Gran Dolina

Location
Fissure fill exposed in a railroad cutting (Trinchera del Ferrocarril) in the Sierra de Atapuerca, some 14 km E of Burgos, Spain.

Discovery
Initially worked by Emiliano Aguirre and students, later by Eudald Carbonell and José María Bermúdez de Castro and colleagues. The TD6 level, containing hominid remains, was reached in 1994.

Material
At this writing, up to 80 hominid cranial fragments and isolated teeth, plus hominid postcranial fragments, are known from TD6. There are faunas and lithics from both TD6 and other levels at the site.

Dating and Stratigraphic Context
Stratified site; the TD6 level has been paleomagnetically dated to the Matuyama reversed Chron, hence before 780 ka (Parés and Pérez-González, 1995). Falguères et al. (1999) have confirmed this date through ESR and U-series studies on teeth. TD6 thus represents the earliest occurrence of hominid fossils yet known from western Europe.

Archaeological Context
TD6 has yielded some 200 lithics comparable to Mode 1 technology in Africa (Carbonell et al., 1999).

Previous Descriptions and Analyses
The TD6 hominid remains include the type material of the species *Homo antecessor*, created by Bermúdez de Castro and colleagues (1997), who claim that this species is ancestral both to modern human and Neanderthal lineages. These specimens and their interpretation are most comprehensively discussed in a series of papers in the journal issue edited by Bermúdez de Castro et al. (1999).

Morphology
(Photographs courtesy of Javier Trueba). Single elements discussed separately; duplicates summarized together.

ATD6-15
Frontal. Very fragmentary and reconstructed, with part of lateral portion of L orbit; also portion of nasal bones and frontal process of maxilla. Thin boned.

Frontal rises well behind brow, is more domed behind posttoral than postglabellar sulcus. Posttoral sulcus long a/p, shallower than postglabellar sulcus (more sunken and concave). R supraorbital "torus" seems continuous over orbit; may have flowed into glabella; does not extend anteriorly as far as glabella; is thinner laterally than medially. At least laterally, "torus" forms crisp edge with orbital roof, from which it angles back. More medial portion of "torus" taller s/i; although not vertical, is not as sloped back as lateral portion. Superior margin of medial portion of "torus" not angled, flows rather smoothly into posttoral sulcus. As seen on the L, some postorbital constriction. Also on the L, a horizontal, moderately developed but crisply defined temporal line courses back from zygomatic process of frontal. Frontosphenoid suture appears to run down and medially, creates inward angulation to alisphenoid. Parietal portion of coronal suture not deeply invaginated or denticulated.

Apparently there was a keel along nasonasal suture. Nasal bones extend much more superiorly than frontal process, taper up to nasion, which was apparently overhung by glabella. Exposed R and L frontal sinuses may have been confluent across midline, with only low septum between them; R sinus extended to midpoint of orbit; L sinus extended slightly more superiorly than the R, but neither went very far up the frontal.

Inferior part of ethmoid preserved, especially at midline. Perpendicular plate broken; its root is preserved. First ethmoid air cell well preserved on the R, does not penetrate upward. As seen especially on the R, frontal sinuses appear to course lateral to field of ethmoid air cells.

Internally, relatively strong and tall frontal crest runs up from foramen caecum, posterior to which a moderate crest may have expanded to become the crista galli. No evidence of cribriform plate. Bone of anterior cranial fossa generally unembellished; hint of corrugation of the bone above orbit.

ATD6-20

L parietal fragment (most of bone missing), preserving much of margin of squamous suture, portion of region of pterion (with small piece of alisphenoid still attached), and part of coronal suture (with outer table extending anteriorly farther than inner table); sphenoparietal suture not fused. Bone thick.

ATD6-38

L zygomaticomaxillary fragment. Large, downwardly pointing infraorbital foramen connected by stout bony tube to infraorbital canal approximately halfway back in orbital floor. Large, shallow fossa lateral to infraorbital foramen. Tight, upwardly concave curve down to inferior margin of zygoma (also in ATD6-19). Maxillary sinus confined, slightly subdivided by thin, creaselike, horizontal ledge in bone (also in ATD6-19, -58).

ATD6-69

Juvenile. Primarily lower portion of facial skeleton and various teeth, with, on the L, inferior orbital region and part of zygoma. RI2–M1 and LP1 and M1 present (RC and P2 still erupting), with LM2 visible in crypt and small LM3 crown positioned on top of it.

Crisp inferior orbital margin oblique, creating "aviator glasses" shape; faces toward medial and inferior "corner." Orbital floor does not flow onto face. Two infraorbital foramina (the larger lying superior to the smaller) point downward. Infraorbital groove occupies much of orbital floor. Anterior root of zygomatic arch originates just above alveolar margin, level with M1; viewed from the front, it arcs steeply up and out and then curves down into maxillary tuberosity. The m/l thin zygoma then curves concavely out and down (also in ATD6-84, which has very obliquely oriented zygomaticotemporal suture). In side view, the somewhat depressed infraorbital region tilts down and back; zygoma flat and vertical with an upwardly arcing inferior margin. As seen in ATD6-58, zygomaticofacial foramina lie above level of inferior orbital margin.

ATD6-84

A fragment of zygomatic arch, including the zygomaticotemporal suture.

A moderate fossa lies lateral to infraorbital foramina and below inferior orbital margin; fossa less pronounced in ATD6-58. Nasoalveolar clivus relatively short but somewhat broad; arcs smoothly from side to side, its surface essentially featureless. In profile, lateral crest of nasal aperture curves posteriorly inward before turning out toward large, elevated, and quite protrusive anterior nasal spine; it bifurcates behind the spine with one branch proceeding downward onto the clivus and the other fading as it joins the spinal crest. Conchal crest faint and oriented downward from near margin of lacrimal groove. Maxillary sinus does not swell into nasal cavity. Floor of nasal cavity stepped down behind rather posteriorly placed, relatively large incisive fossae. Palate shallow anteriorly, slopes gently posteriorly; side walls also slope gently medially. Moderate-sized, not quite vertical incisive foramen lies just behind alveolar margin. Dental arcade diverges somewhat posteriorly. Medial and lateral pterygoid plates (seen on the L) confluent superiorly.

R and LI1 alveoli large, much larger than LI2 alveolus. RI2 (looks more like a L) tall crowned and almost parallel sided; bears deep lingual groove; mesial margocristid larger than distal margocristid. Cross section of RC almost an equilateral triangle; lingual surface very steep, slightly concave, bears margocristae, and dominates occlusal aspect of crown. RP1–2 and LP2 long m/d, lack central pillars, and bear small para- and metaconules with slight indications of foveae. P1–2 buccal surfaces noticeably swollen (beveled) compared with lingual sides. RP1 bears very broad gutter between para- and protocones; buccal side m/d longer than lingual side; cusp apices more mesial than on P2, where they are rather centrally placed. R and L M1s very long m/d, taper slightly distally; protocones very centrally placed, extending into trigon basin between bases of buccal cusps; hypocone and metacone subequal in height and expansiveness. M3 crown small, incomplete, sits atop the larger M2, which appears similar to M1, but with more definitive enamel wrinkling.

ATD6-14

(from photograph). Juvenile. L maxillary fragment to palatal suture, retaining inferior portion of nasal aperture; damage exposes I1–2 visible in crypts and dc and dm1 erupted.

Lateral crest of nasal aperture curves tightly medially toward anterior nasal spine. Nasoalveolar clivus relatively long and somewhat vertical. Palate very shallow and broadly curved out from midline. Incisive foramen well behind the region of I1.

dc very long m/d with slight buccal cingulum, styles mesially and distally, and slight internal pillar mesially with fovea behind. dm1 with large, buccally distended paracone and smaller metacone, each cusp with internal pillar; strong preprotocrista and postcingulum joined in d/l corner (where there is slight hypocone) by horizontal protocone fold coming from mesially placed and b/l compressed apex of large, somewhat centrally placed and lingually swollen protocone.

ATD6-16, -17, -57

R temporal fragments and part of mastoid region (tip of process missing); ATD-17 with attendant portion of sphenoid.

Infratemporal fossa not delineated from temporal fossa by "corner" in bone. Mandibular fossa narrow laterally, expands broadly medially (may have been closed off medially). Articular eminence barely elevated. Parietal notch long, deep, but not too broad; separates process from laterally protruding paramastoid crest. Parietomastoid suture long, thick, not deeply denticulate. Vertical, shallow sulcus midway along suture; low elevation near suture. Parietal notch apparently shallow. Mastoid region thick; process was probably short, with anterior part of occipitomastoid suture lying well below; exposed air cells numerous, small. Occipitomastoid crest difficult to discern. Part of occipitomastoid suture preserved; bone of region thins inferiorly, with its margin "cupping" occipital contribution to suture. Three small foramina would have lain above and behind mastoid process. Moderate-sized foramen ovale enclosed by bone. Internally, floor of middle cranial fossa flat above region of mandibular fossa. Surface of bone below jugal region and lesser wing very slightly corrugated. R optic foramen partially preserved. Single R sphenoid sinus exposed in body of sphenoid. Anterior, vertical portion of sigmoid sinus tall s/i, and very to not very deep; arcs at top; transverse portion shallow.

ATD6-77

R occipital condyle (partly damaged medially). Long, modestly arced in coronal plane. Large anteriorly placed hypoglossal canal.

ATD6-89

Fragment of adult sphenoid. Number of moderate-sized air cells exposed. Basilar part thins markedly down clivus, bears lateral grooves for articulation with petrosals. Internally, posterior part of hypophysial fossa preserved along with base of dorsum sellae. In profile, tuberculum sellae much taller than dorsum sellae; shallow hypophyseal fossa slung like a hammock between the two sellae.

ATD6-18

Fragment of L petrosal. Large single air cell lies lateral to region of arcuate eminence and just inside squamosal. Carotid foramen would have been large and anteriorly facing. Jugular "notch" is shallow and horizontally oriented. Styloid process thick; tip broken; lies lateral to carotid foramen. Thick vaginal process wraps around styloid process; probably did not extend fully laterally along ectotympanic tube. Large stylomastoid foramen lies behind and somewhat away from styloid process. Carotid canal large. No arcuate eminence. Cochlear canaliculus deep.

ATD6-5

Partially crushed fragment of R mandible with M1–2 erupted, M3 in crypt. Relatively gracile. Corpus tall but thin; inferior margin thin, and even thinner below

M_3. Root of ramus originates at distal end of M_2, would have obscured M3. Mylohyoid line very faint.

M1 quite and M2 somewhat worn; M3 roots just beginning to form; all extensively wrinkled. M1 as wide b/l as M2; was probably not long m/d; general disposition of cusps similar to M2, but hypoconulid clearly demarcated from hypoconid by thin crease; presence of centroconid or metastylid uncertain. M2 long m/d and wider b/l than M3; buccal cusps quite internally placed, lingually compressed, and somewhat incorporated into cresting system that encloses talonid basin; centroconid moderate; metastylid large; base of hypoconid runs quite lingually along fairly narrow talonid basin; moderate-sized hypoconulid lies slightly buccally; hypoconid less swollen buccally than hypoconid; paracristid thick, delineated distally by thin, creaselike trigonid basin. M3 narrow and smoothly rounded distally; cusps (especially buccal) quite internally placed, incorporated into low cresting system that encircles long, narrow talonid basin; hypoconulid large, quite centrally placed; centroconid distinct; base of hypoconid extends to middle of talonid basin.

REFERENCES

Bermúdez de Castro, J., et al. 1997. A hominid from the Lower Pleistocene of Atapuerca, Spain: possible ancestor to Neandertals and modern humans. *Science* 276: 1392–1395.

Bermúdez de Castro, J., et al. 1999. Special Issue on Gran Dolina Site: TD6 Aurora Stratum (Burgos, Spain). *J. Hum. Evol.* 37 (3–4): 309–700.

Carbonell, E. et al. 1999. The TD6 level lithic industry from Gran Dolina, Atapuerca (Burgos, Spain): production and use. *J. Hum. Evol.* 37: 653–693.

Falguères, C. et al. 1999. Earliest humans from Europe: the age of TD6 Gran Dolina, Atapuerca, Spain. *J. Hum. Evol.* 37: 343–352.

Parés, J. M. and A. Pérez-González 1995. Paleomagnetic age for hominid fossils at Atapuerca archaeological site, Spain. *Science* 269: 830–832.

Repository

Departamento de Paleobiología, Museo Nacional de Ciencias Naturales, 28006 Madrid, Spain.

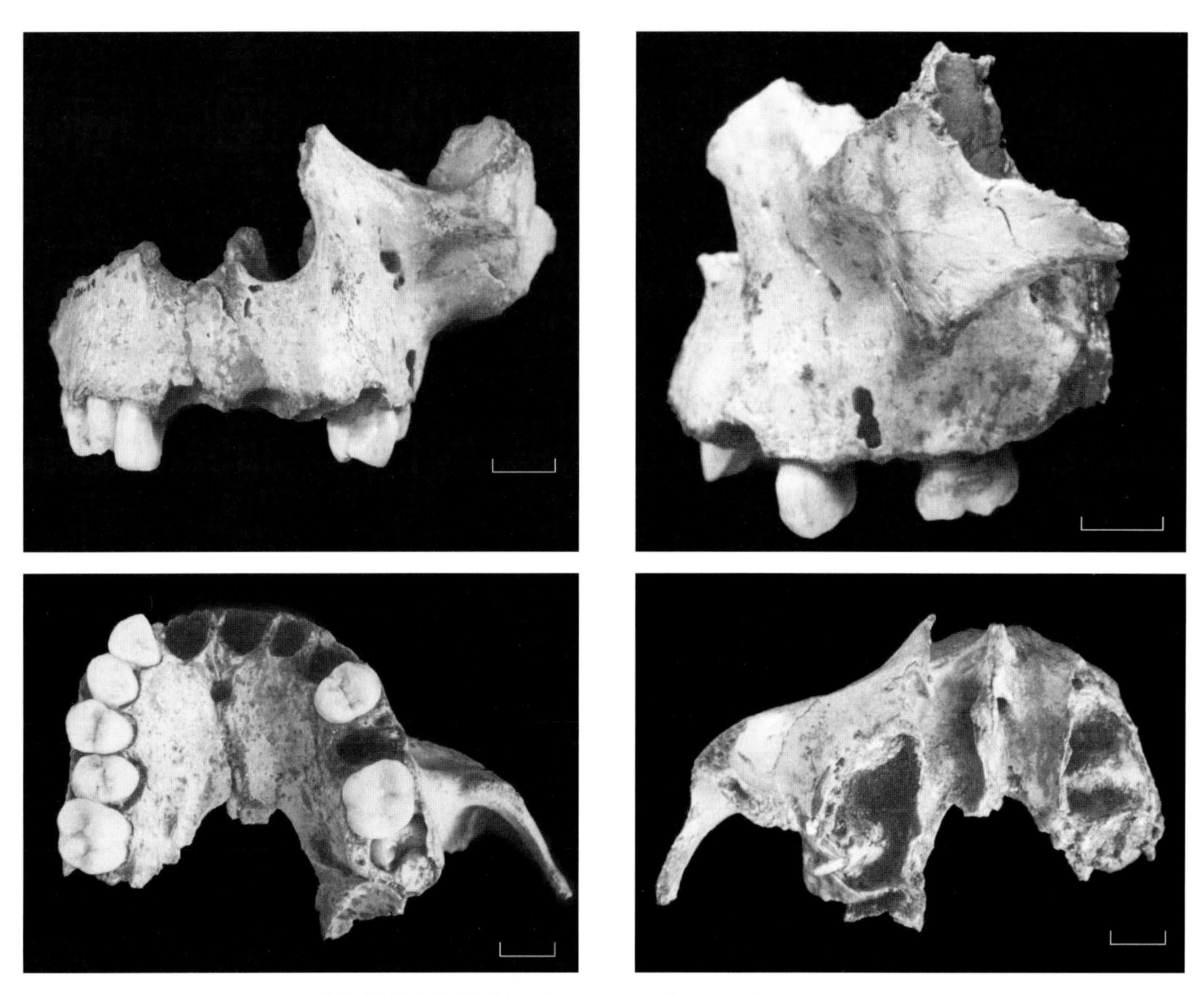

GRAN DOLINA Figure 1. ATD6-69 (scale = 1 cm).

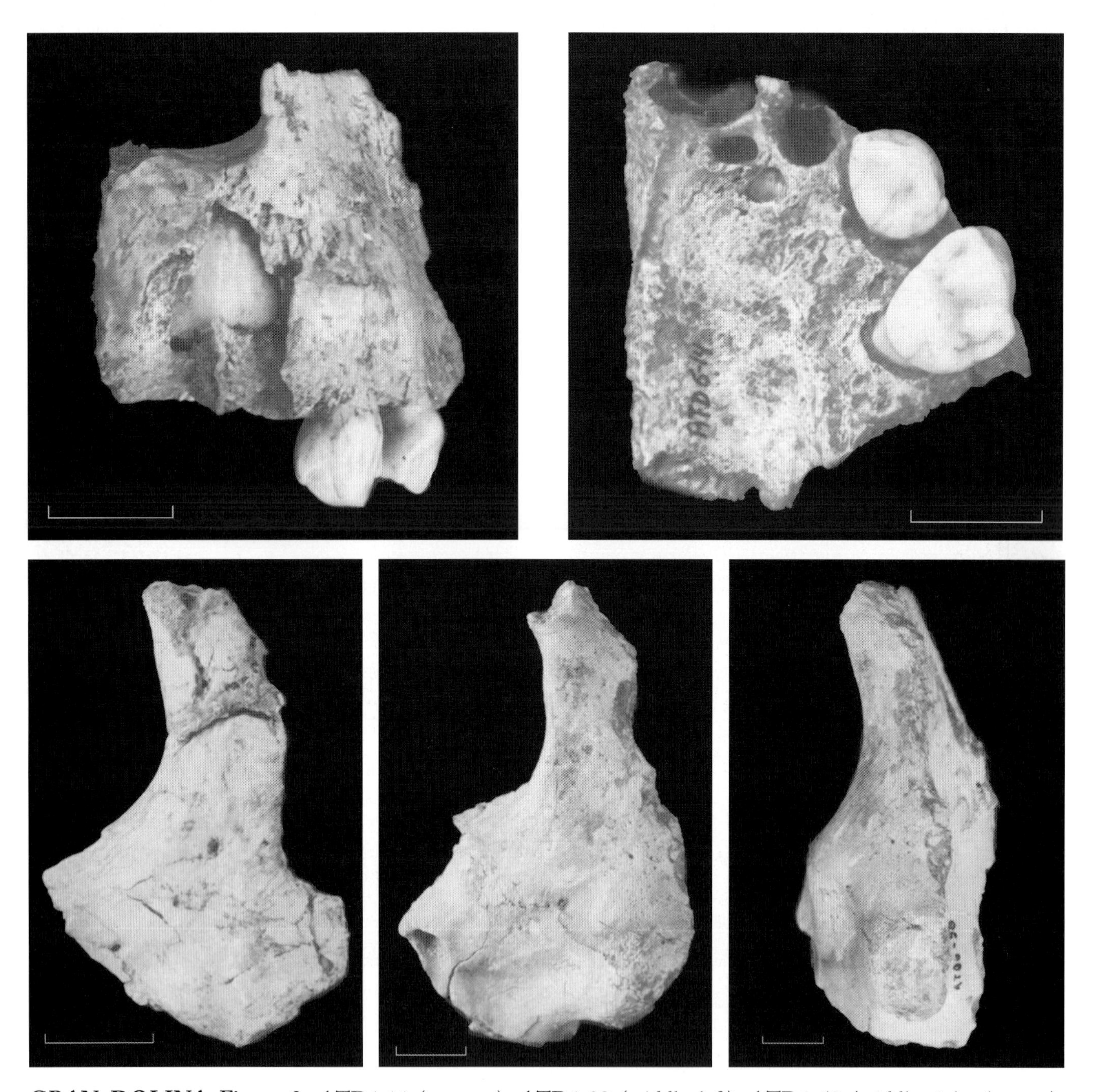

GRAN DOLINA Figure 2. ATD6-14 (top row); ATD6-38 (middle left); ATD6-58 (middle right, bottom), (scale = 1 cm).

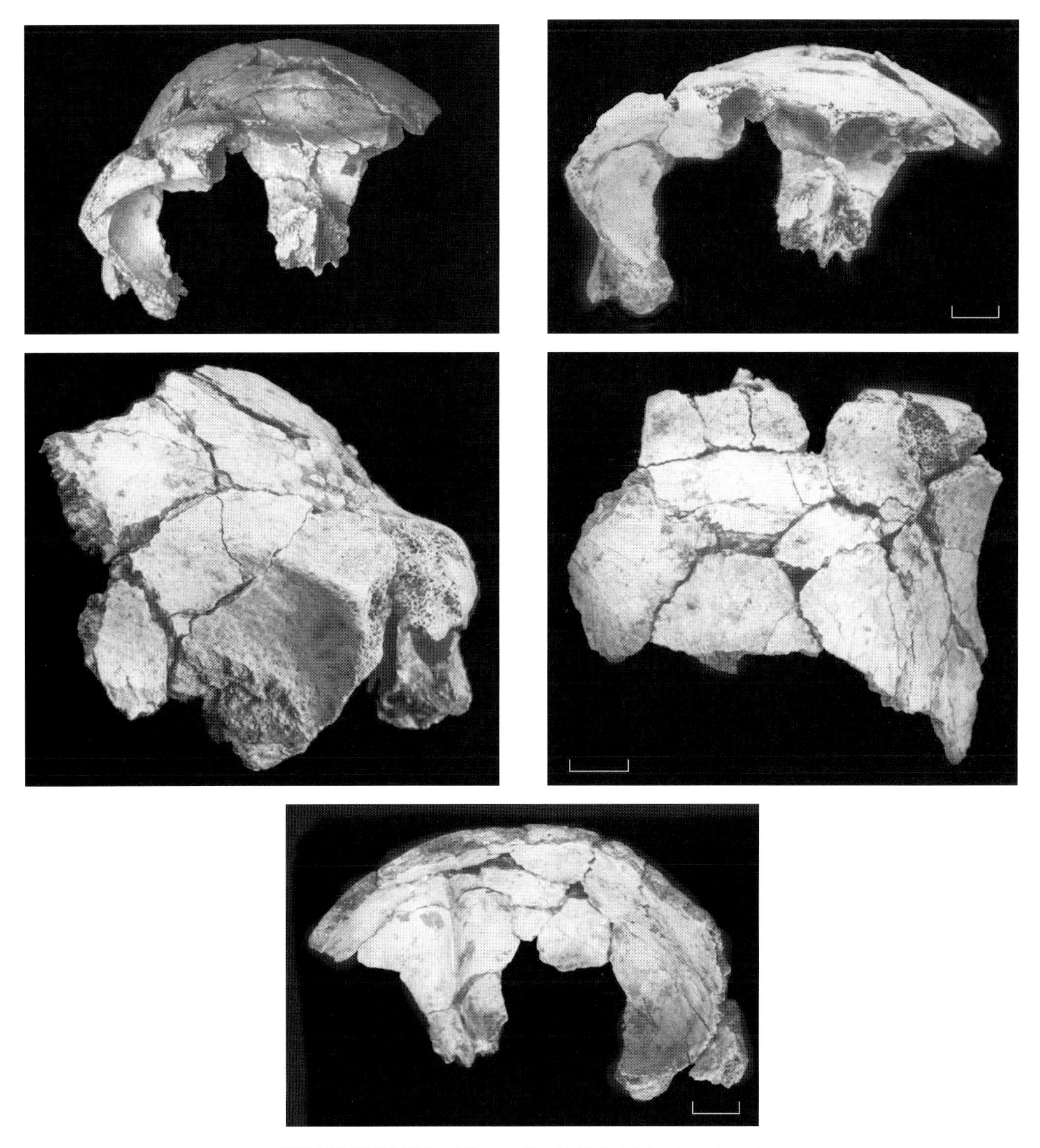

GRAN DOLINA Figure 3. ATD6-15 (scale = 1 cm).

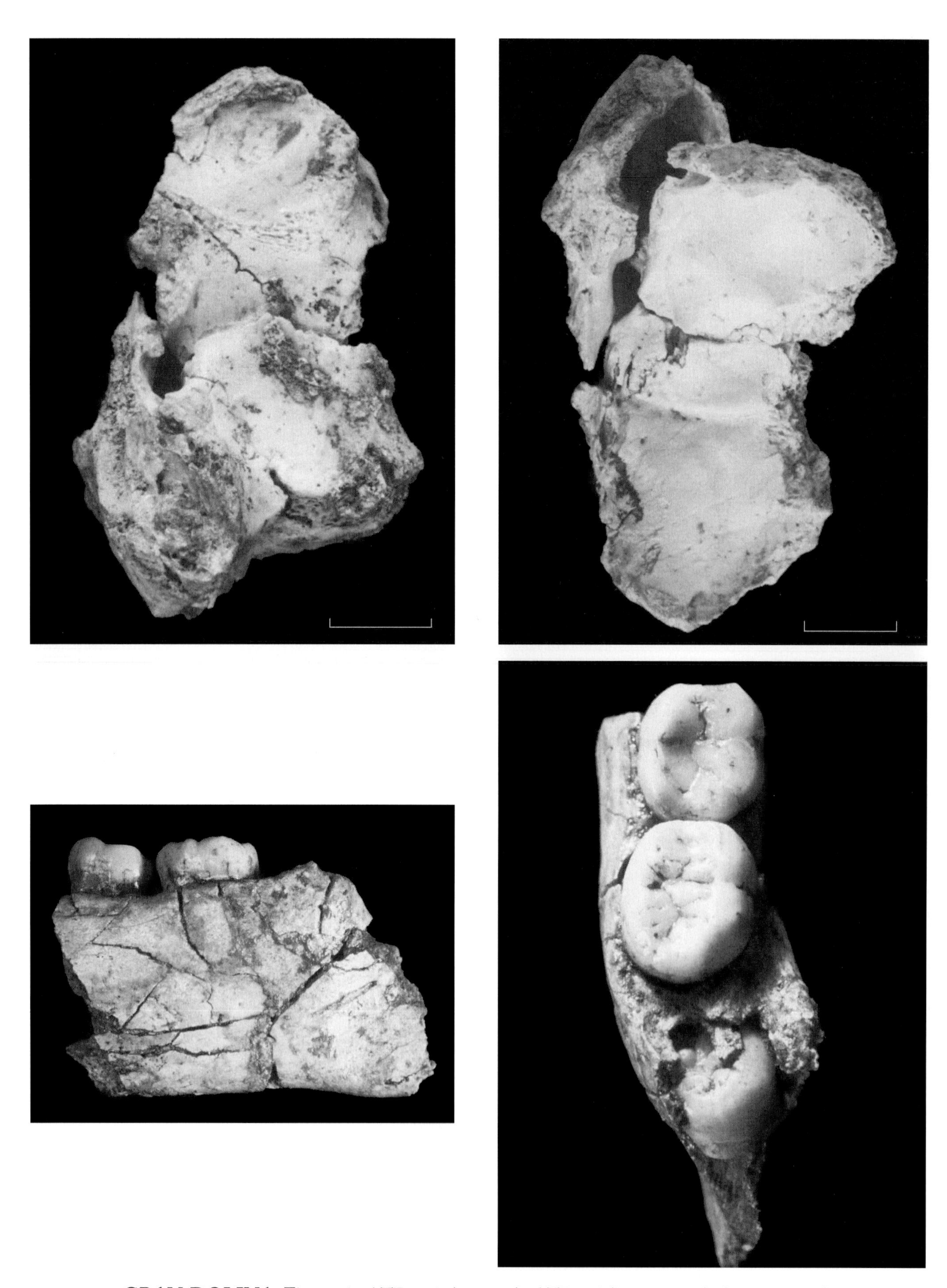

GRAN DOLINA Figure 4. ATD6-17 (top row); ATD6-5 (bottom row), (scale = 1 cm).

Atapuerca: Sima de los Huesos

Location
Vertical shaft some 0.5 km inside the Cueva Mayor-Cueva del Silo complex in the Sierra de Atapuerca, E of Burgos, Spain.

Discovery
By E. Torres and other speleologists, in 1976. Systematic excavation started in 1984, under E. Aguirre, and subsequently continued under J. L. Arsuaga, J. Bermúdez de Castro, and E. Carbonell (see Arsuaga et al., 1997).

Material
The Sima de los Huesos has lived up to its name, having yielded an enormous trove of hominid bones (most recently listed in Arsuaga et al., 1997). This important assemblage is mostly fragmentary but includes an almost complete skull (Skull 5) and pelvis, as well as a series of good calvaria, mandibles, and limb elements. Despite the fragmentary nature of most of the hominid bones, preservation of detail is extraordinarily good.

Dating and Stratigraphic Context
The hominid bones have been found at the bottom of a deep vertical shaft, embedded in a muddy breccia that yields above to sediments containing bear remains. The paleontological layers show evidence of postdepositional disturbance. They exhibit no clear stratification and are capped by flowstones and guano-rich muds. U-series and radiocarbon dates indicate that the flowstone cap dates from about 68 to 25 ka, and U-series analysis of speleothem clasts associated with the hominids shows an equilibrium state, suggesting a date of greater than 350 ka. Bischoff et al. (1997) support the probability that the cave was entered before about 320 ka ago, with a minimum age for entry of 200 ka. How the bones were accumulated in this unusual locality is a matter of controversy, but most investigators lean toward human agency of some kind (e.g., Andrews and Fernandez Jalvo, 1997).

Archaeological Context
None.

Previous Descriptions and Analyses
Numerous descriptions have been published by the excavating team and are well summarized at considerable length in a special journal issue (Arsuaga et al., 1997). These investigators conclude that the Sima hominids retain primitive features of the cranium and face while also showing some features that are "transitional" to Neanderthal morphology. Arsuaga et al. believe that the Sima hominids are early members of the Neanderthal lineage and are also related to other European Middle Pleistocene fossils. Neither group, they find, is affiliated with the ancestry of modern humans. In contrast, Tattersall and Schwartz (2000) concluded that European Middle Pleistocene hominids belong to a coherent but diverse clade, of which the Neanderthals and the Sima hominids form distinct components. They agree that this clade is entirely distinct

from that containing *Homo sapiens*. Arsuaga et al. (1997) estimated the cranial capacity of Sima H 4 at 1390 cc and of Sima H 5 at 1125 cc.

MORPHOLOGY

(Photographs courtesy of Javier Trueba). **Skull 5**: Relatively complete, reconstructed cranium, missing part of R lateral orbit, L parietal near coronal suture, part of R palate and bones within nasal cavity. Although bone feels light, vault quite thick, with substantial diploic layer. Also, reconstructed mandible lacking most of L and part of R ramus. Upper teeth (as preserved in February 2000) include RM1, LM1, and R and L M2–3; lowers include R and L P1–M3. All teeth very worn, both occlusally and interstitially (thus morphologically unrevealing).

Cranial vault of medium size; facial skeleton quite large, robust. Frontal rises from behind supraorbital margins, curves back moderately steeply to peak at bregma. Posterior to this, profile straightens out before descending from point above mastoid process and arcing smoothly and quite steeply across lambda. About halfway down occipital plane, surface becomes vertical as it runs to moderately developed superior nuchal line, where bone distinctly angles. Viewed from above, skull tapers gently forward to modest postorbital constriction. Viewed from rear, cranium has distinct "roofed" or "tented" profile, with short vertical sides and midline peak. Long parietal slope from temporal line to sagittal suture. Parietals themselves broad relative to their length.

As judged from better-preserved L side, supraorbital torus quite well developed, does not project forward markedly. Posttoral shelf longer laterally than centrally; in central region, slight angle delineates superior margin on both sides; descending margins in this area delineate L and R tori from slightly sunken glabellar region. Some superficial pathology, but roof of orbit flows out forward, then "corners" before rolling gently out, around, and up. Almost no postglabellar plane; profile flows smoothly up from glabella to very slightly domed central portion of frontal bone. Interorbital region massively broad, gently curved out.

Orbits modestly "aviator glasses" shape, with somewhat oblique inferomedial corners. Infraorbital foramina moderately large, point down, lie well below inferior orbital margin. Infraorbital groove preserved on both sides; extends almost to inferior orbital margin. Infraorbital canal very short, lies just lateral to midline of orbital floor. Orbital floor flows quite smoothly out onto face, with no marked edge. Within orbit, inferior orbital fissure very long, well marked; superior orbital fissure short, vertical. On both sides, anterior and posterior ethmoid foramina lie in frontal. As seen on the R, lacrimal fossa shallow, with very poorly defined anterior and (especially) posterior lacrimal crests; crests appear to converge superiorly. On both sides, three zygomaticofacial foramina, scattered from below to above inferior orbital margin.

Nasal bones wide superiorly; become even wider inferiorly. In profile, curve out below nasion and down again at their inferior extremities; flexed back from slight midline keel. Nasion lies well below superior orbital margins. Frontal processes of maxilla long a/p; entire superior and lateral nasal region appears somewhat inflated. Breakage reveals maxillary sinus extending up into these areas. Lateral and inferior margins of L side of nasal aperture have been pathologically remodeled; appears that aperture was large, trapezoidal; is very wide both superiorly and inferiorly. As seen on the R, lateral crest of nasal margin crisp; flows down onto nasoalveolar clivus, where it fades out. Also on the R, a crest is preserved that runs from anterior nasal spine back up to join the conchal crest (thus prenasal fossa between lateral and spinal crests opens onto clivus). Nasoalveolar clivus damaged; was quite long, broad across. Within nasal cavity, conchal crest low, fairly horizontal; lies just within entrance to nasal cavity. Floor of nasal cavity quite flat around and behind bulky but not projecting anterior nasal spines. At rear of nasal cavity appears to be voluminous, single-chambered sphenoid sinus. Appears that ethmoid was truly multifocular, with numerous air cells. Posterior superior root of vomer preserved, extends as far posteriorly as superior roots of pterygoid plates; lies well behind preserved posterior border of palate.

Not very wide anterior roots of zygomatic arches originate quite far above and level with M^2. In front view, curve up and out quite steeply (thus, in front view, face narrows inferiorly in stepwise fashion; also anteriorly projecting although not wedge shaped). Infraorbital region substantially swollen by maxillary sinus (as breakage reveals). In top view, anterior root of zygomatic arch goes straight back, flowing into zygoma, which runs straight back.

Temporal lines thick, low, emerge high up and well behind supraorbital tori; trajectory low, running almost directly back; reach highest points above external auditory meati, then recurve strongly toward parietal notch; fade out behind mastoid process. Parietals relatively short a/p, wide m/l.

Squamosal relatively short, not very tall s/i; squamous suture strongly arced. Parietal notch quite vertical, forms angle slightly greater than 90°. Marked supramastoid tubercle lies level with posterior root of zygomatic arch; no suprameatal crest connecting the two structures. Posterior root of zygomatic arch bears lateral swelling posteriorly; arch robust but does not flare laterally. Temporal fossa relatively small, divided into anterior and posterior components by angulation at anterior squamous suture. No sphenoid crease demarcating infratemporal fossa.

Mandibular fossae long a/p at their maximum dimension; taper medially and laterally; not bounded anteriorly by an eminence of any kind (presence of reactive bone anteriorly in Skull 5 suggests that mandible may have articulated in this area); bounded posteriorly by very large postglenoid plate. Plate extends laterally well beyond auditory meatus; is quite massive at its midpoint. Walls of tubular ectotympanic quite thin, especially anteriorly; tubes not fully ossified laterally; bilaterally is ossification of superolateral margin.

Mastoid processes quite thick at bases; as seen on more complete L side, project well below level of occipitomastoid suture. As also best seen on the L, projecting part of mastoid process relatively restricted a/p and narrow m/l (in Skull 4, this structure much bulkier). Parietomastoid suture very long, essentially straight and horizontal (both sides); flows into short anterior lambdoid suture. Mastoid notches on both sides very deep, narrow, posterolaterally oriented; lack digastric fossae posteriorly; are bounded medially by quite tall, robust, longitudinal paramastoid crests. On the L, distinct groove lies between paramastoid crest and flat occipitomastoid suture. On the R, this groove runs virtually along occipitomastoid suture. Waldeyer's crests very markedly distended, thick; run almost straight posteriorly from position medial and close to occipitomastoid suture and almost opposite mastoid processes. Lateral faces of Waldeyer's crests bear laterally oriented, relatively deep fossae. Bilateral short, stout crests lie behind closed-off condylar canals and medial to Waldeyer's crests.

Basicranially, petrosals both damaged; the L is thick m/l, angles only gently forward. It seems that both jugular foramina were relatively small, the R being slightly larger; both occupied double-chambered jugular fossae. Large stylomastoid foramen lies (both sides) in line with mastoid notch and slightly posterolateral to thin styloid process. Styloid processes lie in deep pits; tips broken. Short vaginal processes curve around and peak on lateral sides of styloid processes, fade out laterally before reaching small auditory meati. Carotid foramina relatively large; lie well medial to styloid processes and quite close to margins of jugular fossae. Sphenotemporal sutures bisect large, thick medial articular tubercles. Tubercles well separated from ectotympanic tubes; do not close off mandibular fossae. Very large foramina of Vesalius (both sides) lie between sphenoid and petrosal. Foramina ovales lie, and apparently also foramina spinosum appear to have lain, entirely within sphenoid. Foramina ovales bounded medially by very thick walls of bone that apparently bore on their inner surfaces impression of petrosal (implies there was no foramen lacerum; confirmed in Skull 4).

Occipital quite broad, moderately low; apparently lambdoid suture did not peak at lambda. Suprainiac region appears somewhat pathological (Sima sample shows considerable cribra crania and larger healed legions); hints of modest, poorly rimmed suprainiac depression. Superior nuchal line curves gently from side to side, fades from midline laterally, forms slight thickened torus level with asterion. Torus delineated inferiorly by shallow, paired, scallop-shaped depressions. No definitive external occipital protuberance. External occipital crest distinct, moderately developed.

Foramen magnum ovoid, moderately large. Condyles quite large, fairly strongly arced a/p; are not very anteriorly positioned. Anterior condylar canals large, open. Basiocciput relatively broad, not very long; does not taper much anteriorly; is thin by margin of foramen magnum; thickens markedly anteriorly; surface smooth except for thickened, almost peaked, lateral margins.

Palate broad, not notably long, with slightly U-shaped arcade; is moderately deep, with steep sides and long anterior slope. Medium-sized incisive foramen lies somewhat posterior to I1s, with a distinct groove in front of it. Palatine foramina relatively large; pair of smaller posterior accessory foramina on both sides. Medial and lateral pterygoid plates fade superiorly as they converge to join below region of foramen ovale; are not vertical (incline forward).

Internally, cerebellar lobes face down; are not reflected in external morphology. R depression much broader but shallower than the L. Internal occipital crest quite short. Occipital lobes lie about level with asterion; are separated by sharp, well-developed midline crest that expands into quite broad internal occipital protuberance. Impression of L occipital lobe smaller but deeper than the R. R transverse sinus separates from sagittal sinus high up, about halfway up occipital plane. Less distinct L transverse sinus runs laterally from internal occipital protuberance. Transition between transverse and sigmoid sinuses runs directly from occipital into temporal. On both sides, superior surface of petrosal broad, flat; slight elevation for superior semicircular canal. Better-preserved L side does not appear to bear superior petrous sinus. On both sides, indentation in region of subarcuate fossa.

Mandibular corpora tall, narrow; jaw is only moderately broad across the front. In profile, symphysis slightly tilted back; inferior margin slightly elevated at front. Shallow curve courses across top of symphyseal region. Internally, very short postincisal slope breaks vertically to area of slight swelling in genial region; below this, internal profile curves smoothly down and forward to slight peak between two broad, shallow, posteriorly directed digastric fossae. As seen on the R, mylohyoid line low, roughened; below is long, moderately deep submandibular fossa. On both sides, very large, oval, obliquely oriented mental foramen lies under M_1 and relatively close to inferior margin. Marked retromolar space on each side. Anterior root of ramus takes origin below distal root of M3. On the R apparently was shallow preangular notch, lying well below the a/p fairly long coronoid process. Gonial region severely truncated and oblique; bears some muscle scarring exteriorly. Internally on angle, series of marked muscle scars, including series of medial pterygoid tubercles just below level of ovoid, compressed mandibular foramen (its margin broken). Low, blunt pillar runs from apex of coronoid process down toward internal alveolar crest, which displays some pathological alteration. Sigmoid notch crest broken in midregion; posteriorly, runs just lateral to midline of huge, medially expanded condyle. Laterally below condyle, bone distended into tubercle-like structure.

M^3s probably quite large; roots inclined forward, bifurcate well below the neck, with quite separate lingual root. Lower anterior tooth alveoli quite large. P_{1-2} alveoli were single, quite small; P_1 alveolus bore slight groove on lingual side. M_3 was as long, if not longer, m/d than other molars. Clefts of molar roots lie well below neck; roots stay separate throughout the series. M_3 roots markedly forwardly inclined.

REFERENCES

Andrews, P. A. and Y. Fernandez Jalvo 1997. Surface modifications of the Sima de los Huesos fossil humans. *J. Hum. Evol.* 33: 191–217.

Arsuaga, J. L. et al. 1997. Special issue on the Sima de los Huesos hominids and site. *J. Hum. Evol.* 33 (2/3): 105–421.

Bischoff, J. et al. 1997. Geology and preliminary dating of the hominid-bearing sedimentary fill of the Sima de los Huesos Chamber, Cueva Mayor of the Sierra de Atapuerca, Burgos, Spain. *J. Hum. Evol.* 33: 129–154.

Tattersall, I. and J. H. Schwartz. 2000. *Extinct Humans.* Boulder, CO: Westview Press.

Repository

Departamento de Paleontología, Facultad de Ciencias Geológicas, Universidad Complutense de Madrid, 28040 Madrid, Spain.

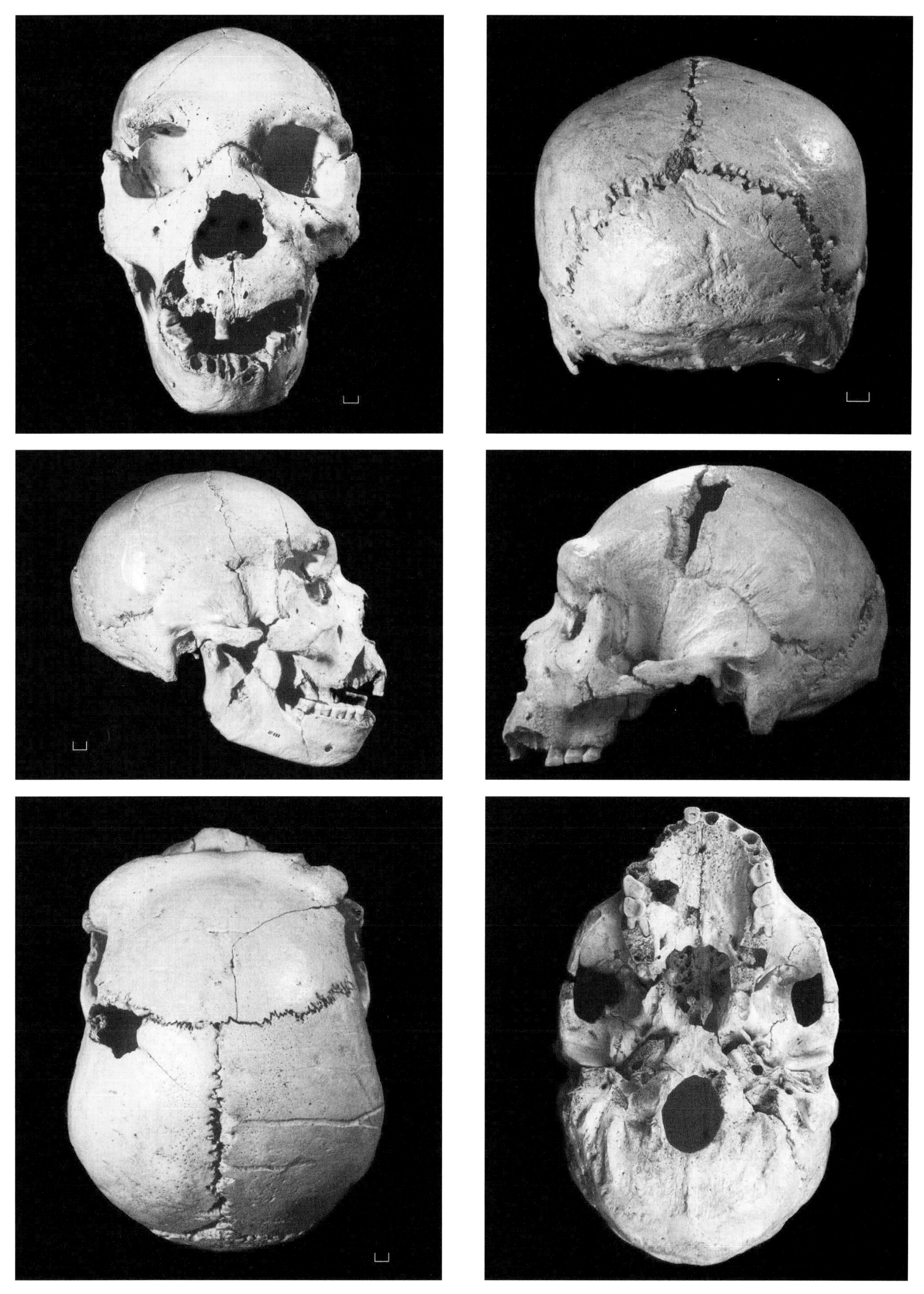

SIMA DE LOS HUESOS Figure 1. Skull 5 (scale = 1 cm).

Biache-Saint-Vaast

Location

River terrace site now incorporated into the Chatillon-Commentry-Biache factory complex, Biache-St-Vaast, midway between Arras and Douai, Pas-de-Calais, France.

Discovery

Salvage excavations directed by A. Tuffreau and J. Piningre, May 1976. Biache 1 found by B. Vandermeersch.

Material

Rear of cranium, palatal fragment, and several isolated teeth were identified during excavation (Biache 1). A partial frontal bone and L temporal with associated part of parietal were identified later among the collections removed from the site (Biache 2).

Dating and Stratigraphic Context

Sands and other deposits of the Scarpe river, overlain by a loess sequence. The river deposits yielded the hominid remains, together with an abundance of lithics and faunal materials. Three principal occupation horizons were distinguished by Tuffreau et al. (1978): some minor levels low in the sequence; Bed IIA, a thin layer composed of several "living floors" and very rich in archaeological remains; and Bed II (base) at the lower limit of the brown paleosol capping the riverine sequence. Biache 1, at least, was discovered just above the top of Bed IIA. Analyses of pollens, invertebrates, and micromammals from the archaeological deposits point to temperate conditions that Tuffreau et al. (1978) assigned confidently to an interstadial of the Saale (penultimate) glacial. This has been confirmed by TL dating of Bed IIA burned flints to around 175 ka (Huxtable and Aitken, 1988).

Archaeological Context

The Bed IIA lithics are strongly Levalloisian in aspect, with no handaxes and many long points that may have been hafted. Stringer et al. (1984) emphasized Mousterian similarities, and Tuffreau (1988) later assigned the Biache industry to the Mousterian of Ferrassie type. Abundant charcoal and other traces of burning indicate the use of fire in Bed IIA times.

Previous Descriptions and Analyses

Biache 1 was briefly described by Vandermeersch (1978), who regarded it as a "preneanderthal" with numerous Neanderthal resemblances but recalling the Swanscombe cranium in certain ways. Stringer et al. (1984) concur in regarding this specimen as essentially an early Neanderthal. Holloway (2000) reported its cranial capacity at 1200 cc. The Biache 2 specimens have yet to be described.

Morphology

Biache 1

Adult. Consists of posterior parts of R and L parietals, occipital and nuchal planes of occiput, L occipital condyle, L petromastoid part of temporal, and mastoid portion of R temporal with partial external auditory meatus and posterior part of petrosal (anterior part missing). Bone not notably thick (even diploe).

Braincase low at rear; "en bombe" in posterior outline; bears chignon. Large infratoral depression delineates broad, horizontal occipital torus. Suprainiac depression broad but shallow and subequally tripartite. No external occipital protuberance; only very modest external occipital crest. Occipital plane is broad, not tall. Lambdoid suture arcuate, does not peak at lambda.

Anterior lambdoid suture long, horizontal, but not straight. Parietal notch lies over midline of mastoid process, which is distinct but very small and points straight downward. Temporal lines faint if observable at all. Effectively no supramastoid crest.

Especially as preserved on the R, thick paramastoid crest lies lateral to occipitomastoid suture; is somewhat posteriorly placed; forms posterolateral margin of broad, shallow mastoid gutter (not notch). Thick, ridgelike occipitomastoid crest lies along occipitomastoid suture medial and parallel to this crest; is broken but was at least as tall as mastoid process (see L side). Low, rugose Waldeyer's crest, preserved bilaterally, lies medial to this crest. Two-thirds of circumference of foramen magnum preserved. Foramen quite elongate; broadest anteriorly, across condyles. Articular surface of L (preserved) condyle extends into postcondylar depression. Anterior condyloid canal very large. Portions of very thick-walled R and L tubular ectotympanics preserved (not evident on every cast); inferior portion of tube only marginally less developed than superior part.

On the L petrosal, arcuate eminence does not protrude; entire superior surface broad, horizontal; no sign of superior petrous sinus. On the R, broken base of petrosal highly pneumaticized. Superior sagittal sinus moderately impressed. R transverse sinus strongly impressed lateral to internal occipital protuberance; fades fast laterally; L transverse sinus thinner and originates lower than R. Sigmoid sinus deep, short.

On R and L sides, stylomastoid foramen lies quite medial to mastoid process. On the L, some of vaginal process preserved; is quite separate from mastoid process. Styloid pit lies quite far medial to mastoid process and in line with mastoid (digastric) gutter; in absence of impressions for posterior digastric fossae, digastric m must have anchored in the gutter. L jugular foramen partly preserved, is not very large and points anteriorly.

References

Holloway, R. L. 2000. Brain. In: E. Delson et al. (eds), *Encyclopedia of Human Evolution and Prehistory*. New York, Garland Publishing, pp. 141–149.

Huxtable, J. and M. Aitken. 1988. Datation par thermoluminescence. In: A. Tuffreau and J. Sommé (eds), *Le Gisement Paléolithique Moyen de Biache-Saint-Vaast (Pas de Calais)*, vol. 1. *Mém. Soc. Préhist. Fr.* 21: 107–108.

Stringer, B. et al. 1984. The origin of anatomically modern humans in western Europe. In: F. Smith and F. Spencer (eds), *The Origins of Modern Humans*. New York, Alan R. Liss, pp. 51–135.

Tuffreau, A. 1988. L'industrie lithique du niveau IIA. In: A. Tuffreau and J. Sommé (eds), *Le Gisement Paléolithique Moyen de Biache-Saint-Vaast (Pas de Calais)*, vol. 1. *Mém. Soc. Préhist. Fr.* 21: 171–183.

Tuffreau, A. et al. 1978. Premiers résultats de l'étude du gisement paléolithique de Biache-Saint-Vaast (Pas-de-Calais). *C. R. Acad. Sci. Paris* D286: 457–459.

Vandermeersch, B. 1978. Etude préliminaire du crâne humain du gisement paléolithique de Biache-Saint-Vaast (Pas-de-Calais). *Bull. Assoc. Fr. Et. Quat.*: 65–67.

Repository

Laboratoire d'Anthropologie, Université de Bordeaux 1, 33405 Talence, France (Biache 1); Laboratoire d'Anthropologie, Faculté de Médécine—Secteur Nord, 13916 Marseille, France (Biache 2).

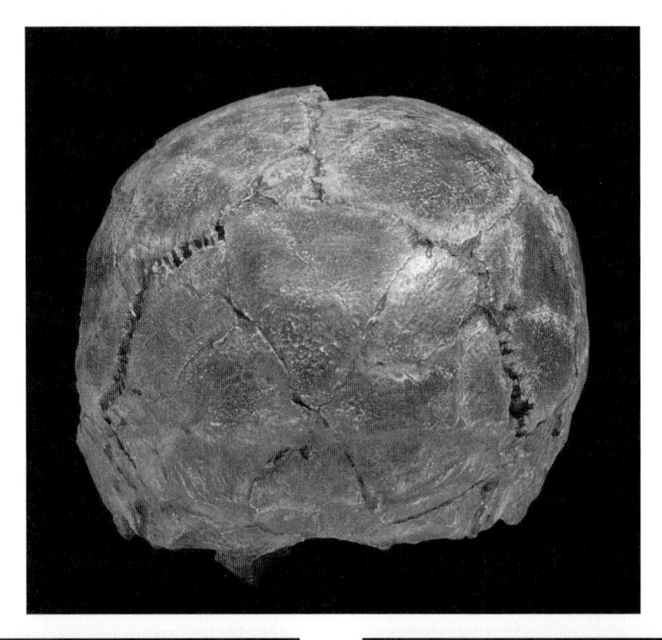

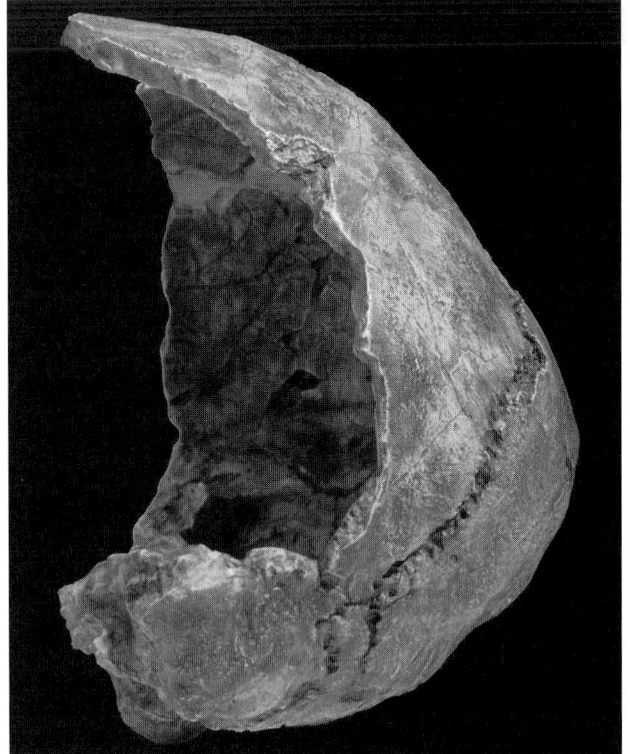

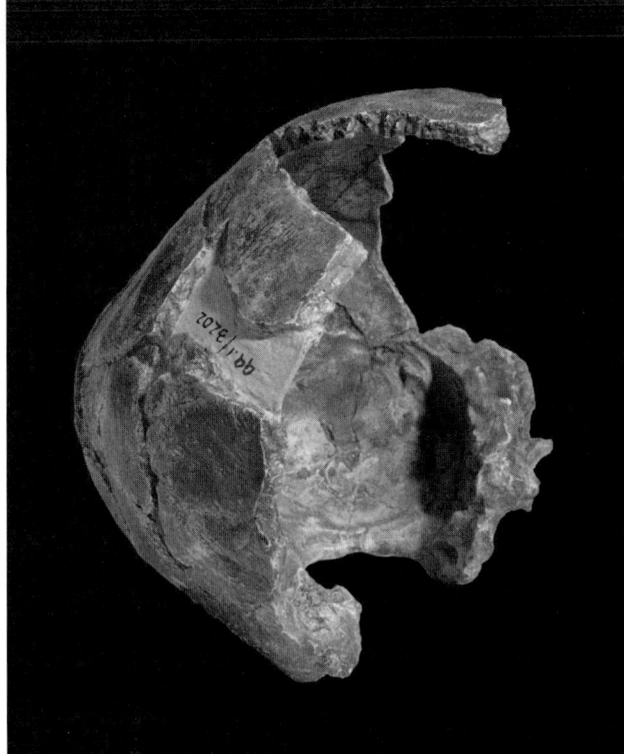

BIACHE-SAINT-VAAST Figure 1. Biache 1 (not to scale).

Bilzingsleben

Location
Open-air locality 1 km S of Bilzingsleben, some 35 km N of Erfurt, eastern Germany, in the Lower Wipper Valley on the northern border of the Thuringian Basin.

Discovery
Site known since the early nineteenth century. All surviving hominid fossils come from ongoing excavations directed by D. Mania. The first of these finds was made in 1972.

Material
Multiple fragments of two skulls (braincases of both have been reconstructed by E. Vlcek) plus an isolated molar from a third individual.

Dating and Stratigraphic Context
The Bilzingsleben site lies near the base of a sequence of Pleistocene travertines built up by a karst spring that flowed into a shallow lake. The archaeological horizon lies in the alluvial deposits by the ancient lake shore. The fauna and flora indicate mild conditions and, together with geomorphological considerations, suggest that the site dates from the stage 9 Holsteinian interglacial (see refs in Mania, 1993), at perhaps 280 ka. However, various U-series dates on calcite samples and ESR dates on teeth have come in considerably higher than this (see summary and refs in Schwarcz et al., 1988), suggesting an age somewhere in the 300- to 400-ka range. At this point, the 280-ka date should probably be regarded as a minimum.

Archaeological Context
Appears to represent a lakeshore campsite, with several concentrations of stone tools and animal bones, and what appears to be a pebble floor (evidence is most recently summarized in Mania, 1997). Some features have been interpreted controversially as the foundations of stone structures, possibly windbreaks or even more complete dwellings, and several hearths and "workshop" areas have been identified (e.g., Mania, 1993, 1997). The lithic assemblage, characterized as "Micro-Clactonian" (Vlcek and Mania, 1987), is rather crude, containing a few large flakes but lacking handaxes and consisting mainly of choppers and rather small flakes, often quite extensively retouched. Some pieces of bone and antler may have served as digging tools, whereas certain bone and ivory plaques bear cut marks that have been interpreted as intentional.

Previous Descriptions and Analyses
The human remains from Bilzingsleben have been described by E. Vlcek and D. Mania (most extensive account in English by Vlcek et al., 1987; most recent update by Mania et al., 1993), who see the affinities of the hominids as lying with *Homo erectus*, most specifically OH9 and some Zhoukoudian and Sangiran individuals. In 1978 Vlcek named the new subspecies *Homo erectus bilzingslebenensis*. In contrast, Stringer

(1981) and Cook et al. (1982) compared them to "archaic *Homo sapiens*," notably Saldanha. However, Stringer later backtracked a little (Stringer et al., 1984), noting that these remains were the "most *erectus*-like and least modern or Neandertal-like" of that group.

MORPHOLOGY

1993+ originals and pre-1993 casts studied. Cast numbers below are Halle numbers; originals have only Bilzingsleben designations. Specimens very fragmented; tend to have worn edges. Were uncovered very scattered around but geographically fall more or less into three major groups (which are discussed below as such):

Set 1

Apparently from same individual, **Individual 1**. A1 (cast, HK 74:206A), partial occipital plane, and A2 (cast, HK:206B), a contiguous, largely nuchal plane fragment with transverse sinus. Also B1 (cast, HK 75: 199), frontal piece including middle supraorbital region plus glabella; B2, probable frontal fragment, with (perhaps) coronal suture (cast, HK 76:529); and B3 (cast, HK 79:1141), said to be fragment of L frontal with temporal line but possibly is R parietal fragment with supramastoid crest. D1 (cast, HK 78:772) was originally associated with this set as R posterior parietal portion but is now associated by Vlcek and Mania with Set 2.

Diploic bone of cranial vault has very tiny air cells; grades indistinctly into thin tabular bone, which only constitutes outside skin.

Apparent frontal bone fragment thick; internal part of preserved stretch of coronal suture appears broken off, but outer portion deeply but uniformly denticulated. Very broad glabellar region only slightly protruding. Nasofrontal suture very wide; nasion lies very high. Very thick, smoothly rolled supraorbital margins slope back gently from glabella. Orbital roof curves smoothly into supraorbital margin. With low frontal crest oriented vertically, frontal presents an a/p long, wide, but quite shallow sulcus behind glabellar and supraorbital regions; more posteriorly, frontal slopes shallowly backward. Frontal lobes lie well behind anterior part of orbital cone. Three separate air spaces in frontal bone; the central one lies in midline and penetrates glabella to an unknown extent. Laterally lie two separate sinuses; the L one extended quite far up and back into supraorbital region, penetrating to below the shallow sulcus; not determinable how far laterally it extended.

What remains of occipital region extremely thick boned in occipital and nuchal planes. Very poorly denticulated and undifferentiated lambdoid suture does not arc broadly across lambda (not peaked either). Occipital plane was apparently very wide from side to side and short, probably with longish, slightly upwardly inclined anterior lambdoid suture. Occipital plane gently arced from side to side; was straighter from top to bottom. Nuchal plane makes distinct inward angle to occipital plane; their border represented by thick, blunt corner (not distinct torus) that bulges out a bit posterior to primary occipital plane and is thus delineated above by a slight depression. Below the corner, in midline, is broad, shallow fossa. Laterally, bone preserved on L as far as posterior aspect of very shallow mastoid notch, which lacks any indication of digastric fossa. Nuchal plane very smooth, with little muscle scarring. Internally, shallow superior sagittal sinus runs down to low internal occipital protuberance, from which the R transverse sinus diverges more strongly. L transverse sinus preserved; fades out laterally toward region of asterion. Cerebellar lobes poorly impressed on inside of occipital.

Set 2

Individual 2. Reassociation of D1 (cast) presumably because meningeal groove lines up with fragment D5 (original). Together these create part of R posterior parietal. D1 seems to have lambdoid suture going up from region of asterion. Front of D1 (cast) articulates at asterion with fragment G1 (Vlcek et al. associate this with temporal fragment G1), with part of its squamosal suture, but D1 might equally be from the L, closer to the parietal notch. G1 is a R temporal fragment, primarily with damaged mastoid and articular regions. There is also a partial L occiput (A3), which includes the L portion of the nuchal plane, with transverse sinus visible internally, a L supraorbital region (B4; cast, HK 87:301), and a R supraorbital region and associated frontal (B7). There are numerous small R parietal pieces: D6, part of parietal with coronoid suture and contact to R parietal fragment (D7), as well as contact farther back on suture with parietal fragment (D8). D8 contacts with D2 (cast, HK 87: 303), another small parietal fragment, which in turn contacts laterally with three pieces sharing the number D10: the D10 piece that articulates with D2 is

479.116; contacting this posteriorly, and continuing along what appears to be the sagittal suture, is 479.110; lateral and inferior to this is 479.106. D6 is a small parietal fragment, probably from the L side, with a small stretch of coronal suture. L parietal fragment D3 (cast, HK 87:302) has the sagittal suture down to lambda plus part of lambdoid suture; articulates on its left break with D4 (cast, HK 87:1205), which possesses part of lambdoid suture. B6 (cast) is a thick, tapering cranial fragment of uncertain identity. D9 is a very thick parietal fragment with part of the temporal line. B5 is part of a frontal with some coronal suture.

Diploic bone of cranial vault has very tiny air cells; grades indistinctly into thin tabular bone, which actually just constitutes an outside skin.

R and L preserved supraorbital regions similar enough to represent same individual (although no contact). Supraorbital margins thick (much less thick than Individual 1), appear not to have tapered until lateralmost part of supraorbital margin. Compared with Individual 1, much sharper corner from orbital roof to external supraorbital margin; superior surfaces of preserved supraorbital margins quite flat, with sharper frontal rises closer to them than in Individual 1. Frontal lobes lay well behind superior orbital region. Preserved frontal sinuses penetrate somewhat up into frontal bone but do not extend laterally beyond midline of orbit. Scarlike temporal ridge emerges from well behind thick zygomatic process, then fairly quickly fans out into low temporal lines that rise up and back. Postorbital constriction was apparently shallow. In superior view, it seems that orbital margins probably receded from glabellar region. Lower part of frontal was apparently somewhat arced from side to side.

Preserved parietal pieces relatively thick, especially posteriorly. Preserved portions of coronal and sagittal sutures were not very deeply denticulated; sagittal suture, at least, was not differentiated. On the L, region of lambda appears not to have been peaked; course of lambdoid suture suggests presence of Wormian bone (not normally curved suture). On the R, it appears that lambdoid suture rose quite steeply from asterion, with no intervening anterior lambdoid suture.

Preserved L occipital fragment preserves lambdoid suture down to asterion; bone thickens remarkably at level of shallow, narrow transverse sinus. Nuchal plane differentiated from occipital plane by thickened, crestlike protrusion that begins to rise quite medial to region of asterion and reaches its peak well lateral to midline, toward which it runs inferiorly, decreasing in height. Superoinferiorly, greatest thickness of occipital bone is achieved at region of peak. This peak also protrudes somewhat backward and inferiorly, being delineated by a shallow sulcus above and below and laterally by a moderately large, lozenge-shaped, and slightly deeper depression.

Temporal bone missing squamosal portion; bone appears to have had moderately long, more or less horizontal parietomastoid suture. Parietal notch oblique, more horizontally than vertically oriented. Entire supramastoid region swells out, with very thick bone, its surface completely smooth above auditory meatus (no sign of suprameatal or supramastoid crests). Broken posterior root of zygomatic arch lies at anterior margin of damaged but moderately large, ovoid, and anteriorly oriented auditory meatus, which is closely approximated to base of broken mastoid process. Inferiorly, along preserved portion of ectotympanic tube, is large, bulky, bony flange (may be vaginal process); internally, this huge process bears impression for moderately sized, rather laterally placed styloid process. Posterior and inferior to, and somewhat separated from, this process is very large stylomastoid foramen. Mastoid process entirely missing; damage reveals presence of moderately sized to slightly larger air cells. Preserved part of mandibular fossa shows it was long a/p, quite deep, and bounded anteriorly only by shallowly sloping articular eminence; in coronal plane, fossa deeply crescentic. No postglenoid plate. Sphenotemporal suture preserved; not swollen into medial articular tubercle. Internally, sigmoid sinus large, deep, and short. Appears that petrosal was quite wide; its posterior origin reveals absence of superior petrous sinus.

Set 3

Teeth. C1 (cast): R upper M1 or 2, lacking roots; crown very worn. Stout postprotocrista rather larger than short preprotocrista. Hypocone very large, swollen, and somewhat distolingually distended. Apices of paracone and metacone relatively close together. Sides of crown straight, with lingual cusps somewhat lingually placed. E3 (cast, HK 87:1214): small, rounded RM_3, very worn, some enamel chipped out, roots missing; talonid basin appears constricted from side to side; small vestige preserved of trigonid basin. F1 (cast, HK87:1213): molar fragment, probably upper LM1 or 2 (listed as lower), very worn, enamel very thick. E1 (cast, HK 88:247): L lower M1 or 2 (listed as lower R), surface eroded; hypoconulid large, centrally placed; trigonid basin long, somewhat constricted from side to side; trigonid basin small. E2 (cast, HK 84:260): possibly a L lower molar trigonid region; very worn. One lower molar found in 1920s now missing.

REFERENCES

Cook, J. et al. 1982. A review of the chronology of the European Middle Pleistocene hominid record. *Yrbk Phys. Anthropol.* 25: 19–65.

Mania, D. 1993. *Homo erectus* von Bilzingsleben–Seine Kultur und Umwelt. *EAZ Ethnogr.-Archäol. Z.* 34: 478–510.

Mania, D. 1997. *Zur Quartärgeologie des mittleren Elbe-Saalegebietes unter besonderer Berücksichtigung der Fundstellen Ehringsdorf und Bilzingsleben.* Jena, Forschungsstelle Bilzingsleben.

Mania, D. et al. 1993. Zur den Funden der Hominiden-Reste aus dem mittelpleistozänen Travertin von Bilzingsleben von 1987–1993. *EAZ Ethnogr.-Archäol. Z.* 34: 511–524.

Schwarcz, H. et al. 1988. The Bilzingsleben archaeological site: New dating evidence. *Archaeometry* 30: 5–17.

Stringer, C. 1981. The dating of European Middle Pleistocene hominids and the existence of *Homo erectus* in Europe. *Anthropologie* 19: 3–14.

Stringer, C. et al. 1984. The origin of anatomically modern humans in Western Europe. In: F. Smith and F. Spencer (eds), *The Origin of Modern Humans.* New York, Alan R. Liss, pp. 51–135.

Vlcek, E. 1978. A new discovery of *Homo erectus* in Central Europe. *J. Hum. Evol.* 7: 239–251.

Vlcek, E. and D. Mania. 1987. *Homo erectus* from Bilzingsleben (GDR)—His culture and environment. *Anthropologie* 25: 1–45.

Repository

Landesmuseum für Vorgeschichte, Richard Wagner Strasse 10, Halle/Saale, Germany (specimens discovered before 1993); Forschungsstelle Bilzingsleben der Friedrich Schiller Universität Jena, 07745 Jena, Germany (specimens collected in 1993 and since).

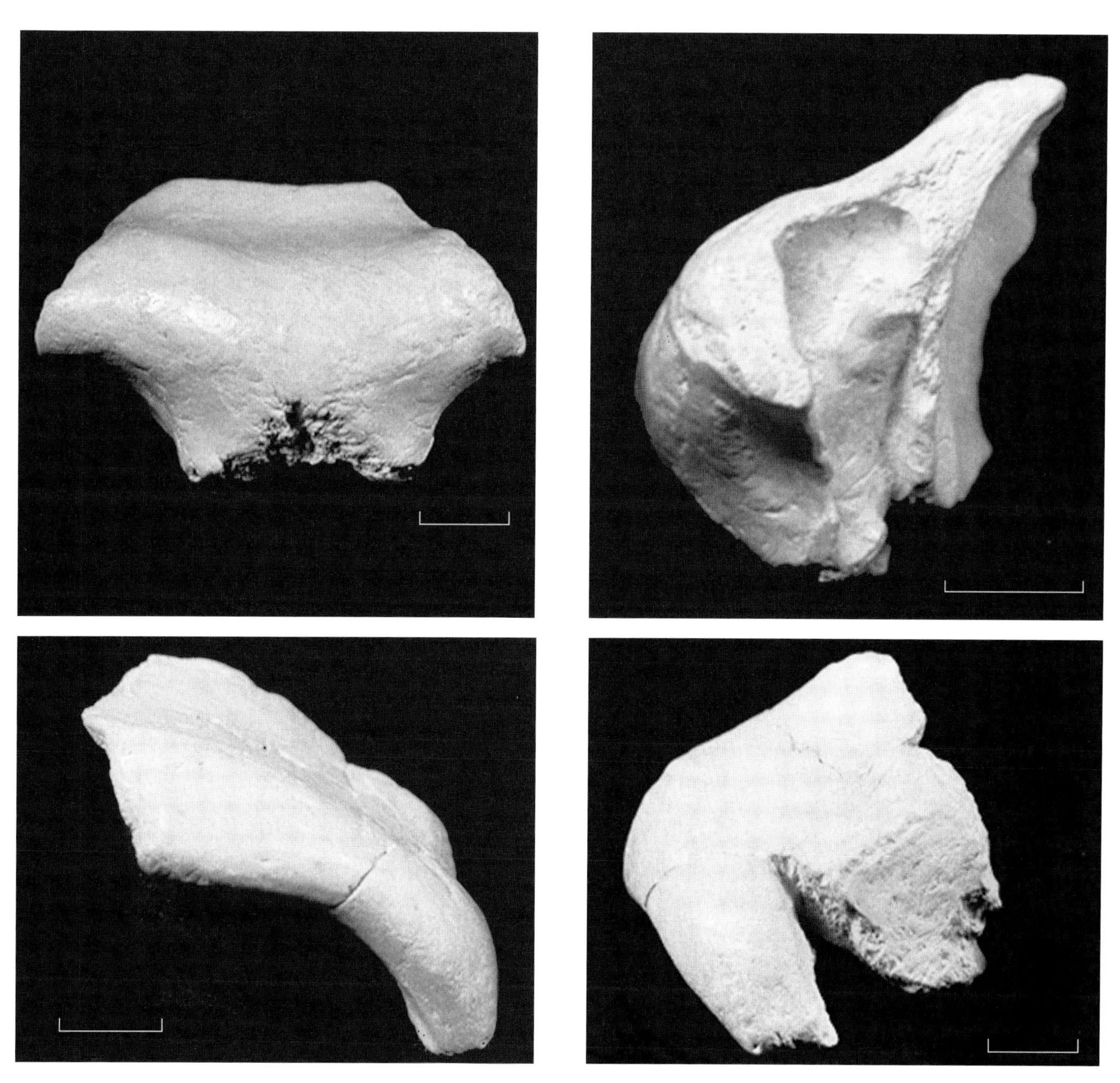

BILZINGSLEBEN Figure 1. Frontals, anterior and lateral: (top left and right) B1; (bottom left and right) B4, (scale = 1 cm).

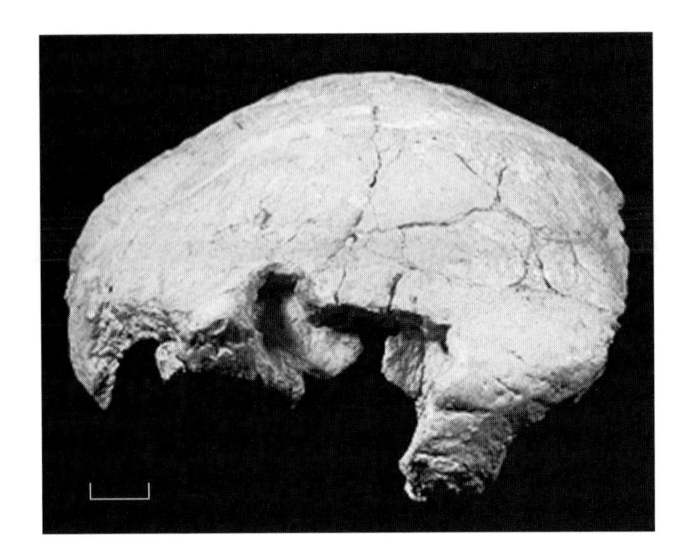

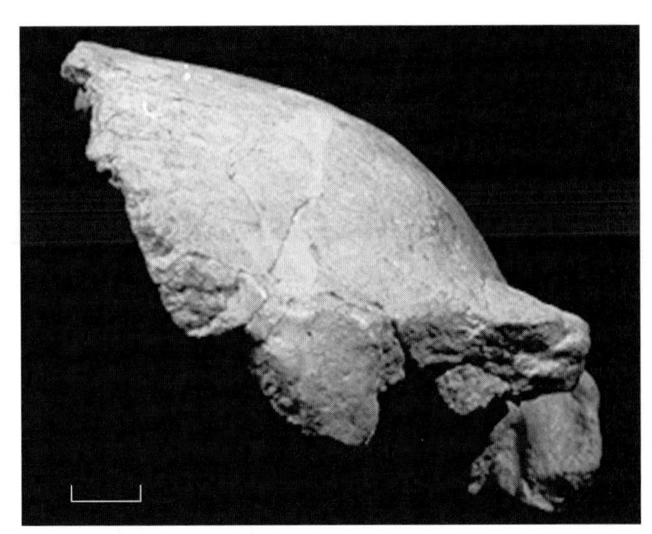

BILZINGSLEBEN Figure 2. H1 frontal (anterior, lateral), (scale = 1 cm).

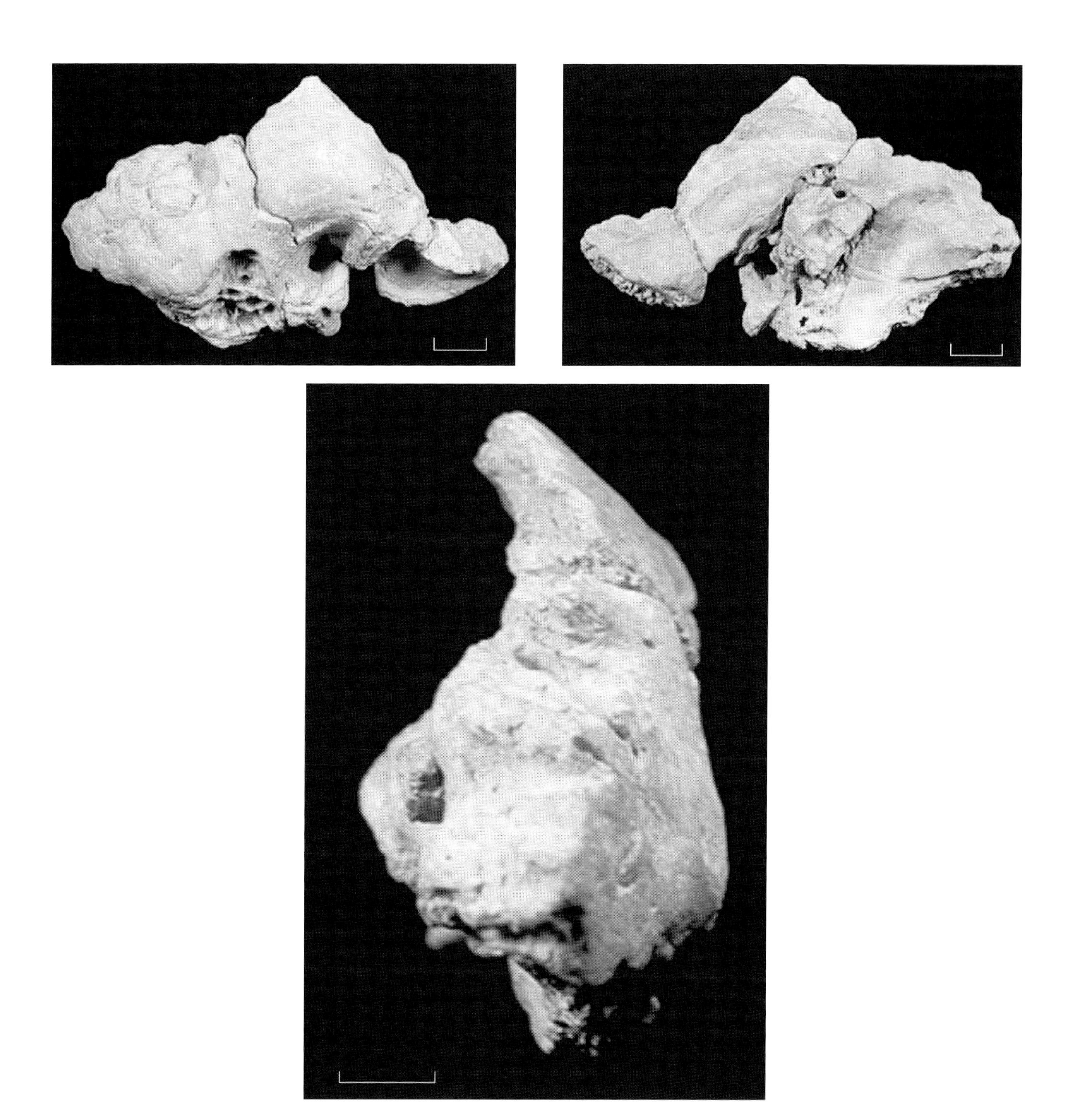

BILZINGSLEBEN Figure 3. G1 temporal (lateral, interior, posterior), (scale = 1 cm).

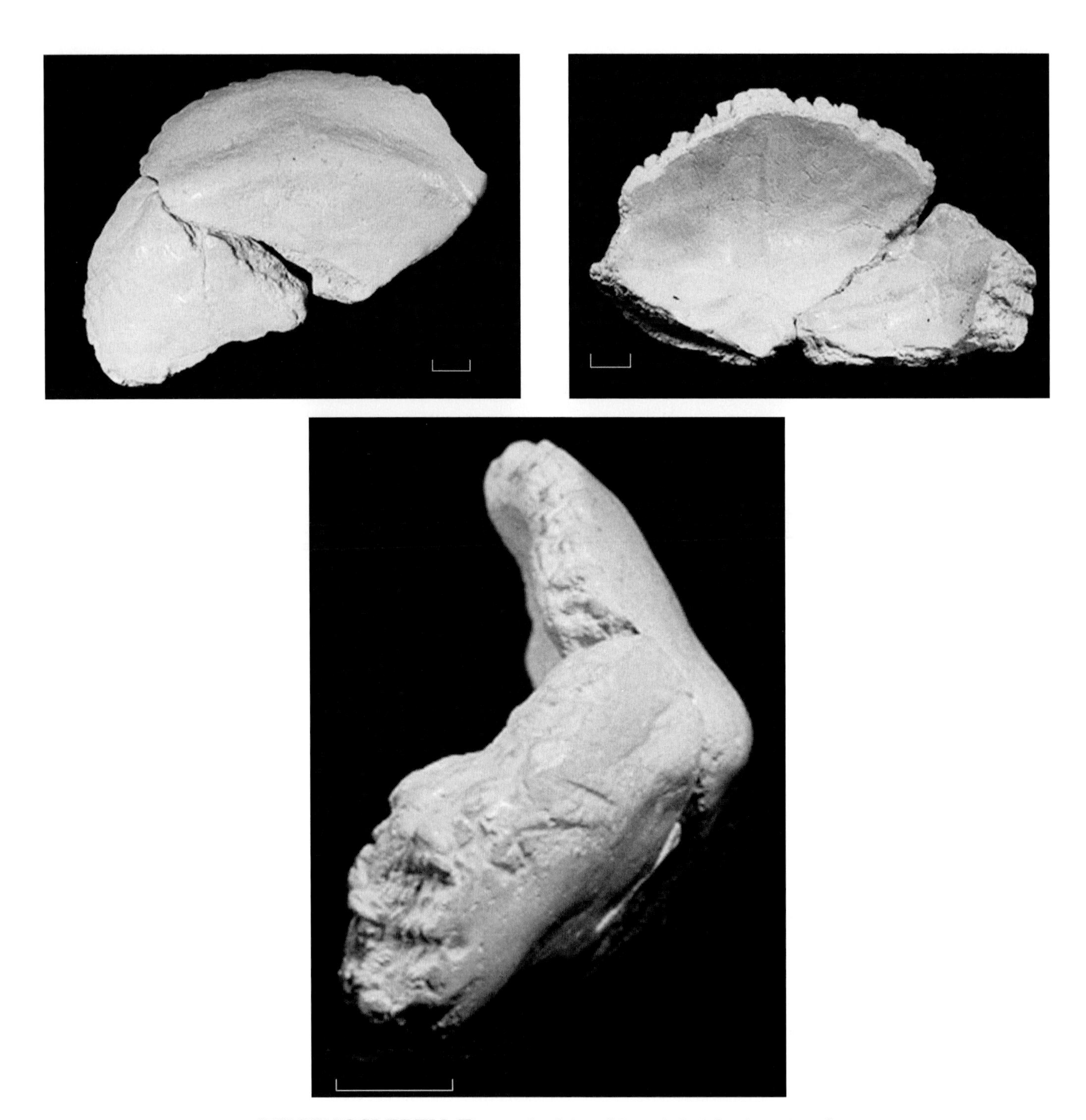

BILZINGSLEBEN Figure 4. A1 + A2 occipital (scale = 1 cm).

Brno

Location

Group of open-air sites in and around the center of the city of Brno, on the Svratka River, Czech Republic.

Discovery

A. Makowsky, 1885 (Brno 1), 1895 (Brno 2); J. Vhylidal and K Trkal, March 1927.

Material

Remains, principally cranial, of three individuals.

Dating and Stratigraphic Context

These sites were dug long ago and have now completely disappeared because of brickyard activities. Stratigraphic provenance of the fossils is poorly understood, but Brno 1 and 2 are currently thought to have been associated with an early Gravettian industry; this fits well with a date of 26.3 ka for the locality that yielded Brno 2 (J. Jelinek, quoted in Schumann, 1995). It is now thought likely that Brno 3, earlier believed to be of Würm II or III age, is post-Pleistocene (Schumann, 1995).

Archaeological Context

Only Brno 2 was found in direct association with artifacts, initially reported as Aurignacian (Makowsky, 1892). More recent work (see Jelinek, 1991) has placed this burial as Pavlovian, part of the eastern Gravettian complex, reflecting the presence of such notable items as bone needles and a multicomponent ivory "puppet."

Previous Descriptions and Analyses

It has always been accepted that the Brno remains represent modern *Homo sapiens* (e.g., Makowsky, 1892; Jelinek et al., 1959), but there has been much discussion of the presence in these specimens, notably the Brno 1 and 2 calottes, of a mixture of "modern" and "archaic" traits (e.g., Vlcek, 1967; Jelinek, 1991, Wolpoff, 1996). The estimated cranial capacity is 1600 cc for Brno 1 (Wolpoff, 1999), 1500 cc for Brno 2 (Vlcek, 1993), and 1304 cc for Brno 3 (Matiegka, 1929; Vlcek, 1993).

Morphology

Brno 1 (A 17 082)

Adult. Calotte plus maxilla lacking RC–I1, LM1; preserved teeth pretty heavily worn. Cranium had been long and narrow, with maximum breadth just above squamosal.

Supraorbital region somewhat prominent; bears glabellar "butterfly" and thin, flattened lateral portions. Frontal sinuses very large; penetrate well into and follow external bulge of superciliary arch; extend laterally almost to midpoint of orbit. Frontal eminences minimal. Forehead fairly low and bulbous. Temporal lines fairly faint, situated quite high, and extending considerably rearward. Lower margin of nasal aperture single; lacks spinal crests. Nasal spine was stout; unclear how far it projected. Maxillary sinuses taper forward as far as P1; no medial bulge or projection into nasal cavity. Nasal floor quite flat. Nasoalveolar clivus moderately long, keeled sagittally, but otherwise forward sloping. Slight depression lies just above peaked lambda. Internally, Pacchionian depressions lie on either side of superior sagittal sinus. Preserved parts of coronal and sagittal sutures appear segmented. Palate very deep with quite vertical sides. Central and lateral Is with small lingual tubercle; no shoveling. Molars decrease in size from M1 to M3; fair-sized hypocone on M1, smaller on M2, absent on M3. Roots of all teeth very short.

Brno 2 (A 17 083)
Adult. Calvaria plus R zygoma; L half mandible with extremely worn teeth.

Skull tall, parallel sided; maximum width occurs just above rather low squamosal suture. In profile, braincase looks "wavy," with depressions behind bregma, at center of sagittal suture, and just above lambda. Strong development of frontal and occipital regions. Bipartite superciliary arch marginally bigger than that of Brno 1. Similar central glabellar "butterfly" and flattened lateral portion. Notch below lateral part of swollen central area. Differs from Brno 1 in apparently lacking frontal sinuses. Preserved trace of maxillary sinus indicates quite limited posterior extent. Three zygomaticofacial foramina, all below orbital rim.

Zygomatic arch quite lightly built, with hardly any muscle scars; fairly sharp angle at malar tubercle; arch does not flare at all (temporal fossa would have been very narrow). Posterior root of zygomatic arch originates over auditory meatus; is confluent behind with the large suprameatal crest and large, upwardly curving supramastoid crest. Vaginal process runs at least to meatal margin. Styloid process lies quite laterally. Mastoid process very broad based, stubby, and only moderately long; points downward. Parietomastoid suture short and horizontal. Lambdoid suture rises sharply from asterion (no anterior lambdoid suture). Mastoid notch deep but only moderately constricted; no digastric fossa behind. Paramastoid crest small. Occipitomastoid suture area broken.

Occipital narrow. Lambdoid suture peaks at lambda, producing squat, triangular occiput. Strong bulge of occipital plane along upper lambdoid suture defined below by distended superior nuchal line; this bulge coincides with deep internal impressions for posterior cerebral lobes. No occipital torus or suprainiac depression. Superior nuchal line takes form of superiorly shelflike, somewhat downwardly oriented crescentic ridge that is part of bow-shaped line extending right across occiput. Sutures in advanced stage of fusion; were differentiated into separate segments. Internally, Pacchionian depressions situated on either side of superior sagittal sinus.

Mandible bears definite mental trigon that "corners" below the Cs. Mental foramen small; lie under P2. Mylohyoid line indistinct; submandibular fossae shallow but quite wide. Mandibular foramen broken. Masseter rugosities present on external aspect of gonial region; posterior part of this region too heavily reconstructed to be reliable. Tooth roots rather short; on M1–2, bifurcate near neck.

Brno 3 (cast)
Adult. Reconstructed cranium; mandible glued to it.

Skull quite gracile, tall and narrow, with feeble supraorbital development and tiny glabella "butterfly." Nasoalveolar clivus short and forwardly inclined. Nasal aperture quite broad (inferior border damaged). Zygomatic arches thin; zygomas rather shallow. On both sides, posterior root of zygomatic arch originates over external auditory meatus; is confluent with low suprameatal/supramastoid crest. Mastoid process broad at base, tapers rapidly, does not project very far downward; tip faced forward. Mastoid notch wide and deep. No digastric fossa. Parietomastoid suture short. A very short anterior lambdoid suture runs back before rising sharply to lambda. Occipital bone narrow and triangular, essentially rounded but slightly puffy below lambda. Weak, low-situated nuchal line runs fully between marked occipitomastoid crests and around back of skull; otherwise, nuchal markings insignificant. Vaginal process ran length of ectotympanic tube to contact mastoid process. Styloid process was stout.

Mandible gracile; gonial angle rather obtuse. Mental trigon small but well delineated. Large mental foramen lies below P2. Masseter insertions quite rugose externally. Low, medial pterygoid attachment surfaces discernible internally, at bottom of gonial angle. Mylohyoid lines faint; submandibular fossae shallow. Roots of upper and lower anterior teeth thin, small.

References

Jelinek, J. 1991. Découvertes d'ossements de la population gravetienne de Moravie. *L'Anthropologie* 95: 137–154.

Jelinek, J. et al. 1959. Der fossile Mensch Brno II. *Anthropos* 9: 1–30.

Makowsky, A. 1892. Der diluviale Mensch im Löss von Brünn. *Mitt. Anthrop. Ges. Wien* 22: 73–84.

Matiegka, J. 1929. The skull of fossil man Brno III and the cast of its interior. *Anthropologie* 7: 90–107.

Schumann, B. 1995. *Biological Evolution and Population Change in the European Upper Palaeolithic*. PhD Thesis, University of Cambridge.

Vlcek, E. 1967. Morphological relationships of the fossil human types from Brno and Cro-Magnon in the European Late Pleistocene. *Folia Morphologica* 15: 214–221.

Vlcek, E. 1993. *Fossile menschenfunde von Weimer-Ehringsdorf*. Stuttgart, Konrad Theiss Verlag.

Wolpoff, M. 1996. *Human Evolution*. New York, McGraw-Hill.

Wolpoff, M.H. 1999. *Paleoanthropology*. New York, McGraw-Hill.

Repository

Anthropos Institute, Moravian Museum, Brno, Czech Republic (Brno 1 and 2; Brno 3 was destroyed during WWII).

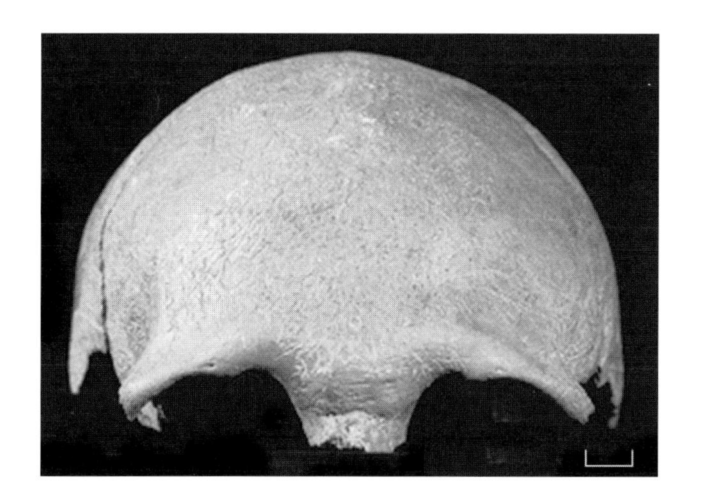

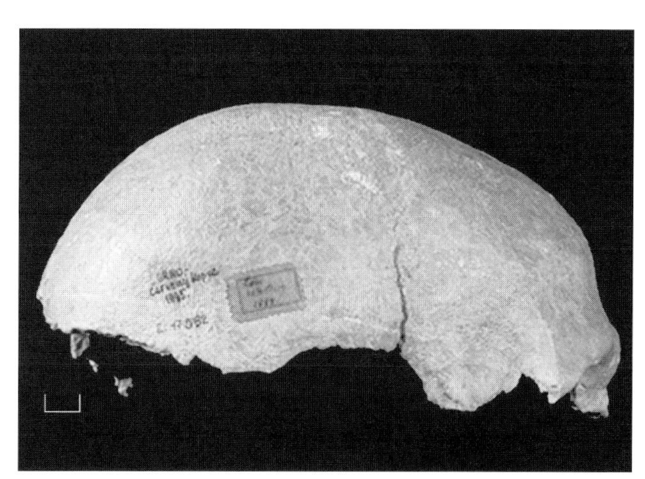

BRNO Figure 1. Brno 1 (scale = 1 cm).

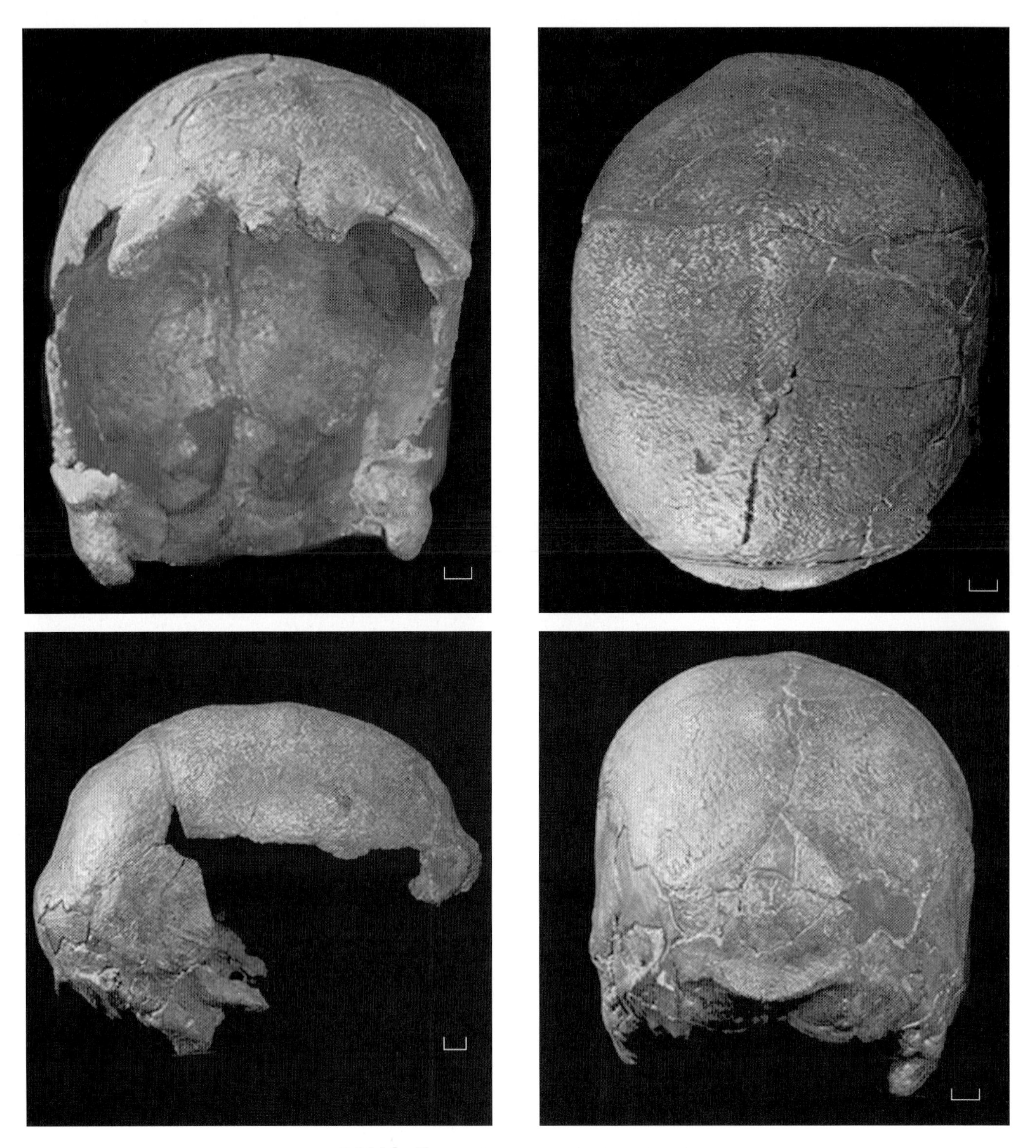

BRNO Figure 2. Brno 2 (scale = 1 cm).

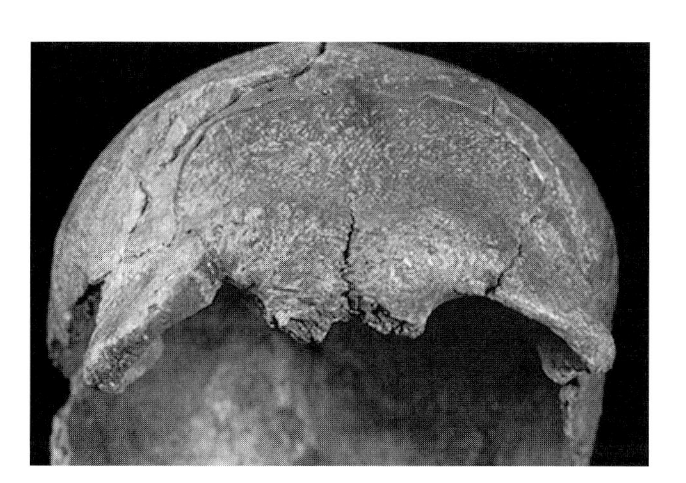

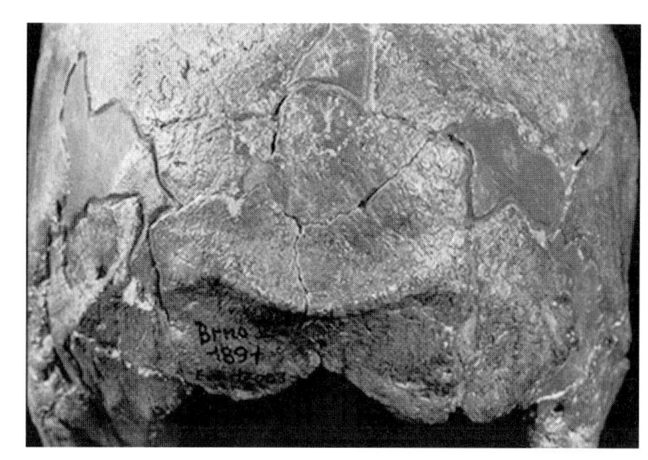

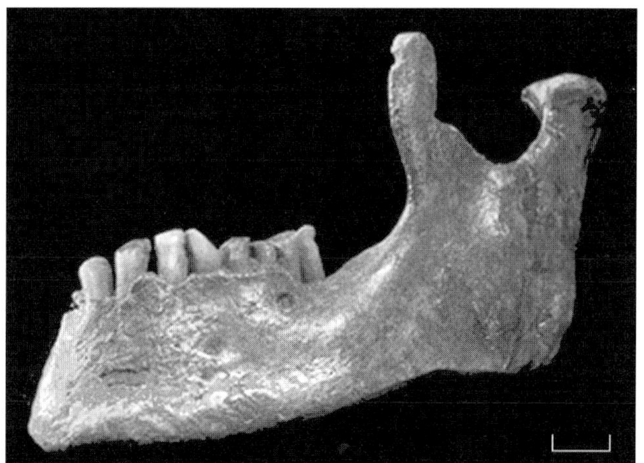

BRNO Figure 3. Brno 3 (scale = 1 cm).

Ceprano

Location
Road cut at Campo Grande, ca. 4 km SW of Ceprano, southern Latium, Italy.

Discovery
I. Bidittu, March 1994.

Material
Fragmentary calvaria.

Dating and Stratigraphic Context
The hominid fossil was found in an otherwise sterile clay stratum, identified as level 9 of the composite stratigraphic column for the Ceprano basin presented by Ascenzi et al. (1996). This level apparently underlies volcanics that have been K/Ar dated at up to 700 ka (Basilone and Civetta, 1975), but it also appears that the hominid was redeposited from still earlier layers (Ascenzi et al., 1996). An age of ca. 800 ka has thus been estimated for the hominid itself by Ascenzi et al. (1996). However, in view of all the associated stratigraphic uncertainties it must be admitted that any age estimate for the fossil must be the roughest of approximations at best.

Archaeological Context
No tools were found in the stratum containing the fossil. Overlying deposits (level 7) with a "lower Acheulean" industry have been correlated by Ascenzi et al. with deposits at another locality dated at ca. 460 ka, but these authors prefer to link the hominid with the underlying layer 13, which contains a very crude chopper-and-flake assemblage.

Previous Descriptions and Analyses
First reconstructed and described by Ascenzi et al. (1996), who compared the specimen most closely to Asian *Homo erectus* while noting certain differences. They rejected affinity with *Homo heidelbergensis* on the rather remarkable grounds that "the Mauer mandible shows absolutely no correspondence to the [maxilla-free] Ceprano calvaria" (p. 422). Later work by Ascenzi and colleagues (2000) leans toward the interpretation of Ceprano as the earliest *Homo erectus* known from Europe, as also advocated by Clarke (2000). Estimated cranial capacity is 1165 cc (Ascenzi et al., 1996).

Morphology
[Description provided by A. Ascenzi, F. Mallegni, and G. Manzi. Photographs by Antonio Solazzi; courtesy of Giorgio Manzi, Italian Institute of Human Paleontology (Rome)]:

Adult calvaria reconstructed from more than 30 pieces, including most of frontal, R parietal, R and L temporals, frontal processes of R and L zygomas, and occipital; L parietal and the sphenoid most incomplete. Large areas of cranial base missing or damaged; tympanic bones lost bilaterally. Cranial bones extremely thick.

Calvaria low but not comparatively long; midsagittal profile curves slightly from massive, protruding glabellar region to angulated occipital that bears a transverve torus. Maximum length between glabella and inion. Considerable latero-lateral expansion creates brachicranic proportions; maximum breadth at level of supramastoid crests. Overall postdepositional deformation produces L/R asymmetry in coronal profiles, with depression of L parietal and clockwise twisting of occipital torus.

Interorbital region wide. Superior border of orbits almost rectilinear. Supraorbital region bulges anteriorly. Tori continuous over orbits but not across glabella, which is depressed. Tori, variably tall s/i, reach maximum height around midline of orbits, and gradually thin laterally. Toral surfaces (bearing foramina) twisted (thus a flat medial part can be clearly distinguished from a more round and bulging lateral part). There is distinct, extended retrotoral (posttoral) depression. Frontal sinuses extend laterally and posteriorly.

Postorbital constriction moderate. Temporal lines marked on frontal and R parietal bones. Frontal squama recedes. No indication of frontal, coronal, or parietal keeling, but signs of slight parabregmatic depression. Parietals squared and relatively flat; strongly angulated in coronal section at level of temporal lines. Angular tori well expressed bilaterally.

Temporal fossa short but wide, without sharp angular distinction between temporal and infratemporal fossae. Mandibular fossa relatively small, deep; bordered by prominent entoglenoid process (medial articular tubercle). Posterior root of zygomatic arch lies above auditory meatus. Temporal squama probably high, arched. Complex parietomastoid sutural pattern in between V-shaped parietal notch and asterion. Mastoid processes massive. Supramastoid crests continuous with moderate suprameatal tegmen; run across both temporal and parietal surfaces.

Occipital squama flat, proportionally large and high; bordered inferiorly by transverse torus with supratoral sulcus. Occipital torus does not reach either asterion. Lambdoid suture peaks at lambda. Partially preserved nuchal plane faces inferiorly and posteriorly. Occipitomastoid crest present along suture. Paramastoid (juxtamastoid) crest absent.

Vault sutures partially fused endocranially. The pattern of identations appears simple and smoothed.

Internally, frontal lobes extend only slightly above orbital cones. Middle meningeal arteriovenous impressions clearly visible on parietal surface. Endinion well below level of inion.

REFERENCES

Ascenzi, A. et al. 1996. A calvaria of late *Homo erectus* from Ceprano, Italy. *J. Hum. Evol.* 31: 409–423.

Ascenzi, A. et al. 2000. A re-appraisal of Ceprano calvaria affinities with *Homo erectus* after the new reconstruction. *J. Hum. Evol.* 39: 443–450.

Basilone, P. and L. Civetta. 1975. The volcanic activity of the Monti Ernici dated by the K/Ar method. *Bull. Soc. It. Min. Petrol.* 31: 175–179.

Clarke, R. J. 2000. A corrected reconstruction and interpretation of the *Homo erectus* calvaria from Ceprano, Italy. *J. Hum. Evol.* 39: 433–442.

Repository

Istituto Italiano di Paleontologia Umana, Piazza Mincio 2, 00198 Roma, Italy.

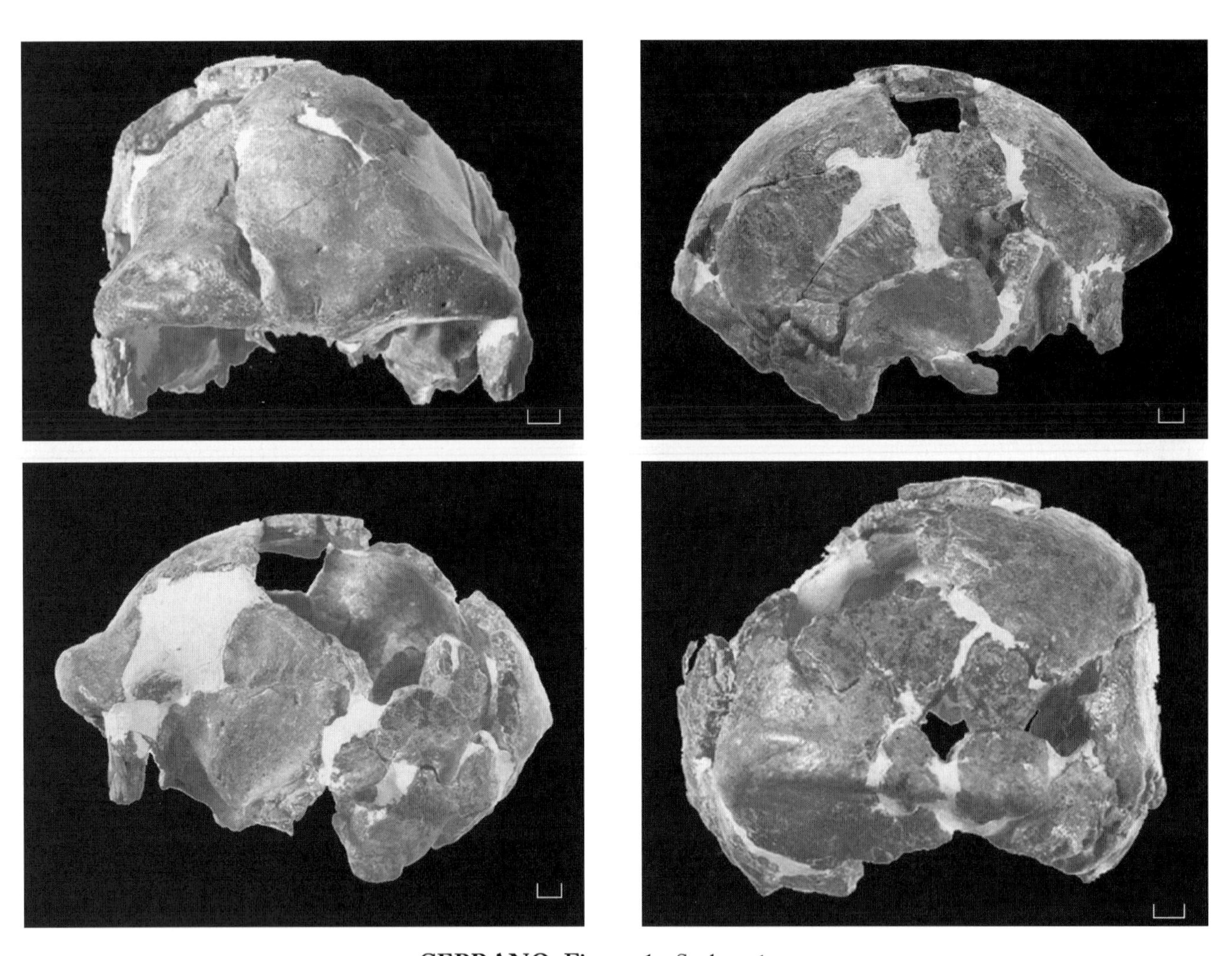

CEPRANO Figure 1. Scale = 1 cm.

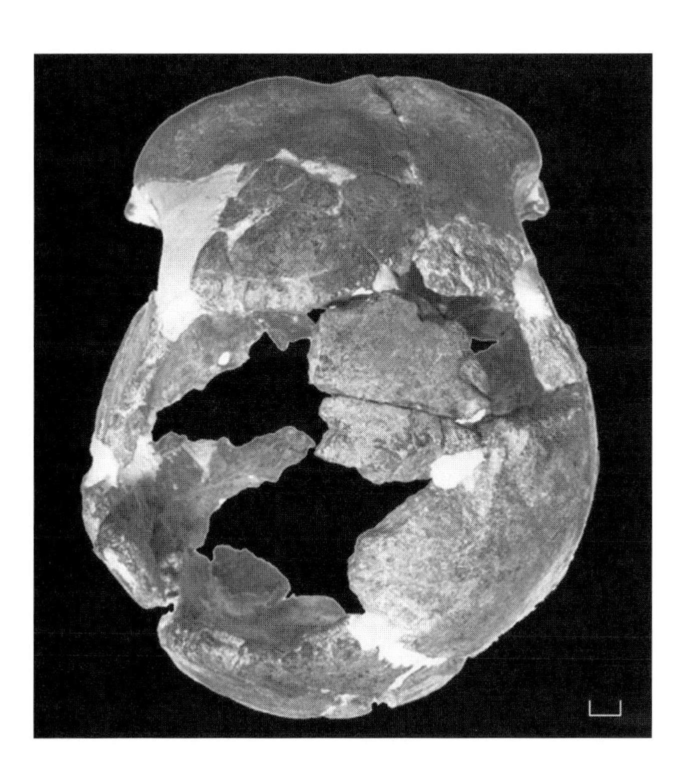

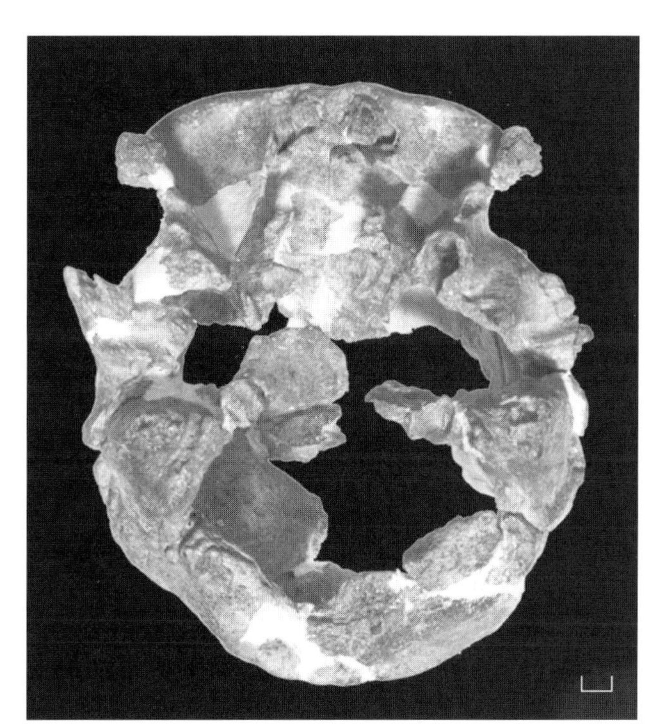

CEPRANO Figure 2. Scale = 1 cm.

Chancelade

Location
Cave site of Raymonden, on the right bank of the Beauronne river, just N of the village of Chancelade, a western suburb of Périgueux, Dordogne, France.

Discovery
M. Hardy and others, October 1888.

Material
Most of a skeleton, including a fairly complete cranium and mandible.

Dating and Stratigraphic Context
The Raymonden cave consists of three Magdalenian levels interbedded with sterile sands and silts. The human skeleton was found buried into the lowest level, about 1.6 m from the surface and resting directly on the original rock floor of the cave (Hardy, 1890). The associated fauna, dominated by reindeer, is indicative of cold conditions, and dating is via the associated industry (see below).

Archaeological Context
Magdalenian III or IV (Sonneville-Bordes, 1959). The burial was highly flexed and was reportedly covered with ochre.

Previous Descriptions and Analyses
The obligatory new species (*Homo priscus)* by Vacher de Lapouge (1899) aside, there has never been any doubt that this specimen represents fully modern *Homo sapiens* (e.g., Testut, 1890). Argument has instead centered on whether Chancelade was an Eskimo of some kind, related to the "Cro-Magnon race," or representative of its own "Chancelade race" (see the historically interesting discussion by Boule and Vallois, 1957). Estimated cranial capacity is 1530 cc (Holloway, 2000).

Morphology
Adult. Fairly complete skull, lacking parts of R and L parietals, part of basicranium, internal and partial external nasal region, medial orbital region. Missing all upper teeth except for root of I2 (antemortem periodontal disease). Somewhat reconstructed; difficult to distinguish parts of reconstruction from bone. Mandible largely complete, missing R condyle and coronoid process and part of L condyle. Lower M3 congenitally missing; all preserved lower teeth (LI2, C, P1, M1–2; RP1–2 and M1–2) extremely worn; RP2 and M1 almost gone.

Cranium large but lightly built, relatively long, tall, broad across neurocranium and zygomas. Glabellar "butterfly" moderately swollen but laterally not very wide; blunt superior margins of the "butterfly wings" are most anteriorly enlarged portions. Very large R and equally deep but less long and wide L supraorbital notch lies under these "wings." Lateralmost extremities of "wings" undercut by medial portion of m/l very wide and somewhat posteriorly sloping lateral plates. Rather vertical frontal arises directly from even more vertically oriented glabellar "butterfly." Especially on the L, frontal bears well-developed eminences. Greatest length of frontal belongs to its almost horizontal posterior part, which curves back sharply above the forehead.

Stout temporal ridges emerge from high up above short, posteriorly directed zygomatic processes of frontal; proceed directly upward about halfway up frontal before being transformed into very indistinct temporal bands that arc up over coronal suture and then arc quite sharply down to asterion. Interorbital region broad; glabella only slightly broader. As better preserved on the L, long oblique slope truncates medioinferior corner of orbit. On both sides, orbital roofs descend to superior orbital margin, then curve tightly back up lateral plates. Inferior orbital margins very thickened and forwardly everted; they greatly overhang infraorbital region, which, in side view, is obliquely angled posteriorly and down.

Nasal bones too damaged for much comment; appear to have been forwardly projecting to some degree. Nasal aperture apparently uniformly wide throughout most of its length; was probably tall and oval. Lateral crest better preserved on the R; is continuous as raised margin to become confluent with apparently fused, quite anteriorly projecting anterior nasal spines. Floor of nasal cavity sunken just below anterior nasal spines; is flat although gently posteriorly sloping downward. On both sides, remnant of conchal crest low and obliquely ascending. Especially on the L, appears that maxilloturbinal had fused to conchal crest and later broken off. On the L, damaged posterior extension of anterior lacrimal crest within nasal cavity quite prominent. Nasoalveolar clivus uncertain (alveolar crest destruction); was not extremely long, and in midline profile, was probably gently concave outward. Difficult to assess original depth of palate; was clearly small and rather horseshoe shaped; it bears low but expansive maxillary torus.

On both sides, anterior root of zygomatic arch apparently arose close to alveolar margin (but above which molar?); in front view, its inferior margin rises steeply before curving out strongly laterally. In inferior view, both zygomas flare out from blunt maxillary tuberosity that sits astride zygomaticomaxillary suture. Thin zygomatic arches reach maximum width just before arcing in to meet their posterior roots on temporal bone; inferior margin of arch oblique upward, with superior margin parallel to it. For size of face, zygoma is a fairly big bone both in height and length.

As seen on the better-preserved L, the straight, fairly oblique anterior squamosal suture is raised. The medially oriented plane this creates runs into an otherwise smoothly and shallowly concave alisphenoid. This raised suture also slightly demarcates between a larger anterior and smaller posterior temporal fossa. Again on the L, sphenoid bears somewhat vertically oblique crest that fades out as bone smoothly curves toward cranial base; no distinct delineation of infratemporal from temporal fossa.

Mandibular fossae quite wide m/l; also quite long a/p because of anterior slope of moderately sized articular eminence. On both sides, very distended, m/l wide postglenoid plate superiorly fused with what remains of ectotympanic tubes, whose posterior walls are completely fused to moderately a/p long, laterally swollen, anteriorly oriented, quite downwardly distended, moderately pointed mastoid processes. Posterior root of zygomatic arch arises anterior to meati; flows into very low suprameatal crest that flows into much more pronounced, ridgelike, upwardly arcing supramastoid crest that is separated from mastoid process by s/i tall, somewhat deep sulcus. Squamosal relatively short a/p; appears to have been tall and deep, with vertical parietal notch that lies anterior to midline of mastoid process. Behind parietal notch is long, horizontal parietomastoid suture. Mastoid notches slitlike anteriorly but broaden slightly toward posterior margin of mastoid process. As preserved more completely on the R, thick, low paramastoid crest separated by deep groove from another crest just medial to it; cannot tell whether occipitomastoid suture ran along this groove. Somewhat medial to latter crest is well-developed, more posteriorly elongate Waldeyer's crest. The R foramen ovale and small foramen spinosum preserved; are well contained within sphenoid. Petrosals short, with large, downwardly directed carotid foramina medially positioned on them. Although damaged, vaginal processes clearly peaked around stout styloid processes; on the R, vaginal process contacts mastoid process. Jugular foramina somewhat forwardly directed; R foramen significantly larger than the L but both relatively small.

Occipital plane appears slightly wider from side to side than tall s/i. Superior nuchal line is bow shaped in inferior and posterior view; has faint lateral origin; becomes rapidly more pronounced towards midline, where it is deeply undercut by horizontal, muscle-scarred sulcus. Discrete external occipital protuberance lacking; central part of superior nuchal line somewhat distended downward in region of midline in shallow crescent. Above its convex border, and to some way on either side, this midline distension bears shallow depression (lacking characteristics of suprainiac depression). In profile, nuchal plane runs forward and slightly downward to apparently round, largish foramen magnum with large, very forwardly positioned occipital condyles. Reconstruction obscures area in front of foramen magnum. Internally, petrosals bear low, domed arcuate eminences; apparently bore superior petrous sinuses.

Lambdoid suture rises fairly steeply from asterion and curves around lambda; has very tightly sinuous interdigitations that are even more elongate along short preserved part of sagittal suture. Similar interlocking, overlapping suture seen in less interdigitated coronal suture. All three sutures show some degree of segmentation.

Mandible quite narrow at symphyseal region. Corpora moderately divergent posteriorly; are relatively deep s/i, becoming shallower posteriorly. Fairly large mental foramina lie under regions of P2. Externally, stout central keel comes down from broken subincisal alveolar region, fans out inferiorly, forming rather tall, in profile straight and inferoanteriorly inclined, triangular structure. Viewed inferiorly, symphyseal region quite thick a/p (much more so than corpora on either side) and rather straight across. Postincisal plane essentially vertical, concave anteriorly. Matrix-filled digastric fossae point somewhat posteriorly and downward; were apparently not very wide m/l. Inferior borders of corpora thinner than bone above mylohyoid line. Mylohyoid line only modestly developed with quite pronounced submandibular fossa below; line runs only to region below P2.

Rami moderately long a/p, quite tall, and fairly vertical. Moderately long, somewhat pointed L coronoid process about same height as m/l moderately wide, coronally peaked mandibular condyle. Sigmoid notch deepest at midpoint. Sigmoid notch crest (as seen on the L) runs toward lateral side of condyle.

Mandibular foramina damaged; the L foramen pointed somewhat up and back. Gonial angle somewhat truncated; inferior border thickened externally, muscle scarred internally. Viewed from below, gonial angle slightly S curved and everted anteriorly. Ramus thickened internally just anterior and superior to mandibular foramen.

Teeth too worn for comment.

REFERENCES

Boule, M. and H. Vallois. 1957. *Fossil Men.* London, Thames and Hudson.

Hardy, M. 1890. Découverte d'une sépulture de l'époque quaternaire à Chancelade (Dordogne). *Int. Congr. Anthropol. Préhist. Archaeol. Paris* 10: 398–404.

Holloway, R. L. 2000. Brain. In: E. Delson et al. (eds), *Encyclopedia of Human Evolution and Prehistory.* New York, Garland Publishing, Inc., pp. 141–149.

Sonneville-Bordes, D. de. 1959. Position stratigraphique et chronologie relative des restes humains du Paléolithique Supérieur entre Loire et Pyrénées. *Ann. Paléontol.* 45: 19–51.

Testut, L. 1890. Recherches anthropologiques sur le squelette quaternaire de Chancelade, Dordogne. *Bull. Anthropol. Lyon* 8: 131–246.

Vacher de Lapouge, M. 1899. *L'Aryen.* Paris, Fontemoing.

Repository

Musée du Périgord, Cours Tourny, 24000 Périgueux, France.

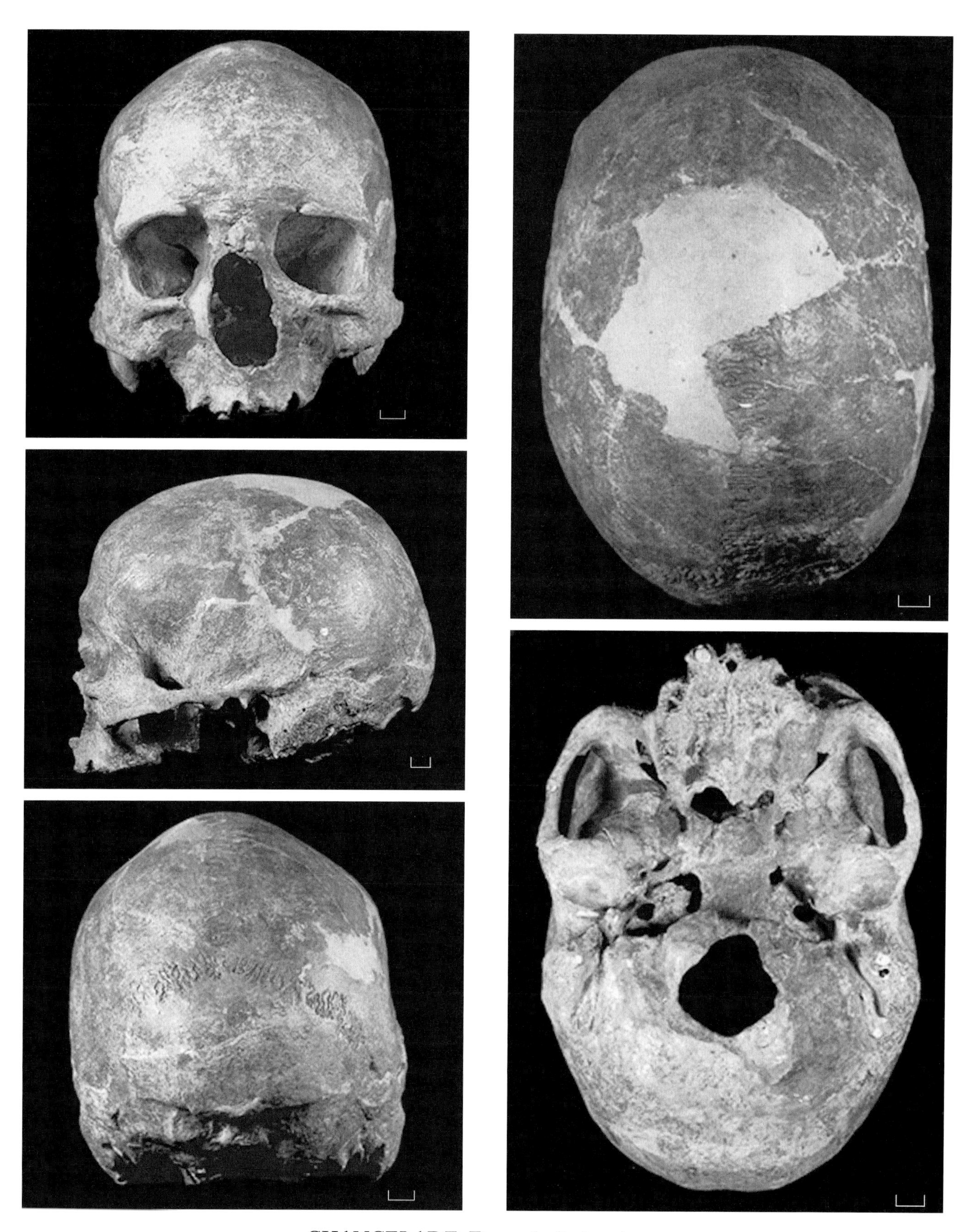

CHANCELADE Figure 1. Scale = 1 cm.

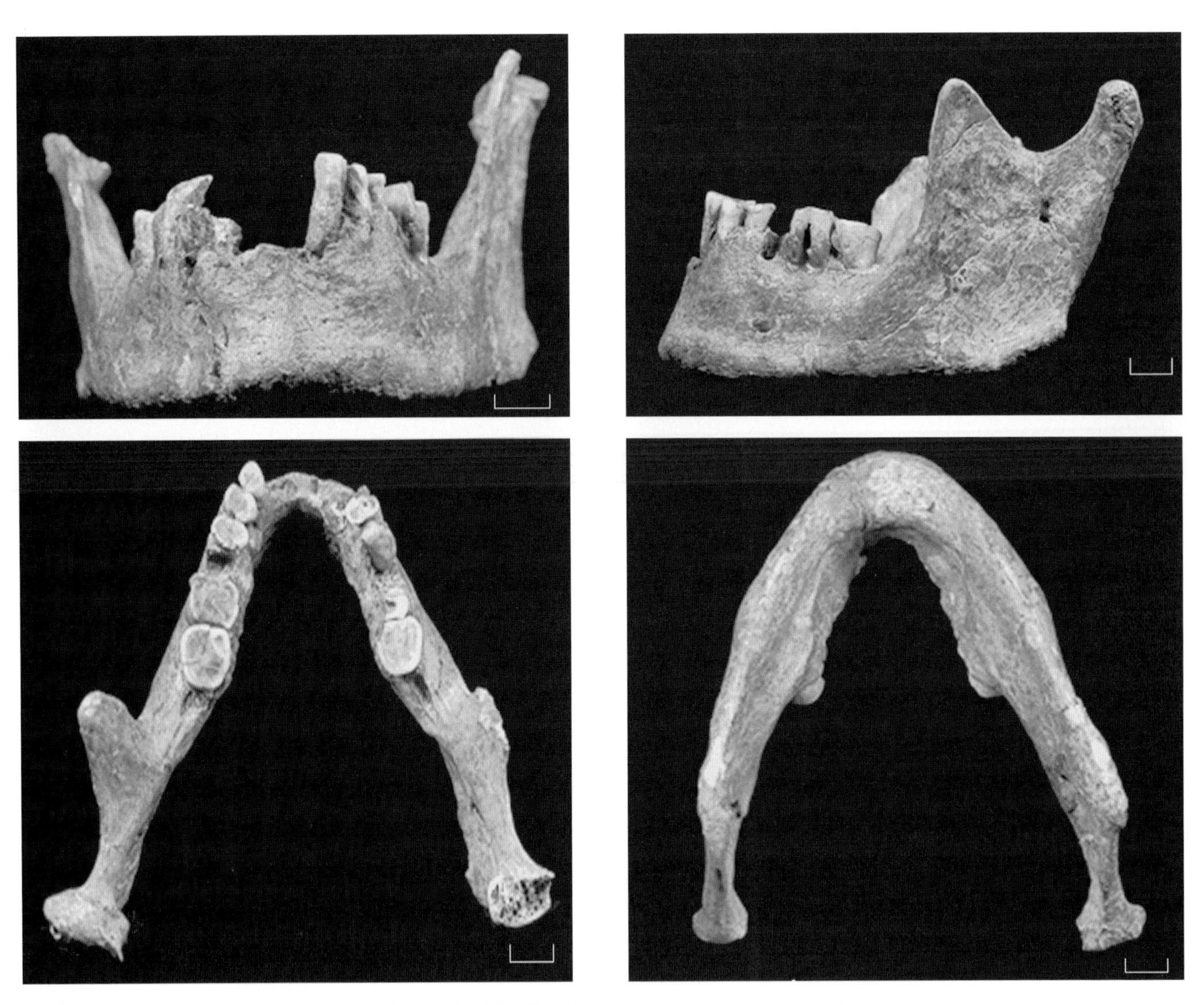

CHANCELADE Figure 2. Scale = 1 cm.

Columbeira (Bombarral)

Location

Cave close to the town of Bombarral, some 60 km N of Lisbon, Portugal.

Discovery

Excavations of G. Zbyszewski and O. da Veiga Ferreira, 1962.

Material

Left dm_2 or M_1 crown.

Dating and Stratigraphic Context

Found in a stratified cave deposit with an upper Pleistocene fauna (Roche, in Zbyszewski et al., 1980–81). Radiocarbon date on associated bone fragments is 28.9 ± 0.8 ka (Antunes et al., 2000).

Archaeological Context

Mousterian with denticulates, rich in Levallois scrapers (Antunes et al., 2000).

Previous Descriptions and Analyses

The tooth was originally described by Ferembach (1964–65), who assumed it was Neanderthal, an assumption more recently borne out by an analysis by Antunes et al. (2000). The date places this among the most recent Neanderthal fossils yet known.

Morphology

Crown, 50% formed toward neck, of either Ldm_2 or LM_1. Overall shape of crown broadly ovoid; metaconid, entoconid, and hypoconulid very peripherally placed. Trigonid and talonid basins large. Unworn cusps tall and pointed and of relatively equal height. Protoconid most massive cusp in area of base area. Sides of protoconid and hypoconid more sloping than other edges of crown. Entoconid just slightly distal to midline of hypoconid. Hypoconulid lies buccal to midline of crown. Hypoconulid separated from hypoconid by deep notch. Downwardly arcuate crests link apices of hypoconulid and entoconid and entoconid and metaconid. Paracristid thick, somewhat mesially arcuate; encloses buccolingually wide and quite deep, somewhat mesially expanded, trigonid basin. Distinct protocristid runs from apex of protoconid to base of metaconid. "Metacristid" runs down distal side of metaconid into large and deep talonid basin. "Centroconid" lies at base of hypoconid.

References

Antunes, M. T., et al. 2000. The latest Neanderthals. In: M. T. Antunes (ed.), *Últimos Neandertais em Portugal: Odontologic and Other Evidence*. Lisboa, Academia das Ciéncias de Lisboa, pp. 269–303.

Ferembach, D. 1964–65. La molaire humaine inférieur moustérienne de Bombarral (Portugal). *Comm. Serv. Geol. Portugal* 48: 185–192.

Zbyszewski, G, Leitao, M, Penalva, C & Ferreira, C. da V. 1980–1. Paleo-anthropologie du Würm au Portugal. *Setúbal Arqueológica*, VI-VII: 7–23.

Repository

Servicos Geologicos de Portugal, Lisboa, Portugal.

COLUMBEIRA Figure 1. Scale = 1 cm.

Combe-Capelle

Location
Rock shelter of Roc de Combe-Capelle on the right bank of the Couze river, just E of St Avit-Sénieur, some 20 km ESE of Bergerac, Dordogne, France.

Discovery
Excavations of O. Hauser, 1909.

Material
Partial skeleton of adult male, including a moderately complete skull.

Dating and Stratigraphic Context
Found in the "red-brown" level of the shelter deposits, apparently buried in a scooped-out grave close to the original rock floor. Dating to Würm II/III is solely by archaeological association (Sonneville-Bordes, 1959); a best guess would be ca. 28–25 ka.

Archaeological Context
Associated industry is said to be Perigordian 1 (Peyrony, 1943), equivalent to a phase of the Gravettian.

Previous Descriptions and Analyses
The Combe-Capelle specimen derives its importance principally from the fact that it was found early on and thus figured in many of the early discussions of the first moderns in France. Described by Klaatsch and Hauser (1910) as the type of *Homo aurignacensis hauseri*, the Combe-Capelle individual has been regarded by most subsequent commentators as a representative of the "Cro-Magnon race" of early *Homo sapiens* (e.g., Boule and Vallois, 1957).

Morphology
Cast of adult skull reconstructed from multiple fragments. Missing part of L frontal, squamosal, and alisphenoid; also bones of pterygopalatine fossa. All teeth apparently present; wear suggests no mirror imaging. Mandible complete except for tip of L coronoid process. Description below from prewar cast that preserves general morphology but almost no detail. Face suspiciously smooth.

Moderately large skull, relatively long and tall. Neurocranium narrow with relatively broad bizygomatic breadth (midface). Judging from the R, braincase was widest at well-developed parietal eminences that lie close behind mastoid process and well above level of squamosal suture.

Frontal bone rises steeply from above glabellar region and begins to dome out. Laterally, frontal bone rises more directly from supraorbital margins to swell into low, distinct eminences. Posterior to short, steep frontal region, profile of braincase changes to more gentle, continuous arc posteriorly, then flattens out at apparently very posteriorly positioned bregma. About halfway along sagittal suture, profile begins to descend to reach most posterior point well down on occipital, where it recurves gently inward toward crescentic, ledgelike, downwardly distended superior nuchal line. This broad, flat, quasi-external occipital protuberance indented by equally wide, s/i short depression. Nuchal plane sharply angled forward. Distinct sulcus separates external impressions of occipital lobes from superior nuchal line.

Glabellar "butterfly" well developed, massive, wide; its pronounced wings undercut lateral to orbital midline by short, slightly posteriorly inclined lateral plates. Both supraorbital margins bear medially placed notches; the L is relatively wide and shallow. Orbital shape symmetrical, indicating orbits were as tall or taller s/i than wide. Orbits obliquely cut off medioinferiorly and significantly distended lateroinferiorly. Thick infraorbital margins project significantly anteriorly (as seen from side), overhang wide, posteroinferiorly inclined, somewhat excavated infraorbital region. Infraorbital foramina were apparently large and situated just under everted inferior orbital margins. Temporal ridges arise from high above downward and slightly laterally oriented zygomatic processes of frontal bone; as seen on the R, ridge arcs steeply up and back a bit before running backward to fade out at region of coronal suture. Very wide glabellar region tapers below to still substantially wide interorbital region.

Nasal bones (original length unknown) flex acutely forward just at or below nasion (indefinable on cast). More or less horizontal roofs of orbits curve thickly up and back into supraorbital region. Nasal aperture was apparently rounded at its inferior corners; true height impossible to determine. Bases of apparently unfused anterior nasal spines thick; what remains of spines has some anterior projection. On the R, small depression in position of prenasal fossa suggests that lateral crest ran medially and inferiorly to join anterior nasal spine; spinal crest lies behind. In the cast, internal walls of frontal processes entirely smooth, suggesting that entire region was reconstructed. Moderately long nasoalveolar clivus fairly vertical; viewed from below, region describes tight, narrow arc from RC to LC, with tooth rows diverging only gently posteriorly.

Anterior roots of zygomatic arches originate above M1–2, when viewed from above, flare strongly. On the L, this root begins to flare close to alveolar margin; on the R, it originates higher up. Viewed from below, zygomatic arches corner somewhat at zygomaticomaxillary suture, then flare to region of zygomaticotemporal suture, where, as seen on the L, the arch runs sharply back toward its posterior root, which appears to lie anterior to auditory meatus. Judging from the R, temporal bone was very long a/p, modestly tall s/i, and apparently curved in long, high arc. Narrow, vertical parietal notch lies over midline of a/p very long, laterally swollen, downwardly pointing, significantly projecting, and blunt-tipped mastoid process. On the R, (the probably more reliable) auditory meatus large, subcircular; its wall was apparently fused to mastoid process. Also on the R, a structure that looks like very long, sheetlike vaginal process increases in height laterally until it fuses with mastoid process. On both sides, appears that the styloid processes were laterally placed. On the more believable R side, what is preserved of mandibular fossa is deep, long a/p, and bounded anteriorly by very pronounced articular eminence. Medially, appears that temporal bone was distended into modest medial projection that only partly closed off mandibular fossa. Mastoid notches very shallow, narrow, and bounded medially by low, thickish paramastoid crests that are separated by grooves from longer, lower, thinner crest (the occipitomastoid crest?).

Foramen magnum was apparently small and subdiamond shaped, with large, very anteriorly placed condyles. Basiocciput broad at least superiorly and strongly upwardly inclined. Posterior root of the zygomatic arch (seen on the R) confluent with low suprameatal crest that expands abruptly into a markedly pronounced, thick, upwardly arcing supramastoid crest that is separated by deep, not very tall sulcus from lateral swelling of mastoid process. Parietomastoid suture long, with apparently many small ossicles in region. Lambdoid suture probably rose steeply from asterion.

Palate was apparently deep, with vertical walls and a steep slope behind the Is. Is quite narrow. Teeth not describable. Ms decreased in size from M1 to M3.

If mandible accurate, is lightly built, relatively short, and tall from alveolar crest to inferior margin. Is tallest at symphysis, in external midline of which, some distance below alveolar margin, low central keel arises and quickly expands laterally into very broad, low triangular swelling. Viewed from below, symphyseal region very narrow and tightly arced; corpora diverge strongly posteriorly. Tall postincisal plane rather vertical. No digastric fossae visible (reconstruction in lower symphyseal region?). Relatively large mental foramina lie under P2s. Well-developed, gently sloping mylohyoid lines lie above long, moderately excavated submandibular fossae. R coronoid process rises modestly above condyle. On both sides, anterior margin of ramus indented slightly and low down by shallow preangular sulcus, which on the R exposes M3 fully. Rami not very tall relative to corpus height; are quite long a/p. Gonial regions form sharp angle. Sigmoid notch appears intact on the R; describes smooth, deep, regular curve that is lowest around midpoint. On both sides, sigmoid notch crest runs to lateral extremity of m/l wide, a/p quite compressed mandibular condyles. Ms cannot be described; appear to have decreased in size from M1 to M3.

REFERENCES

Boule, M. and H. Vallois. 1957. *Fossil Men.* London, Thames and Hudson.

Klaatsch, H. and O. Hauser. 1910. Homo aurignaciensis Hauseri. *Prähist. Z.* 1: 273–338.

Peyrony, D. 1943. Le gisement du Roc de Combe Capelle. *Bull. Soc. Hist. Archäol. Périgord* 70: 158–173.

Sonneville-Bordes, D. de. 1959. Position stratigraphique et chronologie relative des restes humains du Paléolithique Supérieur entre Loire et Pyrénées. *Ann. Paléont.* 45: 19–51.

Repository

The specimen was removed to Berlin, where it was destroyed during WWII. A frontal fragment remains in the Musée National de Préhistoire, 24620 Les Eyzies de Tayac, France.

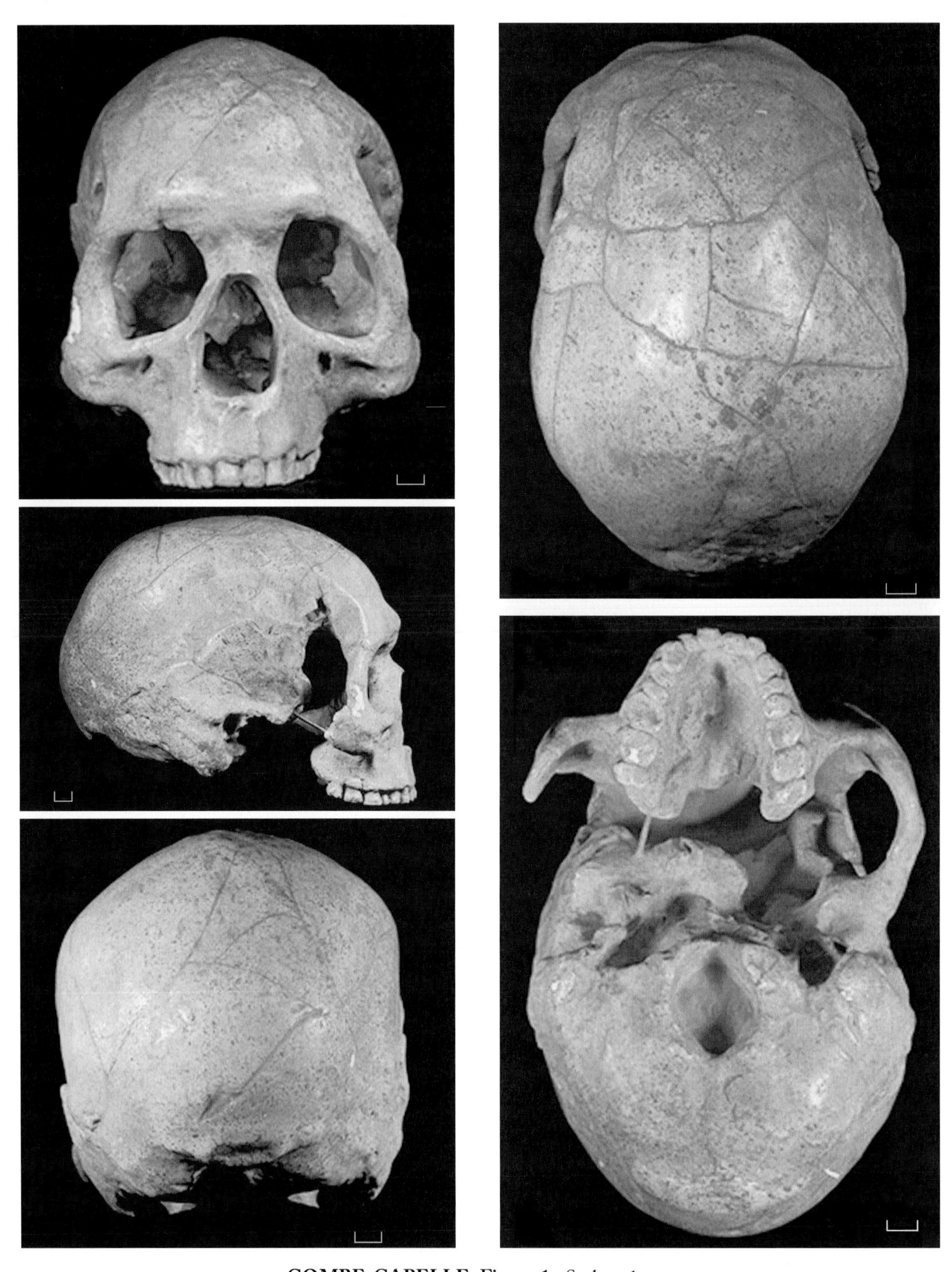

COMBE-CAPELLE Figure 1. Scale = 1 cm.

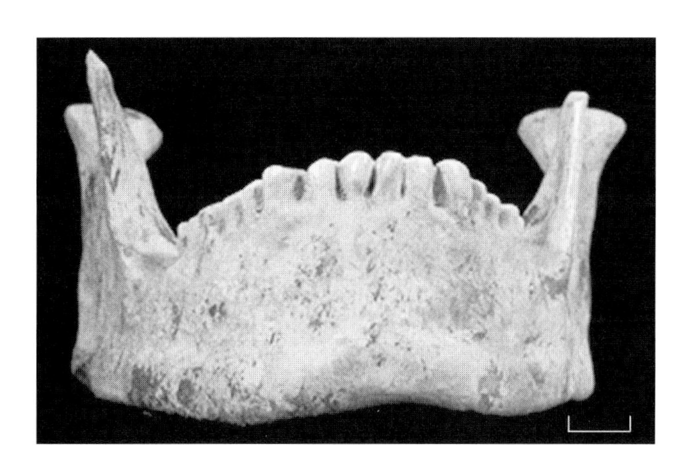

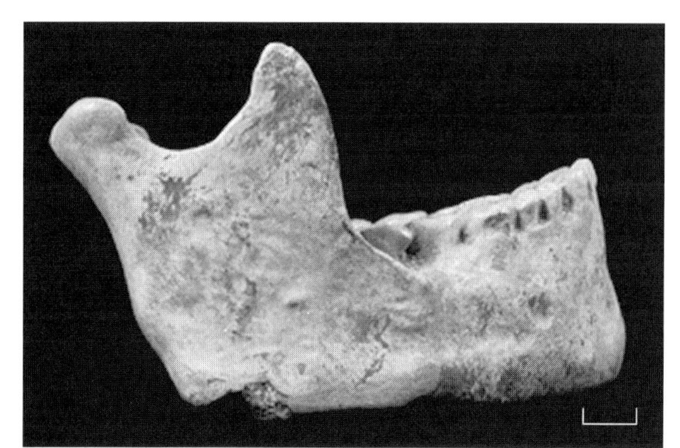

COMBE-CAPELLE Figure 2. Scale = 1 cm.

Cro-Magnon

Location
Limestone rockshelter at the eastern end of the village of Les Eyzies de Tayac, Dordogne, France.

Discovery
Railway laborers, followed by L. Lartet, March 1868.

Material
Skeletons of several individuals, from neonates to aged adults. The best-known fossils are the adult male Cro-Magnon 1, with edentulous skull and many postcranial elements; the adult female Cro-Magnon 2, with calvaria and several postcranial bones; and Cro-Magnon 3, adult male calotte and partial mandible with some long bones. Cro-Magnon 4 and 5 are, respectively, an adult cranial fragment and a partial neonate calvaria with some long bone diaphyses. Numerous unassociated bones were also recovered.

Dating and Stratigraphic Context
The site was partly emptied by laborers before excavation, which was conducted well by the rather primitive standards of the day (most of the work was carried out in 1868–69, with occasional additions up to 1905). Twelve strata (A–L) were identified, containing numerous faunal and archaeological remains. The human fossils came from the uppermost cultural layers, Bed J according to Lartet and Christy (1865–75), although some recent scholars think Bed K more probable. There is no doubt that the hominids were contemporaneous with the cold-adapted fauna found along with them. This contained extinct species such as the woolly mammoth, and was immediately recognized to be of Pleistocene age, probably from the Würm glacial (Lartet, 1868). The lithics and bone tools found in association suggest an age of about 30 ka, or perhaps a bit younger (Movius, 1969).

Archaeological Context
According to de Sonneville-Bordes (1960) the associated industry is of "evolved Aurignacian" type (see also Movius, 1969). No earlier or later industries are represented at the site. Along with diagnostic flint tools, artifacts included split-based bone points. Stone-lined hearths were also identified in the deposits. Some of the human remains bore traces of red ochre, and it is widely believed that they represent burials. Numerous perforated seashells and mammal teeth may be what is left of body ornaments the deceased were wearing when interred.

Previous Descriptions and Analyses
In 1899 the Cro-Magnon remains were referred to the new species *Homo spelaeus* by Vacher de Lapouge, but otherwise doubt has never been expressed that these hominids are fully modern *Homo sapiens* (e.g., Broca, 1868). Indeed, the main claim to fame of the Cro-Magnon remains is that they provided the first generally accepted evidence for the existence of truly ancient anatomically modern humans in association with an extinct fauna; and their name became informally extended to all later finds of Upper Paleolithic

humans in Europe. The main systematic discussion in which these hominids have been embroiled is the largely sterile but long-running debate over whether a "Cro-Magnon race" could be identified in the terminal Pleistocene of Europe. Average estimated cranial capacity is 1600 cc for Cro-Magnon 1 (Vallois and Billy, 1965) and 1730 cc for Cro-Magnon 3 (Wolpoff, 1999).

Morphology

Cro-Magnon 1 (CRM 1)
Aged male skull. Surface of cranium eroded; only minor damage including missing teeth and zygomatic arches. Orbital floors and ethmoidal region lacking. Some encrustation obscures detail in several areas. Mandible lacks all teeth and L ramus, upper part of R ramus, with internal part of surface missing in lower part. Large erosive lesion above glabella and R orbit.

Cranium long with relatively vertical frontal and moderately expressed frontal and parietal eminences. Rear of cranium moderately rounded with straight and quite vertical sides.

Erosion and encrustation do not obscure glabellar "butterfly," which extends laterally to region of supraorbital foramina. "Butterfly" and glabella itself not pronounced. Lateral to supraorbital foramina, supraorbital region more platelike and oriented backward. Orbits rectangular, being much wider m/l than tall s/i. Superior orbital margins thin and blunt. As seen especially on the R, inferior orbital margin rises slightly above floor of orbit; externally, margin appears toruslike because concavity of region below. Infraorbital foramina are moderate in size, lie relatively close to inferior orbital margin. Interorbital region moderately wide. Nasal bones flex outward at fairly strong angle below nasion; are strongly curved transversely; broaden inferiorly and project farther than anterior limit of nasomaxillary suture. Nasal aperture tall, narrow, and tear shaped; encrustations obscure details of inferior margin of aperture; it appears that it was simple, lacking prenasal fossa. Anterior nasal spines missing. Nasoalveolar clivus was moderately long, with distinct anteroinferior slope. Internally, inferior nasal conchae were apparently coalesced with conchal crests, as indicated on the L by vestige of downwardly oriented thin shelf of bone; this shelf runs horizontally from front to back, as does less pronounced counterpart on the R. Vomer extends quite far forward, to margin of nasal aperture. Floor of nasal cavity slopes very gently downward and backward.

Infraorbital plane oriented quite vertically, faces forward. Inferior margin of anterior root of zygomatic arch arcs sharply up and laterally. Lateral to zygomaticomaxillary suture, body of zygoma curves strongly backward (zygomatic arch only minimally flared). Maxillary sinuses were not broad at base. Sphenoidal sinuses were huge; on the L, appear to have extended over roots of pterygoid plates.

Alisphenoid curves inward smoothly (no delineation of distinct infratemporal fossa). Squamosal curves continuously into alisphenoid. Temporal fossa capacious inferiorly but somewhat constricted above. Temporal lines emerge anteriorly at lateral margin of orbit, run up and back; on the L, most posterior part of temporal line can be seen arcing down and forward to meet supramastoid crest. Squamosal appears to have been moderately long and tall and probably with smooth superior arc. Posterior root of zygomatic arch originates just above anterior edge of auditory meatus, flows posteriorly into moderately pronounced suprameatal crest that in turn is confluent with upwardly sweeping supramastoid crest. Distinct concavity below latter separates it from a very bulky, laterally swollen, anteriorly pointing mastoid process. Although broken, auditory meatus evidently large, vertically ovoid; mastoid process closely contacts its posterior wall. Mandibular fossae deep, bounded anteriorly by very stout, downwardly extended articular eminences; are not closed off medially. Carotid foramen relatively large; situated just anterior to sphenotemporal suture. Vaginal process arises from lateral margin of carotid foramen. Although broken, styloid process evidently extended entire length of ectotympanic tube, contacting mastoid process. Thin styloid processes lie quite lateral to carotid foramen as well as to forwardly facing jugular foramina (the R larger and more pocketed than the L). Stylomastoid foramen not identifiable with certainty.

Mastoid processes broad at bases, markedly distended downward. As seen especially on the L, low, relatively thick paramastoid crest delimits long, narrow mastoid notch medially. Occipitomastoid suture not visible on either side; on the L, low, short longitudinal ridge may represent occipitomastoid crest, medial to which lies longer, more elevated Waldeyer's crest. Parietal notch lies over posterior extremity of mastoid process. Region of asterion obscured by bony overgrowth caused by muscle scarring of posterior aspect of mastoid area (cannot determine extent of parietomastoid suture). Lambdoid suture rises quite sharply from back of muscle scar, peaks at lambda. Occipital plane tall and wide at its base (roughly forms equilateral triangle); surface bears highest nuchal line situated considerably above superior nuchal line, which forms boundary between occipital and nuchal planes. No well-developed external occipital protuberance; series of small parallel muscle scars in region. Anterior to superior nuchal line, nuchal plane quite heavily rugose. Anterolateral rim of modestly sized foramen magnum slightly damaged; foramen appears to have been more or less egg shaped. Long condyles, also damaged, were probably quite forwardly positioned.

Palate moderately long, narrow, slightly posteriorly diverging. Even with severe degree of alveolar resorption, palate was clearly quite shallow, modestly sloping at its front and vertical on its sides. Palate was apparently broad and long, with low maxillary torus. Pterygoid plates oblique and parallel, converge at superior and inferior extremities. As seen on the R, foramen ovale lies lateral to lateral pterygoid plate. Superior root of medial pterygoid plate articulated laterally with basisphenoid. Alae of vomer lay well anterior to sphenooccipital synchondrosis, which bears large pharyngeal pit.

All traces of coronal, anterior half of sagittal, and middle segments of lambdoid suture completely obliterated; rest of sagittal and lambdoid sutures indicate differentiation and deep interdigitation.

Internally, petrosals rather narrow, with distinct superior petrous sinuses (particularly well developed on the R); subarcuate fossae closed off, but not filled out. Superior surfaces of petrosals relatively smooth, with only small protrusion above region of superior semicircular canals.

Mandible narrow anteriorly. Corpora thin, especially inferiorly; were originally quite tall; diverge quite strongly posteriorly. Inferior margin of symphyseal region is strongly projecting and straight across; it terminates laterally in well-defined blunt corners that lie below the regions of the Cs, well anterior to the mental foramina lying below the P2s. No midline keel in area immediately below alveolar margin. Apex of mental trigon defined below by "corners" that lie high on symphysis. Internally, symphysis bears pair of well-developed, closely approximated genial tubercles. Digastric fossae well defined, oriented obliquely backward. Mylohyoid lines not well defined. On the R, well-developed preangular sulcus just exposes back of M3. Gonial region weakly reflected outward and fairly smoothly rounded.

Cro-Magnon 2 (CRM 2)

Partial adult (?female) cranium, more lightly built than CRM 1. Lacks R temporal, ethmoid, and L zygoma; most of R parietal, occipital, and sphenoid; and part of L temporal. Lacks all teeth but RM^{1-2}.

Cranium relatively long and quite domed, with slight slope to frontal and well-developed frontal and to lesser extent parietal eminences. Lightly built with thin bone. Rear of cranium smoothly rounded; sides of braincase vertical.

Frontal rises quite steeply. Fairly well-marked temporal ridges emerge quite high behind zygomatic processes of frontal; at about level of coronal suture, ridge separates into quite faint temporal lines that continue to diverge as they arc back along side of skull. Inferior temporal line arcs sharply down and forward from point above asterion to meet low, thick supramastoid crest; superior temporal line fades out above lambdoid suture at point well posterior to asterion. Tiny frontal sinuses (visible from interior) confined to region of glabella.

Low, distinct glabellar "butterfly" extends laterally as far as supraorbital foramina. Glabellar region relatively flat, smooth. Lateral to supraorbital foramina, supraorbital margin thin, platelike, oriented backward. Interorbital region was moderately broad. Nasal bones missing; did not flex at nasion. Partially reconstructed R orbit subsquare, its inferior margin thick, blunt, and somewhat everted. Infraorbital region is relatively vertical, oriented anteriorly. Relatively small infraorbital foramen lies very close to inferior orbital margin. Externally, appears that frontal process of maxilla did not project anteriorly; within nasal cavity, this process bears low, thin, horizontal conchal crest. Lateral crests of nasal aperture did not continue medially to meet anterior nasal spines. Spinal and turbinal crests absent. Nasal cavity floor perfectly flat. Relatively short nasoalveolar clivus angles anteriorly. Inferior margin of anterior root of zygomatic arch originates well above M1, runs horizontally outward before angling sharply posteriorly just behind zygomaticomaxillary suture. In side view, inferior margin of arch appears to slope upward somewhat as it runs back.

Judging from preserved part of L alisphenoid, squamosal was not very long; from sutural scars preserved on the L, appears that squamosal was not very tall. Temporal lacks clearly defined parietal notch. Auditory meatus moderately small, ovoid, vertically oriented. Low, thick suprameatal crest flows into strongly upwardly curving supramastoid crest, below which is broad sulcus delineated inferiorly by strong lateral bulge of massive, anteriorly pointing, not very projecting mastoid process. Very broad, deep mastoid notch separates mastoid process from broken ridge that may be paramastoid crest (cannot tell location of occipitomastoid suture). Partially preserved carotid foramen large, downwardly pointing; posterolateral to it appears to be base of styloid process, adjacent to which lies stylomastoid foramen. Carotid canal large. Jugular fossa discernible (region broken).

Tall occipital plane was probably broad; presents smoothly rounded contour. Superior nuchal line arises quite far laterally at point level with putative paramastoid crest, becomes more distinct as it courses downward and toward midline. Lambdoid suture runs sharply upward from asterion, peaking at lambda, is heavily interdigitated and differentiated into segments, as is sagittal suture. More finely denticulate coronal suture is segmented.

Palate would have been moderately long, relatively broad, and somewhat squared off in front; as seen on the R, it deepened considerably posteriorly and had vertical walls and fairly steep slope at front. Moderately sized incisive foramen very forwardly positioned.

Internally, cranial vault exhibits strong, long frontal crest, well-defined superior sagittal sinus, and very well-defined grooves for meningeal arteries. Poorly defined L transverse sinus runs laterally from what appears to have been large internal occipital protuberance. L petrosal moderately wide, inwardly tapering; bears superior petrous sinus along its preserved length; its superior surface bears expansive, low, domed arcuate eminence; subarcuate fossa closed over, almost completely flat.

Preserved M^1 extensively worn; much of enamel is missing from M^2. M^1 squarish, being distended in distolingual corner by enlarged hypocone; buccal roots separate almost at level of neck. M^2 much narrower tooth m/d; probably had poorly developed hypocone. Preserved alveoli indicate that roots of P^1 were more separated along their length than in P^2.

Cro-Magnon 3 (CRM 3)
Adult. Calotte plus mandible lacking all teeth and rami.

Braincase very long, somewhat domed, with somewhat bulging occipital plane. Forehead fairly steep; frontal and parietal eminences pronounced. Skull broad, more or less parallel sided; widest point lies above squamosal suture. Cranial bone relatively thin.

Very pronounced glabellar "butterfly," more clearly limited on the R in region of lateralmost supraorbital foramen. Lateral portions of both supraorbital regions thin, platelike, angled backward. Glabella somewhat swollen. Interorbital region moderately broad. Frontal sinuses confined to region of "butterfly" laterally, extend little higher into frontal bone. Reasonably pronounced temporal ridges emerge from relatively high up on zygomatic processes of frontal, arc up to peak just posterior to coronal suture, where faint superior and inferior lines diverge slightly and descend toward lambdoid suture, recurving down and forward near level of lambda.

Trace of metopic suture down glabella. Coronal suture highly fused; was finely denticulated and segmented. More visible sagittal and lambdoid sutures highly denticulated, divided into segments. On both sides, remaining parts of squamosal suture suggest that squamosal was long and not very highly arched (on the R, squamosal appears to have been rather unusually peaked at midpoint). Lambdoid suture rises sharply toward lambda; is interrupted in this region by large ossicle; another large ossicle lies at midpoint of L lambdoid suture. Some of apparent bulging of broadly triangular occipital plane below lambda may be an artifact of reconstruction, but area does protrude noticeably. Apparent highest nuchal line crosses occipital plane more or less horizontally a short distance above slightly bow-shaped superior nuchal line; below this, on the L, scalloped area of muscle attachment preserved.

Internally, frontal crest long, strong. Visible anterior part of superior sagittal sinus bears number of Pacchionian depressions. Grooves of middle meningeal vessels large, deeply incised.

Mandible narrow at front; corpora diverge strongly toward back. Single, moderately sized mental foramina lie below single-rooted P_2. P_1 was also single rooted. Symphyseal region forwardly projecting and strongly rounded from side to side. Just below alveolar margin lies the apex of broad, softly defined mental trigon. Lateral to trigon, inferior margin of corpus thickened externally. Digastric fossae well excavated and oriented obliquely backward. Internally, pair of thin, closely approximated genial tubercles lie in midline. No distinct mylohyoid line, although region below is well excavated.

Cro-Magnon Child

Few fragments of braincase of very young child, plus some long bones. Uninformative.

REFERENCES

Broca, P. 1868. Sur les crânes et les ossements des Eyzies. *Bull. Soc. Anthropol. Paris* 3: 350–392.

Lartet, L. 1868. Une sépulture des Troglodytes du Périgord (crânes des Eyzies). *Bull. Soc. Anthropol. Paris* 3: 335–349.

Lartet, L. and H. Christy (eds). 1865–1875. *Reliquiae Aquitanicae: A Contribution to the Archaeology and Palaeontology of Périgord.* London, Williams and Norgate.

Movius, H. 1969. The abri de Cro-Magnon, Les Eyzies (Dordogne), and the probable age of the contained burials on the basis of the nearby Abri Pataud. *Ann. Estud. Atl.* 15: 323–344.

Sonneville-Bordes, D. de. 1960. *Le Paléolithique Supérieur en Périgord.* Bordeaux, Delmas.

Vacher de Lapouge, M. 1899. *L'Aryen.* Paris, Fontemoing.

Vallois, H.V. and G. Billy. 1965. Nouvelles recherches sur les hommes fossils de l'abri de Cro-Magnon. *Anthropologie* 69: 47–74.

Wolpoff, M.H. 1999. *Paleoanthropology.* New York, McGraw-Hill.

Repository

Laboratoire d'Anthropologie Biologique, Musée de l'Homme, Place Trocadéro, 75116 Paris, France.

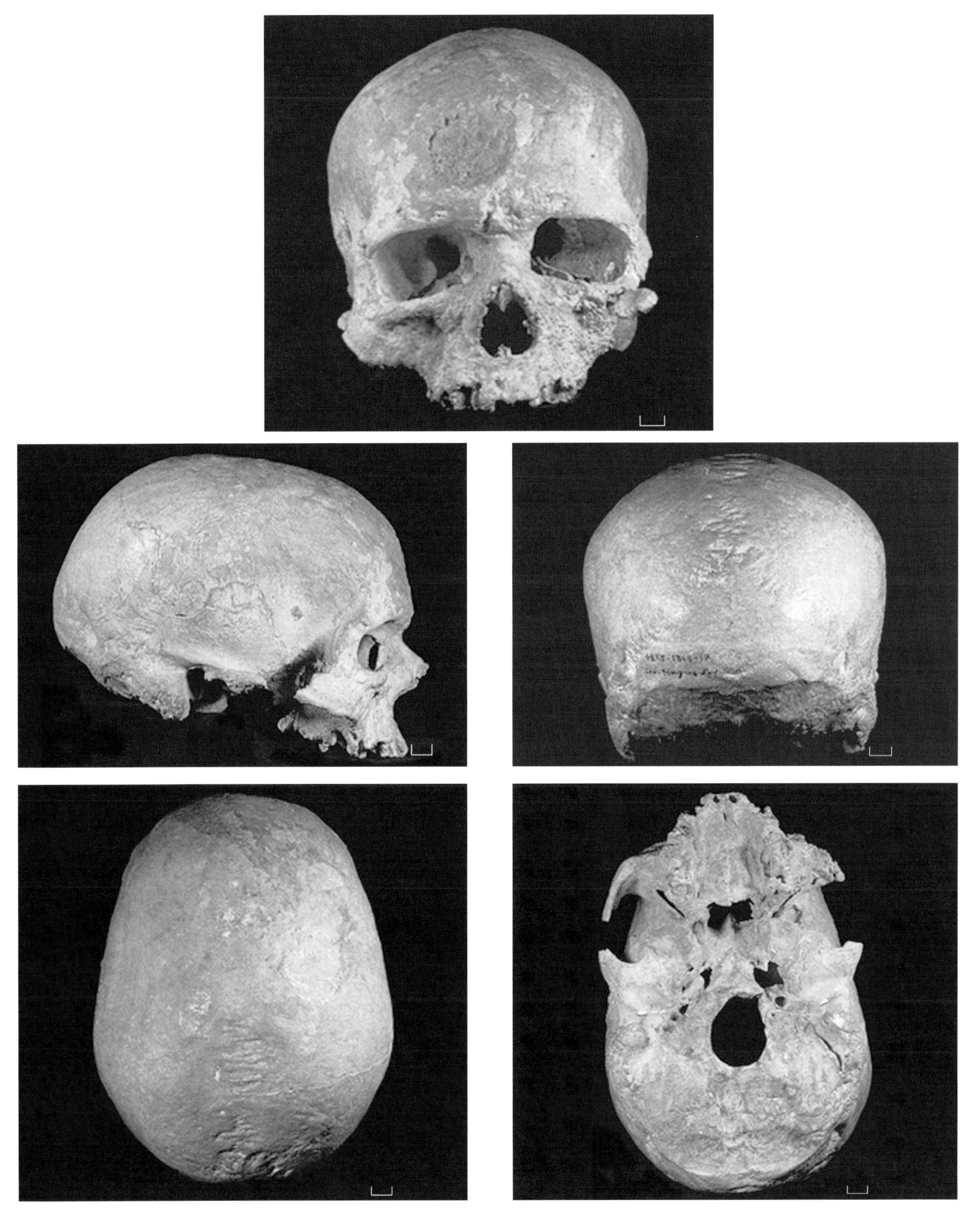

CRO-MAGNON Figure 1. Cro-Magnon 1 (scale = 1 cm).

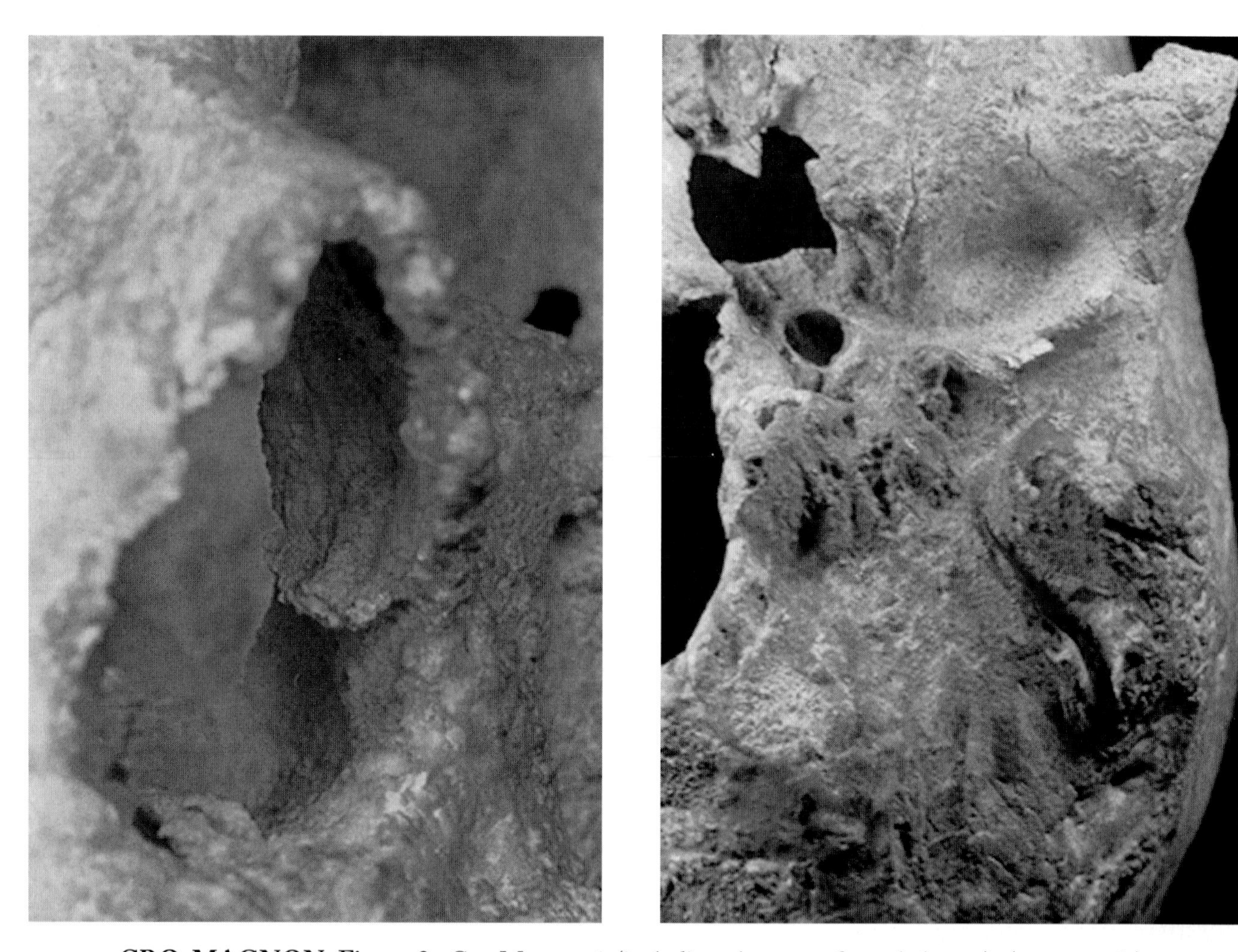

CRO-MAGNON Figure 2. Cro-Magnon 1 (including close-up of conchal crest), (not to scale).

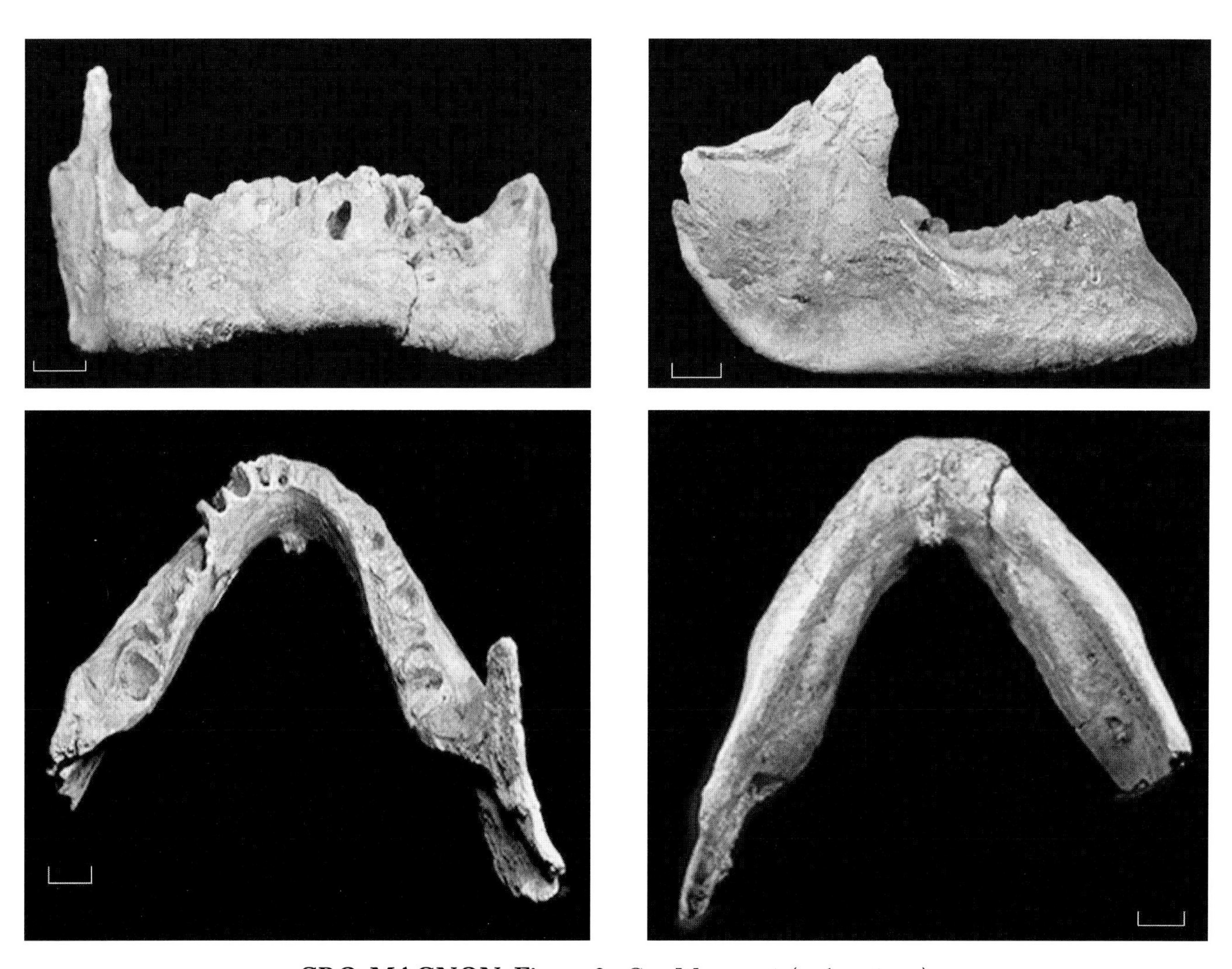

CRO-MAGNON Figure 3. Cro-Magnon 1 (scale = 1 cm).

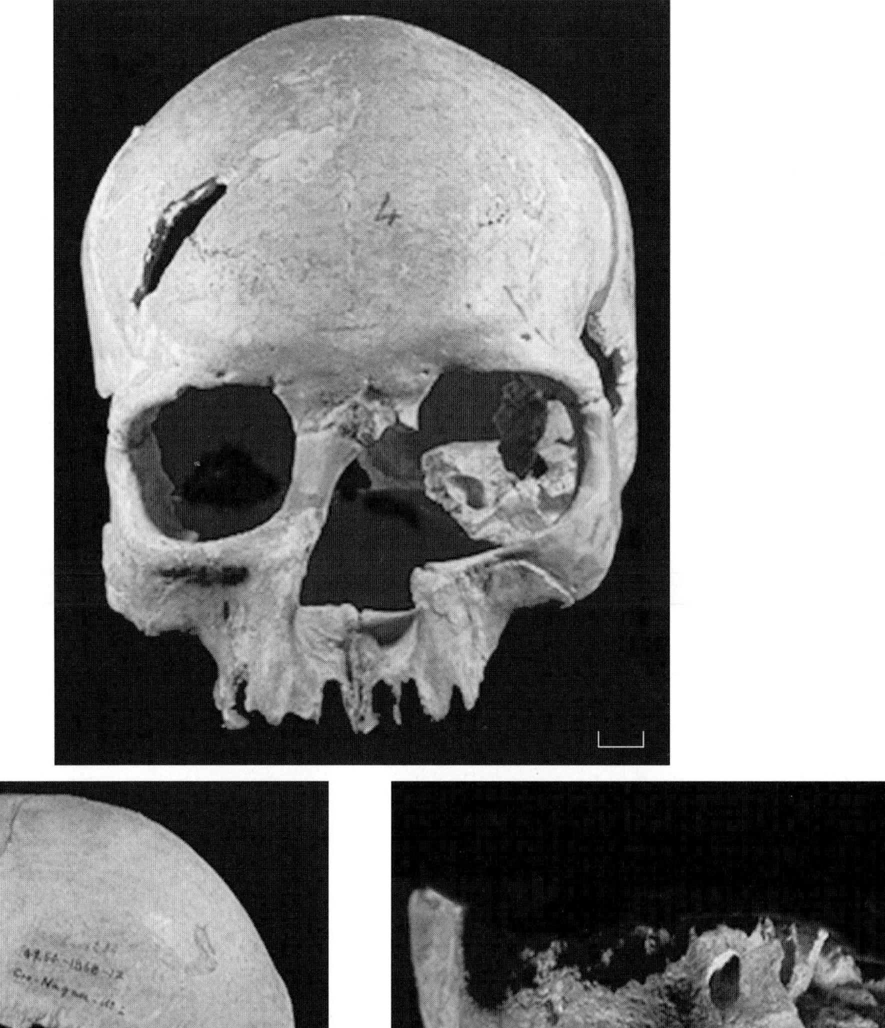

CRO-MAGNON Figure 4. Cro-Magnon 2 (scale = 1 cm).

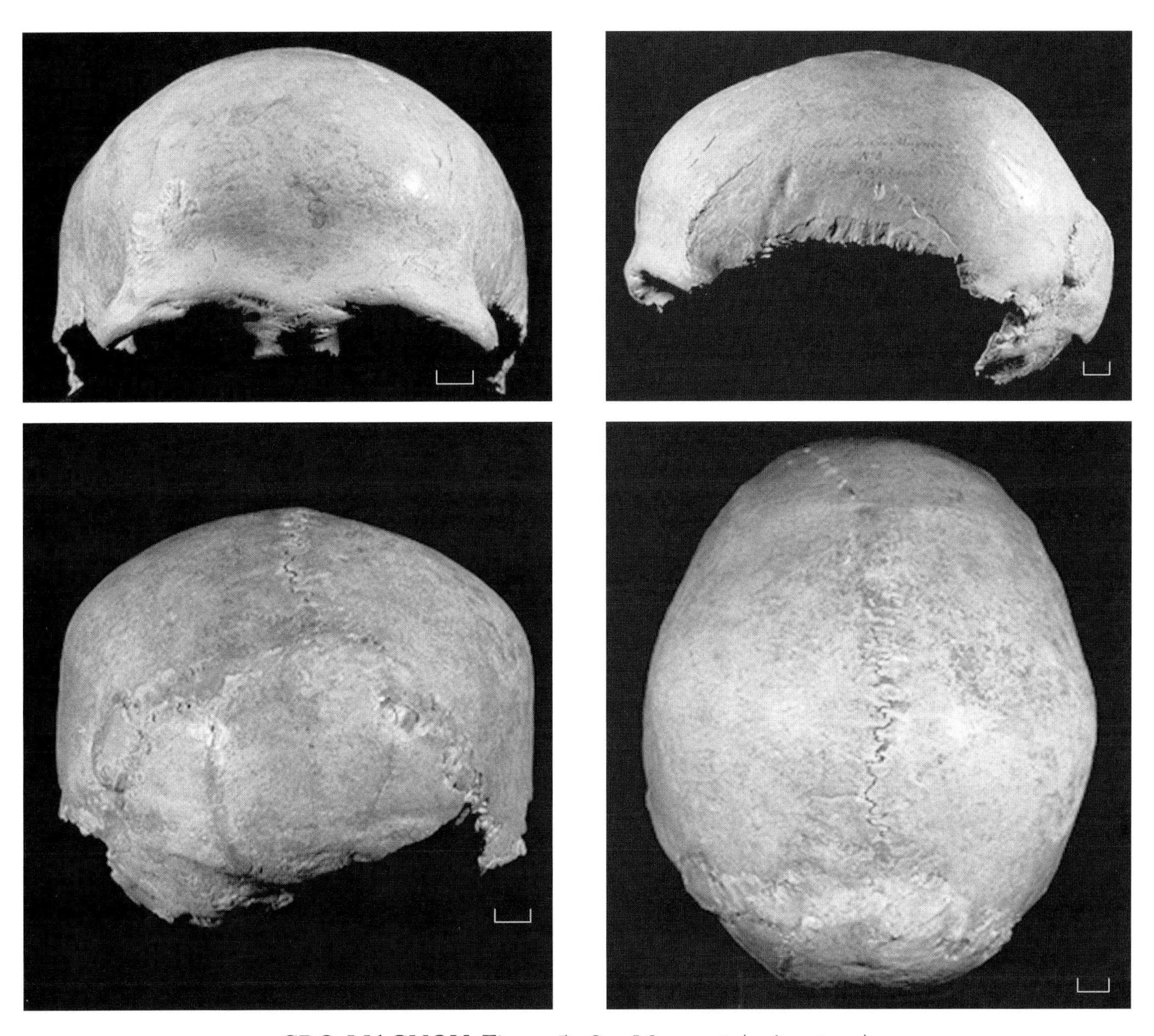

CRO-MAGNON Figure 5. Cro-Magnon 3 (scale = 1 cm).

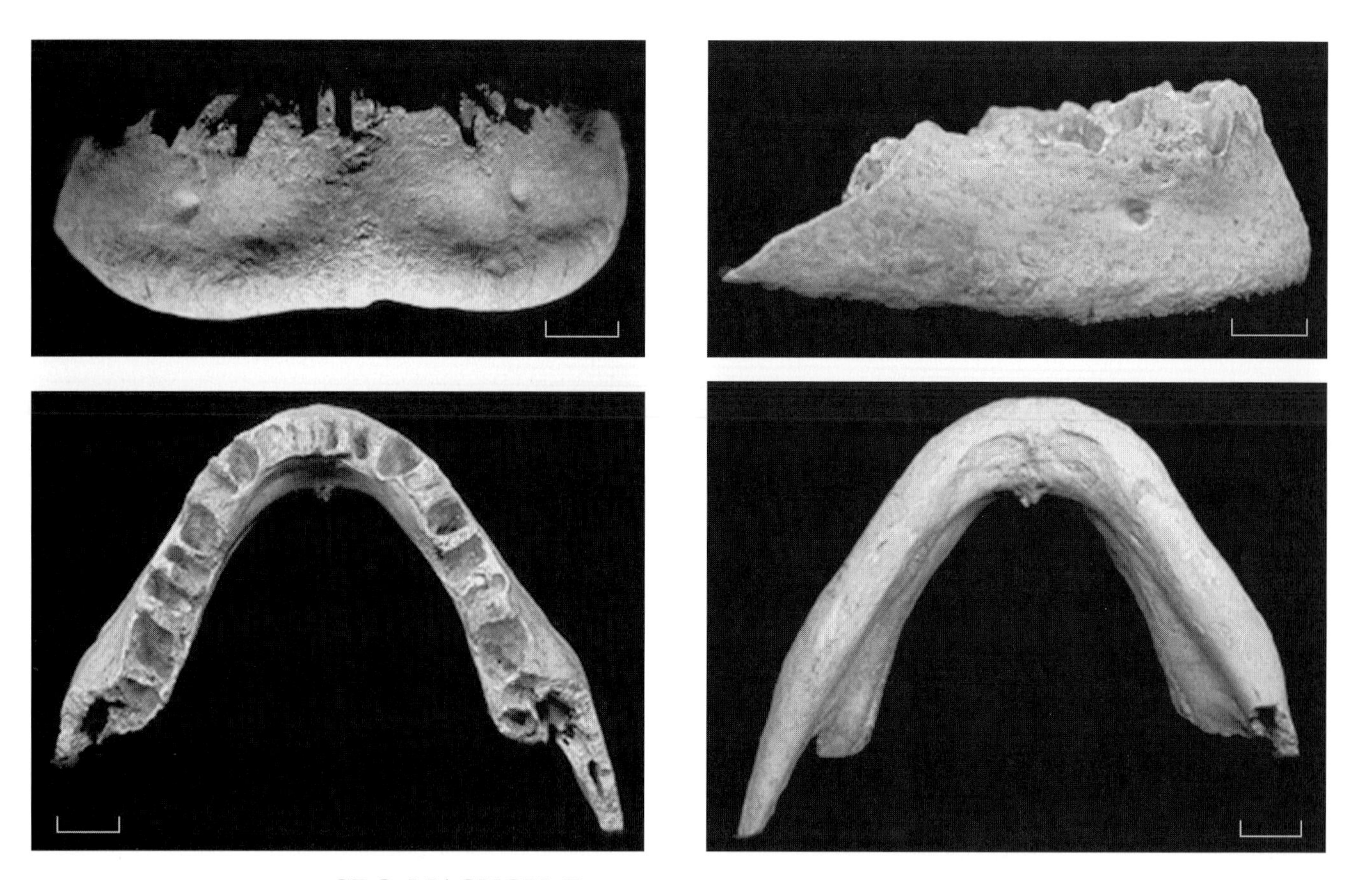

CRO-MAGNON Figure 6. Cro-Magnon 3 (scale = 1 cm).

Dmanisi

Location
Abandoned medieval town of Dmanisi, some 85 km SW of Tbilisi, Georgia, Caucasus.

Discovery
First hominid mandible was found in late 1991 during joint archaeological excavations of the Archaeological Research Center of the Georgian Academy of Sciences and the Römisch-Germanisches Zentralmuseum. Two crania were recovered at the same stratigraphic level in 1999, and a second mandible followed in 2000.

Material
Bilateral mandibular corpus with complete dentition (D211); undescribed mandible with partial dentition; one cranium (D2282), relatively complete; one calvaria (D2280).

Dating and Stratigraphic Context
The hominids were recovered from alluvial deposits, with paleosols, that overlie a thick basalt (Gabunia and Vekua, 1995a,b). The associated fauna is said by these authors to be of Late Villefranchian aspect, which would generally agree with a K/Ar date on the basalt of 1.8 ma (Majsuradze et al., 1991). Near contemporaneity of the basalt and the overlying alluvium is suggested by the fact that both are of normal polarity (presumably the Olduvai subchron), and that the lava is of fresh aspect (Gabunia et al., 2000). It has been argued that the age of the fossils is still in doubt because it is possible that the hominid-bearing sediments may have been reworked and that the normal polarity of the alluvium may be due to later magnetic overprinting. However, the recent review by Gabunia et al. (2000a), incorporating fuller data, indicates that a combination of paleontological, archaeological, geochronological, and paleomagnetic data strongly points to an earliest Pleistocene age of about 1.7 ma.

Archaeological Context
The hominid level and sediments above have yielded Oldowan (Mode 1) stone implements (Gabunia and Vekua, 1995a,b).

Previous Descriptions and Analyses
First description of the hominid jaw was by Gabunia and Vekua (1995a,b), who concluded that it represents "a population of *H. erectus*, possibly foreshadowing (European) early *H. sapiens*" (1995, p. 510). Dean and Delson (1995) agreed with the attribution to *H. erectus*, very broadly defined, while pointing out distinctive features of the jaw. Rosas and Bermudez de Castro (1998) also found a mixture of primitive (mostly in the jaw and anterior dentition) and derived (mostly in the cheek teeth) features, but concluded that the specimen is distinctive enough to be classified as *Homo sp. indet.* (*aff. ergaster*), while also bearing comparison in some features with Asian *H. erectus*. The latest analysis, by Gabunia et al. (2000a), of the assemblage including the first mandible and the new crania, concludes that the closest affinity of these spec-

imens is with *Homo ergaster* from Koobi Fora. Schwartz (2000) pointed out that the Dmanisi mandible does not compare favorably either with specimens from Java that most likely represent *H. erectus* or with the type specimen of *H. ergaster* (KNM-ER 992) and that the skulls (which do not appear similar to skulls from East Africa referred to *H. ergaster*) might represent different morphs. Gabunia et al. (2000b), however, interpret the differences between the skulls as reflecting different ages at death, sexual dimorphism, or individual variation. Gabunia et al. (2000a) estimated the cranial capacity of D2280 as 780 cc and that of D2282 as 650 cc.

MORPHOLOGY

D211

Adult. Partial mandible lacking both rami and inferior surface on both sides from P2 back. Jaw relatively small, narrow at front, not very tall s/i, but quite thick m/l, especially anteriorly and around molars. Tooth rows relatively long, slightly divergent. Symphyseal region bears three depressions below incisor roots; below this, midline slightly and smoothly mounded. Viewed from front, midline of symphyseal region gently peaked downward at its base; inferior margin on either side of this peak is slightly excavated upward. In profile, incisors overhang bone below; region of incisor roots gently concave, with slight bulge below that curves gently back to inferior margin. Appears that inferior margin between region of Cs had been slightly elevated. Below the Cs and P1s, inferior margin noticeably thickened on its outer surface into two small, flat, s/i tall and forwardly pointing tubercles. Large, single mental foramen lies below P1/2 on the R; on the L, moderate mental foramen lies below P1/2 and a small foramen posteroinferior to it. Bone of corpus begins to swell below region of M1. As indicated on the R (where base of ramus is partially preserved), a modest gutter extends back from M2. Appears that anterior root of ramus took origin below M2, and the ramus itself would have obscured M3.

Seen from below, inferior marginal tubercles form broad, blunt, somewhat obtuse corners, between which symphyseal region swells out gently. Both digastric fossae present; are long m/l, shallow, downwardly pointing. L digastric fossa delineated anteriorly by low ridge that runs from tubercle to midline peak of inferior margin. Internally, postincisal slope very long, steep; below it, the bone becomes more vertical. A small genial pit and a small and low single genial tubercle lie at base of postincisal slope. Bilaterally, very faint indication of mylohyoid line lying quite close to the M3s posteriorly. Submandibular fossae seen as shallow depressions; most pronounced under M1s and well below mylohyoid line.

All teeth present; appear relatively small. Is and Cs worn to expose dentine; other teeth minimally worn. I1–2s tall crowned; crown constricted to width of neck for some distance upward (thus the Is, and I1s especially, are seen to flare symmetrically high up). I2s slightly larger, more asymmetric, with a slight lateral flare, than I1s. I1–2 lingual surfaces smooth, slightly concave, bear very slight lingual swellings at their bases. Cs probably did not rise significantly above level of incisor crowns; have very short, almost horizontal mesial slopes; had long, steep distal slopes that terminate in distinct stylidlike swelling that is further delineated on buccal surface by vertical groove. Buccal surface moderately curved; lingual surface bears broad, inferiorly broadening pillar with shallow mesial and posterior foveae adjacent. Cs appear to be slightly skewed in dental arcade.

P1s subtriangular in occlusal outline, with slightly curved buccal surfaces; axis of single root oblique; root very wide b/l, compressed m/d. Protoconid very large, lies just anterior to midline of tooth; has subequal, gently sloping mesial and distal edges that terminate in small stylids that are themselves termini of relatively thick lingual crests. The mesial crest runs obliquely back along protoconid base; is separated by small notch from distal crest, which runs along distal side of protoconid before cornering in to terminate as a small swelling at protoconid base. Mesial crest surrounds thin basin; distal crest surrounds slightly more open basin. Slight swellings lie at base of midline of protoconid (much too small to be regarded as metaconids). P2s slightly wider b/l, but shorter m/d, than P1s. P2 protoconids relatively large, lie slightly mesial to midlines of crowns; metaconids lie opposite protoconids, with distinct m/d crease between them; cusps subequal in height and poorly projecting (metaconid more modest than protoconid). Thin cristid runs smoothly between bases of protoconid and metaconid, delineating b/l narrow but deep anterior fovea. Mesial slope of P2 crown slightly shorter than distal slope. Distal slope terminates in thick cristid that runs along distal side of tooth and slightly lingually around metaconid, delineating b/l narrow but deep posterior fovea (more extensive than anterior fovea). P2 roots wide b/l and somewhat compressed m/d.

All molar crowns subovoid in outline. M1 substantially larger, especially m/d, than M2, which is markedly larger, especially m/d, than M3. M2 and especially M3 preserve markedly deep wrinkling approaching cuspulation; appears M1 would have been similarly wrinkled. M1–3 have distinct and long trigonid basins, with somewhat filled-in talonid basins, and a broad, somewhat buccally positioned hypoconulid. In M1–3 metaconid large and most mesially positioned cusp; grooves at base of metaconid delineate wedgelike structure (a metastylid?). Base of large hypoconid extends lingually beyond midline only in M1. M1 has relatively thick cingulid around buccal side of protoconid; there is also trace of cingulid between bases of hypoconid and hypoconulid. M2 and M3 have slight pit in region between protoconid and hypoconid. M3 has distinct cuspulid in center of tooth, at base of metaconid. Appears that M1 has two roots that bifurcate well below neck. M2–3 roots visible above alveolar crest; appear barrel shaped.

D2280

Adult. Calvaria with upper nasals, missing much of basiocciput and basisphenoid and much of R temporal. Partially reconstructed, with some preparation scratching; numerous pathological lesions of moderate to small size, especially on the frontal and parietal. Bone very fresh. Bone of vault very thin. Cranial volume 780 cc (Gabunia et al., 2000a).

Skull very small, long and not very tall for its size, very lightly built. In profile, long frontal rises right from glabellar region with moderate slope to bregma. Behind bregma, profile flattens out to above mastoid region, behind which it curves gently down to the prominent and peaked external occipital protuberance. Below the protuberance, the nuchal plane (which is longer than the occipital plane in profile) slopes somewhat steeply down and forward. Seen from the rear, skull was broadest across supramastoid regions. As seen on the L, the side of the braincase is short; it slopes gently in; well below the well-marked temporal line, it curves in to slope continuously up to sagittal suture. General region of sagittal suture slightly elevated in midregion.

Seen from the rear, general outline of skull somewhat "hamburger bun" shape but rather high. Seen from the top, somewhat bulging side wall runs anteriorly to deep postorbital constriction; in front of constriction, lateral portions of supraorbital tori flare quite rapidly. Viewed from above, medial portion of supraorbital tori slightly concave lateral to broad, flat glabellar region; lateral portions of tori more convex, curving back laterally (i.e., profile of tori from above not smooth). Seen from the front, appears that face was relatively narrow.

Supraorbital tori moderately tall s/i and continuous laterally, with the medial being flatter and the lateral portion more bulbous. Medial part continues smoothly from broad, flat glabellar region to above very broad, somewhat excavated supraorbital notch. Superior part of flattened medial region continues laterally above notch; superior border of flattened area set off by blunt corner between its anterior surface and the short, steep posttoral plane behind. R supraorbital torus pathologically indented along its lateral anterior surface. Unmodified L torus tapers very little laterally. Concave orbital roofs angle bluntly upward onto anterior toral surface. Interorbital region (as preserved) markedly broad, strongly arced from side to side. As currently reconstructed, region of glabellar and frontonasal fragment contact not sutural. In preserved frontonasal region, frontal processes quite wide and nasal bones relatively narrow. Damage reveals two chambers of frontal sinus on the R and the L; the medial chamber is larger on the R. The lateral sinuses lie behind the medial sinuses, do not extend beyond level of supraorbital notch. Inner surfaces of preserved frontonasal region concave on either side of midline (suggests that region was invaded by sinuses). Frontal lobes extend only modestly over orbital cones. On the L are preserved some of the small sphenoidal sinuses.

Frontal rises moderately steeply from glabellar region; more laterally, curves up from a/p short posttoral plane. Very low midline keel along posterior half of frontal; otherwise, frontal shows little sign of doming. Crisp temporal ridges run close to orbital margin laterally, curve back and up rapidly as they emerge from behind zygomatic processes of frontal, continue to run up and back, parallel to, but well separated from, midline. As seen on the L, and level with auditory meatus, temporal line turns down strongly; at the level of asterion turns in steeply toward (apparently) broad parietal notch. As temporal line parallels lambdoid suture, becomes much more bandlike and somewhat elevated. As seen on the R, temporal fossa deep, moderately long a/p; is separated from infratemporal fossa by very sharp cornering of alisphenoid at level of posterior root of zygomatic arch. Also as seen on the R, anterior squamosal flows smoothly into temporal fossa. Appears that squamosal was long, low, but not very tall s/i. As seen on the L, parietomastoid suture somewhat short, horizontal. Moderate sinuses in R squamous. Hole of uncertain origin in posterior root of L zygomatic arch.

On the L, slightly upwardly angled posterior root of the zygomatic arch lies anterior to auditory meatus, flows into prominent suprameatal crest, which, in turn, continues into what appears to have been a low, moundlike supramastoid crest. On the R, appears that posterior root of zygomatic arch flared out quite prominently as shelf. Mandibular fossae deep, very wide m/l, very long a/p. Large medial articular tubercle partially closes off fossae medially. Anterior wall of fossae quite vertical and curved anteriorly; fossae closed off posteriorly and more laterally by short postglenoid plate; as seen on the L, medial posterior "wall" of mandibular fossa formed by ectotympanic tube slopes back strongly. No distinct articular eminence; bone flat across sphenotemporal region. Breakage exposes sphenoid sinus penetration lateral and down into bases of pterygoid plates. Large foramina ovales lie lateral to, and not much behind, region of lateral pterygoid plate; are bounded medially by moderately thick wall of bone; lie markedly and almost directly behind small foramina rotunda. As seen better on the L, moderate foramen rotundum lies in sphenotemporal suture, which itself lies somewhat in front of medial articular tubercle. Region of matrix on the R appears to be filled foramen rotundum; it lies almost entirely in temporal bone. Preserved portion of basisphenoid very broad, essentially flat.

As better preserved on the L, ectotympanic tube very thick walled, short; extends laterally not much beyond midline of mandibular fossa. As seen on the L, tube appressed to base of mastoid process, from which it is separated posteriorly by shallow, medially widening depression; stylomastoid foramen lies in this depression; auditory meati very small, ovoid. Vestiges of a low vaginal process (preserved on both sides) that appears to have peaked around region of what was probably moderately sized styloid process and not to have extended medial to styloid process; laterally it faded out well before reaching edge of ectotympanic tube. As seen on the L, stylomastoid foramen lies somewhat lateral and well posterior to styloid process. As preserved on the L, moderate-sized carotid foramen points straight back; given position of preserved extremity of sigmoid sinus, was well separated from the jugular foramen. Petrosal appears to have turned anteromedially only slightly.

L mastoid process quite damaged; reveals numerous small air cells within its capacious base as well as within its m/l broad and flat posterior surface. Not possible to tell how prominent mastoid process was. Mastoid notch apparently very narrow anteriorly; broadened strongly into very broad but shallow digastric fossa that is delineated posteriorly by thin, low ridge. Still on the L, low (paramastoid?) crest lies lateral to what appears to be region of occipitomastoid suture, somewhat medial to which is relatively short, thin Waldeyer's crest. Posterior rim of foramen magnum preserved; is strongly curved, suggesting that it was quite narrow, at least posteriorly.

Lambdoid suture runs gently up from asterion; becomes obscured (due to pathology?) well lateral to midline. Appears that occipital plane was very low s/i and extremely wide. Superior nuchal lines very pronounced and ledgelike as they emerge from prominent, downwardly peaked external occipital protuberance; quickly fade out well before region of asterion. Nuchal plane otherwise poorly muscle scarred. Nuchal plane runs forward and down quite steeply. Just behind rim of foramen magnum is pair of broad depressions separated by moderately long but thin external occipital crest.

Internally, frontal crest very strong, prominent; runs more or less vertically down. Only grooves for middle branches of middle meningeal arteries well impressed; as seen on the L, are not highly arborized. As preserved on both sides, superior surface of petrosal very broad; on the R is flat; on the L is raised slightly by superior semicircular canal. On the L, deep superior petrous sinus runs along superior surface of petrosal (not along its margin). Appears that petrosals were quite thin s/i. Internal occipital protuberance broad, prominent; lies well down on nuchal plane. Faint L transverse sinus takes origin well above protuberance. R transverse sinus not really discernible; low ridge between impressions for occipital and cerebellar lobes runs laterally from protuberance. L occipital lobe impression broader than R; reverse applies to cerebellar lobes, which faced back (not down). As preserved on the L, short, deep sigmoid sinus comes off transverse sinus well below asterion; deeply undercuts back of petrosal. Its course suggests a very posteriorly placed jugular foramen (or another exit?). A small groove comes off sigmoid sinus below level of internal acoustic meatus, runs a short distance across posterior part of petrosal (suggesting bifurcating drainage pattern).

Coronal suture visible, particularly between temporal lines; is essentially straight, with only an occasional undulation. Preserved lateral parts of lambdoid suture show broad and short interdigitations. Sagittal suture very shallowly denticulate in region behind bregma; the section behind this has thinner and longer interdigitations.

D2282

Adult. Crushed and distorted partial cranium missing most of orbital and upper nasal regions, most of sphenoid, and the basiocciput. Also missing parts of both temporals. Preserves RP2 and M1, LM1–2, and alveoli for all teeth anterior to these. M3s apparently never erupted.

Skull small, lightly built, with thin cranial bone throughout. In general, face looks large relative to braincase. Even though crushed basally, braincase is relatively long, extremely low, posteriorly broad. In profile, somewhat distorted facial skeleton was apparently well forward on neurocranium; probably quite tall s/i, with slightly dished subnasal region. Low frontal begins to rise from relatively long and flat postglabellar plane; rise is quite short and sloping, then slopes more shallowly back to bulging bregmatic region. Behind this is a long, gentle slope down to lambda, behind which the occipital plane bulges slightly. Profile continues more steeply down occipital plane; apparently then curved smoothly but strongly forward onto nuchal region. Nuchal region highly broken; was probably fairly flat. Seen from behind, skull was very broad across bulging mastoid region (as seen on the L); on both sides, walls of braincase are quite tall, straight, slightly inwardly tilted; profile curves in quite strongly just lateral to temporal lines and runs almost horizontally to sagittal suture (not quite "hamburger bun" shape). Viewed from the top, as better preserved on the less crushed L side, braincase tapers strongly to very deep postorbital constriction. As better preserved on the R, relatively straight supraorbital torus retreats from region of bregma. Seen from the front, brows were at least as broad bilaterally as braincase.

Upper face missing. Brows relatively thin s/i, being thickest over general region of very deep, moderately wide supraorbital notches; thin quite strongly laterally. Interorbital region was extremely broad, at least superiorly. Superior surfaces of supraorbital tori smooth; flow into relatively long posttoral planes that accentuate slope of frontal. Orbital roofs shallowly concave; lateral to notches, angle quite sharply on to anterior toral surfaces. Glabellar region damaged, but postglabellar plane relatively wide and long. Breakage reveals moderately large L and R frontal sinuses that do not penetrate beyond supraorbital notches laterally or posteriorly beyond postglabellar plane. Frontal lobes extended well forward over orbital cones to base of frontal rise. Preserved zygomas suggest that orbits were somewhat rounded laterally and curved into infraorbital margin. In contrast, superior orbital margin much straighter. Zygomas were apparently very deep s/i; had very thick temporal processes. Maxillary tuberosities were apparently quite expansive, oriented downward, and (seen from below) formed corners with the anterior portions of the zygomatic arches. Anterior roots of zygomatic arches take origin well above level of M1. On both sides, grooves coming from region of infraorbital foramina run essentially straight down, delineating anteriorly flat facial pillars laterally. Bone lateral to these deep grooves appears to be forwardly facing (suggesting flat, anteriorly facing infraorbital region). The L infraorbital foramen is small, downwardly facing; was probably situated well below infraorbital margin. Nasal aperture relatively broad; its preserved sides are bluntly rounded. Inferolateral corners of nasal aperture strongly curved; preserved lateral margins turn gently inwardly. As better preserved on the L, nasal cavity floor sunken well below a/p rounded inferior nasal margin. Vomer apparently extended right up to base of inferior margin; apertures of incisive canals were right behind it. Region of anterior nasal spines broken; seems that spines were lacking. As seen on the L, maxillary sinus large; did not swell into nasal cavity. Wall of the R side of nasal cavity preserved higher up; is essentially smooth, lacking signs even of conchal crest. Facial pillars anteriorly flat; extend all the way down region of canine roots to alveolar margin. Nasoalveolar plane between pillars slopes slightly out inferiorly. Nasoalveolar clivus quite tall s/i relative to its size.

As seen on the L, low temporal ridge runs up behind zygomatic process of frontal and over posttoral plane before curving back strongly as thin, low, bandlike temporal line. On both sides, temporal line angles medially as it crosses coronal suture; behind this it arcs back and down. As seen on the R, temporal line descends to lambdoid suture well behind region of asterion, where it apparently faded out. Region of bregma broadly swollen, with shallow sulcus behind. Posterior to postbregmatic sulcus, undistorted region of sagittal suture appears to be slightly keeled. Temporal fossae were deep, apparently quite long. Zygomatic arches seem to have thinned posteriorly; probably flared modestly laterally. Posterior root of zygomatic arch extends laterally from region of mandibular fossa; broadens rapidly into an a/p short shelf. Absence of squamous suture markings on parietals suggests that squamous portions of temporals were quite low. Preserved R anterior squamosal region makes low, blunt corner into temporal fossa; swelling lies along the inferior part of the anterior squamosal suture. Also as preserved on the R, alisphenoid corners smoothly medially, creating infratemporal fossa. As better preserved on the R, mandibular fossa was apparently m/l wide, deep, and bounded by a rather vertical anterior wall that apparently bore a low articular eminence. On both sides, a smooth and moderately well-developed tubercle marks lateral extent of articular eminence. On the R, sphenotemporal plane flat between articular eminence and bases of broken pterygoid plates, which are invaded by sphenoid sinus. Sphenotemporal suture visible on the R; seems that it ran to medial edge of articular eminence and close to foramen ovale. Space between foramen ovale and mandibular fossa not very large. Preserved anterior margin of foramen ovale shows that the foramen lay well within the sphenoid. Foramen rotundum probably preserved just anterior to pterygoid plates and not far in front of foramen ovale. As preserved on the L, ectotympanic tube very thick walled; extends somewhat beyond midline of mandibular fossa. Postglenoid plate absent bilaterally. As indicated on the R, vaginal process probably peaked around thin styloid process. Ectotympanic tube and vaginal process form vertical wall to back of mandibular fossa. Low toruslike structure lies at base of ectotympanic tube anteriorly (as seen on the R).

Mastoid regions quite fractured, but it appears that true mastoid processes were absent. Also appears (from the R) that parietomastoid suture was relatively long and horizontal. On both sides, posterior roots of zygomatic arches flowed into well-developed, slightly sloping suprameatal crest that continued, in turn, into laterally swollen supramastoid region. On the R, occipitomastoid suture preserved; laterally, it runs almost directly medially before turning forward slightly; bone in front of and behind it appears to have been flat; no evidence along suture or on either side of any cresting. On the L side, very shallow groove may represent mastoid notch. Lambdoid suture slopes gently up from asterion; arcs broadly across region of lambda, which also bears some tiny ossicles. Occipital plane very short, smoothly bulging; only on the R is there any hint of superior and inferior nuchal lines. Nuchal plane was apparently very long and essentially devoid of muscle markings.

For size of specimen, palate quite large and deep. Dental arcade forms narrow parabola. Anterior part of palate more sloping than sides, which become more vertical posteriorly. Low but distinct palatine torus posteriorly. Large incisive foramen lies level with P1s; a moderately deep groove runs anteriorly from it to the septum between the I1s. Greater palatine foramina rather small. No indication that M3 was ever present; there is a moderate expanse of bone between M2 and broken pterygoid plates.

Internally, frontal crest stout, quite vertical. Meningeal grooves faint; only posterior branch of middle meningeal artery impressed at all. Sagittal sinus visible posteriorly. No indication preserved of definitive internal occipital protuberance. On the R, very low down on occipital, is distinct transverse sinus that separates small impression of occipital lobe from very anteriorly positioned and downwardly facing cerebellar lobe. This latter lies quite far anteriorly and laterally, right next to the short preserved portion of rim of foramen magnum. Sigmoid sinuses not detectable (region damaged). Superior surfaces of petrosals bear slight arcuate eminences; no sign of superior petrous sinuses. A longitudinal ridge on the R petrosal runs anteriorly along superior surface both in front of and behind arcuate eminence. As seen on the R, superomedial aspect of petrosal is crestlike; a crest descends from it near region of arcuate eminence to margin of internal acoustic meatus; there is a depression on each side of this short vertical crest.

Coronal and sagittal sutures have short interdigitations. Lambdoid suture wildly interdigitated with lots of small islands of bone. None of the sutures is segmented.

Vacant alveolae in general large and deep. I1s would have been much broader than I2s. P1 alveolus wider b/l than the ovoid and quite large canine alveolus. As seen on the L, P1–2 alveoli housed single-rooted teeth; both alveoli compressed m/d, but P1 alveolus slightly pinched in midline mesially. Preserved teeth moderately worn. Lingual side of P2 quite rounded and bulbous, longer m/d than the more gently curved buccal surface. Paracone quite buccally and centrally placed; had short, gently sloping mesial and distal edges. Protocone slightly more internally and mesially placed. Crease between bases of protocone and paracone. Metaconule lies to distal side of paracone, in moderate basin. Thick pre- and postcingula become more expansive as they join the protocone. Preserved R and LM1 and LM2 were probably somewhat wrinkled. All have very expanded hypocones that are distinguished from protocones by groove or crease. M2 hypocone more swollen distally. Buccal cusps of all molars close together; have long internal slopes. Trigon basins relatively small but deep. All molars have thick precingulum that is delineated from paracone by crease. M1s have distinct pit on mesial side of protocone. On all molars, buccal cusps more peripherally placed than protocone.

REFERENCES

Dean, D. and E. Delson. 1995. *Homo* at the gates of Europe. *Nature* 373: 472–473.

Gabunia, L. and A. Vekua. 1995a. A Plio-Pleistocene hominid from Dmanisi, East Georgia, Caucasus. *Nature* 373: 509–512.

Gabunia, L. and A. Vekua. 1995b. La mandibule de l'homme fossile du Villafranchien supérieur de Dmanisi (Géorgie Orientale). *L'Anthropologie* 99: 29–41.

Gabunia, L. et al. 2000a. Earliest Pleistocene hominid cranial remains from Dmanisi, Republic of Georgia: Taxonomy, geological setting, and age. *Science* 288: 1019–1025.

Gabunia, L. et al. 2000b. Taxonomy of the Dmanisi crania (response). *Science* 289: 55.

Majsuradze, G., et al. 1991. Paleomagnetik und Datierung der Basaltlava. Jahrb. Romisch-German. Zentralmus. 36: 74–76.

Rosas, A. and J. Bermudez de Castro. 1998. On the taxonomic affinities of the Dmanisi mandible (Georgia). *Am. J. Phys. Anthropol.* 107: 145–162.

Schwartz, J. H. 2000. Taxonomy of the Dmanisi crania. *Science* 289: 55.

Repository

Georgian State Museum, 3 Purtseladze Street, 380007 Tbilisi, Republic of Georgia.

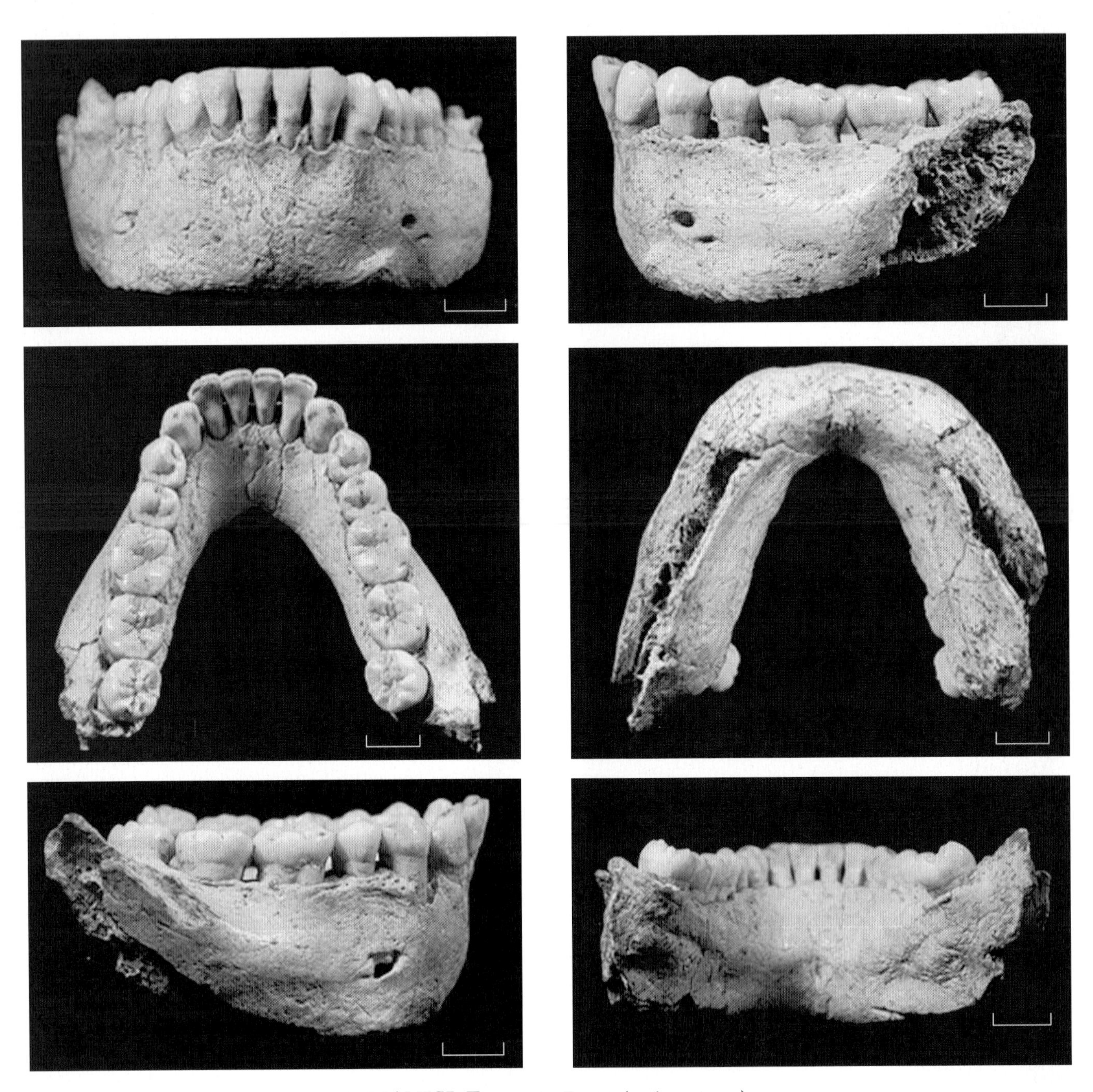

DMANISI Figure 1. D211 (scale = 1 cm).

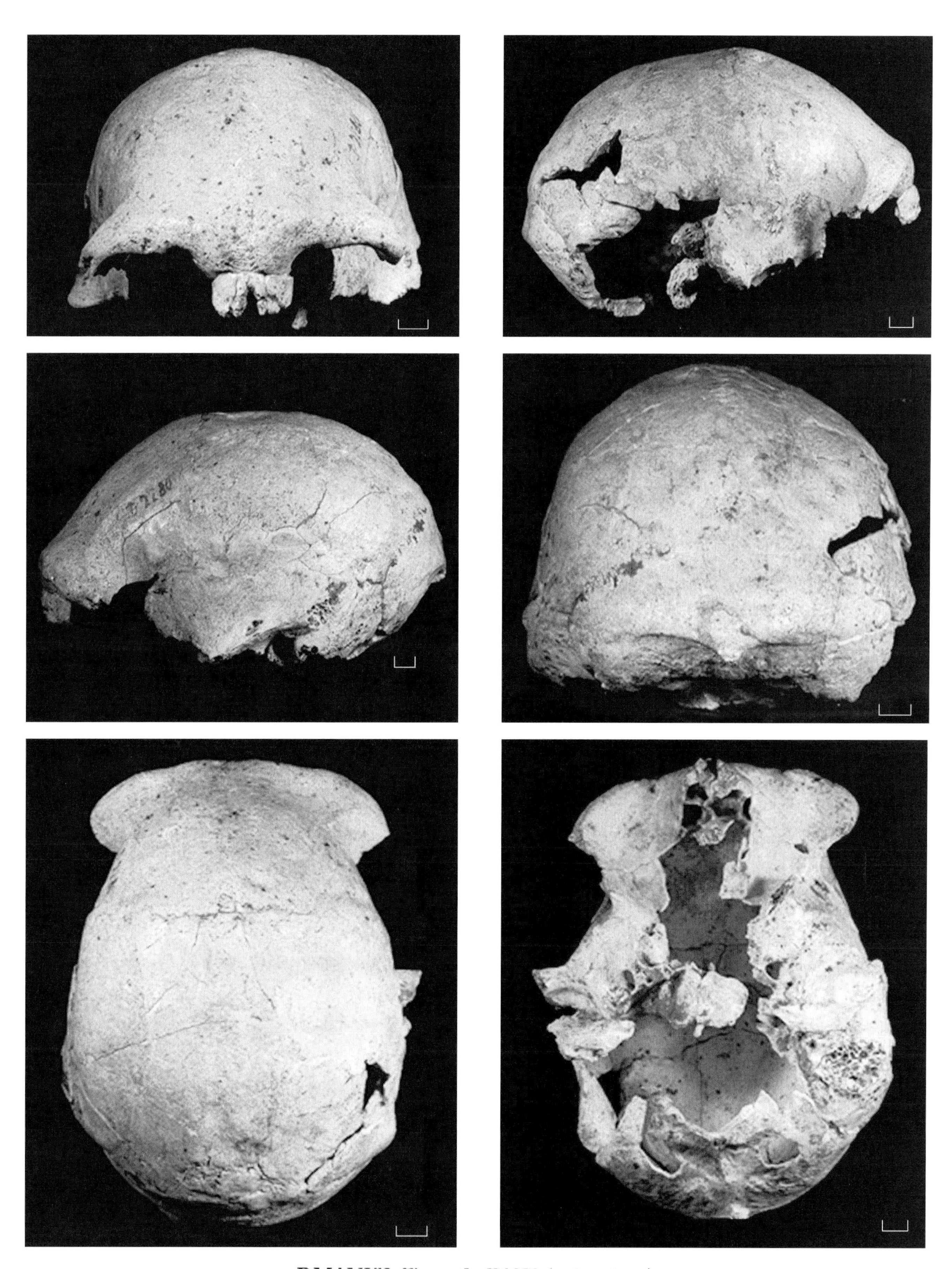

DMANISI Figure 2. D2280 (scale = 1 cm).

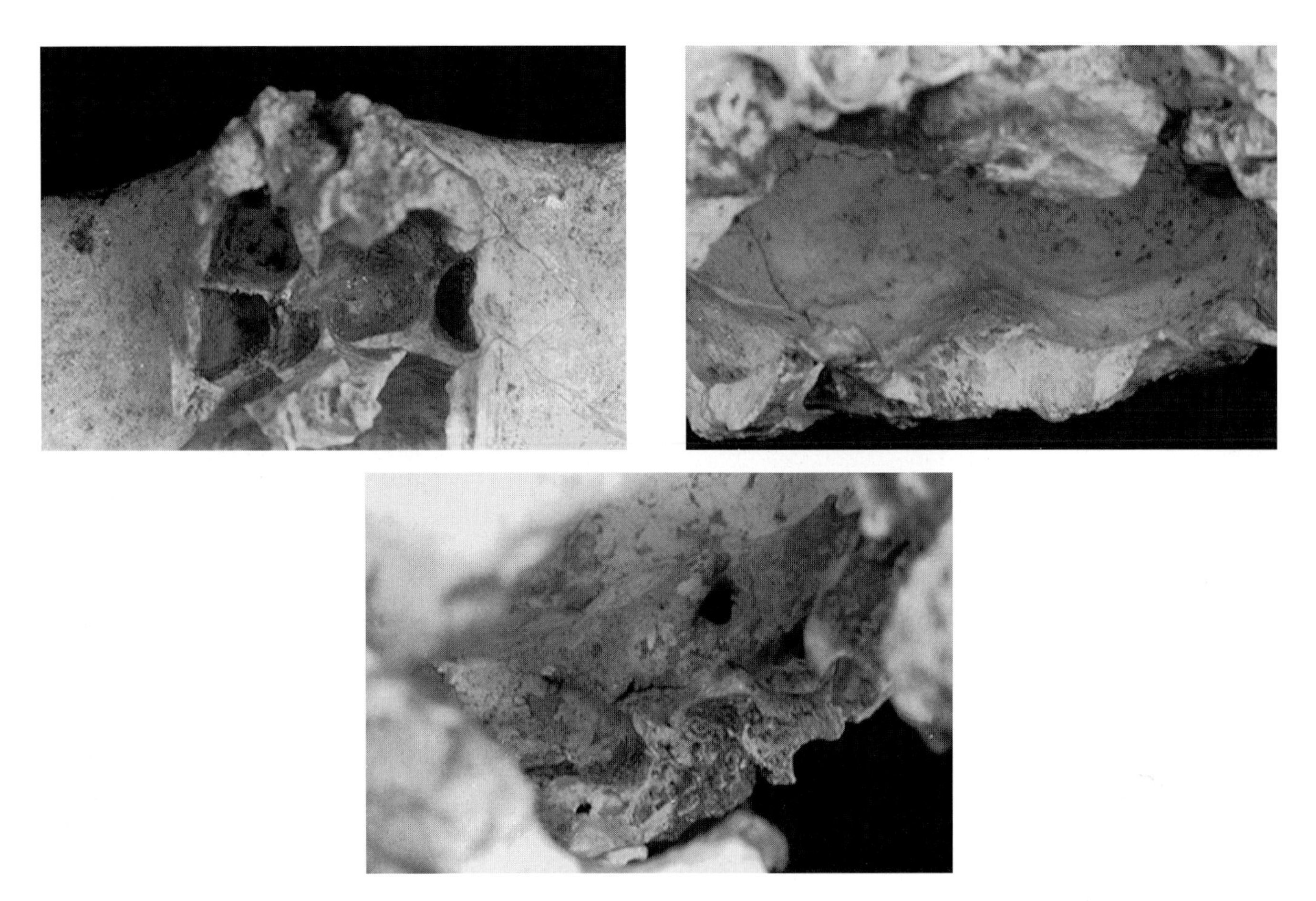

DMANISI Figure 3. D2280: close-ups of frontal sinuses (top left), transverse sinuses (top right), and left petrosal (bottom), (not to scale).

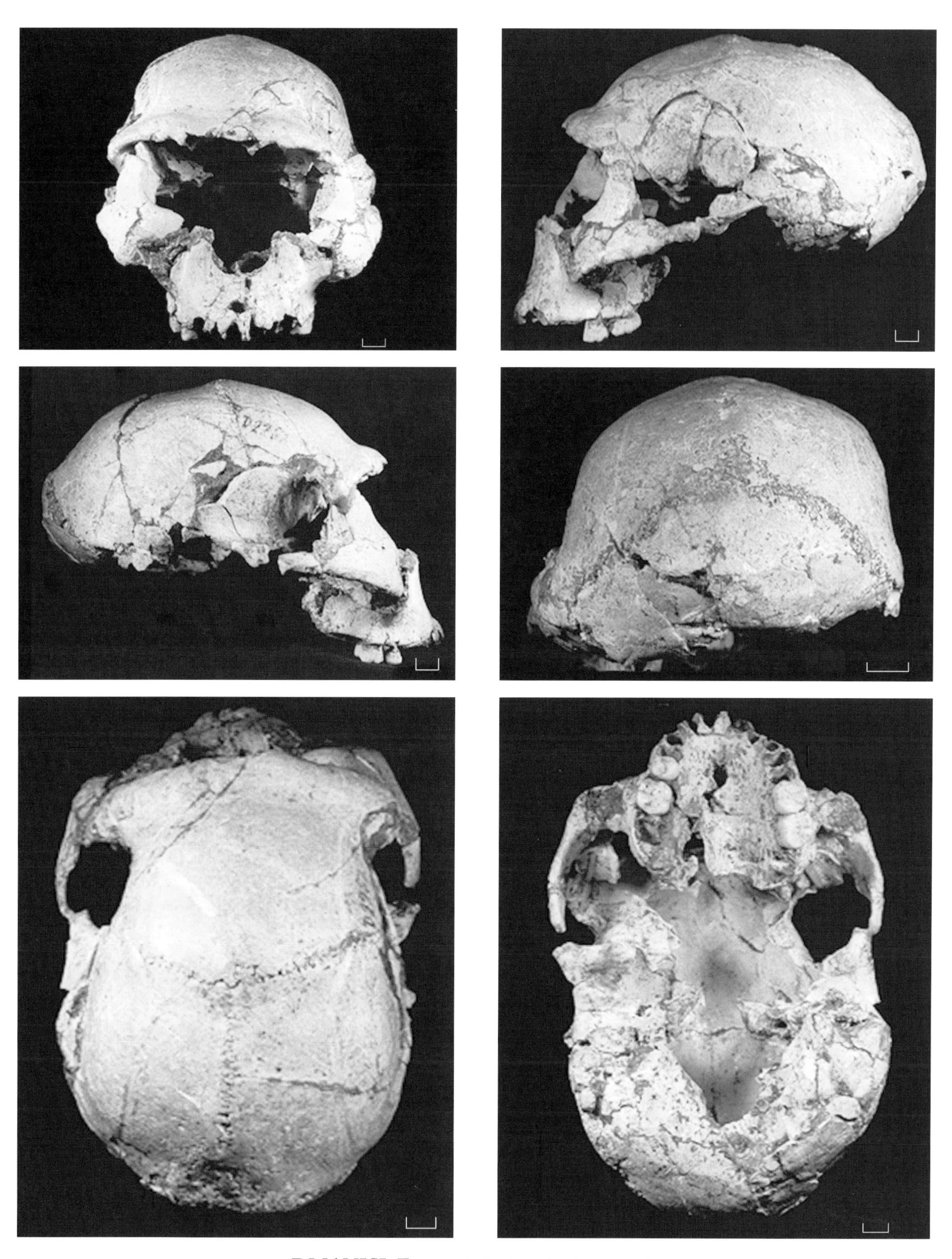

DMANISI Figure 4. D2282 (scale = 1 cm).

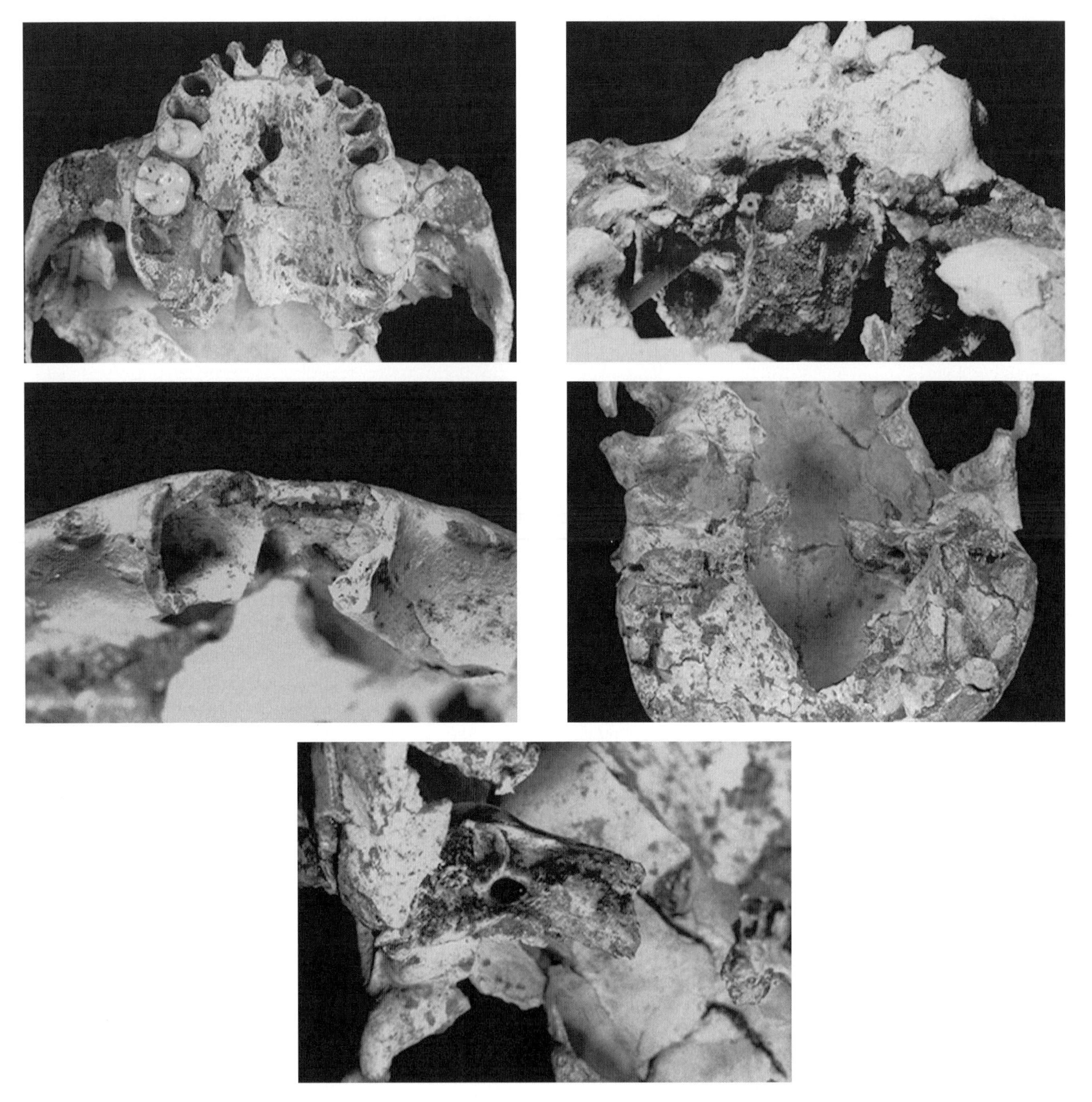

DMANISI Figure 5. D2282: close-ups (including floor of nasal cavity (top right), frontal sinuses (middle left), and petrosal (bottom)), (not to scale).

Dolni Vestonice

Location

Open-air complex of sites on the lower slopes of the Pavlovske Hills below the ruins of the Divci Hrad, about midway between the villages of Dolni Vestonice and Pavlov, some 35 km S of Brno, Czech Republic.

Discovery

Excavations of K. Absolon, May 1925–May 1934 (Dolni Vestonice I–II, IV–VIII); excavations of B. Klima, July 1949–April 1987 (DV III, IX–XVI).

Material

Sixteen individuals in varying states of completeness, from isolated teeth to entire skeletons (there are seven of the latter, plus two quite complete calvaria).

Dating and Stratigraphic Context

Dolni Vestonice is not a single site but rather a cluster of occupation areas, fairly shallowly buried below modern vineyards, that all date from about the same time. Extensive dating by radiocarbon of both charcoal and human bone fragments has yielded dates in the 26.4- to 29-ka range (Vogel and Zagwijn, 1967; Allsworth-Jones, 1986; Svoboda and Vlcek, 1991). Charcoal within the grave of a spectacular triple burial (DV XIII–XV) gave a date of 27.64 ka (Klima, 1988).

Archaeological Context

There are two principal concentrations of human activity at Dolni Vestonice: the "eastern slope" area, whence the bulk of the occupation evidence comes, and the "western slope" containing most of the human remains. The archaeological materials all belong to the Upper Paleolithic Pavlov culture of the Eastern Gravettian complex, the dates of which at other sites agree closely with those obtained here. Some, perhaps all, of the human remains represent burials, of which the most remarkable is the triple interment noted above (Jelinek, 1987; Klima, 1988), in which three young adults were buried together with a variety of grave goods. One, at least, may have been killed (Klima, 1988). Dolni Vestonice also offers evidence of dwelling structures, artistic activity (including figurines fired in kilns that represent the earliest evidence of ceramic technology), and impressions in clay of woven plant fibers that equally provide the oldest evidence of basketry or mat-making (Absolon, 1945; Klima, 1963).

Previous Descriptions and Analyses

The Dolni Vestonice human remains were published piecemeal as they were discovered (from Morant, in Absolon 1938, for DV I, to Svoboda, 1988, for DV XVI). Some have been more fully described than others, but there has never been any suggestion that these remains are anything other than fully modern *Homo sapiens*. Estimated cranial capacities are 1285 cc for DV III (Jelinek, 1954), 1481 cc for DV XIII, 1538 cc for DV XIV, 1378 cc for DV XV, and 1547 cc for DV XVI (Vlcek, 1993).

Morphology

The following composite description derives from the entire Dolni Vestonice sample, which convincingly represents the same population.

Crania thin boned; parallel sided with maximum width high up. Frontal vertical; may be bulbous, with eminences sometimes developed. Bipartite supraorbital region poorly to moderately developed but with discernible to moderate glabellar "butterfly"; lateral portion either indistinguishable or flattened and partly backwardly angled. Frontal sinus confined to midline. Inferior orbital margin protrudes, with single zygomaticofacial foramen below it. Internally, variable number of Pacchionian depressions. Temporal lines broad and low. Nasal bones are downwardly angled. Maxillary frontal processes not forwardly projecting. Nasal aperture small with thin lateral crest that may descend below single inferior margin (no prenasal fossa). Spinal crest incomplete or absent. Conchal crest absent or weak, thin, and horizontal. Nasal cavity floor flat; anterior nasal spine may be large and bifid. Nasoalveolar clivus short and slightly angled forward. Canine fossae small.

Zygomatic arches gracile, lie close to vault wall, arc smoothly or angle back at maxillary tuberosity; anterior root may face forward. Posterior root originates above anterior or middle part of external auditory meatus. Infratemporal fossa not delineated from temporal fossa. Mandibular fossa not closed off medially; is deep with strong articular eminence. Mastoid processes may be small to moderately projecting and thick at base, with tip pointing downward. Mastoid notch narrow to moderately wide, and shallow to deep with small digastric fossa behind. Paramastoid crest distinct. Occipitomastoid and Waldeyer's crests variably developed. Vaginal process courses along length of ectotympanic tube, peaks around possibly stout styloid process, and runs to contact mastoid process. Stylomastoid foramen lies a little lateral and posterior to styloid process. Suprameatal and mastoid crests may be low. Upwardly curving supramastoid crest more robust; is confluent with suprameatal crest. Parietomastoid suture short to long and irregular; may be horizontally oriented.

Triangular occipital bone narrow; protrudes slightly to significantly. Superior nuchal lines weak or not discernible; highest nuchal line may be present. External occipital protuberance may be indistinct or low and broad, or there may be scar or depression in that region. Lambdoid suture courses obliquely upward from asterion (no anterior lambdoid suture); peaks at lambda. Nuchal plane long, narrow; bears weak superior nuchal line. Foramen magnum small, smoothly lozenge shaped. Petrosal bears low, domed arcuate eminence, well-developed superior petrous sinus, and depressed subarcuate fossa that is almost fully closed off. All sutures segmented.

Mandible typically gracile and small; corpus low. Distinct to moderately developed mental trigon bears central keel. Good genial crest in midline. Digastric fossae small to moderate, shallow. Genial tubercles may be crestlike. Mental foramen small, lies under P2. No retromolar space. Gonial region perhaps inflected outward with faint to conspicuous masseter markings; angle widely rounded, not "cut off." Medial pterygoid tubercles low. Mandibular foramen small, quite compressed, and incompletely closed over; opens upward. Sigmoid notch crest runs laterally to very edge of condyle. Weak to strong mylohyoid line angled obliquely down, with small to large submandibular fossa below and depressions above and/or in front of line.

Teeth typically heavily worn. Upper anterior tooth roots short. P^1 larger/broader than P^2. Molars decrease in size from M^1 to M^3 (M^3 may be very small and lack hypocone); talon basins insignificant. All molars have short roots, which bifurcate close to neck. Lower anterior teeth have short, slender roots. P_2 larger than P_1. Molars decrease in size from M_1 to M_3 and bear small anterior foveae (no trigonid basins); only M_1 may have hypoconulid.

DV 16 more robust than others, e.g., in prominence of muscle markings, mental trigon with corners, symphysial keel, zygomatic arches, temporal lines, mastoid processes, and region of unpeaked external occipital protuberance. Relatively extensive sinus development on either side of median septum within glabellar "butterfly"; however, there is little lateral supraorbital development. Otherwise, DV 16 consistent with sample.

REFERENCES

Absolon, K. 1938. *Die Erforschung der diluvialen Mammut-Jäger-Station von Unter Wisternitz an den Pallauer Bergen in Mähren, Arbeitsbericht über das zweiter Jahre 1925.* Brünn (Brno).

Absolon, K. 1945. *Vyzkum diluvialni stanice mamutu v Dolnich Vestonicich na Pavlovskych na Morave. Pracovni zpravna za treti rok 1926.* Brno, Moravska Knihtiskarna Polygrafie.

Allsworth-Jones, 1986. *The Szeletian and the Transition from Middle to Upper Palaeolithic in Central Europe.* Oxford: Clarendon Press.

Jelinek, J. 1954. Nalez fosiliho cloveka Dolni Vestonice III. *Anthropozoikum* 3: 37–92.

Jelinek, J. 1987. A new Paleolithic triple-burial find. *Anthropologie* 25: 189–190.

Klima, B. 1963. *Dolni Vestonice. Vyzkum taboriste lovcu mamutu v letech 1947–1952.* Prague, Nakladatelstvi CSAV.

Klima, B. 1988. A triple burial from the Upper Paleolithic of Dolni Vestonice, Czechoslovakia. *J. Hum. Evol.* 16: 831–835.

Svoboda, J. 1988. A new male burial from Dolni Vestonice. *J. Hum. Evol.* 16: 827–830.

Svoboda, J. and E. Vlcek. 1991. La nouvelle sépulture de Dolni Vestonice (DVXVI), Tchecoslovakie. *L'Anthropologie* 95: 323–328.

Vlcek, E. 1993. *Fossile menschenfunde von Weimer-Ehringsdorf.* Stuttgart, Konrad Theiss Verlag.

Vogel, J. and W. Zagwijn. 1967. Gröningen radiocarbon dates VI. *Radiocarbon* 9: 63–106.

Repository

DV III: Anthropos Institute, Moravian Museum, Zelny trh 6, 65937 Brno, Czech Republic. DV XI–XVI: Institute of Archaeology, Dolni Vestonice, Czech Republic. The other specimens were destroyed during WWII.

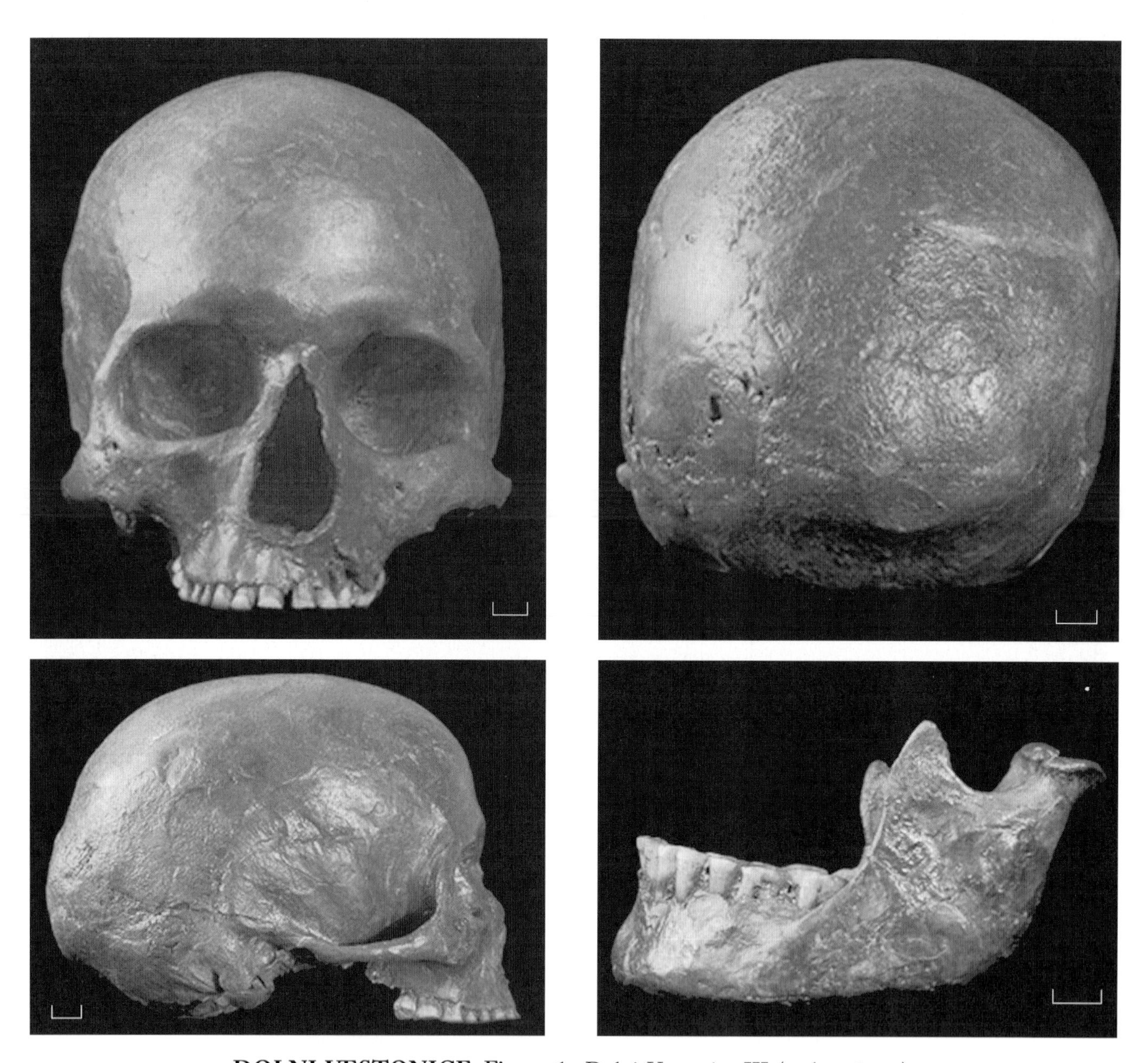

DOLNI VESTONICE Figure 1. Dolni Vestonice III (scale = 1 cm).

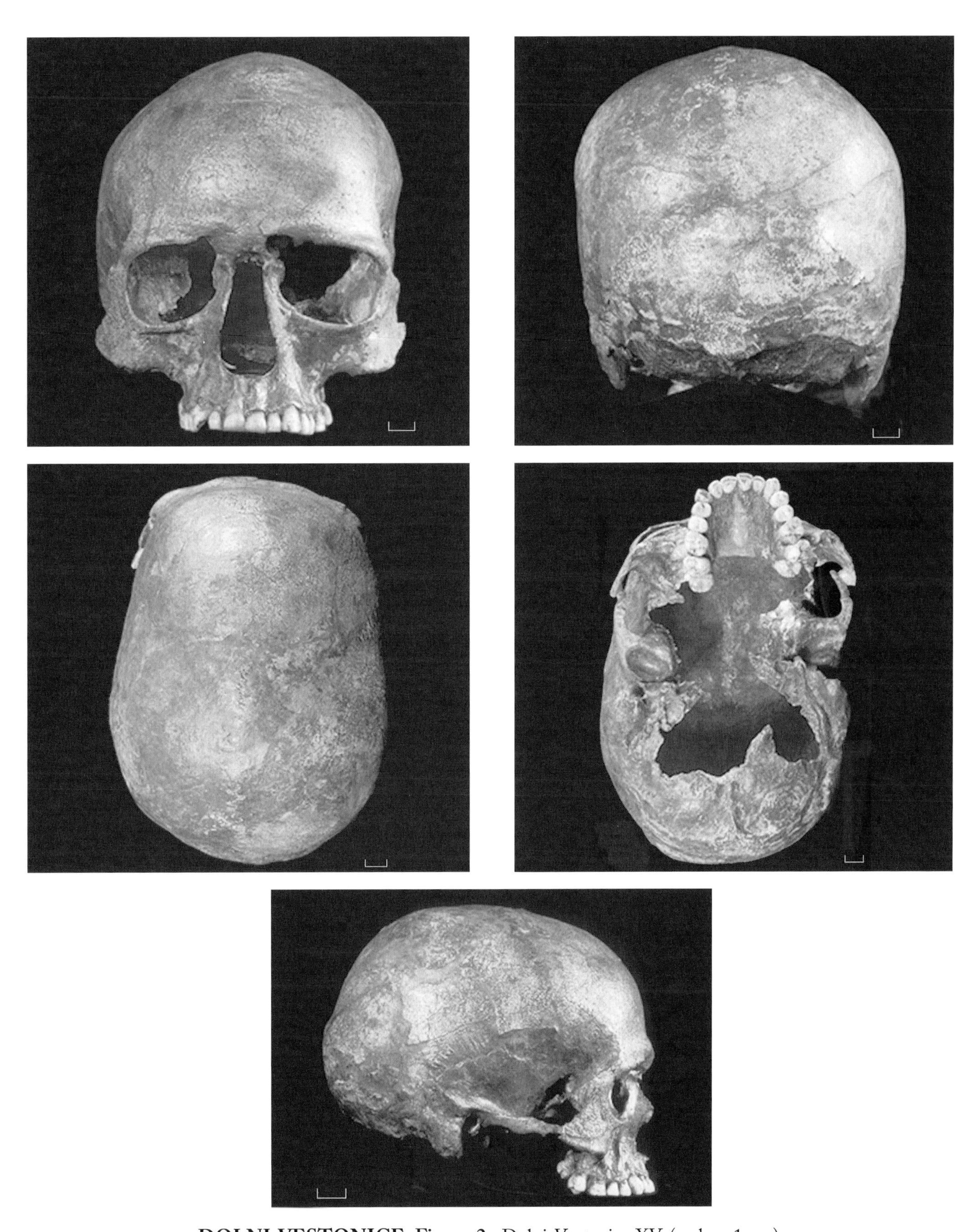

DOLNI VESTONICE Figure 2. Dolni Vestonice XV (scale = 1 cm).

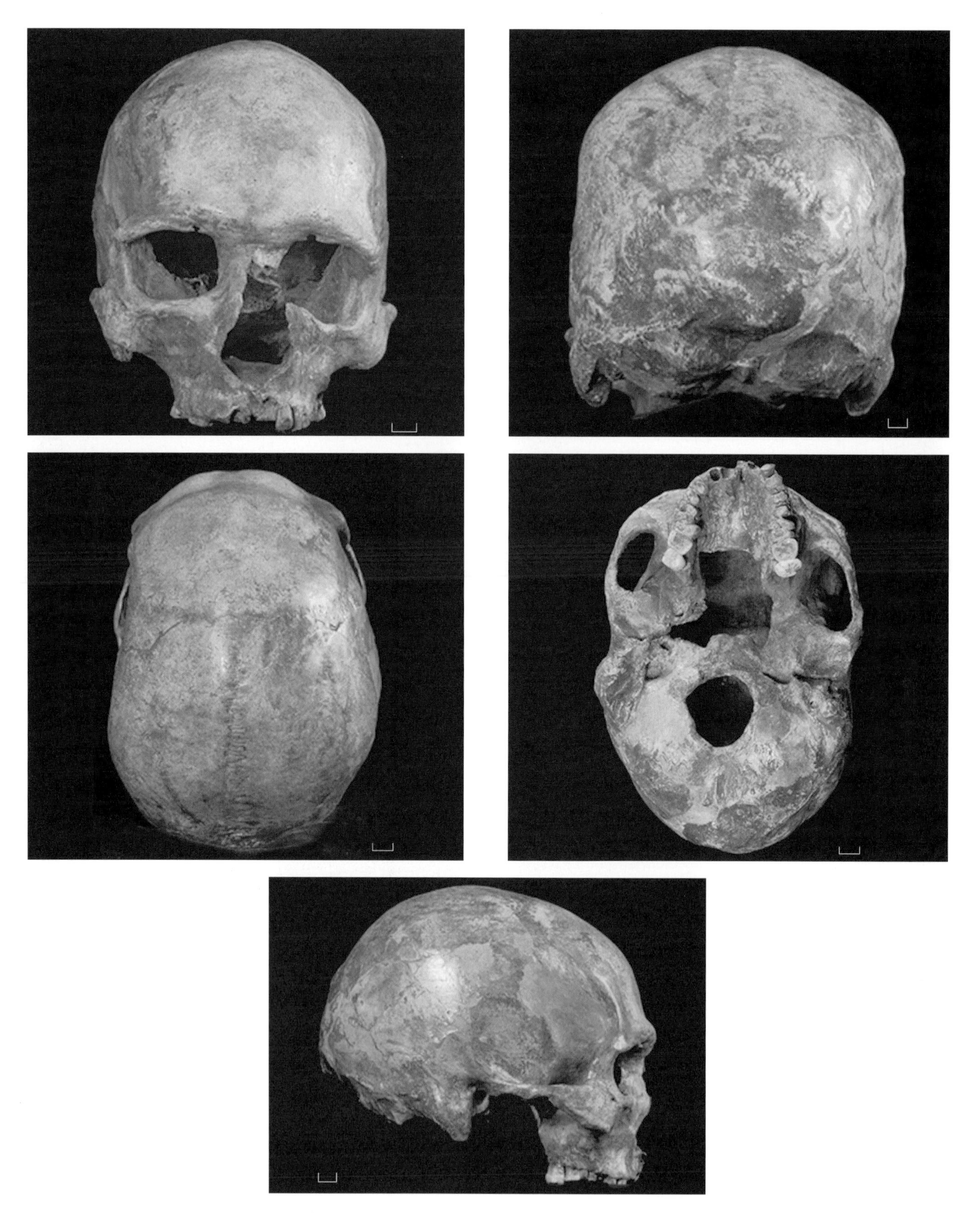

DOLNI VESTONICE Figure 3. Dolni Vestonice XVI (scale = 1 cm).

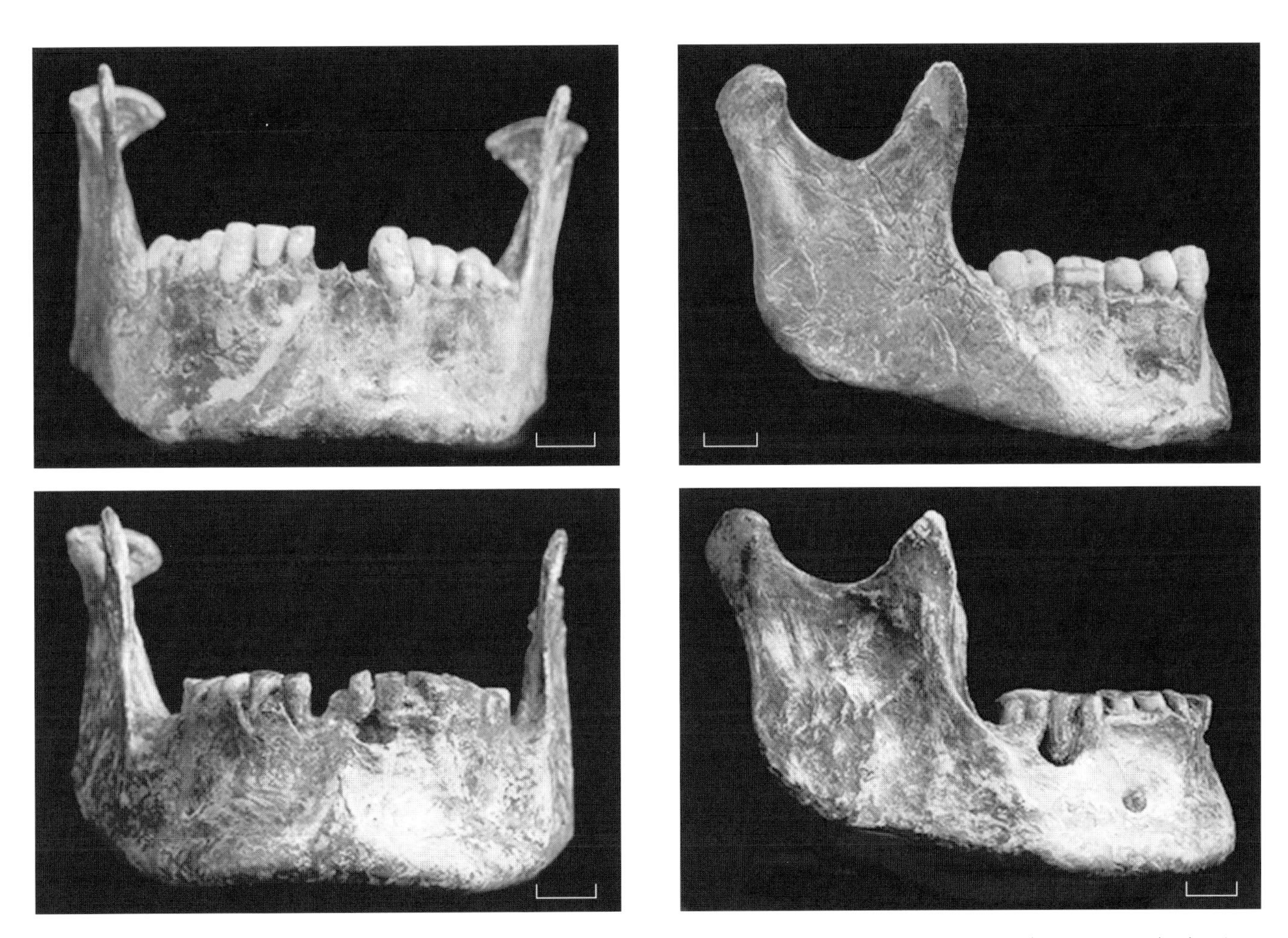

DOLNI VESTONICE Figure 4. Dolni Vestonice XV (top row); Dolni Vestonice XVI (bottom row), (scale = 1 cm).

Ehringsdorf (Weimar-Ehringsdorf)

Location
Fischer's and Kämpfe's quarries, adjacent travertine exploitations in Ehringsdorf, a southern suburb of Weimar, eastern Germany.

Discovery
H. Fischer, E. Lindig, and others, 1908–1925.

Material
About 35 fragments in all, including several cranial pieces, two mandibles, and various postcranial bones. According to Vlcek (1993) nine individuals are represented in this assemblage. Much of the calvaria of one individual (H) can be reconstructed, and there are parietal fragments from at least four others (A, C, B, D), as well as a partial right sphenoid that may belong with individual H. The two partial mandibles, an adult (F) and a juvenile (G), do not appear to be associated with any other skull fragments, although several postcranial elements of individual G are preserved.

Dating and Stratigraphic Context
All Ehringsdorf hominid specimens come from the 15-m-thick Lower Travertine, whose fauna indicates fully interglacial conditions. Which interglacial has, however, been actively debated. A fairly recent analysis of the Lower Travertine and its biota suggested that their traditional assignment to the last interglacial (stage 5e) was appropriate (Steiner, 1979). However, there has always been a suspicion that these deposits were older, and U-series dates by Brunnacker et al. (1983) and Blackwell and Schwarcz (1986) point to the penultimate interglacial, ca. 230 ka (stage 7). This conclusion is supported by ESR dates obtained by Grün et al. (1988), and by the microfaunal studies of von Heinrich (1981).

Archaeological Context
Open-air site adjacent to a warm spring. Numerous lithics have come from the Lower Travertine, mostly bifacially worked scrapers and points (Feustel, in Vlcek 1993). Most workers have identified this industry as a Mousterian variant (e.g., McBurney, 1950; Behm-Blanke, 1960; Feustel, 1983). Numerous hearths have been identified in the lower and middle parts of the Lower Travertine.

Previous Descriptions and Analyses
The Ehringsdorf hominids were described at various times, and the H calotte has been reconstructed in a number of different ways. The most important early studies were those of Virchow (e.g., 1920), and of Weidenreich (in Weidenreich et al., 1928), who reconstructed the calotte. The former found the mandibles to be of Neanderthal type, whereas the latter judged the cranium to be intermediate between Neanderthal and modern in form. The most recent and exhaustive study is that of Vlcek (1993), who concluded that the Ehringsdorf H braincase is of essentially modern architecture, certainly closer to moderns (and to Steinheim and Swanscombe) than to Nean-

derthals. Others, however, would incline more toward the judgment of Smith (1984), who found the most extensive similarities to lie with the Neanderthals. Olivier and Tissier (1975) estimated the cranial capacity of the most complete skull at 1450 cc.

MORPHOLOGY

Given the nature of the sample, elements are described separately, with possible associations suggested.

H1 1024/69

Large frontal fragment with most of coronal suture. Very little of supraorbital margin except tiny medial portion on the L; from medial point of L orbit laterally, all bone missing; tiny lateral portion preserved on the R. Bone moderately thick. Interorbital region was apparently relatively broad. Glabellar region projected anteriorly in front of supraorbital margins, which, judging from the L, would have angled back when viewed from above. With long, fairly low frontal crest held vertical, frontal rises sharply from moderate distance behind orbital margins. Bone describes low, prominent anterior dome, with good curvature in coronal and sagittal planes, and continues rising quite steeply to coronal suture (suggesting profile of braincase was quite high vaulted).

Glabellar region appears to have been flattish; as preserved on the L, supraorbital margin confluent with (i.e., continuous across) it. Supraorbital margins appear not to have been very tall; may have tapered a bit laterally. Missing bone of glabellar region exposes frontal sinuses, which rose to base of frontal dome, i.e., moderately far, but may not have extended laterally as far as midline of orbits. Posteriormost part of R temporal ridge preserved; is quite low, may have faded out posteriorly. Internally, as judged from the R, frontal lobes did not extend fully along length of orbital cone although did extend quite far. Preserved stretches of coronal suture poorly denticulated; suture was probably unsegmented.

This frontal bone fragment probably associated with similarly mineralized L frontal fragment H2 1025/69, R parietal H8 1031/69, and the two parietal pieces associated with the latter, H7 1030/69 and H6 1029/69.

In same box are three parietal fragments, two from the R, one from the L; they apparently belonged to three different individuals. R parietal Ehr.C 1005/69 apparently not associated with frontal above; is fairly complete, its surface rather weathered. The bone thickens medially; is characterized by very dense diploe with thin tabular bone; appears undistorted. Small, relatively short. Bears moderate sagittal curvature, but very strong coronal curvature (it becomes almost vertical as it descends toward squamosal suture). Is thus less "tent shaped" in coronal profile than other two parietal fragments. Widest point on skull appears to have been quite posteriorly and inferiorly placed. Temporal line faint and low on side of skull. Sagittal suture not deeply denticulate or segmented (sinuous curve suggests that suture contained large Wormian bone). Internally, grooves for meningeal vasculature deeply incised, especially posterior branches.

Ehr. H2 1025/69

Part of L frontal including lateral supraorbital margin. Bone quite eroded and surficially damaged. Supraorbital margin very thin s/i, with long superior plane behind it before very steep rise of frontal, which appears to have been quite flattish from side to side. Roof of orbit angles sharply into anterior surface of supraorbital margin. Rather stout temporal ridge emerges from high up behind zygomatic process of frontal and runs quite vertically up. Interiorly, frontal lobe apparently extended just over halfway forward over orbital cone.

Ehr. H3 1026/69

L temporal, almost complete. Squama partially broken. Bone appears to have been short and moderately tall, with tightly arced sutural margin. Posterior root of zygomatic arch originates over middle part of very small, subcircular, anteriorly oriented auditory meatus; diverges strongly from lateral cranial wall. Ectotympanic tube relatively short. Broken, stout vaginal process runs along midline of tube, incorporating quite medially placed styloid process. Posterior extent of tube appressed to base of rather slender, triangular, downwardly pointed mastoid process; vaginal process remains well separated from latter. Large styloid foramen lies laterally and slightly posterior to styloid process. Area medial to mastoid process missing. Mandibular fossa long and very wide from side to side; medial aspect longest a/p; central portion rather deep. Fossa bounded anteriorly by articular eminence that slopes gently anteriorly to descend to level of inferior margin of auditory meatus. Fossa not closed off medially. Suprameatal crest flows back from posterior root of zygomatic arch; reaches minimum salience over auditory meatus and expands posteriorly into tubercle-like supramastoid crest. Lateral surface of mastoid process quite smooth. Parietal notch deep, very vertical; lies over posterior margin of mastoid process. Parietomastoid suture would have been relatively long and horizontal. Internally, long, quite anteriorly tapering petrosal very wide; superiorly, bears two very low swellings, one over region of superior semicircular canal, the other quite far lateral to it. Fairly large superior petrous sinus extends along most of length of petrosal, which bears small depression in region of subarcuate fossa. Only trace of sigmoid sinus.

Ehr. H8 1031/69

R parietal; crushed, somewhat reconstructed; appears to run from just behind coronal suture to lambdoid suture. Surface weathered, especially interiorly. Very thin boned. Was not long a/p; was quite tall s/i and quite strongly curved in coronal plane. A/p curvature difficult to determine. Superior sagittal sinus shallow. Very little indication of meningeal arborization.

H7 1030/69

Portion of L parietal including stretches of coronal and sagittal suture; has match posteriorly with H6 1029/69; together these more or less articulate with H8. When these pieces are held together, appears that superior part of lambdoid suture was broadly arced and did not peak at lambda. On all bones, preserved stretches of coronoid and lambdoid suture are shallowly denticulated but probably undifferentiated.

H5 1028/69, 1001/69, and 1002/69

Various uninformative fragments of cranial bone; 1001 and 1002 relatively thick, with very dense diploic layer sandwiched between very thin tabular layers.

H4 1027/69

Fragment of thin-boned L parietal; shows oblique coronal suture and quite strong side-to-side curvature. Could go with either H7, which goes with H6, or H5 (posterior portion of L parietal with parietal contribution to parietal notch), which is plausibly associated with temporal H3.

H9 1032/69

Large fragment of occipital; probably also goes with above. From below, medial portion quite flat across; only begins to arc forward quite far laterally. Occipital plane strongly arced in midsagittal plane; appears to have been quite broad relative to its height. As this plane curves around inferiorly, moderately large depressions lie bilaterally on either side of midline. Below them, and defined only by them, lies relatively horizontal "occipital torus." Laterally, as seen on the L, "torus" begins to arc down; it fades out rapidly after this point. Below "torus" are two large, shallow, scallop-shaped depressions bilaterally on either side of midline. Below them again, another set of deeper bilateral depressions, which end at posterior margin of foramen magnum. This fragment of margin is very tightly curved, suggesting either elliptical or ovoid foramen. Internally, occipital lobes were fairly small but well impressed on bone. Internal occipital protuberance low, very broad. Ridge along emplacement of transverse sinuses higher on the R than on the L.

Ehr.D 1006/69 (orse E (P) 3/61)

Smaller, more heavily permineralized fragment of R parietal; has "tent-shaped" coronal profile, with strong eminence marking angle between its superior and lateral surfaces. Maximum width would have been posterior, well above squamous suture. Temporal lines low on skull, crossing eminences. Internally, meningeal grooves faint.

Ehr.B 1003/69 (orse E (P) 1/61)

Part of L parietal; thicker (especially at parietal eminence), more heavily permineralized than above. Has somewhat "tent-shaped" coronal profile but less vertical lateral wall. Maximum width would have been posterior, well above squamous suture. Appears occipital plane was wide relative to height and lambdoid suture arced across lambda. Internally, large meningeal grooves well incised. These two specimens probably do not belong to the same individual.

1039/69

Fragment of what appears to be R sphenoid, with tiny part of central body, anterior part of middle fossa, posterior orbital wall, and alisphenoid. Sphenoid sinus was apparently confined to body. Gracility, coloration, mineralization suggest that this specimen is associated with the gracile individual.

Ehr.F 1009/69

Partial mandible missing L and part of R ramus, as well as R incisors; all teeth very worn. Very narrow, with corpora diverging only slightly. Preserved subalveolar bone projects far in front of preserved inferior region of symphysis. Inferiorly, symphysis only lightly curved from to side to side, as well as in sagittal plane. When placed on flat surface, inferior margin elevated relative to corpora behind. Digastric fossae very laterally wide and deep. Bone of region below missing RIs greatly remodeled; is depressed relative to region below it. Slight mound above inferior margin of symphysis slightly offset to the L. Postincisal plane very long, oblique; delimited below by fairly pronounced, bilaterally asymmetrical fossae, below which, in midline, lie faint genial scars. Internally, moderately developed mylohyoid lines; substantial submandibular fossae below. On the L, huge, single mental foramen lies below M1; on the R, large mandibular foramen and small one lie in same position. Appears that lingual surface of LC was somewhat excavated. On P1, mesial crest runs from side of protoconid to base of lingual swelling, from which is separated by crease. RM3 lies well in front of root of largely missing ramus. M3s small; hypoconulid-bearing M1–2 subequal in size.

Ehr.G(?) 1010/69

Partial mandible missing most of R side, part of L corpus and ramus. M3 in crypt; M2 just erupted. All incisors in; RC well erupted; RP1 in course of eruption. Corpus light boned. Symphyseal region appears to have been relatively broad, somewhat arced from side to side; rather vertical in profile, with only slight overhang of incisor alveolar region. On the R, digastric fossa weathered away; was probably wide (given width of mandible). Postincisal plane moderately long and steeply descending; is delineated below by shallow depression. Inferior margin was also somewhat elevated. Gonial margin cut off and somewhat deflected inward; bears huge medial pterygoid tubercle. Coronoid process may have risen above condyle; is very long a/p at its base. Maximum depth of sigmoid notch lies close to base of missing condyle. Sigmoid notch crest would have run to midline of condyle. Low, broad pillar runs down internal midline of coronoid process and flows into internal alveolar crest. I2s, especially, have well-developed lingual tubercles. Lingual surface of rather short RC is shallowly concave. Buccal surface of RP1 bulges out; low, very well-developed tubercle lingually; anteriorly, a crest comes down to just below tubercle's base; distally, a stouter crest runs to apex; both crests delineated from tubercle by vertical fissures. M2–3 enamel deeply invaginated. M2 has distinct trigonid basin and large hypoconulid just buccal to midline; hypoconid extends into center of talonid basin (even more so on M3).

References

Behm-Blanke, G. 1960. Altsteinzeitliche Rastplätze im Travertingebiet Taubach, Weimar, Ehringsdorf. *Alt Thüringen* 4: 151–200.

Blackwell, B. and H. Schwarcz, 1986. U-series analysis of the Lower Travertine at Ehringsdorf, DDR. *Quat. Res.* 25: 215–222.

Brunnacker, K. et al. 1983. Radiometrische Untersuchungen zur Datierung mitteleuropäischer Travertinvorkommen. *Ethnogr.-Archäol. Z.* 24: 217–266.

Feustel, 1983. Zur Zeitlichkeit und kulturelle Stellung des Paläolithikums von Weimar-Ehringsdorf. *Alt Thüringen* 19: 16–42.

Grün, R. et al. 1988. ESR dating of spring-deposited travertines. *Quat. Sci. Rev.* 7: 429–432.

McBurney, C. 1950. The geographical study of the older Palaeolithic stages in Europe. *Proc. Prehist. Soc.* 16: 163–183.

Olivier, G. and H. Tissier. 1975. Determination of cranial capacity in fossil men. *Am. J. Phys. Anthropol.* 43: 353–362.

Smith, F. 1984. Fossil hominids from the Upper Pleistocene of central Europe and the Origin of Modern Europeans. In: F. Smith and F. Spencer (eds), *The Origins of Modern Humans.* New York, Alan R. Liss, pp. 137–209.

Steiner, W. 1979. *Der Travertin von Ehringsdorf und Seine Fossilien.* Wittenberg, Neue Brehm-Bücherei.

Vlcek, E. 1993. *Fossile Menschenfunde von Weimar-Ehringsdorf.* Stuttgart, Konrad Theiss Verlag.

Virchow, H. 1920. *Die Menschlichen Skeletreste aus dem Kämpfe'schen Bruch im Travertin von Ehringsdorf bei Weimar.* Jena, Gustav Fischer.

von Heinrich, W.-D. 1981. Zur stratigraphischen Stellung der Wirbeltierfaunen aus den Travertinfundstätten von Weimar-Ehringsdorf und Taubach in Thüringen. *Z. Geol. Wiss.* 9: 1031–1055.

Weidenreich, F., F. Wiegers and E. Schuster. 1928. *Der Schädelfunde von Weimar-Ehringsdorf.* Jena, Gustav Fischer.

Repository

Museum für Vor und Frühgeschichte Thüringens, Weimar, Germany.

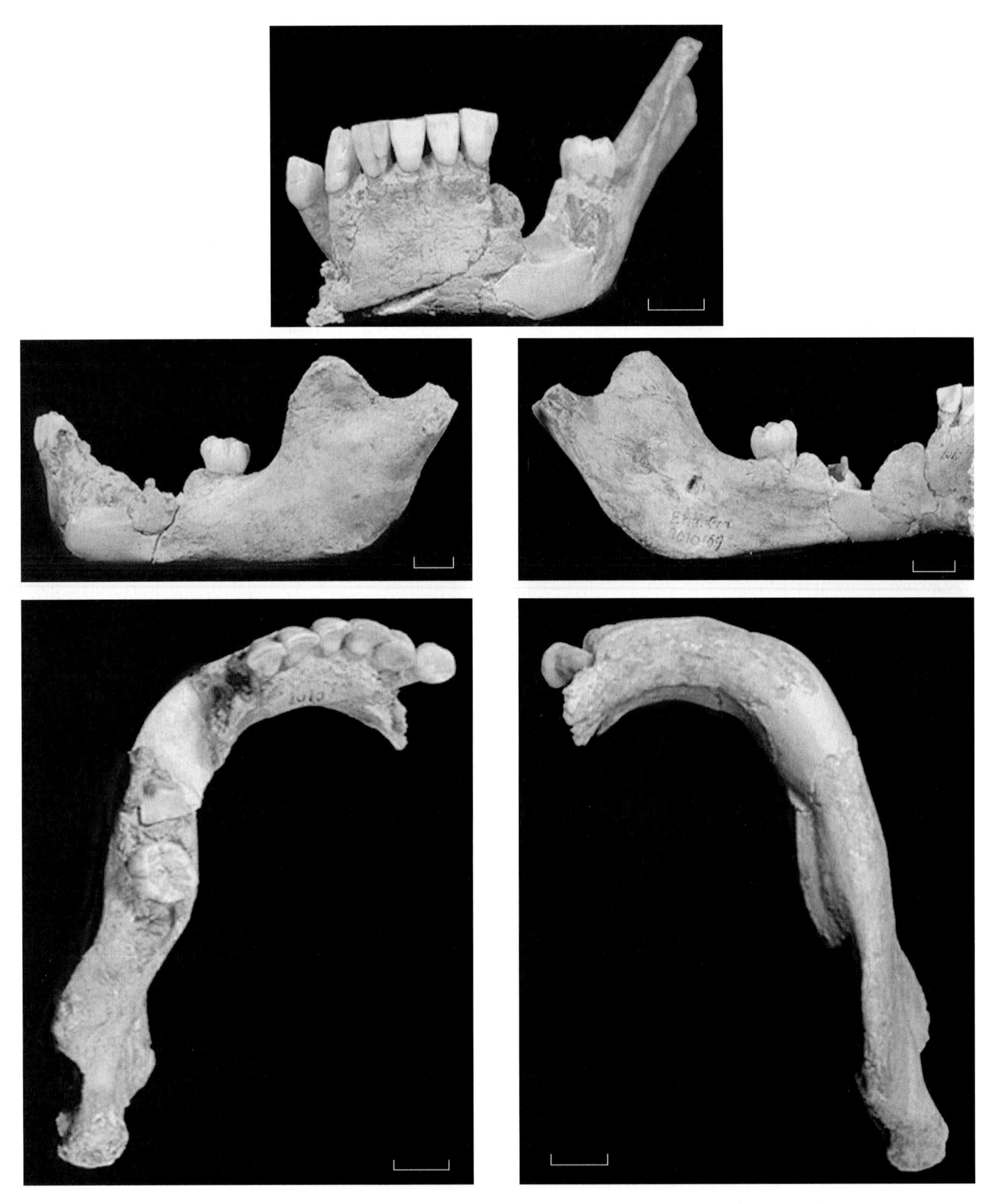

EHRINGSDORF Figure 1. Ehr. 1010/69 (scale = 1 cm).

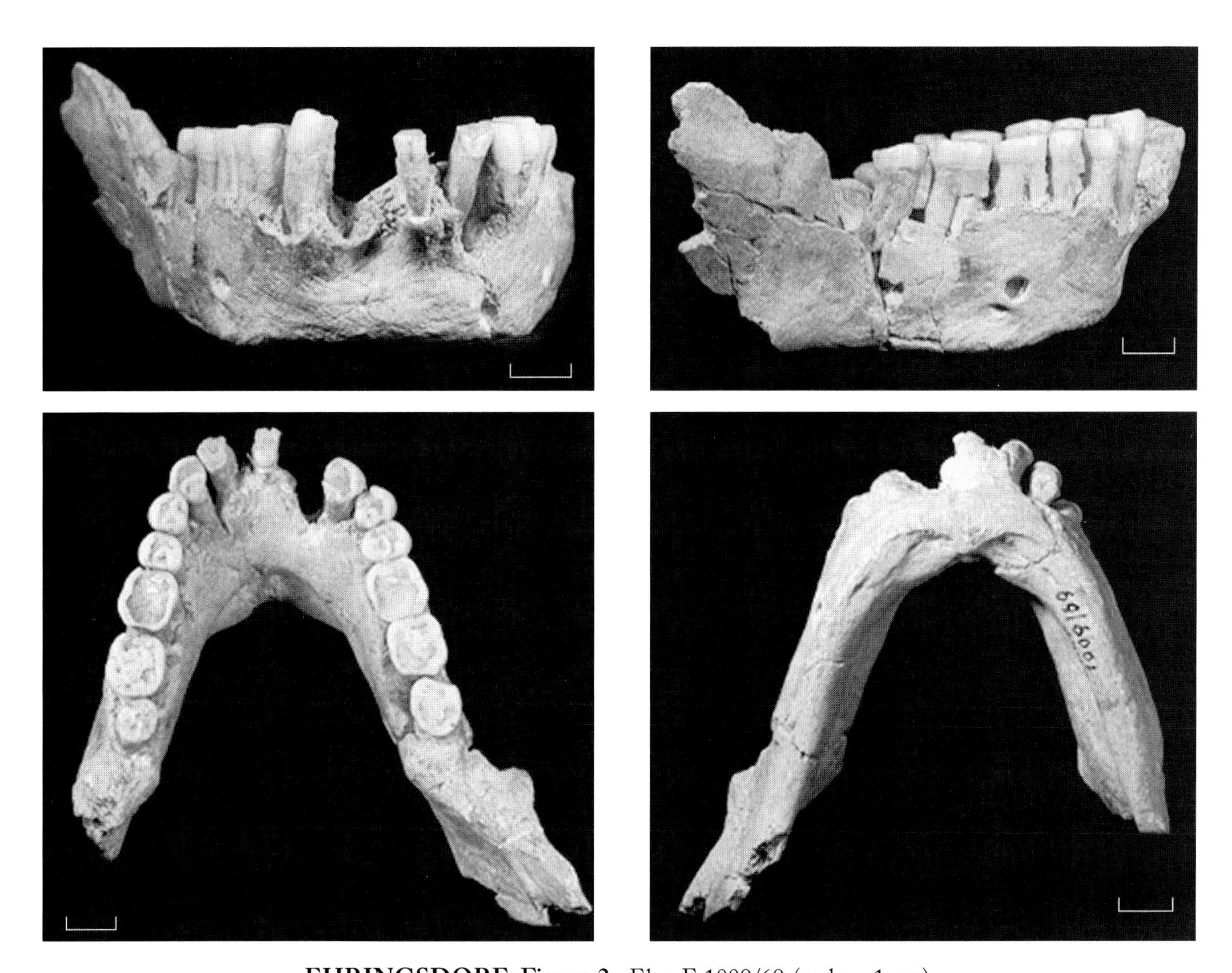

EHRINGSDORF Figure 2. Ehr. F 1009/69 (scale = 1 cm).

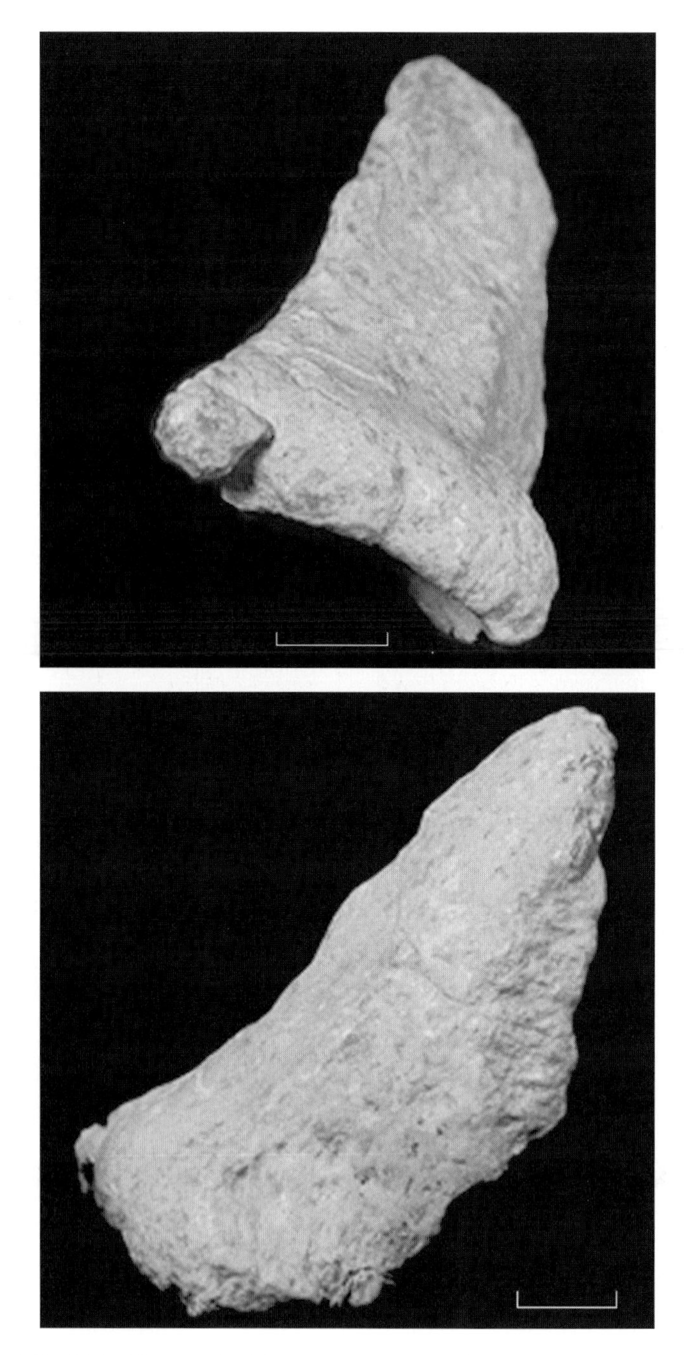

EHRINGSDORF Figure 3. Ehr. H2 1025/69 frontal fragment (scale = 1 cm).

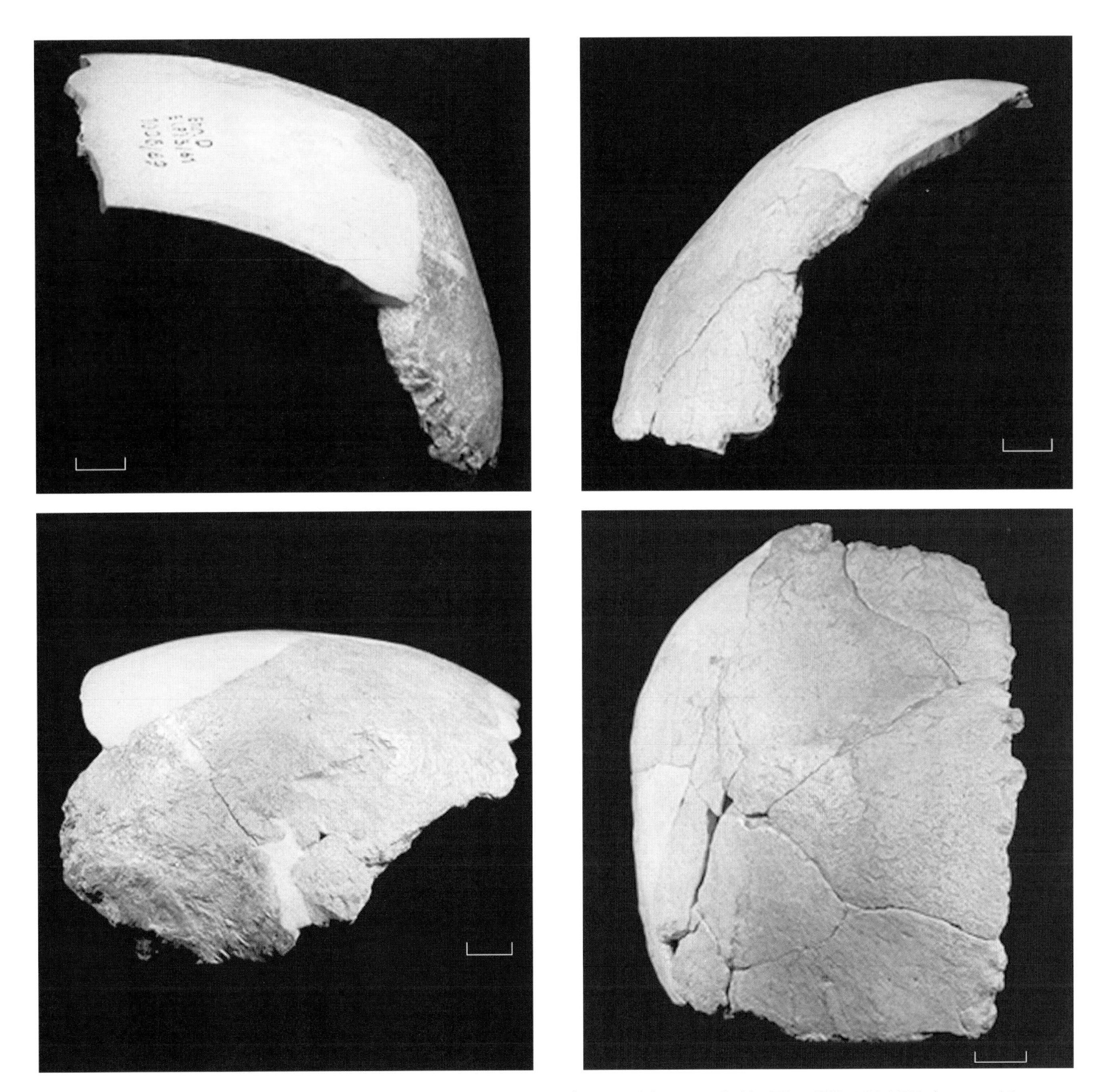

EHRINGSDORF Figure 4. Parietals: Ehr. D 1006/69 (top and bottom left); Ehr. H8 1031/69 (top and bottom right), (scale = 1 cm).

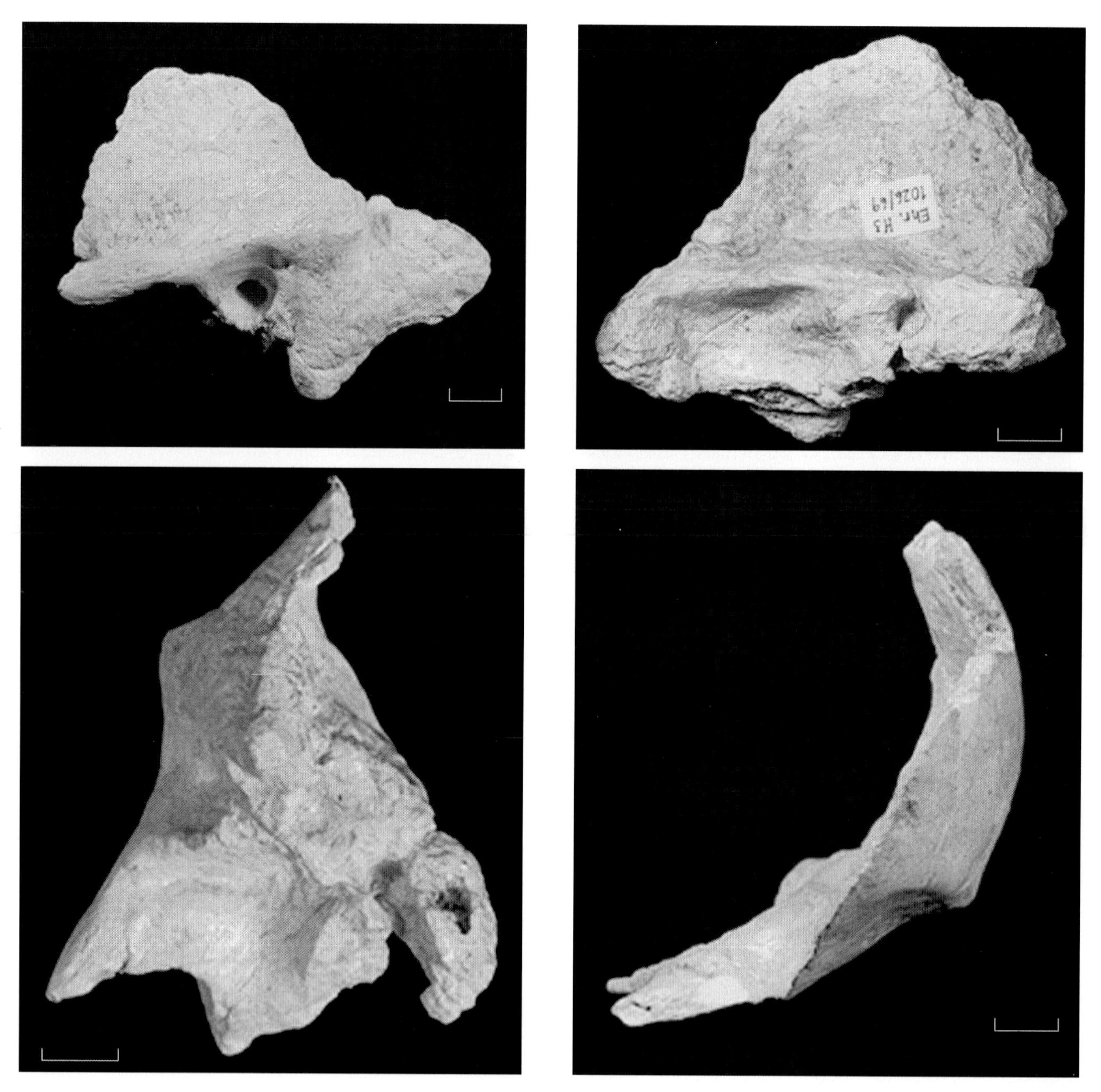

EHRINGSDORF Figure 5. Ehr. H3 1026/69 temporal (top row and bottom left); Ehr. H9 1032/69 occipital (bottom right), (scale = 1 cm).

Engis

Location
Former cave site of Awirs (the "second cave," long since destroyed by mining activities) just N of Engis village, near Liège, Belgium.

Discovery
Excavations of P.-C. Schmerling, 1829 (Engis 1–3); excavations of E. Dupont, 1872 (Engis 4).

Material
Almost complete adult neurocranium (Engis 1); partial child's cranium with separated maxilla (Engis 2); minor cranial and postcranial fragments, now missing (Engis 3); ulna (Engis 4).

Dating and Stratigraphic Context
The hominids were recovered from stratified cave deposits. C. Fraipont (1936) summarized the results of successive excavations by Schmerling (1829–30), E. Dupont (1872) and J. Fraipont (1885), the last of these being undertaken just before the definitive destruction of the cave by miners. Fraipont showed convincingly that the Engis 2 child's cranium was contemporary with the Mousterian level 3 "upper ossiferous layer" in which it was found; the Engis 1 adult calvaria, on the other hand, appears to have been buried into the underlying Mousterian from a higher level. According to C. Fraipont this specimen is Aurignacian ("Brno race"), an attribution supported by the presence of a variably thick Aurignacian layer (level 2) overlying the Mousterian (C. Fraipont, 1936); alternatively, however, it may date from the latest Pleistocene, a conclusion supported by its condition of preservation and the associated fauna (E. Poty, personal communication).

Archaeological Context
Mousterian (Engis 2); Upper Paleolithic (Engis 1).

Previous Descriptions and Analyses
As early as 1833 Schmerling clearly recognized that he had evidence from Engis of truly ancient humans. Not until C. Fraipont's review in 1936, however, was it recognized that these humans were of two kinds. The Engis 1 calvaria, the adult, is clearly that of a modern *Homo sapiens*; but Fraipont demonstrated that the Engis 2 child's cranium was that of a Neanderthal, a conclusion subsequently supported by other authors, including Tillier (1983), who affirmed this allocation on the basis of extensive comparisons with other juvenile Neanderthals. The cranial capacity of Engis 2 is approximately 1362 cc (Fenart and Empereur-Buisson, 1970).

Morphology

Engis 1
Adult. Almost complete neurocranium, missing R temporal, parts of R parietal, face, nasal bones, ethmoid, sphenoid, basiocciput, part of R squamosal.

Skull moderately long, somewhat domed. Frontal eminences low and moderately prominent. Parietal eminences marked. Superior to glabellar "butterfly," frontal rises steeply to level of eminences, where, in profile, it curves back strongly, then arcs downward yet more strongly at point just behind moderately marked parietal eminences. This steep, slightly posteriorly oriented profile continues into the occipital, which reaches its posteriormost extent just above the low, horizontal, crestlike external occipital protuberance.

Brow bipartite with quite prominent glabellar "butterfly" (glabella is most swollen part). Frontal sinus medially situated. From medially placed supraorbital notches, the flat, backwardly directed lateral potion undercuts "butterfly" wing. Orbital roofs form blunt but rather acute angle with lateral part of supraorbital area. Minimal expansion of the ethmoidal sinuses into wide interorbital region. Frontal crest huge.

Squamosal was moderately long and its superior margin gently arced. Parietal notch lies over midline of anteriorly directed mastoid process, which is broader at its base than at its blunt, anteriorly pointing tip. Although distended downward greatly, tip of mastoid process lies noticeably below preserved mastoid crest. These two structures define a narrow, relatively shallow mastoid notch, which lacks digastric fossa behind. Lateral surface of mastoid process very rugose; relatively small air cells visible through broken tip of R process. Moderately long parietomastoid suture relatively horizontal. On the R, preserved portion of suprameatal crest low and blunt, flows posteriorly into upwardly curving supramastoid crest of similar proportions.

Tubular ectotympanic fused to base of mastoid process. Auditory meatus relatively small and circular. Vaginal process peaks around relatively thin, somewhat laterally placed styloid process, at base of which lies small stylomastoid foramen; vaginal process contacts base of mastoid process. Lateral to styloid process, vaginal process broken; evidently extended entire length of ectotympanic tube. Jugular fossa/foramen appears to have been quite large. Moderately sized, backwardly directed carotid foramen lies quite far laterally on petrosal.

Occipital bone tall, narrow, triangular; bulges slightly from side to side. External occipital protuberance present.

As preserved on the R, petrosal only moderately wide internally; bears low, distinct arcuate eminence, internal to which runs deep superior petrous sinus. Lateral to arcuate eminence, superior petrosal surface flat. Internal auditory meatus very large and long and anteromedially directed. Subarcuate fossa fully filled in. Sigmoid sinus tall and curved. Posterior fossae greatly excavated for cerebellar lobes. Foramen magnum was probably small.

All cranial vault sutures noticeably segmented; coronal suture most finely and sagittal suture most deeply denticulated. Lambdoid suture rises steeply from asterion, peaks at lambda. Nuchal plane curves gently into posterior margin of foramen magnum. Superior nuchal line bow shaped.

Engis 2

Partial cranium with separated maxilla, plus some isolated upper and lower teeth. Missing most of L side of skull (from orbit into parietal region), nasal bones, posterior nasal cavity, ethmoids and sphenoids, parts of occipital, and R parietal and temporal. Auditory and mastoid characters suggest an alleged age of 3–4 years; small foramen of Huschke present bilaterally.

Cranium relatively long, with quite vertical frontal and broad, flattish top. Cranial bone very smooth, showing pitting only in suprainiac region. Braincase quite broad posteriorly and ovoid in posterior profile. Facial skeleton narrow compared to more posterior regions of skull; would have tapered inferiorly. Relatively steep frontal bears eminence on preserved R side; posterior to this, plane of frontal becomes more horizontal and relatively flat. This relatively planar surface continues into parietal region. Regions of parietal eminences quite broadly swollen. Below eminences, braincase tapers inferiorly (maximum width quite high up on cranial wall).

Interorbital region broad, with faint beginnings of supraorbital and glabellar swelling that diminishes toward lateral margin of orbit. Missing nasal bones would have been broad across region of nasion; their exposed articulation with the frontal is very long superoinferiorly. No frontal sinus development; already small air cells along ethmofrontal articulation. R orbit preserved sufficiently to indicate that it was ovoid and very tall but perhaps narrower inferiorly. Frontal process of zygoma preserved on the R; external surface oriented quite laterally, corresponding with narrow, forwardly oriented posterior zygomatic arch root.

Mastoid processes just becoming distinct (more so on the R than on the L); inferiormost extent lies well above small paramastoid crest, medial to which is more pronounced occipitomastoid crest. Area of mastoid notch effectively an almost vertical sulcus occupying space between mastoid process and paramastoid crest. Preserved R jugular foramen oriented anteriorly. On both sides, medially placed, rather small carotid foramina point almost directly backward. Styloid pits quite medially placed. Vaginal processes run along incompletely ossified floor of ectotympanic tube; peak above styloid pits. Rather small stylomastoid foramina situated quite lateral, slightly posterior to these pits.

R mandibular fossa quite wide and fairly well excavated, with long anterior slope (but not distinct articular eminence). Neither the wide, well-developed postglenoid plate nor the stubby medial articular tubercle contacts ossifying ectotympanic tube. Posterior root of zygomatic arch originates well in front of incomplete auditory meatus, above which is low, upwardly curving crest that does not continue over mastoid region behind it. Preserved on lateral surfaces of both mastoid processes is roughened mastoid crest that runs almost vertically down toward process's tip. Parietal notch lies above midline of mastoid process. Long parietomastoid suture relatively horizontal. As preserved on the R, squamosal portion of temporal moderately long; its sutural margin quite tall and arcuate. No remnant of sphenoid. Preserved portions of coronal, sagittal, and lambdoid sutures relatively finely and uniformly denticulate.

Posterior to their eminences, relatively flat parietals project somewhat posteriorly as they run towards lambda, below which occipital continues the posterior projection. Just below posteriormost extent of occipital is preserved an area of what would have been a relatively large, elliptical, grossly pitted suprainiac depression. Below this depression is slight undercutting along superior nuchal line, inferior to which the slightly muscle-scarred nuchal plane runs fairly directly anteriorly.

Foramen magnum long and elliptical. Occipital condyles contained entirely on lateral parts of occiput, which were not fused completely to basiocciput. Basiocciput broad; flexes upward somewhat relative to plane of foramen magnum. On both sides, anterior lambdoid suture long and relatively horizontal although sinuous. Lambdoid suture proper rises rather steeply, then runs more horizontally to form poorly defined peak at lambda.

Internally, petrosals quite wide; bear distinct, moderately domed arcuate eminence, lateral to which is another swelling of superior surface. Subarcuate fossa closed over on the L but represented by thin fissure on the R. Neither side exhibits superior petrous sinus.

R and L partial maxillae with complete L frontal process. Anterior nasal spines bulky, projecting, separate. Nasal aperture incomplete; may have been broad at base and superiorly. L lateral crest preserved; is crisp and flows inferiorly around to become lateral margin of the anterior nasal spine. Also on the L, spinal crest takes origin at base of anterior nasal spine, courses laterally and internally, diverging slightly from lateral crest (creates narrow prenasal fossa laterally). Frontal process preserved superiorly; is quite long a/p. L nasomaxillary suture slopes gently forward and down. Distinct medial projection on nasal cavity side of frontal process, whose surface, especially anteriorly, is quite strongly pitted. Superior nasal crest, emerging from medial projection, broken in its superior extremity. Lacrimal groove covered only minimally by nasal cavity extension of lacrimal crest. No evidence indicating articulation of inferior nasal concha over lacrimal groove. Maxillary sinus extends only as far forward as lacrimal groove; is very tall in area of frontal process; medial wall broken far anteriorly and inferiorly; medial and lateral walls diverge strongly inferiorly (suggesting at least some encroachment into nasal cavity). Floor of nasal cavity descends quite steeply behind anterior nasal spines. Nasoalveolar clivus tall and vertical. Front of maxillae strongly arced from side to side.

Preserved, erupted upper teeth: Rdi1, Ldi2, R and Ldc; visible through breaks in bone, crowns of LI1-2 and P1 and RP1. Isolated teeth include R upper and lower dm1, dm2, and M1. dis worn; quite shoveled with relatively strong margocristae and moderate lingual tubercles. Distal edge of di2 strongly truncated. dc crowns broadly triangular in outline; low down on lingual side, moderately sized crista surrounds base of tooth, also subtends small anterior and posterior foveae. Upper dm1 has centrally placed cusps with bulbous sides. Strong preprotocrista arcs mesially and buccally up side of paracone; weaker postprotocrista courses to small, buccolingually compressed metacone from which stout postcingulum courses lingually before turning to run straight up side of protocone. Cingula enclose large and deep basins. Buccal cusps of upper Rdm2 more compressed and peripherally placed; lingual cusps more bulbous and their apices more centrally placed. Metacone almost as large as paracone. Postprotocristae and postcingula enclose large, somewhat crenulated basins. Hypocone large.

Small pit mesially and another distally at base of protocone. Upper M1 cusps bulkier, puffier than on upper dm2. Upper M1 has greatly swollen, distolingually distended hypocone; preprotocrista that encloses small basin that lies mesial to bases of protocone and paracone; stout, segmented, postprotocrista that fills in middle of crown; deep fissure on the lingual side that separates hypocone from metacone; and protocone that bears small pit mesially on its base.

Rdm_1 somewhat exodaenodont on buccal side; enamel extends low down over anterior root. Tiny protoconid rather centrally placed and slightly mesial to level of metaconid. Well-defined paracristid descends from protoconid to base of metaconid, enclosing quite deep, narrow basin. Cristid obliqua runs directly forward from somewhat compressed hypoconid to protoconid; hypoconid not connected to small entoconid (which leaves the long, deep talonid basin open distally). Rdm_2 basally bulbous, broadly ovoid in occlusal outline with slightly internally placed cusp apices. Protoconid and metaconid lie opposite one another; low, stout paracristid in front of them encircles deep, narrow basin. Short cristid obliqua runs directly between protoconid and hypoconid. Small hypoconulid at base of hypoconid not connected to anteriorly placed entoconid (creating long, deep talonid basin that opens distally). RM_1 more bulbous on sides and more subcircular tooth than dm_2; bears relatively larger, slightly more centrally placed hypoconulid that, along with enamel wrinkling, occludes talonid basin. Centroconid-like island of enamel in middle of talonid basin; anterior to it is small conulid-like structure filling in trigonid basin.

REFERENCES

Dupont, E. 1872. Sur une nouvelle exploration des cavernes d'Engis. *Bull. Acad. R. Sci.* 33: 504–510.

Fenart, R. and R. Empereur-Buisson. 1970. Application de la methóde 'vestibulaire' d'orientation au crânes neanderthaliens. *Arch. Inst. Paleontol. Mem. Hum.* 33: 89–148.

Fraipont, C. 1936. Les hommes fossiles d'Engis. *Arch. Inst. Paleontol. Hum.* 16: 1–52.

Fraipont, J. 1885. Nouvelle exploration des cavernes d'Engis. *Ann. Soc. Géol. Belg.* 12: 187–191.

Schmerling, C.-P. 1833. *Recherches sur les ossemens fossiles découvertes dans les cavernes de la Province de Liège.* Liège, P.-J. Collardin.

Tillier, A.-M. 1983. Le crâne d'enfant d'Engis 2: Un exemple de distribution des caractères juvéniles, primitifs et néanderthaliens. *Bull. Soc. R. Belg. Anthropol. Préhist.* 94: 51–75.

Repository

Service de Paléontologie Animale et Humaine, Université de Liège, Allée du 6 Août, Bâtiment B18, 4000 Liège (Sart-Tilman), Belgium.

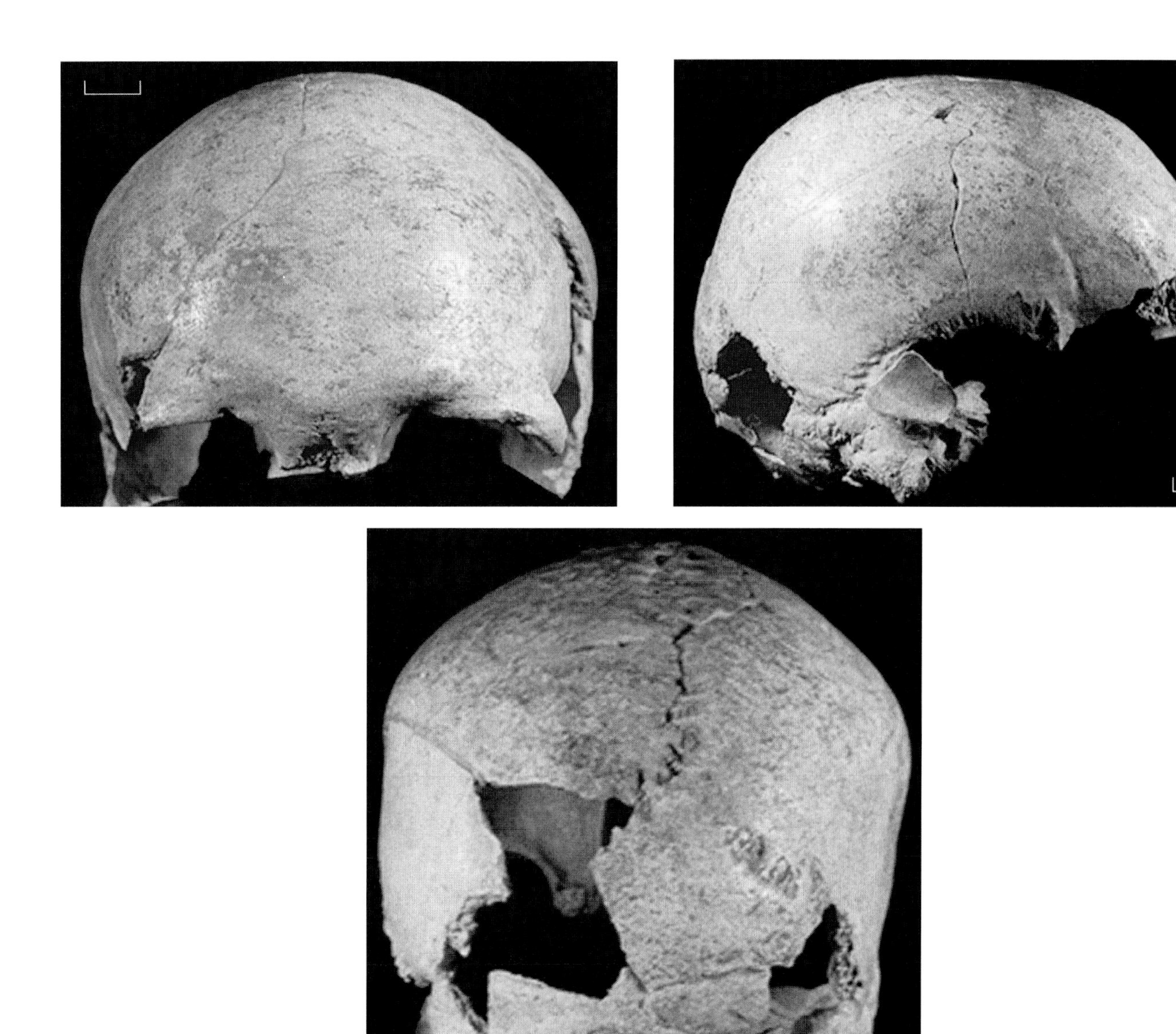

ENGIS Figure 1. Engis 1 (scale = 1 cm).

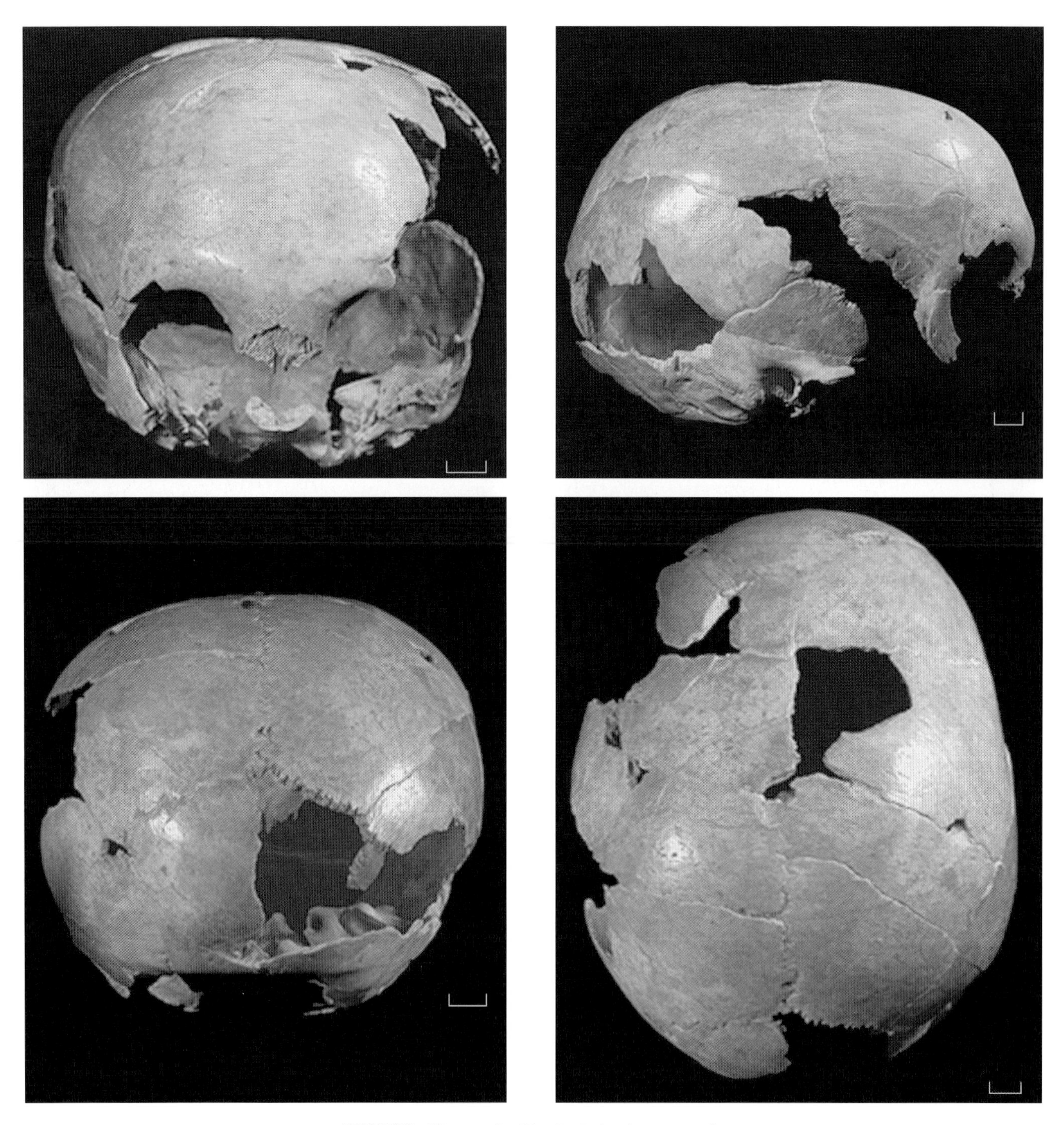

ENGIS Figure 2. Engis 2 (scale = 1 cm).

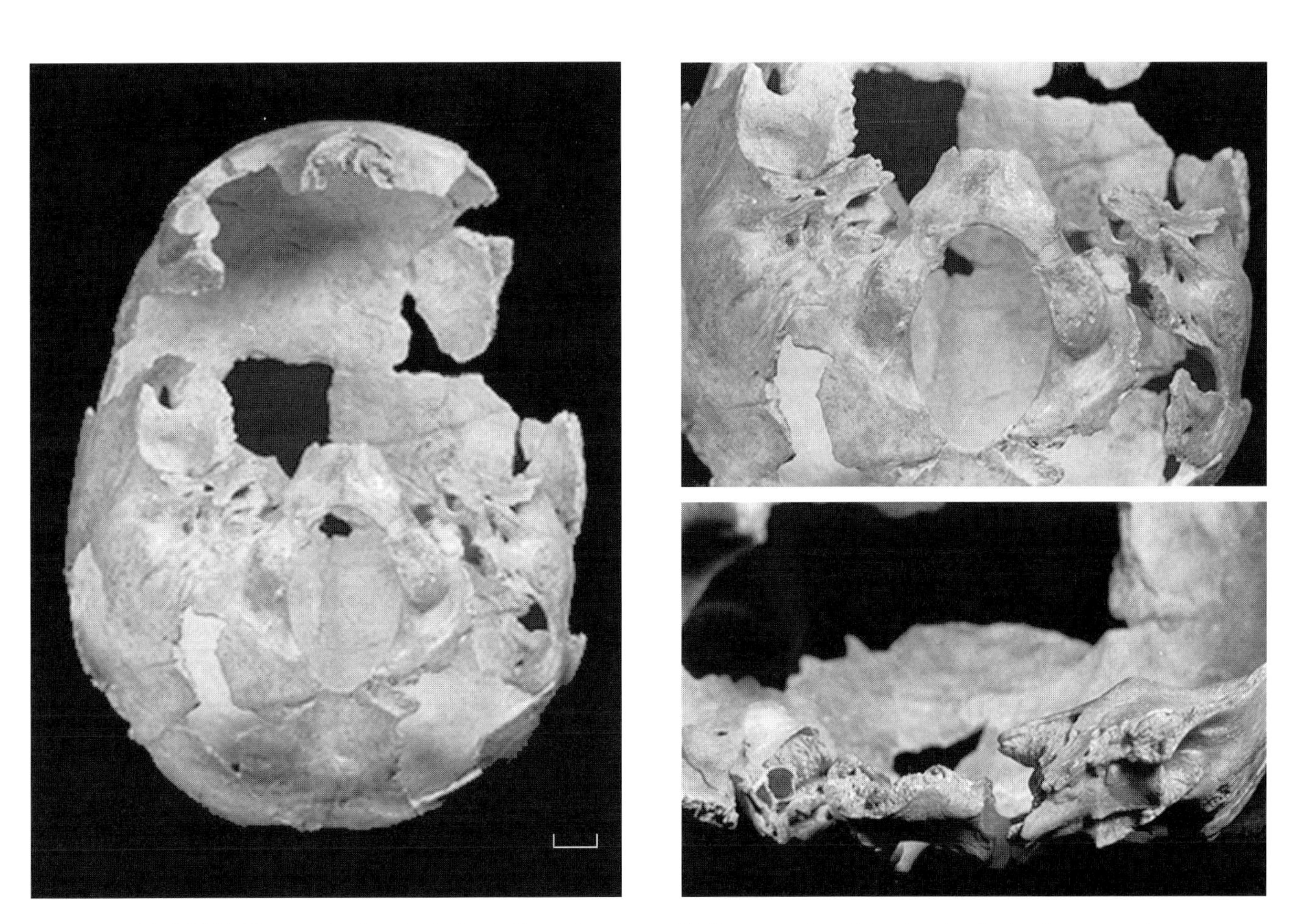

ENGIS Figure 3. Engis 2 (including close-up of left petrosal and basiocciput), (scale = 1 cm).

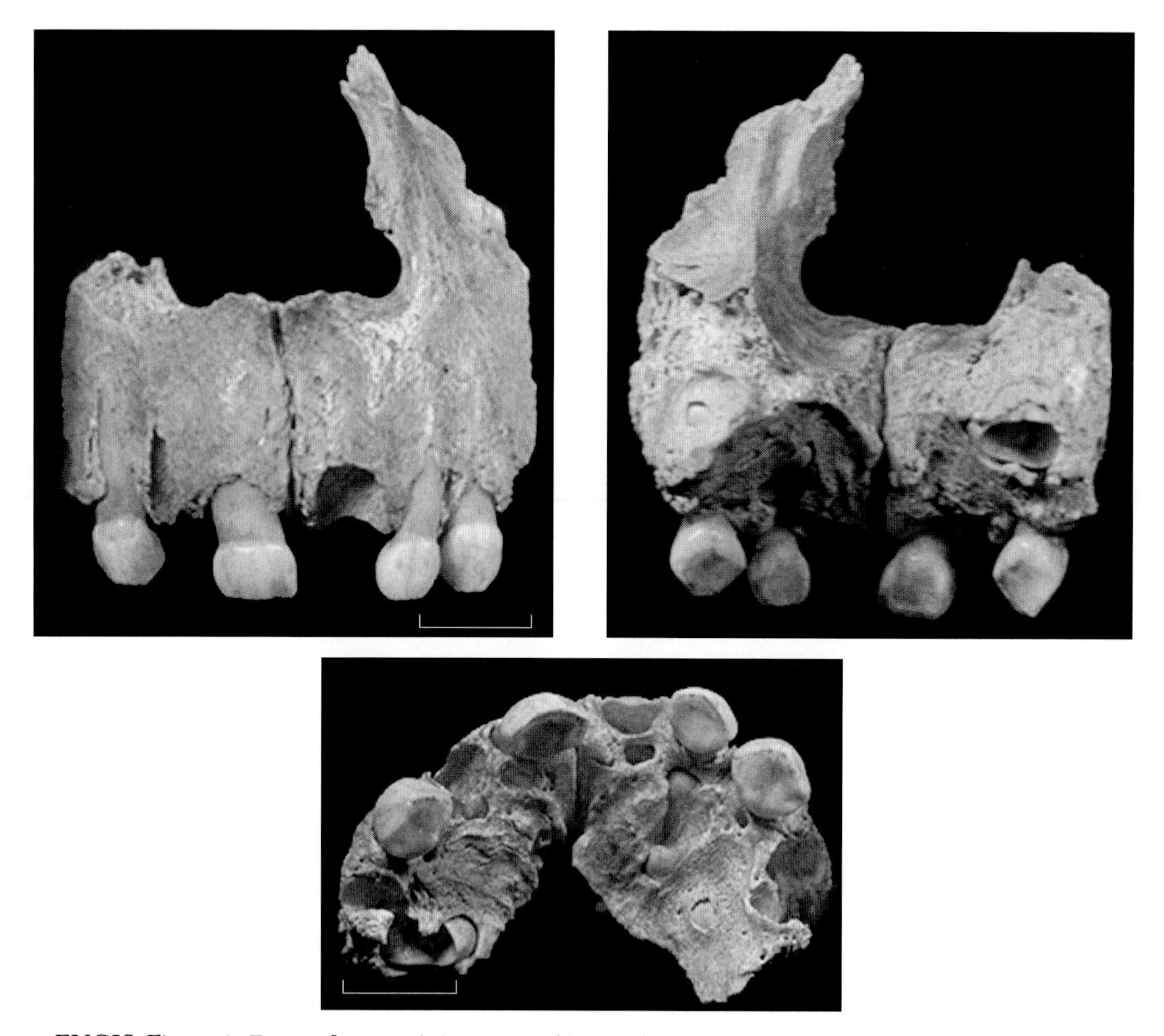

ENGIS Figure 4. Engis 2 [note medial projection (damaged) on left side of nasal cavity wall], (scale = 1 cm).

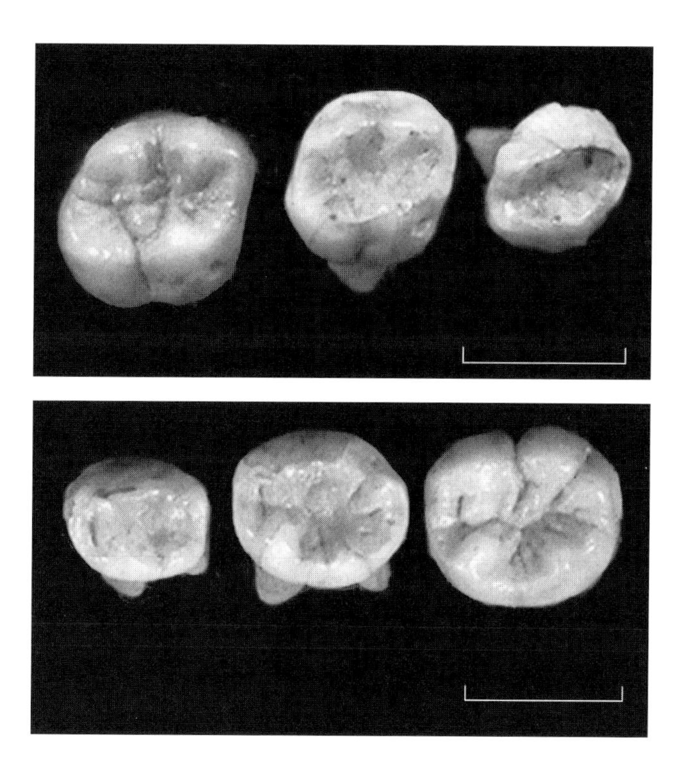

ENGIS Figure 5. Engis 2: Rdm^1-M^1 (top); Rdm$_1$-M$_1$ (bottom), (scale = 1 cm).

Feldhofer Grotto (Neanderthal)

Location

Limestone cave, now destroyed by mining, in the valley of the Düssel river between the towns of Erkrath and Mettmann, some 12 km E of Düsseldorf, Germany.

Discovery

Miners, August 1856. Identified by C. Fuhlrott.

Material

Adult calotte and partial postcranial skeleton. Fragments of a second individual (possibly the same one as reported in 1895 and subsequently lost) have been informally reported (Schmitz and Thissen, 1999) but not seen by us.

Dating and Stratigraphic Context

The original skeleton, possibly initially complete, was unearthed by miners in cave deposits that they subsequently destroyed. There is no published associated fauna (although cave bear fossils were found nearby), and as yet no stratigraphic or archaeological context to help with dating. Informal reports of direct ^{14}C AMS dating on fragments of bone from a second Feldhofer Neanderthal individual, recovered from the miners' dumps, indicates an age of around 40 ka (Schmitz and Thissen, 1999).

Archaeological Context

Schmitz and Thissen (1999) recently reported finding stone tools and broken faunal bones in the rediscovered miners' dumps, but details are not yet available.

Previous Descriptions and Analyses

As the first kind of extinct human to be recognized, the Neanderthaler occupies a special place in the history of paleoanthropology. The controversies it unleashed are too well known to require rehearsal here (but see reviews in Trinkaus and Shipman, 1993, and Tattersall, 1995a); however, it must be noted that as early as 1864 King made this specimen the holotype of *Homo neanderthalensis*. It is thus Neanderthal by definition, and it is generally agreed that morphologically it fits well among the western European "classic" Neanderthals of the last glacial (see Stringer and Gamble, 1993; Tattersall, 1995b). The Feldhofer Grotto individual has lately furnished the first mtDNA to be extracted from any extinct human (Krings et al., 1997). Unsurprisingly, this specimen proved to be an outlier relative to all living human populations; moreover, the age of divergence of the Neanderthal lineage was estimated to be four times that of the common ancestor of all living humans. Holloway (1985) calculated the cranial capacity as being approximately 1525 cc.

Morphology

Calotte, plus various postcranial bones. In general, cranial bone not very thick. Cranium long and low, with gradual rise in profile; fairly smooth curve down to suprainiac depression; very little trace of an occipital bulge. Frontal rise most steep immediately behind posttoral region. Posteriorly, slope gentler throughout anterior one-third of sagittal suture; curve then begins an almost equally gentle descent to farthest preserved point (= above suprainiac depression). In coronal profile, skull presents low curve from side to side; reaches maximum width just above level of squamous suture and quite far posteriorly, above parietal notch. Rear profile en bombe. Chignon slight.

Frontal begins rise quite far posterior to supraorbital region (creating posttoral plane, not sulcus). Double-arched supraorbital tori confluent across smooth, broad, only modestly protrusive glabella, which in profile slopes down and in. Tori quite thick superoinferiorly as they emerge from glabella; taper a little beyond midline of orbit. Toral anterior surfaces smoothly rolled. Slight transverse peak along their superior margins, which arc somewhat higher medially than laterally. Roofs of orbits flow smoothly into upward curve of anterior surface of tori. Bone of tori vermiculate and penetrated by numerous small to tiny foramina. Viewed from above, tori angle back slightly from glabella. Laterally on front of R supraciliary arch is an indentation of uncertain origin. At least four supraorbital/frontal foramina lie on each side; on R is huge supraorbital notch. Interorbital region quite broad. Frontal part of frontonasal suture exposed; is quite extensive, tall, and peaked. Nasal bones would have flexed below nasion. Frontal sinuses (as seen especially on the R) extend well up into glabella, as well as posteriorly and laterally at least to midline of orbit.

Postorbital constriction minimal. Modest temporal ridges arise along side of short zygomatic processes of frontal; from these ridges, poorly marked temporal lines run back in shallow curve, with inferior and superior lines diverging posteriorly. Just above widest point of braincase, temporal lines recurve tightly down and forward. Although temporals missing, preserved squamosal suture on parietal indicates that squamosal portion was long, not notably curved superiorly, and that parietomastoid suture (of undetermined length) was horizontal.

Most of nuchal plane of occiput missing. Tall, broad occipital plane preserved above superior nuchal line ("occipital torus"), which undercuts suprainiac depression. "Torus" quite horizontal but dips somewhat in midregion (lowest at inion). Suprainiac depression centrally small, but is continued very laterally by asymmetrical patches (which help define "torus" below) that are also asymmetrical vertically (much more expansive superiorly on R). Surface of depression somewhat roughened throughout. Enough of lambdoid suture preserved to indicate there was horizontal anterior lambdoid suture of some extent; behind, lobate, not segmented lambdoid suture "proper" runs steeply upward to arc broadly across region of lambda.

Internally, frontal crest stout but not salient. Frontal lobes of brain extended no further than midway along tops of orbital cones.

REFERENCES

Holloway, R. L. 1985. The poor brain of *Homo sapiens neanderthalensis*; see what you please. In: E. Delson (ed), *Ancestors; The Hard Evidence.* New York, Alan R. Liss, pp. 319–324.

King, W. 1864. The reputed fossil man of the Neanderthal. *Q. J. Sci.* 1: 88–97.

Krings, M. et al. 1997. Neandertal DNA sequences and the origin of modern humans. *Cell* 90: 19–30.

Oakley, K. 1964. The problem of Man's antiquity. *Bull. Brit. Mus. Nat. Hist. Geol.* 9: 1–65.

Schmitz, R. W. and J. Thissen. 1999. New human remains and the first archaeological finds from the rediscovered site of the Neanderthal type specimen. Abstract, *Workshop on Central and Eastern Europe from 50,000 to 30,000 BP.* Neandertal Museum, Mettmann.

Stringer, C. and C. Gamble. 1993. *In Search of the Neanderthals.* London, Thames and Hudson.

Tattersall, I. 1995a. *The Fossil Trail.* New York, Oxford University Press.

Tattersall, I. 1995b. *The Last Neanderthal.* New York, Macmillan.

Trinkaus, E. and P. Shipman. 1993. *The Neandertals.* New York, Knopf.

Repository

Rheinisches Landesmuseum, Colmanstrasse 15, Bonn, Germany.

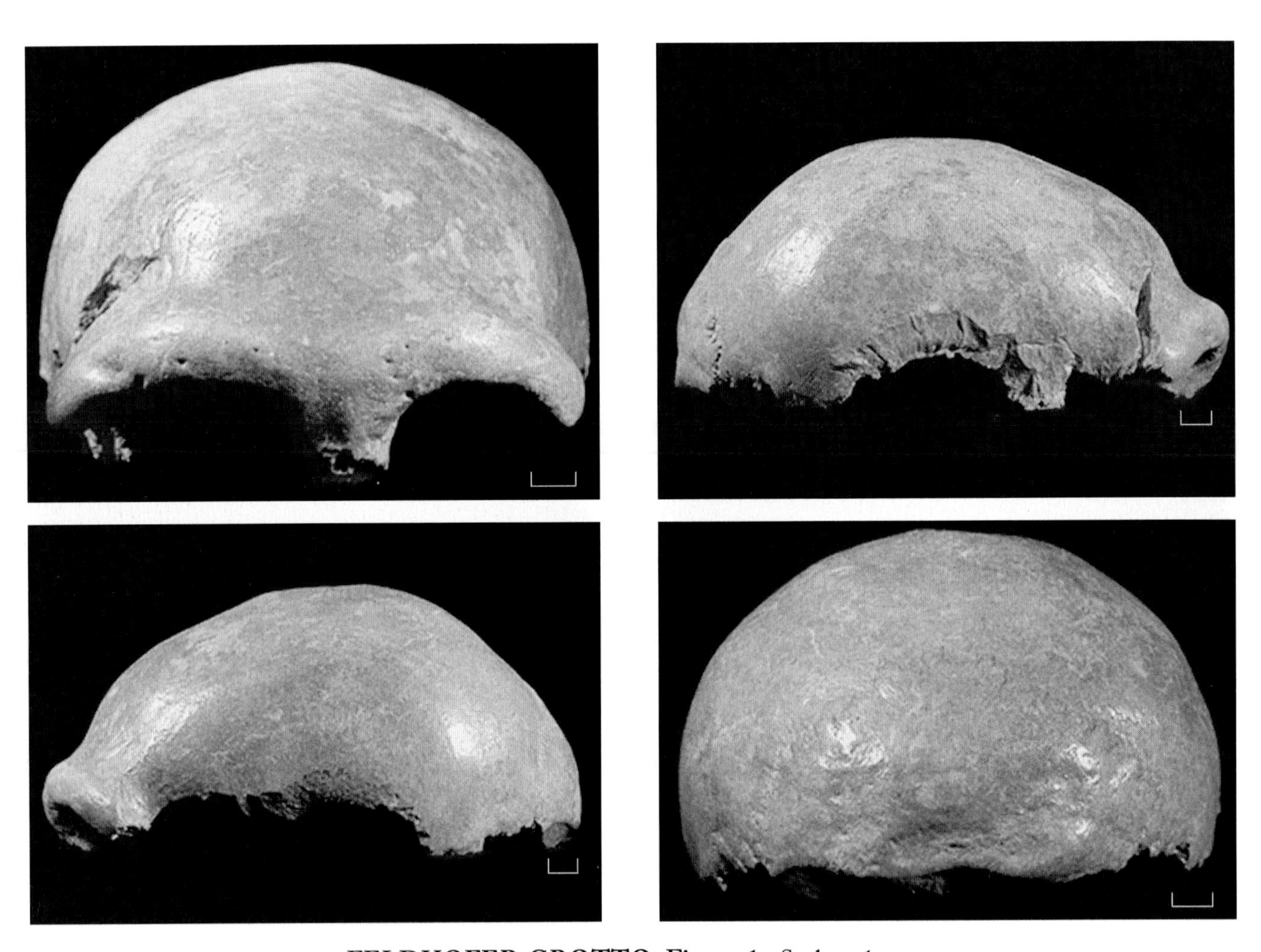

FELDHOFER GROTTO Figure 1. Scale = 1 cm.

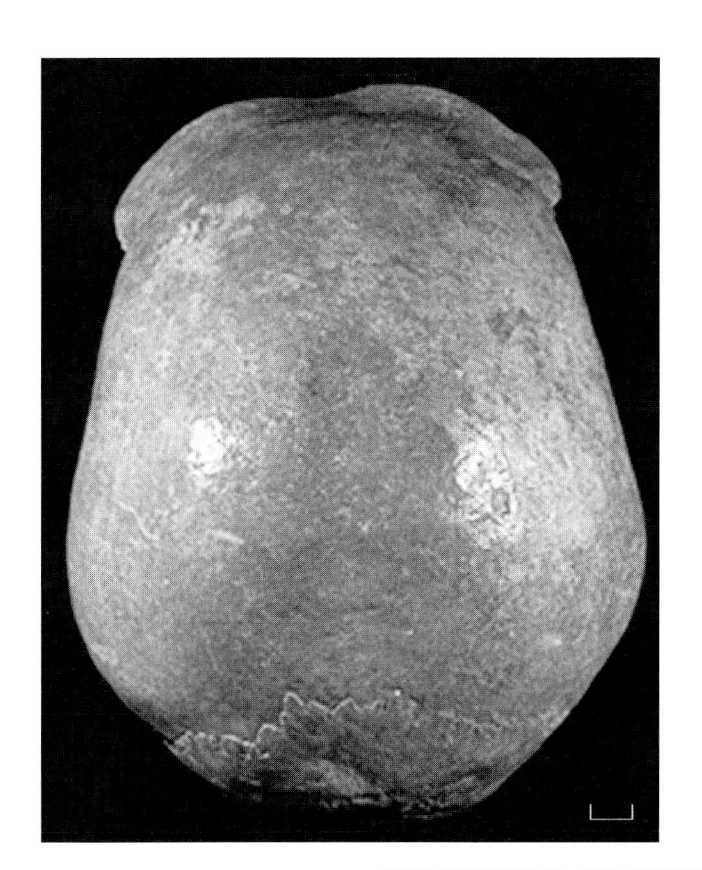

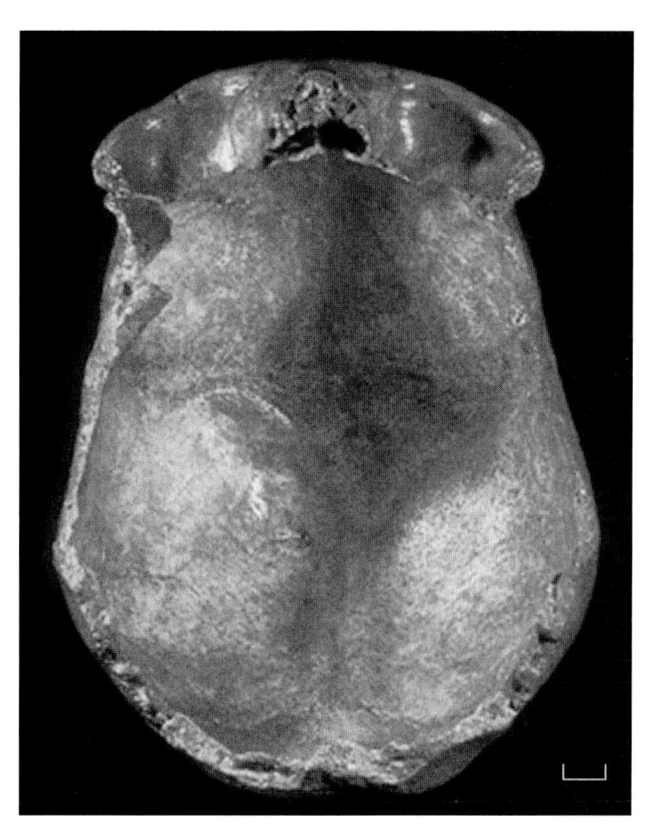

FELDHOFER GROTTO Figure 2. Scale = 1 cm.

Figueira Brava

Location

Limestone cave on the coast of Portugal about 35 km SE of Lisbon, about 400 m SW of the Printho fortress on the Setubal Peninsula.

Discovery

Excavations coordinated by M. Telles Antunes, September 1987 (phalanx) and June 1988 (tooth).

Material

Isolated LP^2; phalanx.

Dating and Stratigraphic Context

The human remains were recovered from Bed 2 of the stratified cave sequence. Radiocarbon date on mollusk shells from Bed 2 is 30.93 ± 0.7 ka, U-series (Th/U) on a cervid tooth from the same deposit is 30.56 ± 10.7 ka (Antunes, 1990–91; Antunes and Cunha, 1992). These dates are mutually supportive.

Archaeological Context

Bed 2 yielded numerous tools ascribed to a denticulate Mousterian tradition. There is no suggestion of burial, and the human remains were found among what appeared to be food refuse (Antunes et al., 2000). Some vertical scratches on the medial and distal surfaces of the tooth appear to represent postmortem modifications (Antunes, 1990–91; Antunes and Cunha, 1992).

Previous Descriptions and Analyses

The human remains from Figueira Brava were first described by Antunes (1990–91), who regarded them as Neanderthal because of their archaeological context. A more recent analysis by Antunes et al. (2000) has confirmed the Neanderthal morphology of the tooth. This finding is rendered of the greatest interest because of the extremely recent date of the fossils.

Morphology

LP^2 buccolingually broad. Single root bears distinct mesial and distal longitudinal grooves; is very long relative to crown height; two distinct apical foramina perforate two low root apices. Crown somewhat worn, particularly on paracone and slightly more so on protocone. These two cusps subequal in size, but paracone may have been taller. Apex of paracone was more peripherally placed than that of metacone, the lingual side of which is somewhat more bulbous. Both paracone and metacone apices worn; when fresh, would not have been much higher. Thus angle formed by inner surfaces of the cusps would have been very open and shallow. Anterior fovea fairly buccolingually wide and deep but narrow. Noticeably larger posterior fovea, which bears some thick crenulation, rather centrally placed, intervening somewhat between junction of bases of cusps. Distal moiety of crown (delineated by axis between the two cusps) larger and more arcuate along its margin than anterior part of tooth. Lingual side of tooth slightly more vertical than the buccal, but both quite steep.

Regions of both mesial and distal interproximal facets bear what appear to be incisions, running down from edges of occlusal surface. It is hard to say what these mean.

References

Antunes, M. 1990–91. O homen da gruta da Figueira Brava (ca. 30,000 BP). *Mem. Acad. Sci. Lisboa* 31: 487–538.

Antunes, M. and A. Santinho Cunha. 1992. Neanderthalian remains from Figueira Brava Cave, Portugal. *Géobios* 25: 681–692.

M. T. Antunes et al. 2000. The latest Neanderthals. In: M. T. Antunes (ed), *Últimos Neandertais em Portugal: Odontologic and Other Evidence.* Lisboa, Academia das Ciéncias de Lisboa, pp. 269–303.

Repository

Centro de Estratigrafia e Paleobiologia, Universidad Nova de Lisboa, 2825 Monte da Caparica, Portugal.

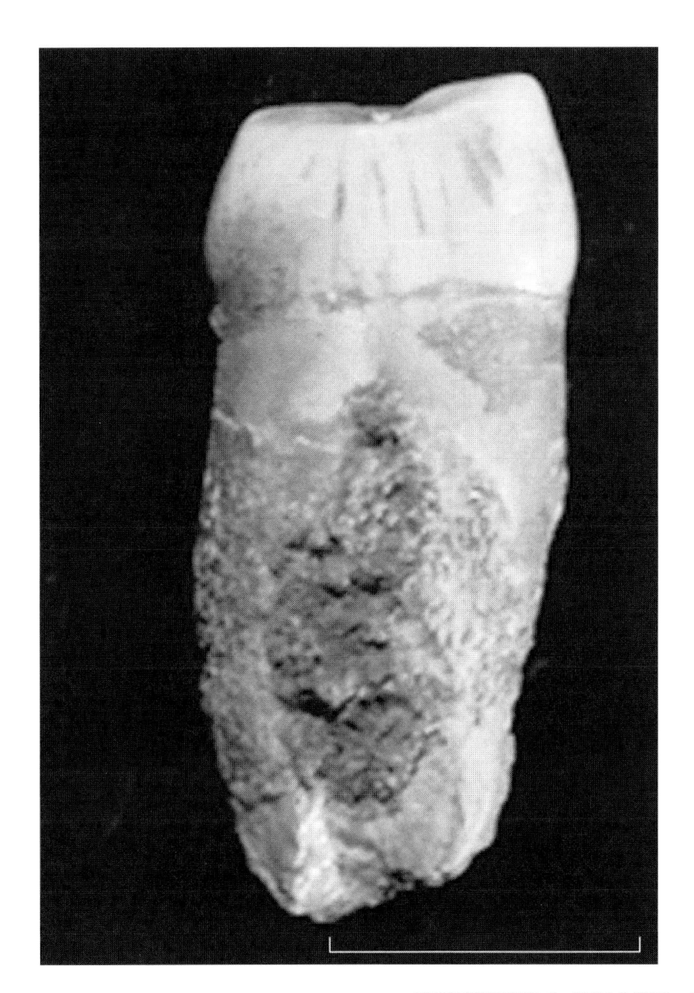

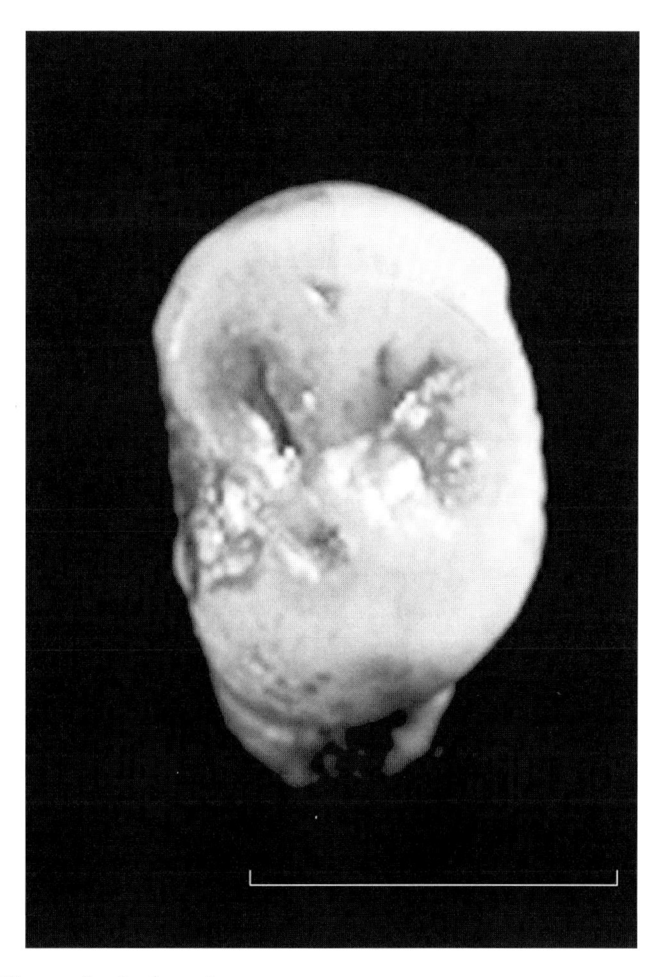

FIGUEIRA BRAVA Figure 1. Scale = 1 cm.

Fontechevade

Location
Cave site in the Tardoire Valley, near Orgedeuil, east of Angoulême, Charente, France.

Discovery
Excavations of G. Henri-Martin, August 1947.

Material
Partial calotte (F1) and probably unassociated frontal fragment (F2) (MNHN/MH 24.448/447–1959).

Dating and Stratigraphic Context
The fossils were found beneath a stalagmitic floor underlying Upper Paleolithic and Middle Paleolithic layers. They are faunally dated to the last interglacial. This age estimate has not been significantly refined since Oakley and Hoskins (1951) used fluorine analysis to demonstrate the contemporaneity of the hominids with the warm to temperate interglacial fauna.

Archaeological Context
Tayacian tools are said to occur right below the stalagmitic layer, apparently in association with the hominid fossils; lower in the section was found a Clactonian industry. McPherron et al. (1999) believe on the basis of new excavations that the cave had a very complex taphonomic history that calls into question the archaeological context of the fossils.

Previous Descriptions and Analyses
Vallois (1949) considered the calotte to resemble *Homo sapiens* and used this specimen to underpin his "presapiens" theory of human evolution, excluding the Neanderthals from modern ancestry. Brace (1964) later argued that the Fontéchevade materials did not represent pre-Neanderthal *Homo sapiens*. More recently, Trinkaus (1973) concluded that preservation was inadequate to determine the affinities of either Fontéchevade specimen, most notably the frontal, which he felt might or might not have borne a frontal torus. Corruccini (1975) used multivariate metrical techniques to deny any special resemblance of the calotte to modern humans. Olivier and Tissier (1975) estimated cranial capacity as 1350 cc.

Morphology
Adult. Calotte (F1), being part of R and most of L parietal, and an indeterminable amount of a frontal bone that is somewhat distorted and eroded. Also several other fragments, most notably part of a frontal containing some of the glabellar region into L upper orbit.

Bone of calotte appears to have been moderately thick. L side better preserved; is long from front to back; was apparently not strongly curved in the a/p axis, but had strong transverse curvature. Damaged but denticulated lambdoid suture rises steeply to lambda, where it is not strongly peaked. At the front, the bone turns downward rather steeply.

Frontal fragment (F2) not eroded; may be some porotic hyperostosis in region of glabella. Bone much less thick than larger piece of calotte and other fragments; is differently textured; also apparent differences in mineralization between it and rest (doubtful association of this fragment with other pieces?). Internally, frontal crest sharp, well defined. R frontal sinus almost completely preserved; was quite tall s/i; apparently did not extend very far laterally. Glabellar region low, essentially undistinguished. In the preserved L part of medial orbital region, very low mound extends laterally from glabellar region. Given possible ways of orienting this region, the preserved frontal could not have sloped backward very sharply. Interorbital space was probably quite broad.

References

Brace, L. 1964. The fate of the "Classic" Neanderthals: A consideration of hominid catastrophism. *Curr. Anthropol.* 5: 1–41.

Corruccini, R. 1975. Metrical analysis of Fontéchevade II. *Am. J. Phys. Anthropol.* 42: 95–97.

McPherron, S. et al. 1999. The Fontéchevade fossils: A reanalysis of their archaeological context based on new excavations. *Am. J. Phys. Anthropol.* Suppl. 28: 199.

Oakley, K. and Hoskins, C. 1951. Application du "Test de la Fluorine" aux crânes de Fontéchevade. *Anthropologie* 55: 239–242.

Olivier, G. and H. Tissier. 1975. Determination of cranial capacity in fossil men. *Am. J. Phys. Anthropol.* 43: 353–362.

Trinkaus, E. 1973. A reconsideration of the Fontéchevade fossils. *Am. J. Phys. Anthropol.* 39: 25–36.

Vallois, H. 1949. L'origine de l'*Homo sapiens. C. R. Acad. Sci. Paris* 228: 949–951.

Repository

Laboratoire d'Anthropologie, Musée de l'Homme, Place Trocadéro, 75116 Paris, France.

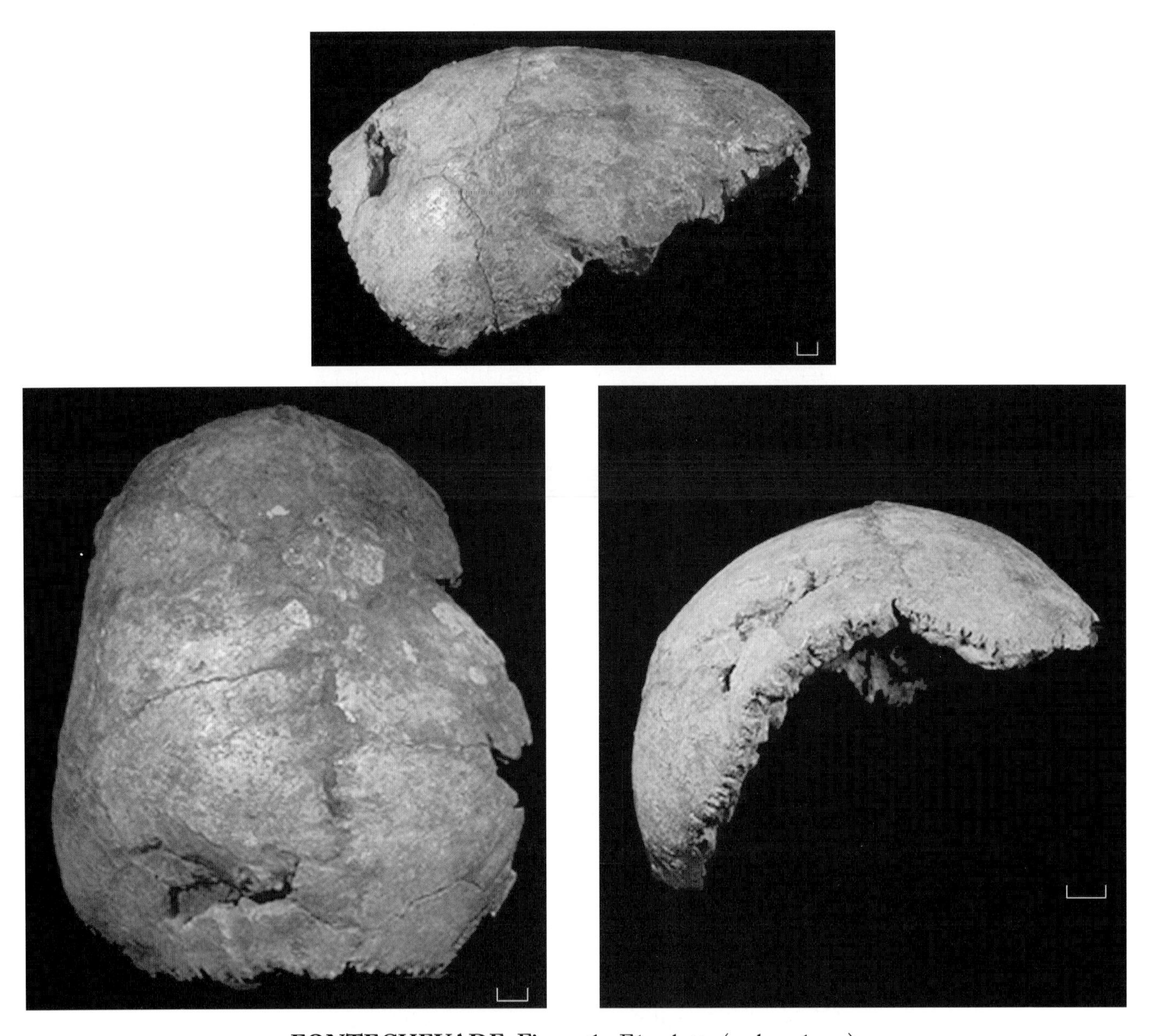

FONTECHEVADE Figure 1. F1 calotte (scale = 1 cm).

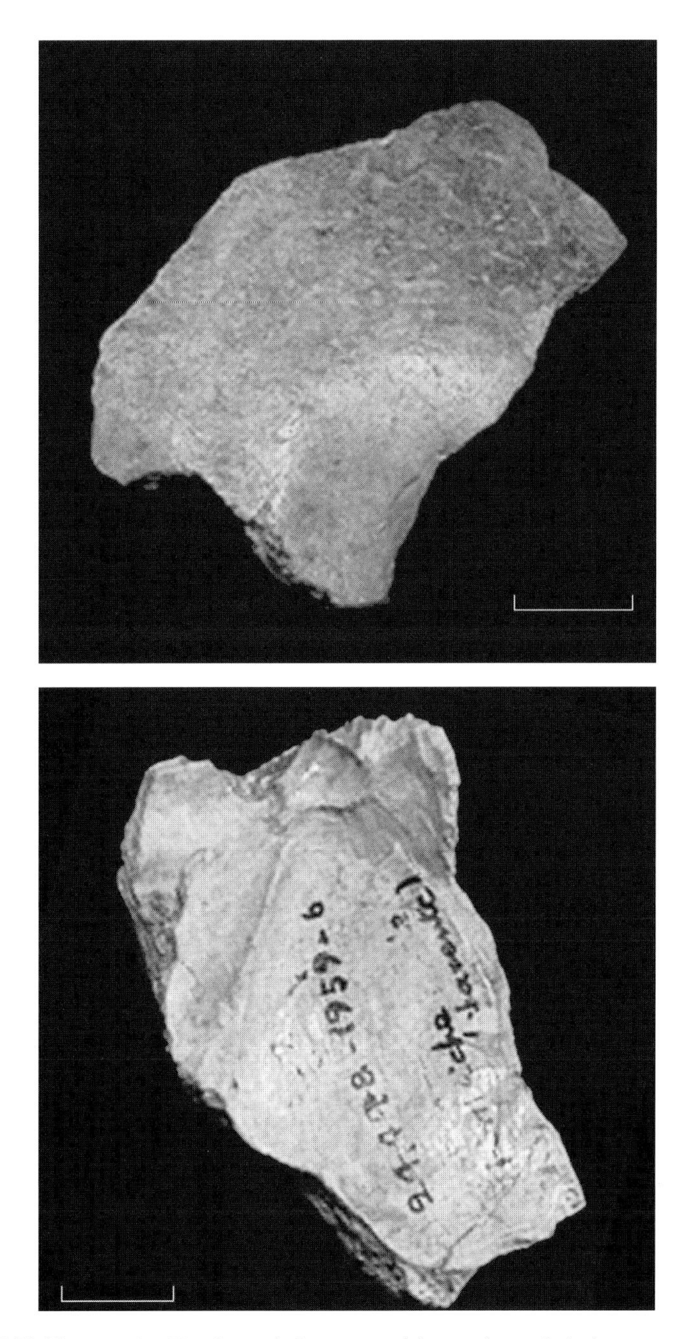

FONTECHEVADE Figure 2. F2 frontal fragment (above: lateral; below: internal), (scale = 1 cm).

Gibraltar: Devil's Tower

Location

Limestone rockshelter site at the western end of the North Front, Gibraltar.

Discovery

D. Garrod, June/October, 1926.

Material

Partial juvenile cranium, ca. 5 years old (Gibraltar 2).

Dating and Stratigraphic Context

Excavated from calcified aeolian sands containing an abundant fauna said by D. Bate (in Garrod et al., 1928) to be of Würm age. Oakley et al. (1971) claim an age of ca. 50 ka by extrapolation from charcoal with similar archaeological context at the nearby Gorham's Cave.

Archaeological Context

"Upper Mousterian" (D. Garrod, in Garrod et al., 1928).

Previous Descriptions and Analyses

Described by Elliot Smith (in Garrod et al., 1928) as Neanderthal, an attribution uncontested since. A recent three-dimensional digitized reconstruction has been reported by Zollikofer et al. (1995). Estimated cranial capacity is 1400 cc (Dean et al., 1986).

Morphology

Partial cranium; mandible with Ldm_{1-2} and RM_1 erupted, LM_1 in crypt; isolated RI^1. Skull would have been low and ovoid in posterior view. Somewhat domed frontal shows very broad interorbital region and prominent glabella. Hint of low, rolled, uniformly thick brow ridges. No frontal sinus development. Huge frontal crest internally. R parietal has short temporal contact and squamosal suture; coronal suture undifferentiated. Palate very shallow. Inferior surface of anterior root of zygomatic arch recedes upward as well as backward. Maxillary sinus visibly swelling out posteriorly to constrict nasal cavity.

Mandibular fossa broad, shallow; already bounded medially by bony elevations. Posterior root of zygomatic arch does not extend over auditory meatus. Developing mastoid process tiny; occipitomastoid crest huge by comparison. Mastoid notch shallow. Parietomastoid suture long and horizontal. Anterior lambdoid suture goes straight back before lambdoid suture proper, turns sharply upwards. R petrosal has large, domed arcuate eminence; closed-over, filled-in subarcuate fossa; no sign of superior petrous sinus; and patent foramen of Huschke. Jugular foramen would been large and multichambered (orientation impossible to tell). Carotid foramen would have pointed down.

Mandible essentially smooth across and broadly arced. Symphyseal region relatively tall s/i and not quite vertical. Viewed from below, symphyseal region thin a/p throughout; bone noticeably thicker posteriorly along corpora. Mylohyoid line indistinct; submandibular fossae placed high up inside of corpora.

Medial pterygoid tubercle present high up internal gonial region.

RI^1 shovel shaped, very high crowned. dm_1 with large talonid, tiny trigonid, and very discrete cusps; paracristid does not close off trigonid. dm_2 distinguished by large trigonid and very large talonid; protoconid and metaconid very closely approximated. M_2 has large trigonid and large, open talonid; large crest ("deflecting wrinkle") runs from metaconid into talonid basin.

REFERENCES

Dean, M.C. et al. 1986. A new age at death for the Neanderthal child from Devil's Tower, Gibraltar and the implications for studies of general growth and development of Neanderthals. *Am. J. Phys. Anthropol.* 70: 301–309.

Garrod, D. et al. 1928. Excavation of a Mousterian rock-shelter at Devil's Tower, Gibraltar. *J. R. Anthropol. Inst.* 58: 1–113.

Oakley, K. et al. 1971. *Catalogue of Fossil Hominids, Part II: Europe.* London, British Museum (Natural History).

Zollikofer, C. et al. 1995. Neanderthal computer skulls. *Nature* 375: 283–285.

Repository

The Natural History Museum, Cromwell Road, London SW7 5BD, England.

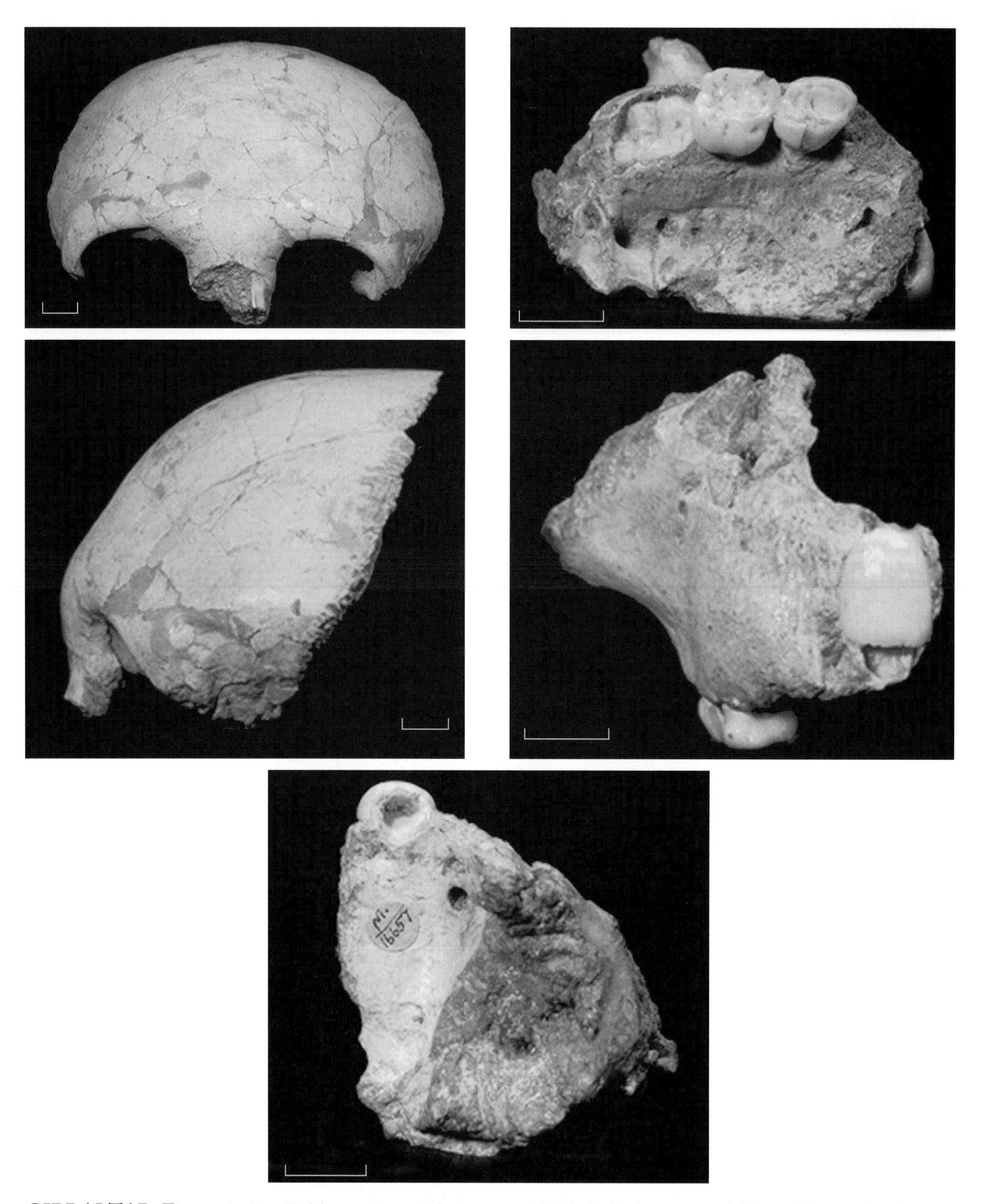

GIBRALTAR Figure 1. Devil's Tower: Frontal (left: top, middle); R (right: top, middle) and L maxillae (scale = 1 cm).

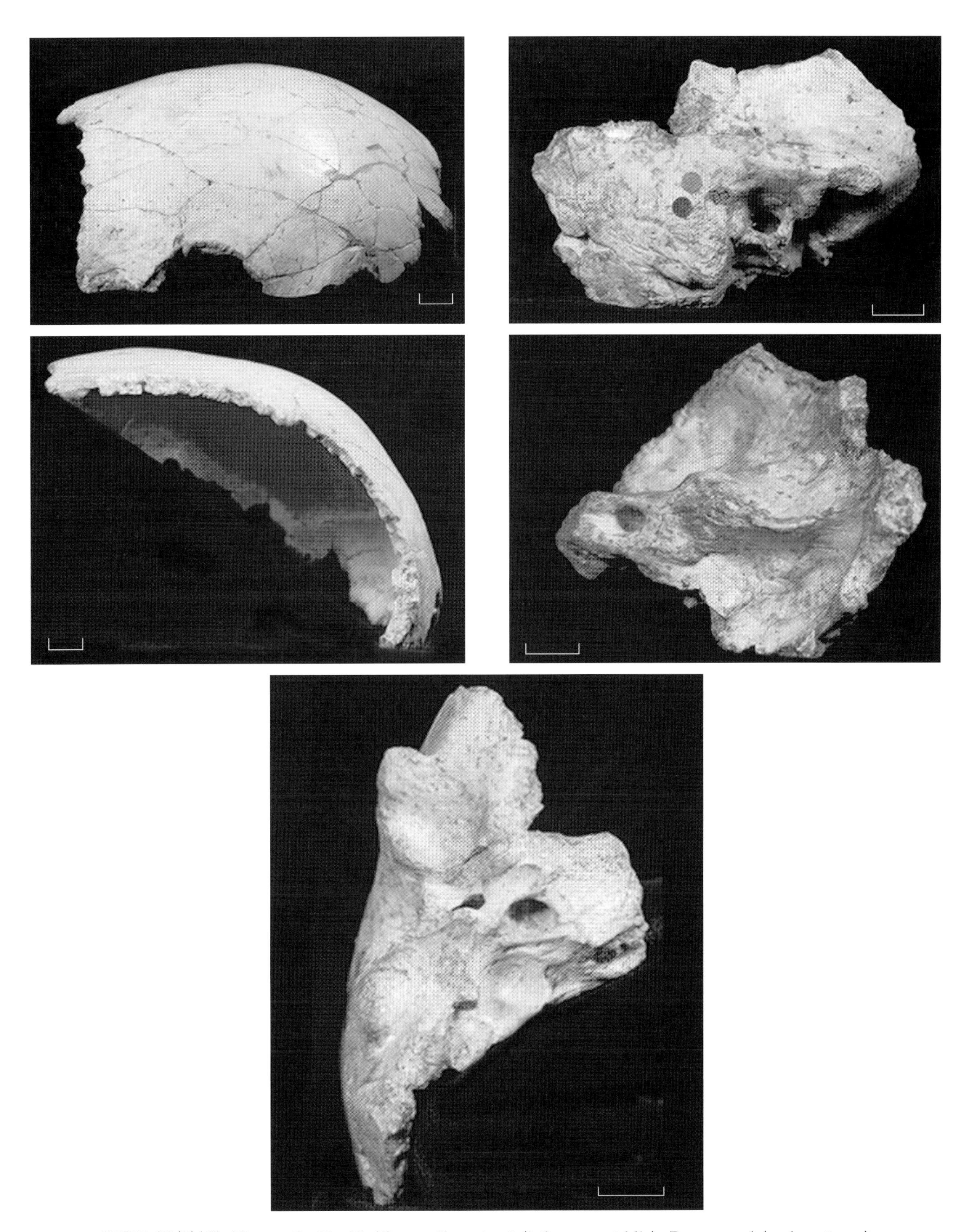

GIBRALTAR Figure 2. Devil's Tower: L parietal (left: top, middle); R temporal (scale = 1 cm).

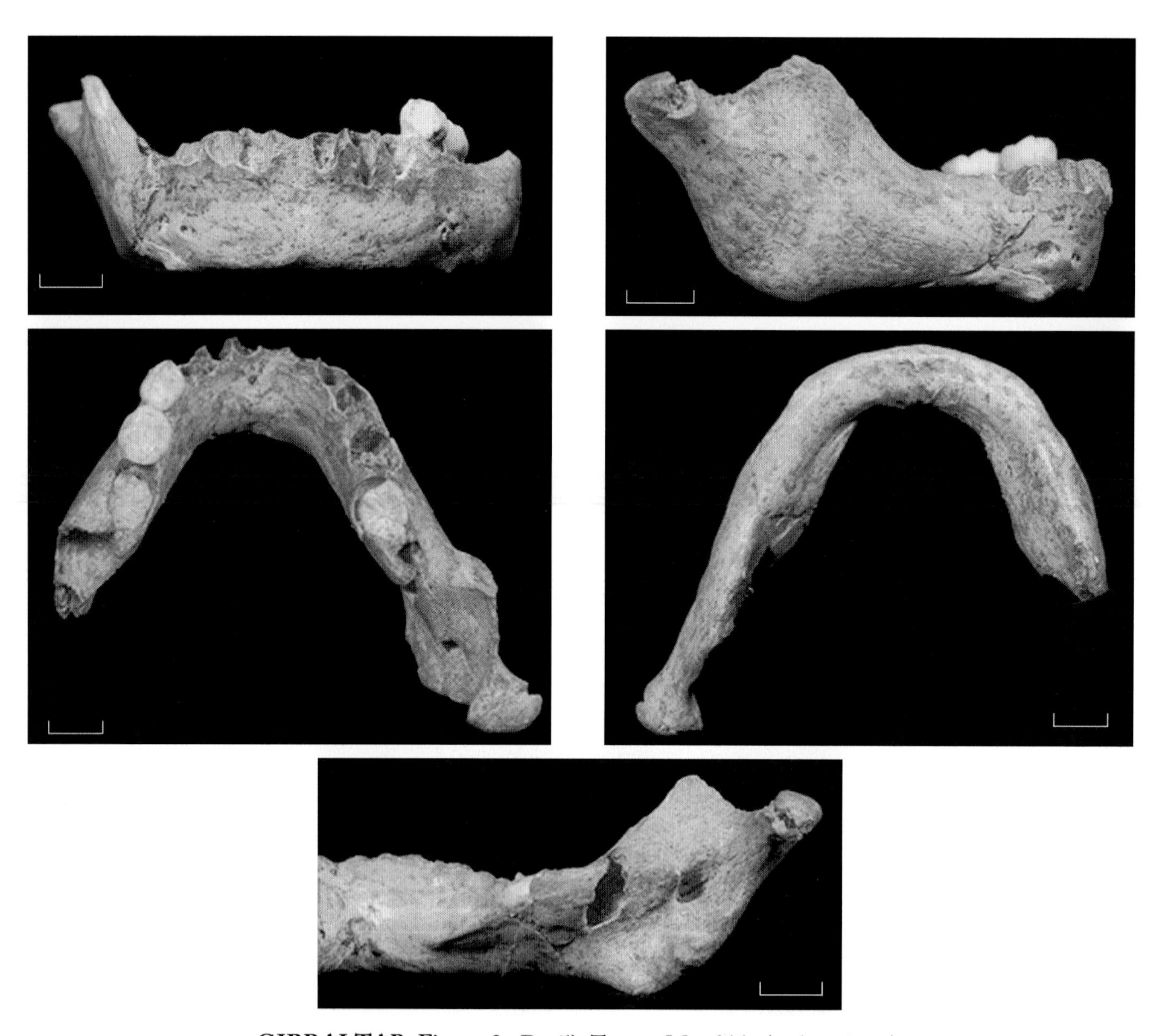

GIBRALTAR Figure 3. Devil's Tower: Mandible (scale = 1 cm).

Gibraltar: Forbes' Quarry

Location
Limestone quarry site close to sea level at the western end of the North Front, Gibraltar.

Discovery
Exact discovery unknown. Reported by a Lt. Flint to the Gibraltar Scientific Society in March 1848 and by George Busk to the British Association for the Advancement of Science in 1864.

Material
Adult cranium lacking part of the vault (Gibraltar 1).

Dating and Stratigraphic Context
Unknown.

Archaeological Context
None.

Previous Descriptions and Analyses
Recognized by Busk (1865) as belonging to the same kind of early human as the Feldhofer Neanderthal, this specimen was made the type specimen of the new species *Homo calpicus* by Keith (1911) and of *Homo gibraltarensis* by Battaglia (1924). Subsequent authors have been virtually unanimous in regarding this specimen as a (gracile female) Neanderthal, but although frequently cited it has never been monographed. Keith (1915) estimated cranial capacity as 1200 cc.

Morphology
Consists of face and most of R braincase; base partly missing. I2–P4 present on both sides; extremely worn both buccally and lingually; P4 has completely bifid root. Cranial bone not remarkably thick in general, although occipital relatively thick.

Cranium quite long. In profile, preserved relatively tall frontal rises moderately steeply from moderate posttoral plane (not sulcus) that flows from the thick supraorbital tori. Frontal curves back strongly anterior to region of bregma. In midsagittal profile apparently would have descended moderately steeply toward lambda. Seen from front, upper face large with posteriorly retreating zygomas and anterior zygomatic roots but lower face relatively very narrow. Posteriorly, occipital bulges somewhat, especially toward where nuchal region is undercut. Modestly concave orbital roofs curve smoothly onto anterior surfaces of s/i tall supraorbital tori that arc slightly over orbits. Tori smoothly rolled from front to back and confluent across glabella. When viewed from above, tori retreat slightly from glabellar region. Glabella very prominent. Long, wide, essentially parallel-sided nasal bones flex strongly and smoothly below nasion. Interorbital space very wide. Orbits deep and roundish, their floors distended by swollen maxillary sinuses beneath; inferior margin blends smoothly into puffed-out face below without an angle or ridge. Infraorbital foramina large, open downward. On both sides is series of small frontal process foramina. Continuous infraorbital canal (not part canal, part groove) runs lateral to midline in orbital floor. Suture above infraorbital foramen

(= premaxillary/maxillary suture) visible bilaterally. Lacrimal fossa appears to have been a/p narrow.

Nasal region swollen and projects in midline. Nasal aperture very large. Lateral crests well defined and continuous all the way around to large, fused, ledge-like, forwardly projecting anterior nasal spine. Prenasal sulcus below spine runs entire breadth of inferior nasal margin. Nasoalveolar clivus relatively short and indented horizontally below anterior nasal spine; bulges anteriorly to accommodate backwardly curving incisor and canine roots.

Large medial projections intrude bilaterally into nasal cavity immediately behind capacious entrance; are continued upward by (superior nasal) crest. No discernible conchal crest. Floor of nasal cavity sunken behind ledge just within nasal entrance, delineating anterior and posterior nasal cavities. Behind medial projections is bilateral swelling out of lateral nasal cavity walls (enlarged maxillary sinuses intrude into cavity); much matrix (but broken) in region; curvature visible posteriorly on R and more anteriorly on L.

Zygomatic arches lightly built; anteroinferior margins sweep both upward and backward. Anterior roots of zygomatic arches originate just above M1s and angle gently laterally. Oblique ridge runs from squamosal to sphenoid; delineates temporal from infratemporal fossa. The latter has distinct bony roof. Anterior squamosal suture angles medially, constricting temporal and infratemporal fossae posteriorly.

Mandibular fossa fairly shallow; is closed off internally by very pronounced medial articular eminence. Posterior wall of fossa quite steep, but articular eminence barely discernible. Sharp vaginal crest runs along midline of tubular ectotympanic to midpoint of carotid foramen. Auditory meatus small and round; above, space between posterior root of zygomatic arch and supramastoid crest. Posterior root of zygomatic arch originates as lateral extension of mandibular fossa.

Mastoid process damaged; is tiny even at base. Occipitomastoid crest very large and probably directly overlies occipitomastoid suture. Parietal notch lies far back. Wide angle between parietomastoid and lambdoid sutures. Lambdoid suture lacks interdigitations; ossicle lies near lambda.

Occipital "torus" more or less horizontal, delineated by sulcus below. Above, "torus" defined only by extremely well-defined, pitted suprainiac depression ("torus" more a ledge than torus).

Palate broad and quite deep, slopes gently down to alveolar rim on all sides. Incisive foramen deep and subcircular, with raised rim at rear. Slight bifid, rounded maxillary torus lies at midline.

Medial and lateral pterygoid plates separated by deep, narrow fossa; converge superiorly and inferiorly. Alae of vomer broad; separated by U (not V)-shaped notch. Preserved part of basiocciput broad and flat. Carotid foramen lies at midpoint of petrosal and points straight down. Foramen ovale broken; appears to lie between pterygoid plates. Jugular foramen well separated from carotid foramen, its fossae small and shallow.

Internally, carotid canal incompletely ossified; carotid sulci large and deeply invaginated. Superior petrous sinus weakly marked. Arcuate eminence huge. No subarcuate fossa, but small foramen is present on the petrosal about 1 cm anterior to internal carotid foramen. R and L transverse sinuses symmetrical. Hypophyseal fossa large, broad. Frontal crest very large; crista galli narrow and tall, with elongate cribriform plate on either side and behind but not extending in front. Tuberculum sellae tall; its surfaces relatively smooth. Middle clinoid processes barely discernible; posterior clinoid processes large (even in broken state) and delineated by very large sulci. Jugum and lesser wings of sphenoid quite long, very deep a/p.

REFERENCES

Battaglia, R. 1924. Osservazioni su l'uomo fossile di Broken Hill. *Boll. Soc. Adriatica Sci. Nat.* 28: 314–321.

Busk, G. 1865. On a very ancient cranium from Gibraltar. *Rep. Br. Assoc. Adv. Sci.* (Bath, 1864): 91–92.

Keith, A. 1911. The early history of the Gibraltar cranium. *Nature* 87: 313–314.

Keith, A. 1915. *The Antiquity of Man.* London, Williams and Norgate.

Repository

The Natural History Museum, Cromwell Road, London SW7 5BD, England.

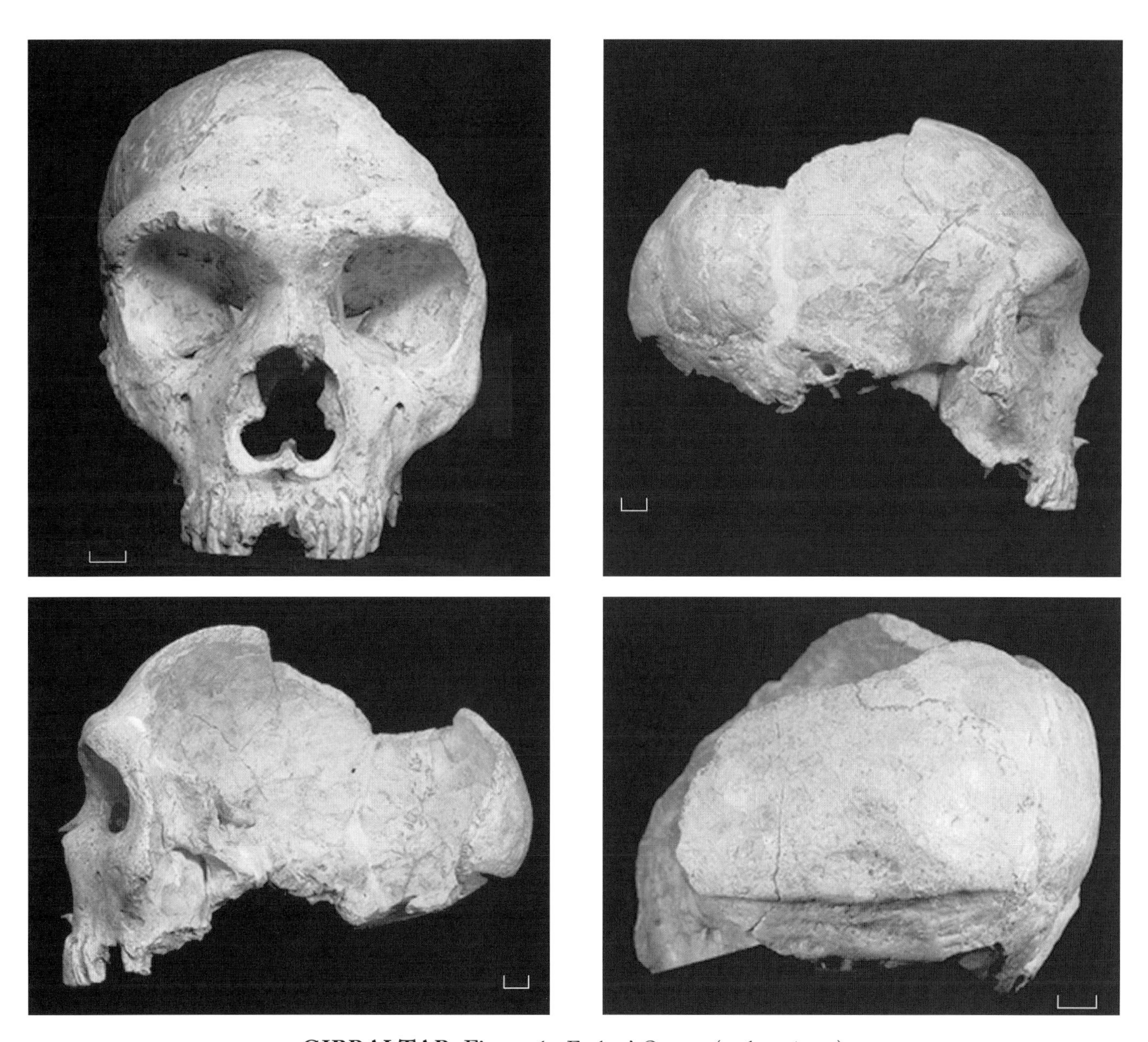

GIBRALTAR Figure 1. Forbes' Quarry (scale = 1 cm).

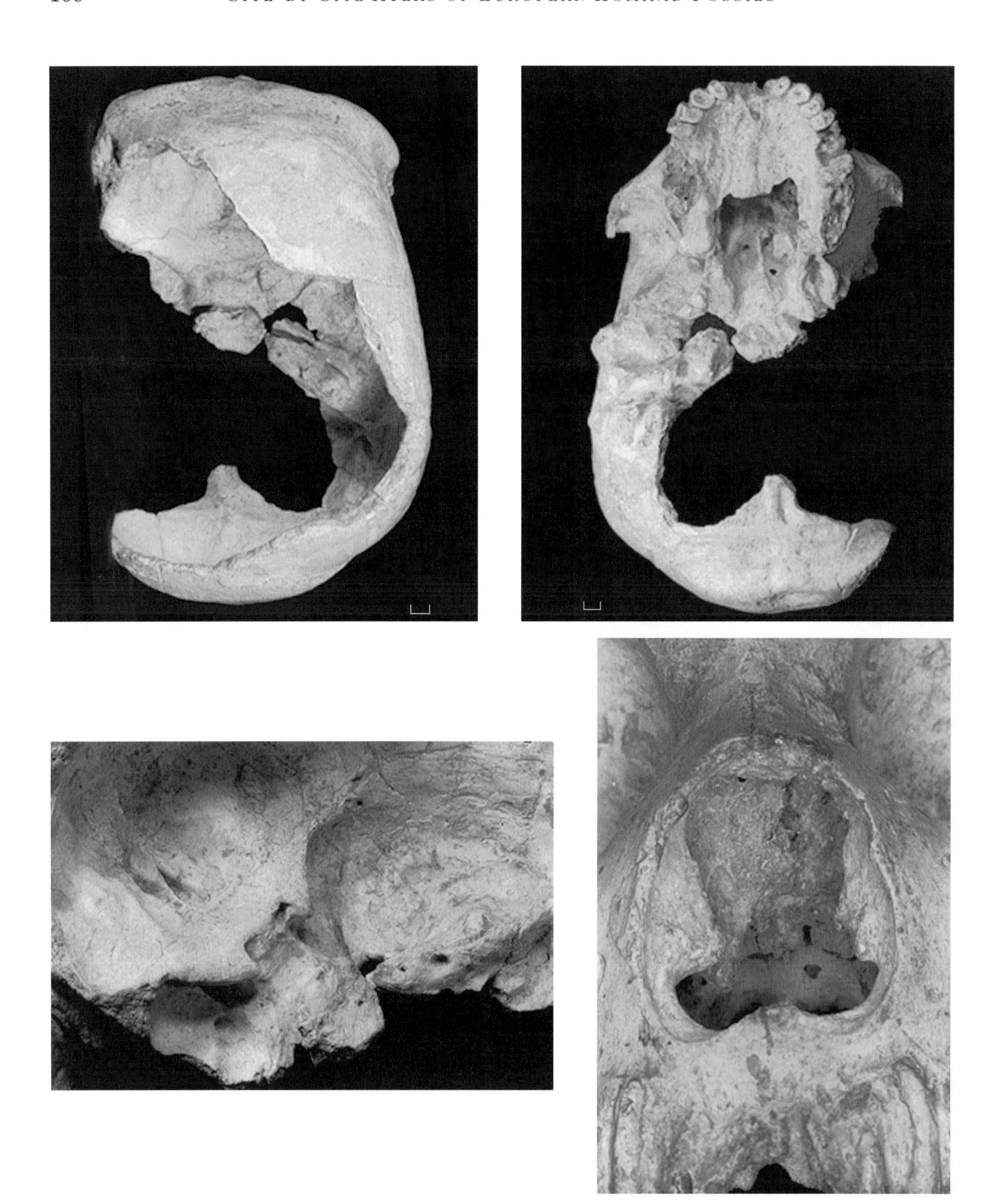

GIBRALTAR Figure 2. Forbes' Quarry (including close-ups of middle cranial fossa and nasal cavity medial projections), (scale = 1 cm).

Grimaldi Caves: Grotte des Enfants (Fanciulli), Barma Grande, Baousse da Torre, Barma del Caviglione

Location

Series of limestone caves, some now destroyed, at the base of the Balzi Rossi (Baousse Rousse) cliffs, just E (Italian side) of the border post between Ventimiglia, Italy, and Menton, France.

Discovery

E. Rivière, 1872 (Barma del Caviglione 1–3), 1873 (Baousse da Torre 1–3); E. Rivière, 1874 (Grotte des Enfants 1), 1875 (Grotte des Enfants 2); L. Julian and S. Bonfils, 1884 (Barma Grande 1); F. Abbo, 1892 (Barma Grande 2–4), 1894 (Barma Grande 5–6); L. de Villeneuve, 1901 (Grotte des Enfants 4–6).

Material

The Grimaldi caves have yielded a remarkable series of burials, most consisting of more or less complete adult, adolescent or infant skeletons. See section above for a count (the missing Grotte des Enfants 3 individual is regarded as intrusive from the Mesolithic; only the Barma del Caviglione 2 and 3 and Barma Grande 6 skeletons are substantially incomplete).

Dating and Stratigraphic Context

Much, although not all, of the early excavation of the Baousse Rousse caves was of indifferent quality, leaving many stratigraphic questions unanswered. Direct radiocarbon dating on the Grotte des Enfants 4 individual has recently yielded a date of 10.25 ka (Binant, 1991), but this is almost certainly a drastic underestimate given the Aurignacian context of this burial (Cartailhac, 1912). Along with the Aurignacian (Cardini, 1931) Barma Grande triple burial (Barma Grande 2–4), this individual is generally considered by archaeological association to be about 25 ka old or more (see discussion by Schumann, 1995). The Grotte des Enfants 1–3 individuals were also found in an Aurignacian context (Cartailhac, 1912; see discussion by Boule and Vallois, 1957), as were the Baousse da Torre skeletons (Rivière, 1887) and the "Homme de Menton" (BC 1) from Barma del Caviglione (Cartailhac, 1912). Archaeological considerations thus place the ensemble of Grimaldi burials in the period between about 30 and 25 ka ago.

Archaeological Context

As noted above, the human skeletons recovered from all four hominid-yielding Grimaldi caves are of Aurignacian context. Virtually all of the complete skeletons appear to have been intentionally buried, some of them with copious quantities of ochre, and perforated teeth and invertebrate shells (e.g., Rivière, 1887; de Villeneuve, 1906; Cartailhac, 1912).

Previous Descriptions and Analyses

The Grimaldi skeletons figured importantly in early scenarios of *Homo sapiens* evolution, principally because while Verneau (1906) considered most of the burials to represent the "Cro-Magnon race," he considered the Grotte des Enfants burials 5 and 6 (recovered from the lowest occupation layer of the site) to be of a distinct "Grimaldi race," possessing "Negroid" affinities. According to Verneau, these two distinct types may have had a relationship of descent, and indeed this author and others subsequently discovered "Grimaldi" features in a variety of Upper Paleolithic skeletons from other European sites—and even in modern European populations (see discussion in Boule and Vallois, 1957). More recent discussions (e.g., Barral and Charles, 1963) have discounted this typological view, focusing instead on variations in robusticity among the Grimaldi skeletons. Among others, Lapouge (1899: *Homo grimaldii*) and Sergi (1911; *Notanthropus eurafricanus recens*) early on applied new nomina to various Grimaldi specimens, but, such instances apart, rarely has any of these skeletons ever been claimed to represent anything other than modern *Homo sapiens* (Schumann, 1995).

Morphology

Barma Grande

Specimens seen in exhibit, although cover of glass case removed.

Barma Grande Burial 2. R individual. Adult male; most of skeleton preserved in original (not destroyed in war). Very tall; large long bones with huge muscle scars. Skull relatively small.

Skull thrown together for exhibit. Visible are one piece with R mastoid; posterior part of R parietal, and adjacent occipital, all connected along sutures; also part of R facial skeleton, smaller part of L facial skeleton, most of frontal, posterior parts of R and L parietals, all joined along sutures. In sand below, part of R mandibular corpus from C through M3; adjacent to it, separate piece, probably L posterior corpus with M1–2. Cranial bone not very thick.

Face relatively short; not very long s/i. Glabellar region through part of R supraorbital reconstructed. Only very lateral portion of R lateral orbital plate preserved: is thin and slopes back from rather thin orbital margin. Maxillary sinus does not extend laterally into zygoma; in region of frontal process, sinus divided into three chambers by bony septa. Temporal ridge defines lateral border of short zygomatic process of frontal, above which it courses up before recurving sharply to run across frontals and on to parietal, where it is transformed into flat temporal lines. Frontal rises only moderately vertical; gentle curve behind in profile. R orbit apparently very wide m/l but only moderately tall s/i; is subrectangular with rounded corners. Inferior orbital margin thickened, somewhat everted. Large infraorbital foramen lies in deepest part of broadly concave infraorbital region and fairly close to inferior orbital margin; faces slightly downward. On the L, within nasal cavity, is low, horizontal conchal crest. As seen on the R, lateral crest of nasal aperture curves inward (not clear whether as far as anterior nasal spine). Anterior root of zygomatic arch takes origin close to M1; inferior margin ascends more steeply than it flares laterally until quite high up, when it arcs laterally quite strongly. From below, inferior margin of zygomatic arch at first anteriorly facing; then, just behind zygomaticomaxillary suture, it curves back strongly to flare a little as it begins to define lateral margin of temporal fossa. Missing posterior to this.

Coronal suture visible in parts; is finely denticulated, segmented. Sagittal suture more deeply denticulated. Huge, downwardly pointing mastoid process very long a/p, swollen laterally at its base; above is pronounced, strongly upwardly arcing supramastoid crest that is confluent with much lower suprameatal crest that lies over compressed, ovoid, moderately large, slightly anteriorly tilted auditory meatus whose posterior wall is fused to mastoid process. Parietal notch apparently shallow. Long parietomastoid suture quite horizontal. Lambdoid suture rises steeply from asterion. More medial morphology not discernible.

Sides of braincase vertical; occipital plane gently curved. Almost flat, minimally muscle-scarred nuchal plane angles slightly and steeply from very distended superior nuchal line, which becomes more prominent and ledgelike as it courses medially and downward to terminate in prominent V-shaped external occipital protuberance. Quite laterally along the superior nuchal line, above level of and slightly posterior to asterion, is downwardly pointing tubercle-like swelling. Mastoid notch relatively deep, narrow; bounded medially by low, thick paramastoid crest immediately medial to which is much more pronounced, downwardly distended occipitomastoid crest. Somewhat medial to latter is less elevated, posteriorly more extended Waldeyer's crest. No crest descends as far as the elongate, downwardly pointing mastoid process.

Palate unanalyzable. All teeth extensively worn. RC–M3 and LC–part M1 preserved. C root and base of crown very large, robust. M1 larger than M2, which is larger than M3.

R mandibular corpus fragment tall s/i, not very deep from side to side (judging from break behind and in top view). Mandible quite thick across symphyseal region. Symphyseal region preserved from lower extremity of anterior tooth roots, from which it swells out anteriorly and laterally quite strongly toward inferior margin. In inferior view, inferior margin straight across in this region; preserves blunt lateral corner on the R. Preserved R digastric fossa is wide m/l, well defined, posteriorly directed. Apparently large mental foramen was under P2. M3 just exposed in front of preserved anterior border of ramus. Steeply descending mylohyoid line moderately well developed; mandibular fossa below is rather shallow. What remains of postincisal plane rather vertical. Genial tubercles thin, weakly developed. Preserved lower RC root and crown are large; P1 is missing; P2–M3 are very worn, but M3 is less than expected. L corpus fragment comparable to the R, except the missing M3 may have been partially hidden by ramus.

Barma Grande Burial 3. Middle individual; subadult. Heavily reconstructed adult skull missing R and L zygomatic regions, all sphenoidal and nasal cavity structures; various teeth, incorrectly placed in upper jaw. L side of mandible buried in sand; what is visible is complete; M3s unerupted, visible in crypts.

Skull moderately long, high vaulted, quite high-domed profile. Both face and neurocranium very narrow. Face also rather tall, although nasoalveolar clivus does not protrude much.

Supraorbital region reasonably intact; low glabellar "butterfly" slightly undercut by damaged lateral plate, which appears to be quite vertical (not posteriorly inclined). Frontal plane only gently sloping; rises directly and steeply from glabellar "butterfly" and flexes backward about halfway up bone to describe fairly flat curve that becomes posteriorly directed halfway along length of sagittal suture. This more gentle slope carries on around occipital plane to region of external occipital protuberance, at which point the somewhat flatter, poorly muscle-scarred nuchal plane angles quite sharply anteriorly. Faint temporal ridge emerges from just behind and above downwardly directed zygomatic process of frontal; behind this, temporal lines obscured by damage and reconstruction.

Glabellar region very broad. Mediosuperior corner of orbit m/l wide but s/i not very tall. R orbit cut off obliquely. L side buried laterally; medial part of orbit exposed, showing similar conformation. Interorbital region broad superiorly, narrows markedly inferiorly. Superior orbital margin thin; orbital roof angles acutely into plane above it. Inferior orbital margin was apparently not distended. Large, downwardly oriented infraorbital foramen lies moderately close to infraorbital margin; appears to penetrate rather flat, slightly inferoposteriorly sloping infraorbital surface.

Nasal bones were apparently rather long and expanded a bit inferiorly; in preserved superior portion, nasal bones somewhat arcuate from side to side. In nasal cavity, on the L, wall of frontal process essentially smooth in region of conchal crest. Lateral margins did not contact relatively large, anteriorly projecting pair of anterior nasal spines. Nasal aperture moderately wide; was apparently not very tall. Moderately long nasoalveolar clivus essentially vertical; its bony surface bears strong impressions of anterior tooth roots.

Anterior root of zygomatic arch appears to take origin quite far above M1; its inferior margin curves strongly laterally before flexing strongly backward at zygomaticomaxillary suture. Posterior root of R zygomatic arch preserved; does not protrude laterally away from skull; its orientation suggests that arch did not flare laterally; takes origin well in front of moderately sized, ovoid, vertically oriented auditory meatus, whose thick bony posterior wall is closely appressed to base of a/p long, stubby, anteriorly oriented, not laterally swollen or downwardly very protrusive mastoid process. Vaginal process apparently peaked around moderately thick styloid process; descends to lateral edge of ectotympanic tube but does not quite contact mastoid process. Moderately large stylomastoid foramen quite far lateral and posterior to styloid process. Reconstruction obscures morphology medial to mastoid process. On the R, squamosal long, tall, its superior margin strongly curved. Anterior squamosal region flows smoothly into deeply excavated alisphenoid. No clear margin between posterior and anterior temporal fossae; temporal fossa broadens considerably anteriorly. Parietal notch was apparently not distinct. Palate damaged. Reconstruction covers most of detail of rear of skull.

Mandible rather tall s/i in midline, decreasing drastically in height to level of M2s. Apex of moderately pronounced triangular swelling on external surface of symphysis arises in subalveolar depression lying across incisor roots. This triangular region most pronounced inferiorly. From below, inferior margin of symphysis straight across; turns backward at bilateral blunt corners. Internally, postincisal plane essentially vertical; genial tubercles lacking, but small midline depression lies in that region with small elevation below it. Ramus not much taller s/i than it is long a/p. Moderately tall, pointed coronoid process relatively long a/p at its base; is undercut by distinct preangular sulcus. Sigmoid notch uniformly curved downward; is deepest at midpoint. Sigmoid notch crest terminates just lateral to midpoint of damaged R condyle. Internally, coronoid process bears pronounced vertical pillar that runs down to become confluent with blunt internal alveolar crest. Ovoid, only slightly compressed, lingula-free mandibular foramen points upward and back; strong mylohyoid groove descends steeply anteriorly from it. Mylohyoid line and submandibular fossa beneath very faint.

Upper teeth in good shape; many in wrong places; may not all be from same individual. I^1s tall crowned, wide, shoveled to some extent. I^2s deeply shoveled. C^1s (reversed in jaw) smooth on their sloping lingual surfaces. Premolars in wrong place, including two teeth in this position that could be pathologically small M_3s. In the place of RM^1 is LM^1; in position of LM^1 is LM^2. M^{1-2} have large hypoconids (quite swollen distolingually on LM^1), distinct, centrally placed talon basins, and subequal pre- and postprotocristae.

In the lower jaw, incisors not shoveled. C_1 lingual surfaces bear slight pre- and postmargocristae. P_1s bear moderately developed lingual swellings that are well separated from protoconid; on either side of this cusp, thin anterior and posterior foveae; mesial edge of this cusp much shorter than distal edge. Preserved RP_1 has subequal protoconid and metaconid connected by low cristid; on either side of this cristid are small foveae; mesial edge of protoconid much shorter than distal edge. R and L M_1s have small, centrally placed hypoconulids; M_2s do not. No sign of M_2 trigonid basin; this basin lacking on exposed crown of RM_3. At least on RM_1, base of hypoconid extends lingually beyond midline of crown; on the M_2s and the unerupted RM_3s, cusps all meet in midline of talonid basin. As seen best on the R, M_1 slightly longer m/d and wider b/l than M_2; both molars are subrectangular in outline and only show minimal enamel wrinkling.

Barma Grande Burial 4. L individual. Adult cranium without mandible. Skull crushed, heavily reconstructed; still distorted; lower face bears a lot of matrix. Preserved is most of R side of skull minus zygomatic arch, L temporal, and some of L lower face. M3 crowns formed; still deep in crypts; all sutures patent: individual quite young. Bone not very thick.

Overall skull shape obscured by distortion. Glabellar region not interpretable. Only part of supraorbital region preserved is lateral to midline; is platelike, quite vertical, and makes sharp angle into orbital roof. As seen on the R, frontal probably rose steeply to moderately pronounced eminence, beyond which slope gentler. Coronal suture moderately strongly denticulated; is segmented, as are preserved sagittal and lambdoidal sutures. Low temporal ridge forms posterior border of thin, backwardly inclined zygomatic process of frontal. This ridge curves back sharply from upper part of zygomatic process; is rapidly transformed into faint temporal lines that lie low on skull.

The R orbit was apparently very wide m/l but only moderately tall s/i (producing wide rectangle with rounded corners); superomedial corner truncated. Infraorbital margin thick, slightly everted; rises vertically above orbital floor. Internally, within nasal cavity on L frontal process is low, horizontal conchal crest. The L lateral crest of nasal aperture does not course around medially along inferior nasal margin. Nasoalveolar clivus only moderately long and only slightly forward sloping. Much matrix in this area. Palate partially reconstructed; appears not to have been deep. On matrix-encrusted R side, seems that anterior root of the apparently not flared zygomatic arch takes root high above M^1; its inferior margin arcs strongly laterally before curving backward into the arch.

Squamosal long, tall; its superior margin strongly curved. Anterior squamosal region flows quite smoothly into shallow alisphenoid. Narrow, vertical parietal notch lies above posterior margin of low, very blunt, a/p long, slightly laterally swollen, anteriorly directed mastoid process. Posterior root of zygomatic arch arises well in front of ovoid, moderately large, slightly anteriorly tilted auditory meatus, whose thick bony wall is fused posteriorly to base of mastoid process. Mandibular fossa moderately wide m/l, long a/p, and relatively deep; is bordered anteriorly by well-developed articular eminence and posteriorly by very thick, m/l wide, somewhat downwardly distended postglenoid plate that lies almost entirely lateral to edge of auditory meatus, from which separated by small gap. Apparently there was no medial articular tubercle. Vaginal process peaks significantly around thick, broken styloid process; former is low medial and lateral to latter. Laterally, vaginal process runs to edge of meatus and contacts mastoid process. Relatively large stylomastoid foramen somewhat posteriorly and laterally separated from styloid process. Base of mastoid process visible on L; like the R, no mastoid notch. Bone is flat medial to this process.

Parietomastoid suture relatively long and horizontal; lambdoid suture rises quite steeply from asterion; damage in region of lambda obliterates sutural configuration there. In spite of crushing, occipital plane was apparently not very wide; it smoothly curved down into long a/p but shallow sulcus along superior part of apparently smooth nuchal plane.

M^3s visible in their crypts; were not erupting. Other teeth quite well worn, as especially seen on lingual and occlusal surfaces of the Cs and Is. M^1s markedly larger than M^2s, which are larger than M^3s. M^1s have large hypocones, which are much smaller on M^2 and absent on M^3. On all Ms, preprotocrista more pronounced than postprotocrista; talon basin lies quite centrally on crown.

Barma Grande Burial 5 (1894). Adult. Skull largely externally complete but heavily reconstructed. Mostly missing internal nasal structures, palate, R zygomatic arch, L half of mandible; inferior border of entire symphyseal region present. RI^2 and M^1 missing (recently broken off); LI_1–M_3 missing.

Very long, high-rising skull; broad, with very vertical frontal that even bulges somewhat anteriorly because of marked and forwardly expansive frontal dome. Glabellar "butterfly" moderately pronounced; extends only to about midline of orbits; is undercut medially by lateral plane. Long, wide, moderately deep supraorbital notch just medial to midline. Lateral plate fairly vertical; angle between it and orbital roof fairly acute; its supraorbital margin is thin. Internal breakage shows that L frontal sinus does not extend beyond region of glabellar "butterfly." Very stout, rugose temporal ridges extend vertically well above short, backwardly tilted zygomatic processes of frontal, then they curve back quite strongly along sides of frontal; their continuation posteriorly obscured by damage and reconstruction. Glabellar region quite wide and rather flat from side to side, although, as seen in profile, the surface is inclined inward so that an angular glabella overhangs nasion. Nasal bones wide; were probably parallel sided; arced strongly from side to side; flexed outward strongly ca. 1 cm below nasion. Lower parts of nasal bones broken toward tip; preserved bones project almost horizontally and were probably quite long. Particularly on the L, inferior orbital margin thickened and somewhat everted; overhangs wide but shallowly excavated infraorbital surface that inclines somewhat posteroinferiorly.

Nasal aperture was apparently not wide or tall. Lateral crest was seemingly confluent with very large, anteriorly projecting, probably tip-fused anterior nasal spines. Both frontal processes bear low, ridgelike, horizontal conchal crests on their nasal cavity surfaces. Anterior portion of nasal floor preserved; is flat, although it lies below level of inferior nasal margin. Nasoalveolar clivus very short and steeply inclined; on the L, bears impressions of anterior tooth roots. Palate may have been shallow (as indicated by long, gentle slope from behind region of Is). Maxillary sinuses do not proceed laterally into zygomas. Although broken, lacrimal canals were clearly roofed within nasal cavity by posterior extensions of anterior lacrimal crests.

On both sides, anterior root of zygomatic arch arises relatively close above M1, from which point it arcs strongly far laterally to region of long, rugose inferior malar tubercle, at which point it arcs backward, flaring only slightly to terminate at very anteriorly placed articular eminence. Short zygomatic arch encloses narrow temporal fossa that, on the L, appears to be subdivided by raised anterior squamosal region into small anterior and much longer posterior temporal fossae. Damage obscures any vertical division of the temporal fossa, if there was any. Posterior root of zygomatic arch takes origin anterior to large, ovoid, anteriorly tilted auditory meatus; meatal wall fused behind to base of a/p moderately long, laterally quite swollen, anteriorly directed, somewhat downwardly projecting mastoid process. Only on the L is any morphology of a mandibular fossa discernible; seems that the articular eminence was very downwardly projecting (and mandibular fossa thus relatively deep).

As seen on the L, vaginal process was apparently low, except where it peaked around laterally placed styloid process; vaginal process appears to have continued laterally to contact mastoid process. Medial surface of mastoid process quite flat; faces onto narrow V-shaped mastoid notch that is bounded medially by moderately pronounced paramastoid crest, medial to which is more downwardly projecting, thicker, longer occipitomastoid crest.

Squamosal long a/p; appears to have been tall s/i. On the L, parietal notch very obtuse and opens onto very long, relatively horizontal parietomastoid suture; lies over midline of mastoid process. Posterior root of zygomatic arch flows into pronounced, thick suprameatal crest that, in turn, flows into a more pronounced, ridgelike, steeply ascending suprameatal crest that is separated from mastoid process below by s/i tall sulcus. Lambdoid suture rises steeply from asterion before becoming lost in reconstruction. Occipital bone largely reconstructed but preserved in midline. At base of what appears to have been gently rounded occipital plane, and overhanging more rugose and planar nuchal plane, is shelflike, crescentic, downwardly, outwardly projecting external occipital protuberance.

Mandible as preserved not tall s/i; is taller at symphysis than below M3. Corpora not remarkably wide; maximum width occurs at symphysis and at rise of ramus. Superior part of symphysis missing. Viewed from below, inferior triangular swelling thickens symphyseal region. Symphyseal region narrow and gently arced from side to side, and L corpus diverges strongly from it. Digastric fossae poorly marked; face posteriorly. Postincisal plane was apparently vertical; pair of strongly developed genial tubercles lie in midline inferiorly. Mylohyoid line poorly marked; submandibular fossa well excavated. Gonial angle gently obtuse; its inferior margin strongly reflected outward; external part of inferior margin bears larger and smaller tubercle-like muscle scars. Internally on gonial region, three muscle scars; superiormost is largest. Mandibular foramen quite compressed and upwardly and backwardly pointed; is notched inferiorly in line with downwardly and anteriorly descending mylohyoid groove. Coronoid process quite a/p long, moderately tall s/i, and pointed; is somewhat taller than m/l very wide mandibular condyle; is undercut by moderately excavated preangular sulcus, which fully exposes distal part of M3; pillar on its internal surface. Sigmoid notch crest runs to lateral extremity of mandibular condyle; deepest point of notch lies quite close to condyle's base.

Preserved upper teeth extensively worn; except on M^3s, variable amounts of dentin exposed. Anterior teeth worn down almost to their necks. Ms appear to have decreased in size posteriorly; M^1 appears to have been only tooth with developed hypocone. M^3 talon basins centrally placed; short preprotocristae more prominent than even shorter postprotocristae.

Preserved LI_1–M_3 quite worn, to dentin; even LM_3 shows significant wear. Ms decrease in size from M_1 to M_3; M_{2-3} apparently lacked hypoconulids.

Grotte des Infants

Adult skeletons on display; only skulls were removed for study. All skulls similar in having somewhat parallel sidewalls, bipartite supraorbital region with distinct glabellar "butterfly" and platelike lateral parts; steeply vertical domed frontal rising from above glabellar "butterfly"; maximum width of skull high and above region of squamosal suture; anterior root of zygomatic arch originating just above M^1, with posterior root anterior to auditory meatus; and mandible with distinct mental trigon and central keel, as well as somewhat narrow symphyseal region and divergent corpora. Variation within samples described below.

Burial 4. Externally almost complete but heavily reconstructed skull and mandible; lacking internal orbital and nasal features, part of sphenoid and some of nasal region. Mandible fixed to skull, obscuring mandibular fossae and details of teeth, which appear to have been highly worn. Bone thin; surface quite weathered, obscuring temporal lines and major sutures.

Skull massive and broad faced; mandible large. Preserved traces of muscle scarring surprisingly weakly developed. Skull moderately long and almost flat topped, with strong frontal doming and occipital curvature and relatively straight middle profile.

Fairly strongly developed glabellar "butterfly" incorporates very broad, moderately swollen glabella with pronounced wings that are undercut by strong medial continuation of platelike lateral portion of supraorbital region. In this medial extension lie, on the L, a large supraorbital foramen and, on the R, a deeply incised notch. Temporal ridges originate from high above somewhat horizontally oriented zygomatic processes of frontal, which has vermiculate surface. Interorbital region would have been quite broad. Orbits and infraorbital and zygomatic planes slightly asymmetrical.

Because the L side was apparently damaged and reconstructed, following description from the R. Orbit very wide m/l, not very tall s/i, and subrectangular in shape. Orbital roof angles acutely back into anterior supraorbital margin. Nasal bones lacking for almost all of their length; appear to have arced forward from just below nasion. Judging by apparent a/p length of R maxillary frontal process at level of infraorbital margin, nasals may have been relatively long. Superiorly, they were only gently curved from side to side. At nasal aperture, lateral crests continuous to anterior nasal spines; faint spinal crests behind create shallow prenasal fossa. Nasal aperture only modestly broad at base; was probably not very tall s/i given probable length of nasals. Internally on the R, appears that maxilloturbinal had been fused to conchal crest and subsequently broken off; L side still covered in matrix. Preserved anterior portion of floor of nasal cavity rather flat. Bases of anterior nasal spines thick; spines were evidently not very forwardly projecting. Relatively long nasoalveolar clivus quite vertical; bears impressions of anterior tooth roots.

Infraorbital foramen preserved on the L; is quite close to infraorbital margin. On the less distorted R side, in front view, inferior margin of anterior root of zygomatic arch curves smoothly but sharply upward and laterally to slight inferior maxillary tuberosity. In side view, infraorbital plane slopes slightly backward beneath thickened inferior orbital margin. Viewed from below, inferior margin of zygomatic arch corners back to run slightly laterally as well as posteriorly to somewhat laterally expanded posterior root. In side view, inferior margin of modestly robust zygomatic arch slopes up; superior margin more or less horizontal. Arch defines quite small temporal fossa (cranial wall missing on both sides).

Squamosal was apparently moderately long and tall. Shallow parietal notch located more anteriorly over forwardly pointing mastoid process on the R than on the L. Auditory meatus moderately large, ovoid, and slightly anteriorly tilted, its posterior wall fused to mastoid process. Posterior root of zygomatic arch flows into well-developed suprameatal crest that is confluent with somewhat more pronounced supramastoid crest, which arcs gently upward. Sulcus below supramastoid crest separates it from very laterally swollen mastoid process.

On the L, broken base of styloid process very laterally placed; posterior and somewhat lateral to it lies relatively large stylomastoid foramen. On the L, basally very long, thick, but not markedly downwardly projecting mastoid process borders relatively shallow, narrow mastoid notch that is bounded medially by thick, low paramastoid crest; on the R, virtually no notch and no crest. On the R, long, well-developed Waldeyer's crest visible.

Occipital and nuchal planes rather smoothly curved; the latter is not very rugose; in midline is fairly small, downwardly convex, somewhat ridgelike structure in region of external occipital protuberance. Region around foramen magnum entirely reconstructed.

Palate broken. Is only moderately deep, with modest anterior slope, almost vertical sides, but only moderately diverging tooth rows. Is relatively small for size of skull and rather narrow at front.

Mandible quite tall s/i; is dimensionally large although bone overall rather thin. Symphyseal region strongly curved; corpora diverge strongly posteriorly. A low central keel originates below the I_1s and quickly fans out into swollen region that becomes more pronounced as it approaches inferior margin. From below, anterior surface of this swollen region short and straight across; bears blunt corners bilaterally. Symphyseal region thickest part of front of jaw. On the R, evidence of posteriorly facing digastric fossa. Mylohyoid lines poorly defined; submandibular fossae below shallow. Internally, gonial regions bear three rugosities bilaterally, which increase in prominence superiorly; highest is in position and form of medial pterygoid tubercle. Mandibular foramina asymmetrical in configuration; both strongly compressed and point backward and up. Coronoid processes low, thin a/p, slightly undercut by shallow preangular sulci. As preserved better on the L, sigmoid notch long and very shallow.

Burial 5. Skull thin boned, externally very weathered. Lacks nasal cavity structures, most of sphenoids; on the R, mandibular condyles, coronoid and gonial regions missing. In upper jaw, teeth too worn to be informative; some missing. In lower jaw, all molars lost antemortem; remaining teeth worn to roots.

Rather small skull; for its size, is relatively long, very high domed, and not very narrow. In profile, strongly vertical frontal curves more gently backward, then more sharply down to midpoint of occiput, at which point bone curves forward along nuchal plane to flatten out as it approaches cranial base. Frontal may have borne low eminence on the R (L damaged). Glabellar "butterfly" very low; terminates laterally at about midline of orbits, from which point the flat lateral part is moderately strongly directed posteriorly.

Glabellar region quite broad. Interorbital region appears to have been broad throughout its length. Supraorbital notch very wide on R, shorter on L; both very medially placed on margin of very a/p wide, subrectangular orbits. In profile, appears that temporal ridges rose high above zygomatic process of frontal, then turned backward to become low-lying temporal lines; damage prevents further comment. Inferior orbital margins thickened and everted strongly anteriorly. Orbital roofs descend to superior orbital margin, where bone arcs back acutely onto anterior surface. Region below everted infraorbital margins broadly and shallowly excavated.

Nasal bones apparently were broad, moderately arced from side to side, flexed forward well below nasion. As preserved best on the R, frontal process of maxilla appears to have been somewhat elongated a/p at level of infraorbital margin (nasal aperture would probably not have been very tall, but nasal bones would have been very long). Nasal aperture was relatively broad inferiorly; lateral (margin) crests were not continuous to midline. Bulky bases of anterior nasal spines situated just at base of nasal aperture. Internally, frontal processes rather featureless; on the L is evidence that lacrimal groove had been to some extent covered by anterior lacrimal crest. Floor of nasal cavity appears to have been rather flat. Moderately long nasoalveolar clivus strongly inclined forward and upward, narrow from side to side.

In front view, infraorbital region faces forward although slightly tilted back. Inferior margin of thin zygomatic arches arc strongly laterally to thickened maxillary tuberosity. Seen from below, zygomatic arch curves strongly backward from tuberosity and arcs gently out toward posterior root. In side view, inferior margin of zygomatic arches oriented upward. Arches enclose rather small temporal fossae. Anterior squamosal curves gently into alisphenoid, which on the L is broadly angled in toward cranial base (weakly defining infratemporal fossa).

Auditory meatus moderately large, ovoid, vertically oriented. Its posterior wall fused to mastoid process. Posterior root of zygomatic arch flows into low suprameatal crest, which is confluent with slightly more prominent, weakly upturned supramastoid crest. As seen on the L, vaginal process extends full length of ectotympanic tube; peaks around large, laterally placed styloid process; terminates at base of mastoid process. As seen on both sides, moderately sized stylomastoid foramen lies lateral to base of styloid process. On both sides, mandibular fossa wide m/l, moderately long a/p, quite deep; is bordered anteriorly by strongly developed articular eminence. Low, short postglenoid plate extends laterally just a bit beyond auditory meatus. Squamosal relatively long; was probably tall. Downwardly pointed L mastoid process preserved; is not very elongate, but quite broad, stubby at base, and swollen laterally; medially, is bounded by very shallow, extremely narrow mastoid notch, whose medial border is formed by low, thick paramastoid crest.

Occiput was narrow and tall; bone of nuchal plane lacks much detail. Lambda was very high; suture across region may have been arced.

Palate appears to have been very shallow, with very long anterior slope; is almost V shaped in outline. Pterygoid plates preserved superiorly; are confluent at their bases. Foramen ovale lies posterolateral to this confluence. Foramen magnum was apparently relatively long and wide. Sutural detail obscured almost throughout, but sagittal suture was strongly denticulate and apparently segmented.

Much bone resorption of mandible due to loss of molars bilaterally. Even so, bone of corpora was evidently never very thick, nor the rami tall. Some matrix still on symphyseal region; below roots of I_1s is apex of triangular swollen area in external midline that broadens laterally to inferior margin. In profile, this margin elevated anterior to the P_1s. In inferior view, bone of corpora thickest in midline; corpora diverge quite strongly from very narrow symphyseal region. M/l wide, quite deep digastric fossae point strongly backward. Bone modification evidently affected more posterior features of mandible.

Burial 6. Externally fairly complete but weathered skull; severely crushed, heavily reconstructed, lacking part of L frontal and parietal region, L orbital and zygomatic region, part of basicranium, nasal cavity, and associated structures, R zygomatic arch. Bone thin. Mandible lacks L condylar region; is permanently fixed to cranium (obscuring mandibular fossa). Difficult to assess age: M^2s fully in, M^3s unerupted, with minimal root formation, RC^1 in place; slightly incompletely erupted LC^1 may have been impacted (perhaps more teeth impacted). Closure of coronal suture difficult to assess; ectocranially, sagittal and lambdoid sutures clearly patent. Four parallel rows of small shells (6 or 7 per row) adherent to posterior part of L parietal.

Viewed from the front, skull generally narrow with considerable subnasal prognathism. In profile, cranium moderately long and very highly vaulted. Frontal rises steeply behind orbits; profile continues in shallower curve throughout most of its length and around to point level with asterion. From here, curve becomes stronger, recurving toward cranial base. Frontal bears distinct but low eminences. Course of temporal lines not visible.

Minimally swollen glabellar "butterfly" terminates laterally just medial to midline of orbits. Platelike lateral parts flat, slightly posteriorly directed. Interorbital region moderately broad superiorly but narrows somewhat inferiorly. Roof of orbit angled downward toward superior orbital margin, where bone rather acutely angles up and back over orbits. As preserved on the R, orbit very wide and subrectangular; inferomedial corner very rounded and medially extended. Shallow supraorbital notches lie quite medially. Most of inferior orbital margin thickened and distended outward somewhat. Region below infraorbital foramen, which lies moderately below inferior orbital margin, deeply excavated. Nasal bones displaced but strongly arced from side to side; were probably narrow, and, in preserved superior part, were probably parallel sided. Frontal processes were probably not elongate a/p. Nasal aperture was quite narrow; lateral crests do not continue around to anterior nasal spines. Anterior nasal spines represented by thick bases quite well forward relative to aperture. Almost imperceptible horizontal conchal crest palpable on R side. Although damaged, continuation of anterior lacrimal crest into nasal cavity was clearly shelflike. Floor of nasal cavity rather flat. Moderate nasoalveolar clivus strongly inclined forward and upward.

In front view, anterior root of zygomatic arch arcs strongly out; in side view, it arcs back from anterior plane of zygoma to lateral plane of arch; in vertical view, zygoma itself slopes backward very slightly. R side broken; appears that inferior border of zygomatic arch was almost horizontal and did not flare out away from cranial wall (temporal fossa thus small). Anterior squamosal flowed relatively smoothly into alisphenoid, which arced gently toward basicranium (thus distinct infratemporal fossa). Squamosal was apparently relatively long and tall. On the L, obtuse parietal notch lies over midline of anteriorly directed mastoid process. Auditory meatus ovoid, vertically oriented, thick walled; solidly appressed to base of mastoid process. Gracile posterior root of zygomatic arch flows smoothly into suprameatal crest that is more pronounced on the L than on the R; both suprameatal crests flow into low, upwardly curving supramastoid crests. Tips of mastoid processes broken (especially on the L). Mastoid processes were quite strongly projecting, quite long a/p, laterally expanded at bases; appear to have faced medially upon rather shallow, narrow mastoid notches. Damage makes unclear whether medial borders of mastoid notches were formed by paramastoid or occipitomastoid crests; however, this area did not project nearly as far as mastoid processes.

On the R, vaginal process broken; was clearly well developed; appears to have extended length of the ectotympanic tube to contact mastoid process. Damaged styloid process quite laterally placed. Medially placed, moderately sized carotid foramen faces backward; small jugular foramen faced down.

Most of region of foramen magnum missing. Preserved anterior margin narrow; preserved, moderately sized, arcing occipital condyle quite forwardly placed. Basiocciput appears to have flexed strongly upward. Somewhat downwardly sloping parietomastoid sutures rather long. Strongly denticulate lambdoid suture rises moderately steeply directly from asterion to peak at lambda. Narrow and tall occipital bears little evidence of nuchal scarring; lacks external occipital protuberance. Internally on the L, rather narrow petrosal bears strongly domed arcuate eminence; otherwise unreadable. Roots of anterior upper teeth, between Cs, strongly impressed on surface of bone.

Palate quite narrow, very long, shallow, with long, gentle anterior slope behind the Is; may have been horseshoe shaped; posteriorly, sides quite short s/i. Relatively small, single incisive foramen lies level with the Cs.

Major cranial sutures segmented and denticulated.

Mandible narrow, quite strongly curved across symphyseal region; corpora diverge quite strongly posteriorly. In side view, bone along roots of anterior teeth slopes forward, even beyond level of small mental trigon below. In front view, low central keel rises at tip of roots of I_1s; fans out strongly to somewhat thickened anterior border. As preserved better on the R, large, shallow mental fossa lies to side of keel and above inferior margin. Small mental foramen visible on the R lies below P_{1-2}. In inferior view, bone at symphysis appears to be only slightly thicker than bone more lateral to it. Part of R digastric fossa visible; was probably quite wide m/l and faced quite strongly posteriorly. Corpora thickest in region below

M_2; rami much thinner behind. Gonial angle only slightly obtuse and minimally inflected along inferior border. Ramus s/i short, a/p long; bears relatively broadly based, pointed coronoid process, below which is tall, shallow preangular sulcus. Anterior root of ramus hides unerupted M_3 completely. Internally, mylohyoid line very faint; submandibular fossa rather shallow.

I^{1-2}s very broad, high crowned, quite shoveled on their lingual surfaces, which bear moderately swollen but subdivided lingual swellings. Fully erupted upper RC^1 not as concave on its subdivided lingual surface. Erupted P^1s bear small conules in region of anterior foveae. On the L, very worn dm^2 retained; on the R, P^2 impacted against P^1. Somewhat worn M^1s have large, distolingually swollen hypocones and very centrally placed small trigon basins; enamel was somewhat wrinkled. Apparently newly erupted M^2s have smaller hypocones and quite extensive enamel wrinkling; moderately sized trigon basin lies centrally; preprotocrista well developed but postprotocrista much less so. As judged by unerupted M^3, enamel also wrinkled, preprotocrista well developed; hypocone lacking.

REFERENCES

Barral, L. and R. Charles. 1963. Nouvelles données anthropométriques et précisions sur les affinités systématiques des "Negroides de Grimaldi." *Bull. Mus. Anthropol. Préhist. Monaco* 10: 123–129.

Binant, P. 1991. *Les Sépultures du Paléolithique.* Paris, Errance.

Boule, M. and H. Vallois. 1957. *Fossil Men.* London, Thames and Hudson.

Cardini, L. 1931. Il paleolitico superiore della Barma Grande ai Balzi Rossi. *Archaeo. Antrop. Etnol.* 60/61: 461–476.

Cartailhac, E. 1912. *Les Grottes de Grimaldi: Archéologie.* Monaco, Imprimerie Nationale.

Lapouge, V. de. 1899. *L'Aryen: Son Role Sociale.* Paris, Fontemoing.

Rivière, E. 1887. *De l'Antiquité de l'Homme dans les Alpes-Maritimes.* Paris.

Schumann, B. 1995. *Biological Evolution and Population Change in the European Upper Palaeolithic.* PhD Thesis, University of Cambridge.

Sergi, S. 1911. *L'Uomo Secondo le origine, l'Antichita. Le Variazioni e la Distribuzione Geografica.* Turin.

Verneau, R. 1906. Les grottes de Grimaldi. *L'Anthropologie* 17: 291–320.

Villeneuve, H. de. 1906. *Les grottes de Grimaldi: Historique et Description.* Monaco, Imprimerie Nationale.

Repository

Musée d'Anthropologie Préhistorique, Monte Carlo, Monaco (Grotte des Enfants 4–6); Musée de Préhistoire Régionale, Menton, France (Barma Grande 1); Museo Preistorico dei Balzi Rossi, Ventimiglia, Italy (Barma Grande 2–5); Musée des Antiquités Nationales, St Germain-en-Laye, France (Baousse da Torre 1–3, Grotte des Enfants 1–2); Musée de l'Homme, Paris, France (Barma del Caviglione 1–3).

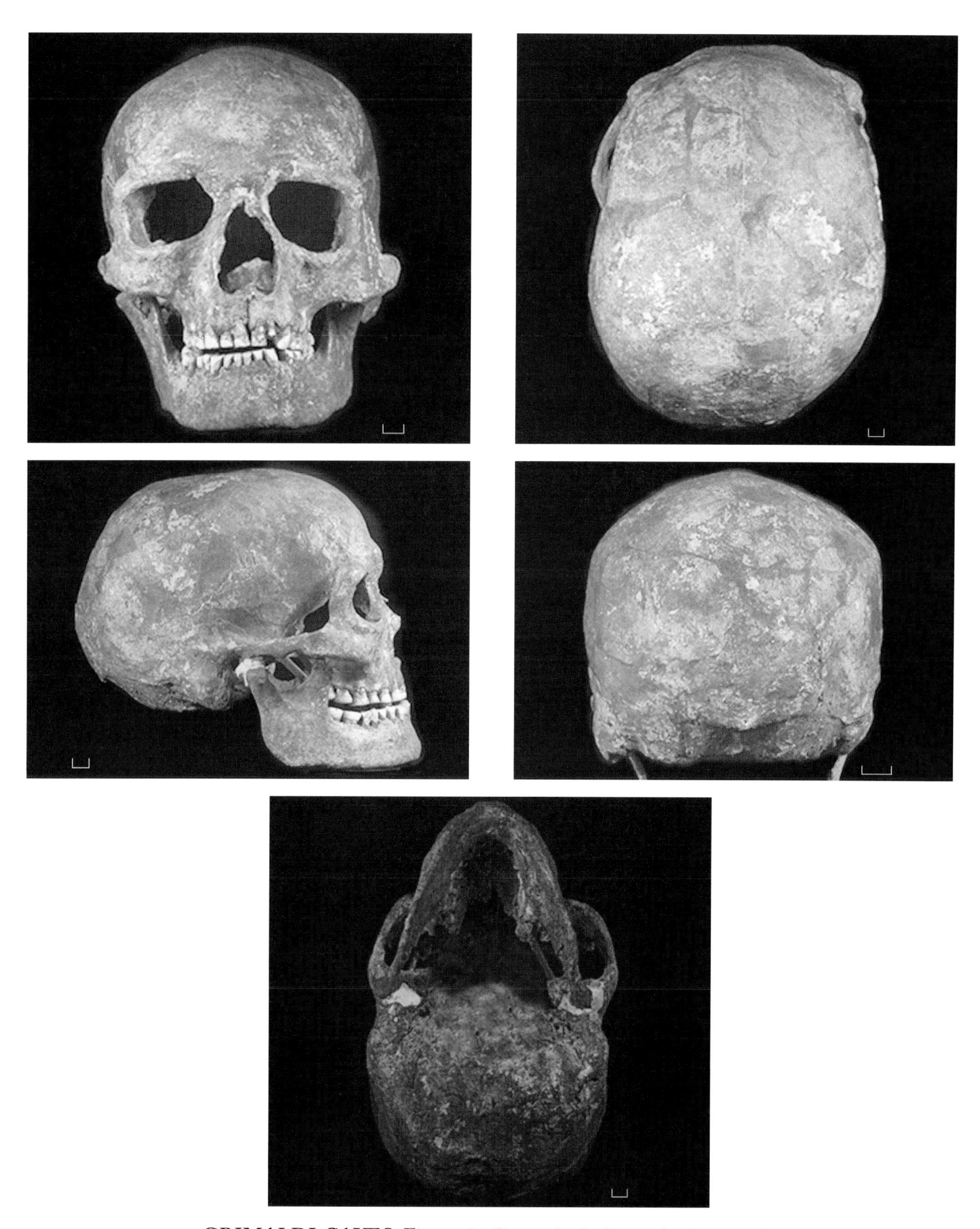

GRIMALDI CAVES Figure 1. Grotte des Infants 4 (scale = 1 cm).

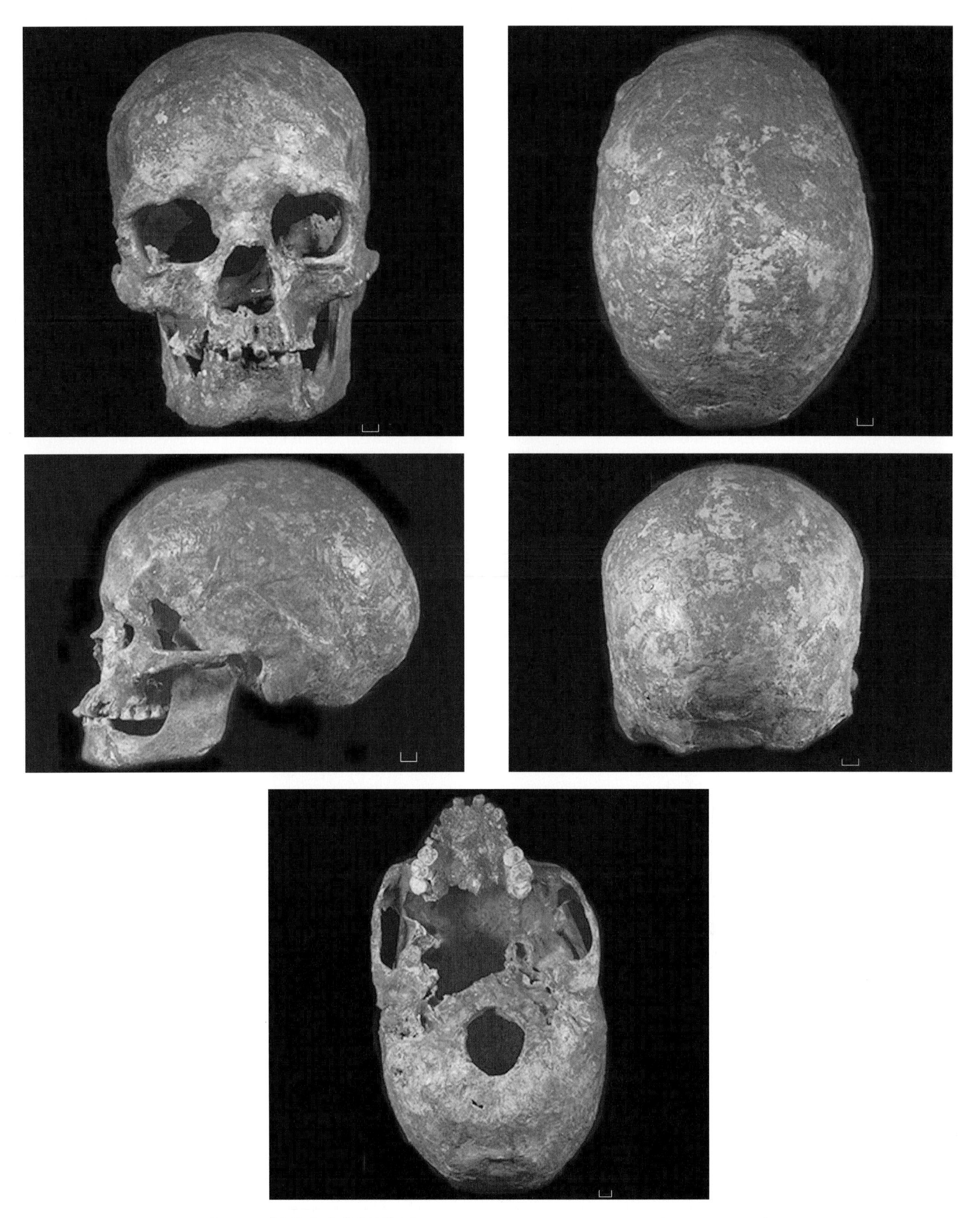

GRIMALDI CAVES Figure 2. Grotte des Infants 5 (scale = 1 cm).

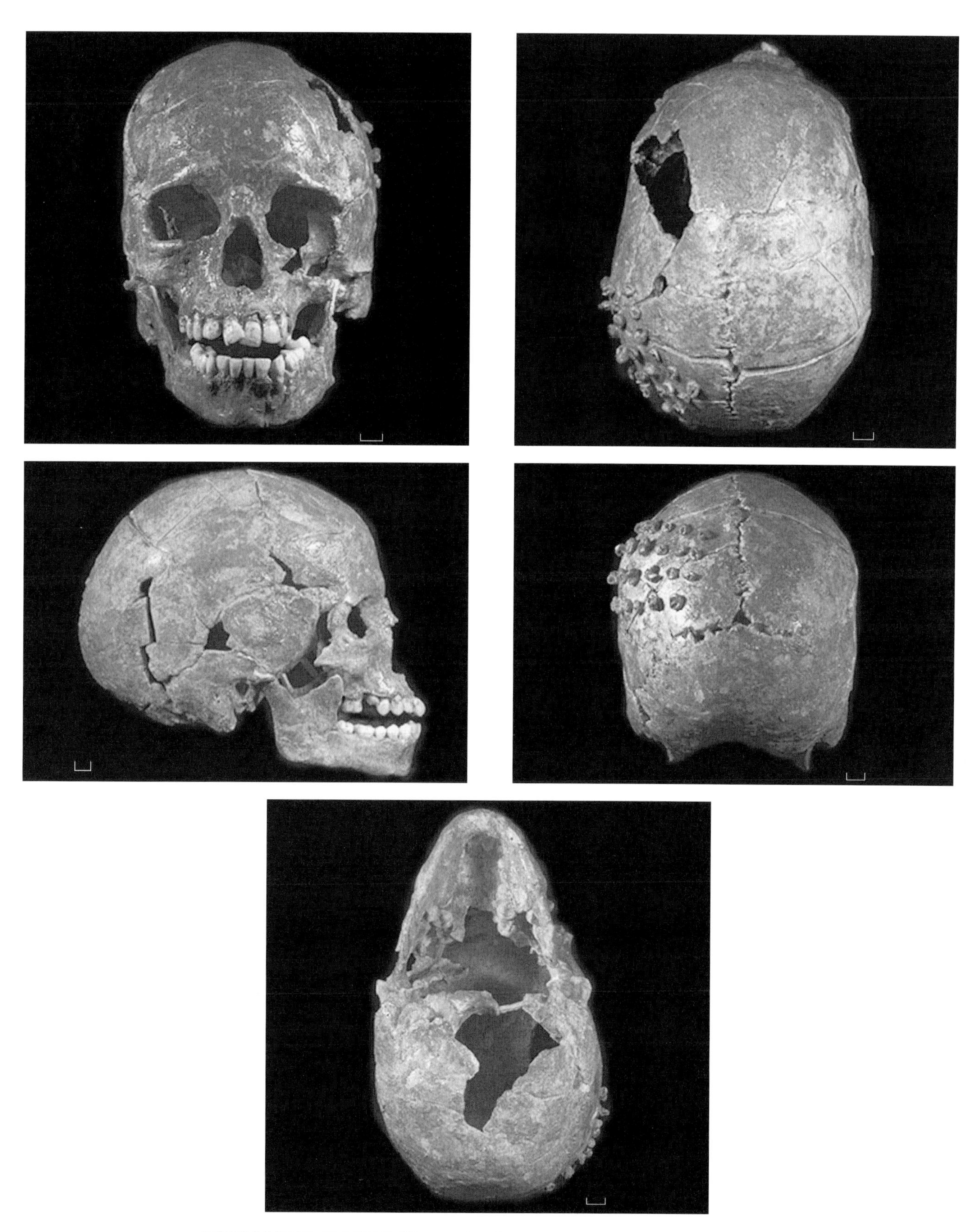

GRIMALDI CAVES Figure 3. Grotte des Infants 6 (scale = 1 cm).

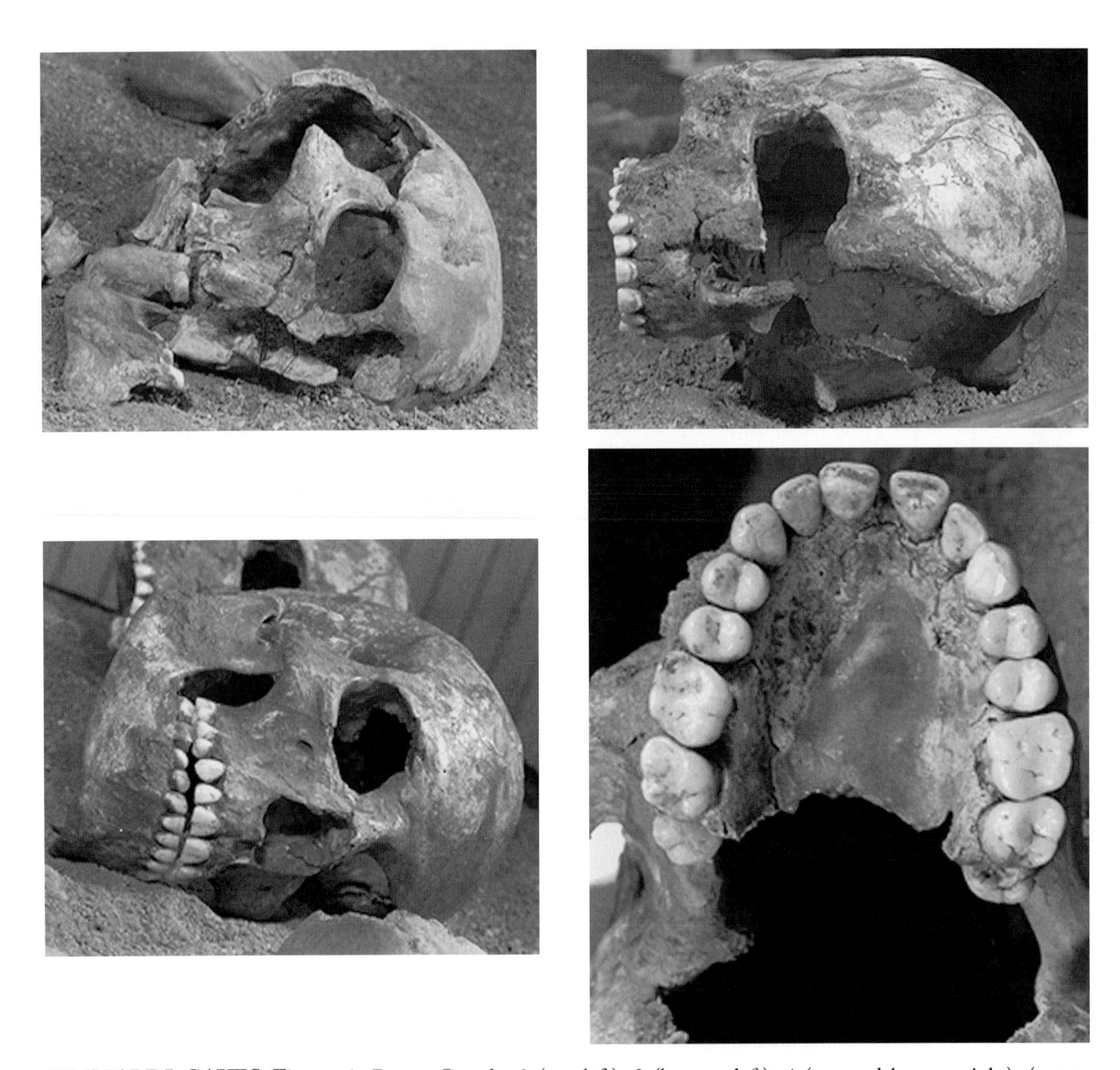

GRIMALDI CAVES Figure 4. Barma Grande: 2 (top left); 3 (bottom left); 4 (top and bottom right), (not to scale).

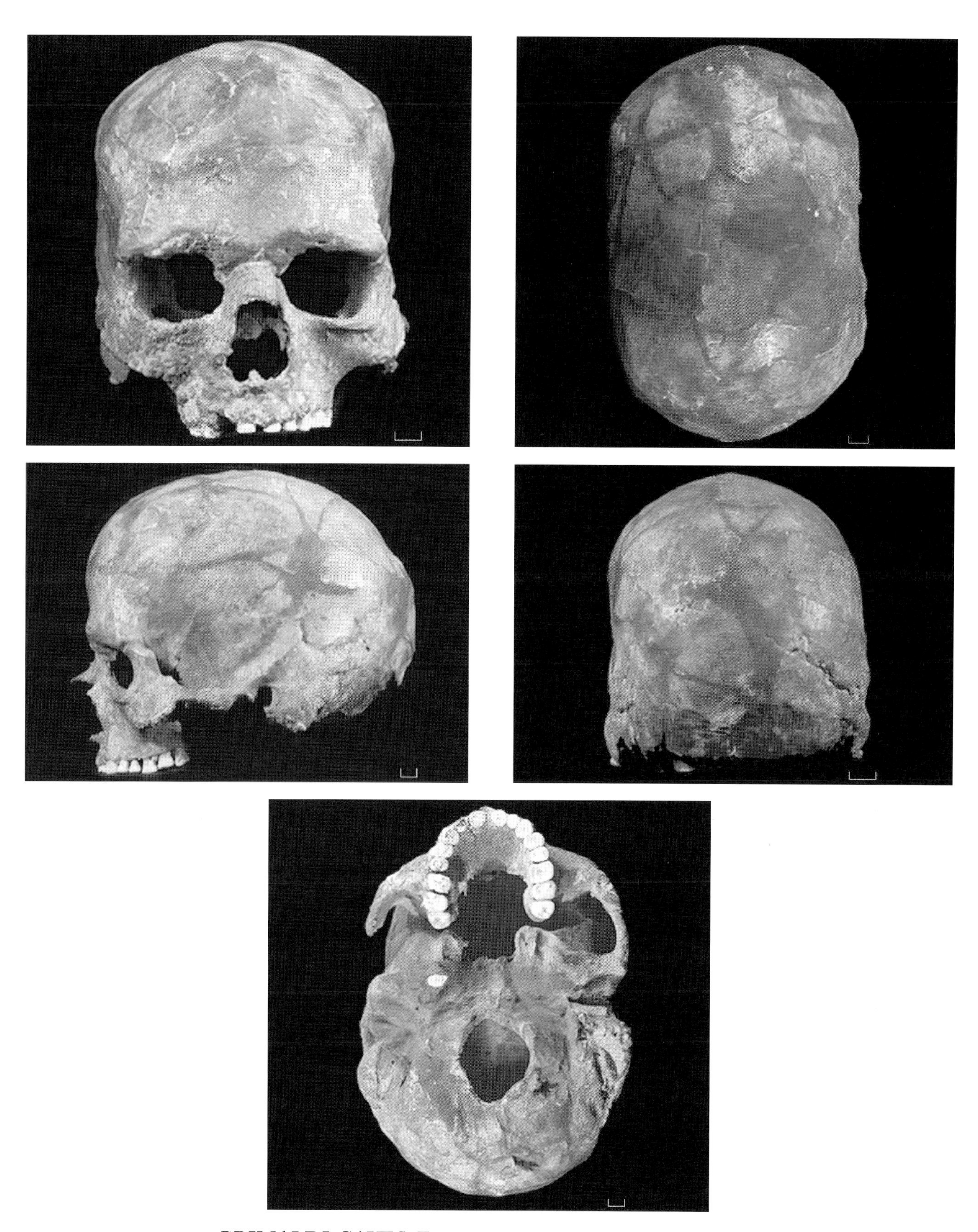

GRIMALDI CAVES Figure 5. Barma Grande 5 (scale = 1 cm).

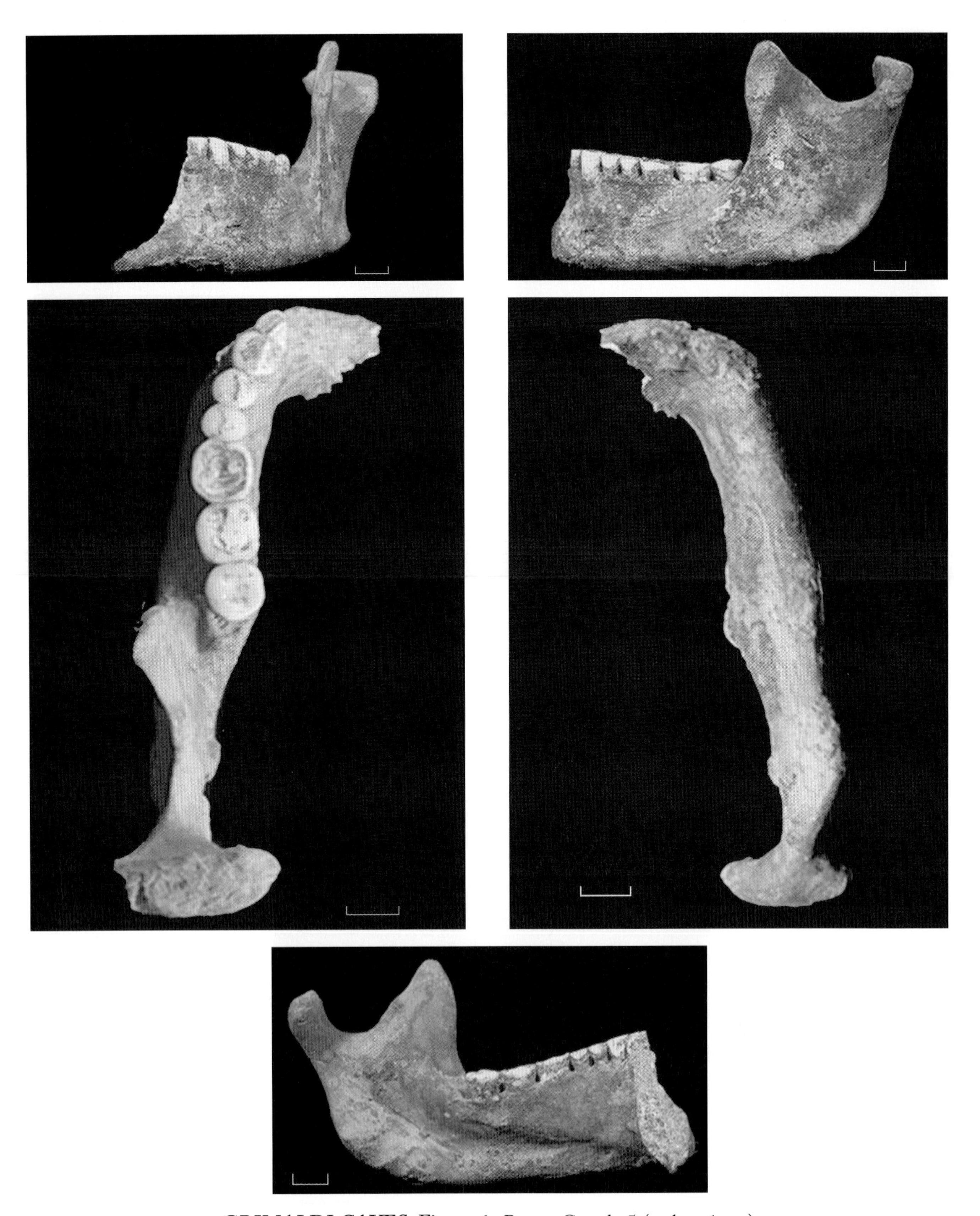

GRIMALDI CAVES Figure 6. Barma Grande 5 (scale = 1 cm).

Guattari (Monte Circeo)

Location

The Grotta Guattari is a small, low-roofed cave site just above present sea level and a few hundred yards E of the village of San Felice on the eastern flank of Monte Morrone, the eastern spur of Monte Circeo, Italy. Circeo is a limestone massif that rises where the Pontine marshes meet the Tyrrhenian Sea, about midway between Rome and Naples.

Discovery

A. Guattari and M. Palombi, February 1939 (Guattari 1 and 2 respectively); A. Ascenzi and G. Lacchei, August 1950 (Guattari 3). See accounts by Ascenzi (1991) and Stiner (1995).

Material

Cranium with damaged base, presumed male (Guattari 1); two partial mandibles (Guattari 2 and 3).

Dating and Stratigraphic Context

Guattari 1 and 2 were surface finds on a cave floor littered with bones; Guattari 3 was found in the collapsed rubble of one of the cave entrances. Four strata containing Mousterian tools and animal bones lie below the surface deposit, which is rich in hyena coprolites and carnivore remains (Blanc and Segre, 1953). U-series dating on the earliest calcite coatings of surface bones from the cave has yielded dates of 51 ± 3 ka, whereas ESR on teeth from the surface deposit has given dates of 44.0 ± 5 ka to 62.6 ± 6 ka, depending on uptake model (Schwarcz et al., 1991). Teeth from the layer immediately underlying have ESR dates of 44.0 ± 6 and 54.4 ± 4 ka. Most probable dates for Guattari 1 and 2 thus lie between 60 and 50 ka, whereas Guattari 3 is plausibly somewhat earlier, perhaps between 60 and 74 ka (Grün and Stringer, 1991).

Archaeological Context

Pontinian Mousterian (Taschini, 1979). The cave's first investigator, A. C. Blanc, believed that the Guattari 1 cranium, with its broken base, had been deliberately placed in the center of a ring of stones and was the victim of a ritual sacrifice (e.g., Blanc, 1961). However, more recent studies suggest that the cave served at least intermittently as a hyena den and that the surface accumulations date from such a time. See discussions by Stiner (1995) and Kuhn (1995).

Previous Descriptions and Analyses

Initial description of Guattari 1 and 2 was by Blanc (1939), who characterized the specimens as Neanderthal and entrusted Sergio Sergi with their study. Sergi followed up with a number of small publications (e.g., 1939, 1954) but did not complete his planned monograph on the cranium, a project finished after his death in a magisterial team effort edited by Piperno and Scichilone (1991). Guattari 3 was described by Blanc in 1951 and by Sergi and Ascenzi in 1955. That all three specimens are Neanderthal has never been questioned. Holloway (1985) estimated the cranial capacity of Guattari 1 at 1550 cc.

MORPHOLOGY

Guattari (Monte Circeo) 1

Adult. Partial cranium lacking large portion of base, R sphenoorbito-maxillary region and zygomatic arch, nasal bones, and much of finer internal nasal cavity detail. Some of external surface, and much of internal, covered with matrix; surface damage to some of bone. Bone not very thick.

Moderately sized, relatively long cranium, with low profile and some chignon at rear, broadest posteriorly; quite low just above squamosal sutures. Face broadly, not acutely wedge shaped. Low rise of frontal starts well posterior to supraorbital margins and slopes back gently to about midpoint of sagittal suture, at which point it descends rather flatly to lambda, beyond which occipital plane bulges in moderate chignon. Most posterior point of skull lies very low on occipital plane, just above area of suprainiac depression (this area has skim coat of matrix on it).

Supraorbital tori moderately thick medially; taper gently laterally, are confluent across very broad, only moderately swollen glabella. Broad, blunt cornering from roof of orbit onto anterior surface of smoothly but somewhat acutely curved back brow (not rolled in constant arc); patches of vermiculate bone across brow. In front of very gentle frontal rise, and behind glabella, is small oblique planar surface; laterally, this plane broadens a/p. On the L, somewhat medially placed shallow notch in superior orbital margin; on the R, longer, much shallower supraorbital notch. Viewed from above, glabella projects anteriorly beyond level of supraorbital margins; lateral to glabella, margins slightly concave before becoming slightly convex laterally (overall profile retreating). Breakage of lateral part of R brow reveals that sinuses did not extend beyond midline of orbit. Nasion low and minimally overhung by glabella. Entire frontonasal-frontomaxillary suture broadly peaked and somewhat depressed. Orbital shape "aviator glasses," with inferomedial corner cut off by inflation of a/p long maxillary frontal process, which raises profile of projecting process above level of orbital floor; superomedial corners of orbits also strongly affected by inflation of frontal bone. As seen on the R, maxillary sinuses continuous with pneumatization of supraorbital region. Breakage in L medial orbital wall reveals multiple small to medium air cells. As also preserved on the L, lacrimal fossa bounded anteriorly by a/p thickened, blunt anterior lacrimal crest that fades out well before reaching frontal suture. Posterior lacrimal crest barely expressed. Relatively shallow lacrimal fossa not very tall s/i; opens superiorly, draining inferiorly into round, not very large lacrimal foramen/canal.

Nasal bones narrow superiorly, quite strongly curved from side to side; missing inferior parts would have been quite broad at nasal aperture margin. Internally, as seen on the L, nasals were invaded by sinuses that apparently derived from ethmoidal complex. Nasal aperture was probably rather squarish, with rounded corners. Lateral crest continuous right down sides of aperture all the way around to midline. Region of anterior nasal spines broken. Vomer present anteroinferiorly; it comes forward quite far. Fairly close behind inferior nasal margin, nasal cavity floor becomes quite depressed. Internally, walls of nasal cavity preserved anteriorly; on the L, wall of nasal cavity behind nasal aperture encrusted with matrix (what appears as series of rugosities that might correspond to conchal crest is simply matrix surface). On the R, where matrix appears to have been largely removed, vertical medial projection situated well back within external margin and protrudes somewhat into nasal cavity. Superior to medial projection, nasal cavity surface of frontal process strongly inflated; on the R, damage exposes undivided airspace. Posterior to medial projection, thin edges of broken bone reveal between them long, vertical gutter that appears to be continuation into nasal cavity of lacrimal fossa. Behind this and parallel to it, is similar long gutter that lay within lacrimal bone itself. Internally, maxillary sinus (as seen on the L) continues well up into frontal process; toward lateral side of this sinus is long bony tube that originates far back in orbital floor as continuation of broad lacrimal groove and opens anteriorly onto somewhat downwardly oriented infraorbital foramen. As seen on the L, maxillary sinus does not extend very far laterally into zygoma; as preserved posteriorly, medial wall of this sinus does swell out noticeably into nasal cavity. Looking toward roof of nasal cavity, plethora of small to medium air cells (visible through broken bone) extend all the way back to region of large R and L sphenoidal sinuses. These air cells form part of a sinus complex that includes nasal bones, as well as ethmoid, sphenoid, and possibly also maxillary sinuses (origin of frontal sinuses possibly not necessarily from ethmoid alone). General inflation of area would also have ensured not much room for development of superior and middle ethmoturbinals.

Nasoalveolar clivus moderately long, somewhat vertical. Most of alveolar crest was destroyed antemortem by periodontitis. Overall arc indicates that smallish palate was evenly and tightly curved anteriorly and somewhat narrow, with only slightly divergent tooth rows. Repaired damage on the R makes original contour of anterior root of zygomatic arch problematic; some encrustation present on the L; judging from the latter, anterior zygomatic root may have originated quite close to original alveolar margin, whose straight inferior border rises sharply outward and slightly backward toward zygomatic arch. Body of zygoma curves backward toward short, s/i tall, slightly outwardly angled zygomatic arch. Inferior border of arch quite horizontal; superior border parallels it about halfway before descending posteriorly.

On the L, short, moderately pronounced temporal ridge originates along posterior border of short frontal process. Almost immediately after curving backward, ridge disappears in area of damaged bone surface; more posteriorly, intermittent traces of faint temporal lines. On the R, superior temporal line traceable posteriorly; describes low curve that recurves and fades out well before reaching parietal notch. On the L, posterior surface of zygoma faces backward upon relatively deep anterior temporal fossa, which is defined posteriorly by blunt, pronounced cornering of anterior squamosal suture into markedly vertically concave alisphenoid. In the vertical plane, alisphenoid curves gently to region of pterygoid plates, with no differentiation of infratemporal fossa.

As seen better on the L, relatively small but heavily muscle-scarred squamosal as tall s/i as long a/p; its superior sutural margin gently curved. Postmortem processes have somewhat displaced squamosals from parietals; seems, as seen better on the L, that posterior margin of squamosal, in region immediately above vertical parietal notch (which lies above midline of mastoid process) is thickened and protrudes laterally, well beyond level of parietal behind it. Thick posterior root of zygomatic arch originates in front of small, ovoid, posteriorly inclined auditory meatus, over which is moderately thick suprameatal crest that is slightly keeled along its midline. On the L, this crest fades out posteriorly after rising into area of thickened posterior squamous border. No supramastoid crest. On the R, keeled area continues over supramastoid region to terminate at raised border of posterior squamosal.

On the L, mandibular fossa shallow, not very long a/p; is very broad m/l and bounded medially by prominent, downwardly projecting medial articular tubercle. Low, backwardly sloping postglenoid plate extends laterally beyond lateral end of ectotympanic tube, to which it is fused medially. No definable articular eminence; region is continuation of articular concavity (not salient projection). Tubular ectotympanics fused to bases of basally stout, apparently strongly tapering, downwardly pointing mastoid processes (both damaged); on the L, vertical swelling may be interpreted as mastoid tubercle. Low, damaged vaginal process apparently faded out laterally before reaching auditory meatus; peaked slightly around small, medially placed styloid process, from which moderately sized stylomastoid foramen is displaced markedly posteriorly and somewhat laterally. Both vaginal processes course along midlines of ectotympanic tubes; would not have contacted mastoid processes. Medially placed, relatively large carotid foramina point somewhat backward. What appears to be broad, deep mastoid notch is probably artifact of cleaning. Medial to this "notch," occipitomastoid suture runs in gutter along midline of raised prominence that may or may not have descended further than broken tip of mastoid process. On the L, putative evidence along broken edge of occiput of low Waldeyer's crest quite far medial to occipitomastoid crest and extending quite far posterior to it.

Preserved anterior portion of thin basiocciput (which bears on its external surface pair of lateral tubercles as well as short median keel) flares out widely posteriorly. Preserved tips of long, thin petrosals extend m/l beyond small foramina lacera that lie on either side of basiocciput. Basiocciput would not have been very strongly flexed upward anterior to foramen magnum. Damaged pterygoid plates were apparently parallel to one another throughout their lengths; may have converged both superiorly and inferiorly. Large foramina ovales lie behind midline between plates; large foramina spinosa lie quite far m/l to foramina ovales. Posterior root of vomer appears to lie just in front of sphenooccipital synchondrosis.

On the L, appears that ossicle at asterion is bordered below by relatively long, horizontal anterior lambdoid suture, from which lambdoid suture proper rises to lambda, where it appears not to peak. L anterior lambdoid suture is direct extension backward of relatively long, horizontal parietomastoid suture. As seen on the L, superior nuchal line arises from juncture of anterior and proper lambdoid sutures; thickens as it curves strongly medially. Almost horizontal nuchal plane quite strongly undercuts superior nuchal line, which appears to be lowest at its midpoint (contour obscured by matrix in inferior view). Matrix-covered region above superior nuchal line appears to bear very wide but not very tall suprainiac depression that defines upper border of "occipital torus."

Coronal suture finely denticulate; straightens out just lateral to bregma. Sagittal suture more deeply denticulated; straightens out in region adjacent to bregma. Lambdoid suture bears ossicles at both asterions, as well as in middle of L side, and possibly also in region of lambda; elsewhere suture is denticulated (more so than the coronal and less than the sagittal).

Internally, dorsum sellae lies well below level of tuberculum sellae and jugum; the latter appears to have been rather long a/p. Moderately wide petrosal bears two low swellings on its superior surface; medial swelling lies slightly posterior to lateral one; no superior petrous sinus. Region of subarcuate fossa closed over and slightly depressed. Internal auditory meatus directed quite medioanteriorly. Long frontal crest runs down to very tall s/i, apparently pointed crista galli.

Guattari 2 and 3

Adults. Both Guattari 2 and 3 mandibles not very broad at front of jaw, but corpora strongly divergent posteriorly. Digastric fossae imprinted underneath "hint" of inferior transverse torus. Mylohyoid lines descend steeply. Long retromolar space. Bone not very thick.

Guattari 2

Quite heavily damaged mandible; retains all of the L, most of the R ramus; all teeth except for RM2.

Corpora uniform in thickness all around. Front of jaw smoothly curved. Symphysis would have been quite tall s/i; is quite straight and vertical in profile. External surface of symphyseal region more or less featureless (chipped bone near midline not depressions). Inferior margin featureless and smoothly curved from side to side. Well-developed digastric fossae wide from side to side, oriented downward. Internally, very short, almost vertical postincisal plane; below, surface becomes broadly concave; toward base of concavity is broad roughened area corresponding to region of genial tubercles. Mylohyoid line on the R more rugose and ridgelike than on the L; below well-developed submandibular fossae. As preserved on the L, inferior border of corpus rises posteriorly, suggesting that there would have been upward curvature to gonial region; inferior border sinks markedly behind, where M3 would have been, and is also slightly inflected.

Remaining M3 very worn; enamel would have been quite highly crenulated; there would have been a large, centrally placed hypoconulid; in occlusal outline, crown would have been rather rounded.

Guattari 3 (III)

Damaged mandible lacking posterior part of L corpus, all of L ramus, region of gonial angle and condylar area of R ramus. Teeth present: R and LI2s, R and LCs, LP1, and R and LM1–3. All teeth quite worn; LM2–3 damaged.

Jaw quite deep s/i, especially toward front. Symphyseal region strongly curved. In front view, region between roots of Cs and I2s, and continuing around roots of I2s, deeply excavated (makes bone below appear to be in relief). In profile, distinct subalveolar depression discernible; above it, roots of anterior teeth somewhat displaced anteriorly, with crowns somewhat reflected back on them. Below this subalveolar depression, symphyseal region gently convex. Viewed from below, rugose, broad, well-excavated digastric fossae point somewhat back and down. Internal surface of symphyseal region in general somewhat vertical; lacks distinct postincisal plane. Lower two-thirds or so of internal surfaces of corpora broadly and shallowly concave bilaterally to mylohyoid lines; submandibular fossae quite steeply descending. On the R, although angle missing, gonial area thin; would have been inwardly inflected. Part of large medial pterygoid tubercle preserved. Moderately sized mandibular foramen compressed; points up and back, its inferior border incomplete posteriorly because of descending mylohyoid fissure. Internal pillar preserved at base of damaged coronoid process; pillar curves down and forward to become confluent with thick internal alveolar crest. Preserved course of rather steep, concave anterior ramal margin hints at presence of preangular sulcus. External surface of ramus bears few vertical muscle markings. On both sides, moderately sized plus smaller mental foramen lie below P2–M1.

Lower I2s shallowly convex lingually, minimally swollen on their lingual bases. Cs not very large; internally, shallowly concave surfaces bordered distally by thick margocristids. P1 quite broad m/d along buccal side; its long mesial and even longer distal edge run directly from protoconid apex. Internally, from the protoconid base, a short, stout crest swells out into lingual tubercle. From extremity of mesial buccal ridge, short, stout crest runs to base of lingual tubercle, from which is separated by thin crease; crest encloses small fovea. From distal buccal ridge, stout crest runs to become confluent with lingual tubercle, enclosing slightly larger posterior fovea. P1 has very strong buccal slope from the rather centrally placed protoconid apex out to the cervix.

M2 slightly larger than M3, which is larger than M1. All three molars with hypoconulids lying buccal to midline of tooth and separated from hypoconids by shallow buccal notch. Judging from the R, all three molars had distinct trigonid basins as well as talonid basins that were compressed somewhat by relatively internally placed buccal cusps (buccal sides of teeth somewhat swollen). Enamel of at least M3 appears to have been somewhat wrinkled.

REFERENCES

Ascenzi, A. 1991. A short account of the discovery of the Monte Circeo Neandertal cranium. In: M. Piperno and G. Scichilone (eds), *The Circeo 1 Neandertal skull: Studies and Documentation.* Rome, Istituto Poligrafico e Zecca dello Stato.

Blanc, A. C. 1939. L'uomo fossile del Monte Circeo. *Atti Acad. Naz. Lincei Rc.,* Ser. 6, 29: 205–210.

Blanc, A. C. 1951. Rinvenimento di una mandibola fossile nella breccia ossifera esterna della grotta Guattari a San Felice Circeo. *Boll. Soc. Geol. Ital.* 70: 590–591.

Blanc, A. C. 1961. Some evidence for the ideologies of early man. In: S. L. Washburn (ed), *The Social Life of Early Man.* Chicago, Aldine, pp. 119–136.

Blanc, A. C and A. Segre. 1957. Excursion au Mont Circé. *INQUA IV Cong. Int.,* Roma, Pisa.

Grün, R. and C. Stringer. 1991. Electron spin resonance dating and the evolution of modern humans. *Archaeometry* 33: 153–199.

Holloway, R. L. 1985. The poor brain of *Homo sapiens neanderthalensis*; see what you please. In: E. Delson (ed), *Ancestors; The Hard Evidence.* New York, Alan R. Liss, pp. 319–324.

Kuhn, S. 1995. *Mousterian Lithic Technology.* Princeton NJ, Princeton University Press.

Piperno, M. and G. Scichilone. 1991. *The Circeo 1 Neandertal skull: Studies and Documentation.* Rome, Istituto Poligrafico e Zecca dello Stato.

Schwarcz, H., A. Bietti, W. Buhay, M. Stiner, R. Grün and A. Segre. 1991. Absolute dating of sites in coastal Lazio. *Quaternaria Nuova* (n.s.) 1: 51–67.

Sergi, S. 1939. Il cranio neandertaliano del Monte Circeo. *Atti Acad. Naz. Lincei Rc.* 41: 305–344.

Sergi, S. 1954. La mandibola neandertaliana Circeo II. *Riv. Antrop.* 41: 305–344.

Sergi, S. and A. Ascenzi. 1955. La mandibola neandertaliana Circeo III. *Riv. Antrop.* 42: 337–403.

Stiner, M. 1995. *Honor Among Thieves.* Princeton, NJ, Princeton University Press.

Taschini, M. 1979. L'industrie lithique de Grotta Guattari au Mont Circé (Latium): Définition culturelle, typologique et chronologique du Pontinien. *Quaternaria* 12: 179–247.

Repository

Guattari 1: Museo Nazionale Preistorico Etnografico "L. Pigorini," P. le G. Marconi 14, 00144 Roma (Eur), Italy; Guattari 2 and 3: Istituto Italiano di Paleontologia Umana, 2 Piazza Mincio, 00198 Roma, Italy.

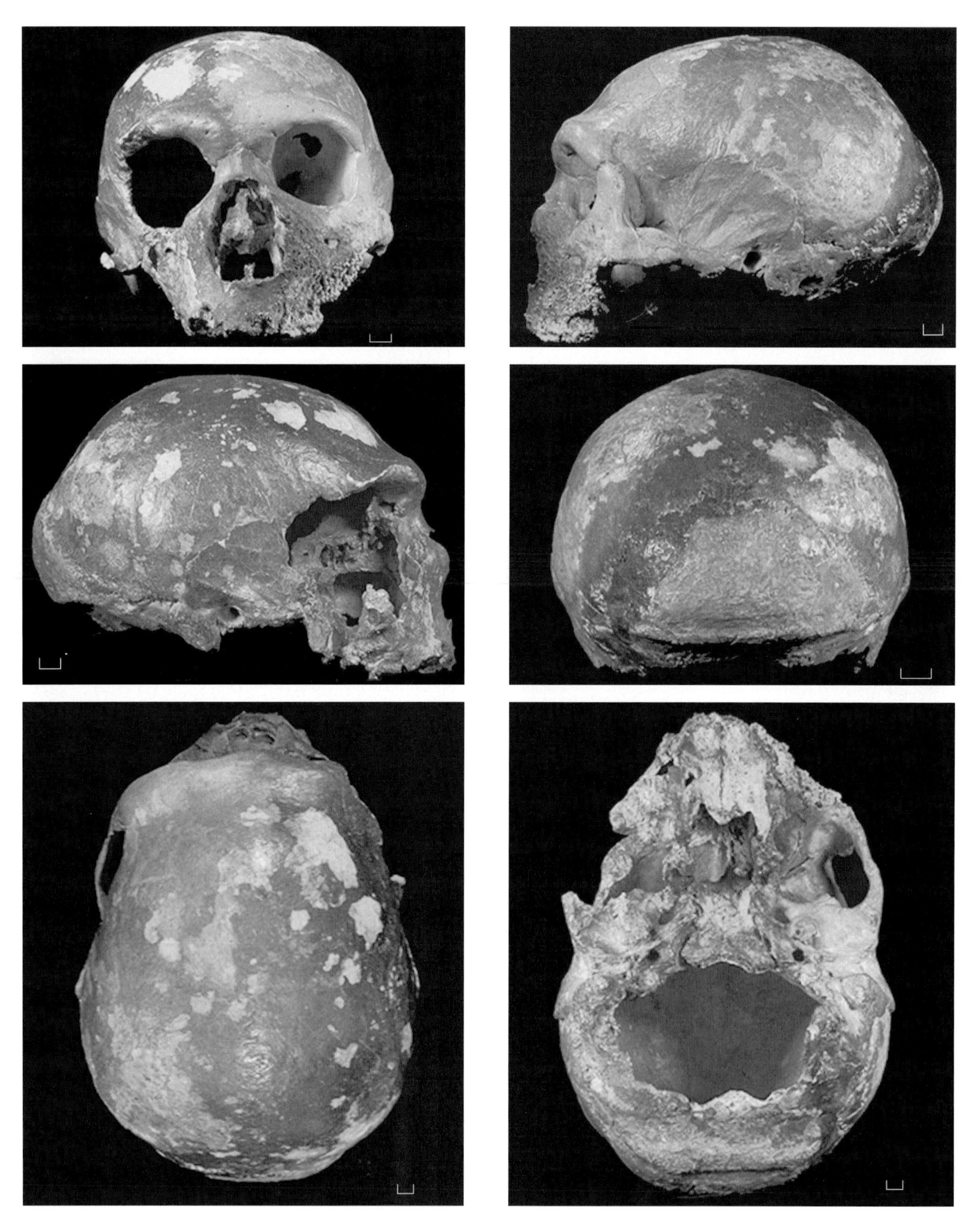

GUATTARI Figure 1. Guattari 1 (scale = 1 cm).

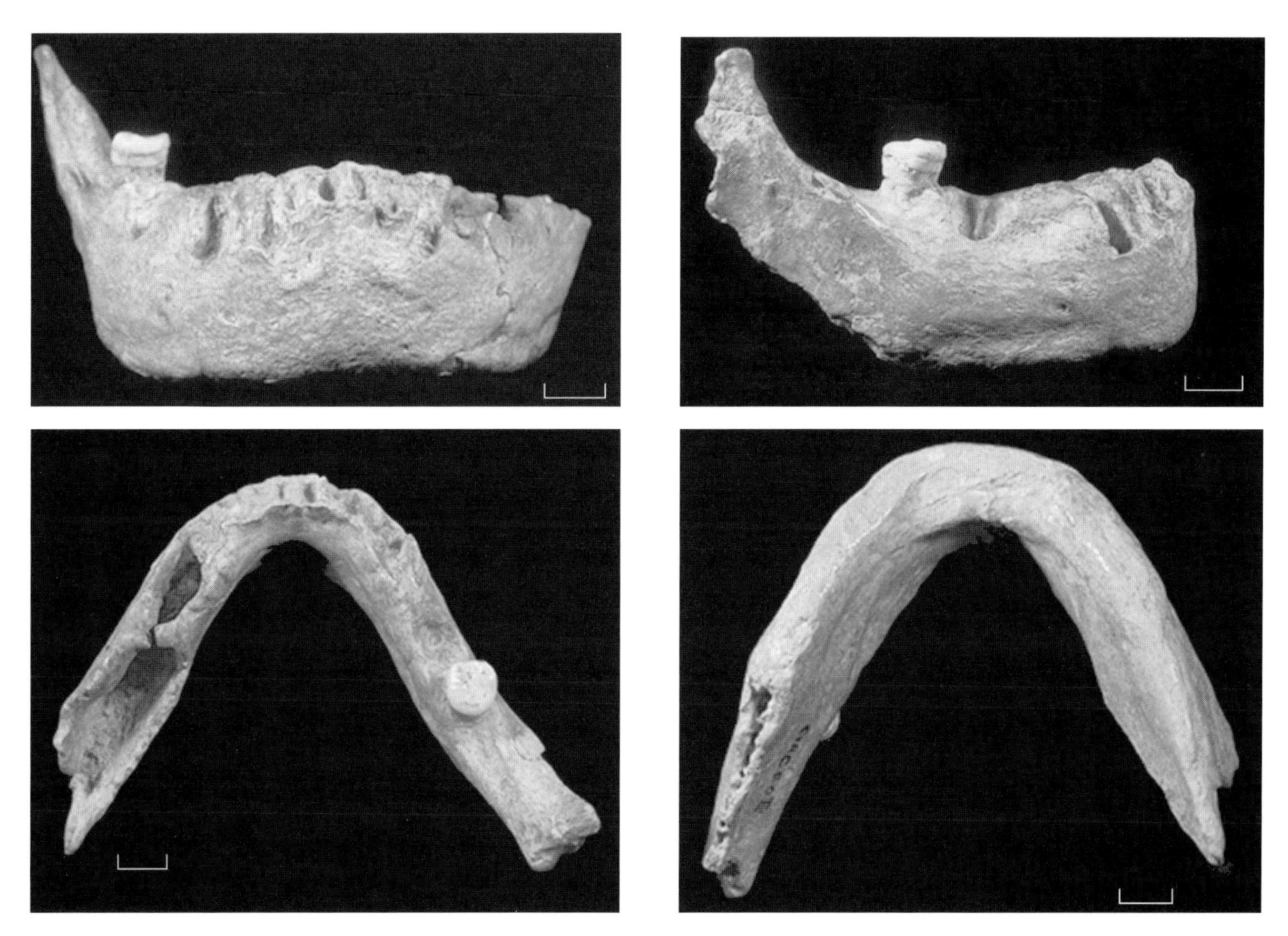

GUATTARI Figure 2. Guattari 2 (scale = 1 cm).

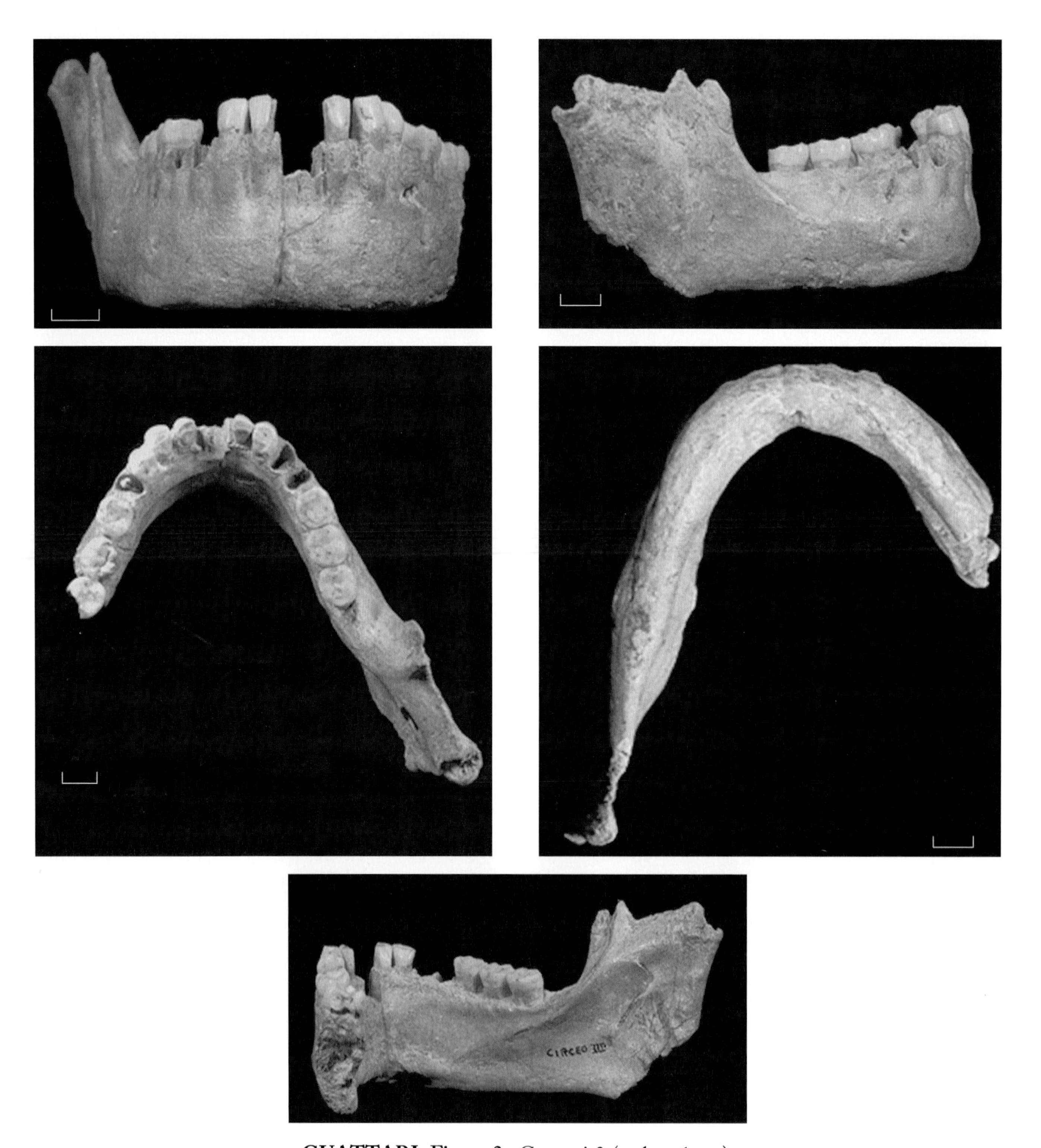

GUATTARI Figure 3. Guattari 3 (scale = 1 cm).

Hahnofersand

Location
S bank of the Elbe River near the western limit of metropolitan Hamburg, Germany, about 1.5 km E of the eastern tip of the island of Hahnöfersand.

Discovery
Found by H. R. Labukt, March 1973.

Material
Much of an adult frontal bone, with heavily damaged supraorbital region.

Dating and Stratigraphic Context
The fossil was found at a depth of ca. 2 m in glacial sands derived from the river bed (Bräuer, 1980a). Direct dating by R. Protsch yielded 36 ka by amino acid racemization and 36,300 ± 600 yr B.P. by conventional radiocarbon (Bräuer 1980a, b; 1981).

Archaeological Context
None.

Previous Analyses
Shows evidence of both Neanderthal and modern human affinities (Bräuer 1980a, b; 1981).

Morphology
Frontal missing parts of supraorbital region including glabella; large portion of R side removed for amino acid racemization and ^{14}C dating. Bears cut marks running obliquely across L supraorbital region and shorter ones just below L frontal eminence. Bone very thick, with thick inner and outer tables as well as relatively wide diploe.

Coronal suture preserved on L segmented; poorly denticulated in medial half but deeply denticulated laterally. Lateral extremity lost from the point where bone starts to thin. L and R superior orbital margins preserved medially; the R bears large depression probably of traumatic origin. On both sides, large, quite medially placed supraorbital notch. Supraciliary areas somewhat prominent, with flattish anterior surfaces. Superior orbital margin peaks just above supraorbital notch. Laterally, damage on both sides; preserved on the L, small portion of orbital roof, which makes sharp angle with brow above. Lateralmost part of supraorbital region clearly preserves change in orientation from mounded medial supraorbital portion to more planar lateral portion. Medially, mounded supraorbital portion flows into what would have been swollen, quite projecting, probably broad glabellar region.

Capacious, multifocular frontal sinuses visible on both sides; extend quite high up into frontal bone and laterally past midpoint of superior orbital margin. On the L, frontal sinus extends almost as far as departure point of temporal ridge. On both sides, plane immediately behind supraorbital region has shallow slope; behind it, frontal rises steeply and arcs over low-placed frontal eminences bilaterally. In midline, lower portion of frontal bears low, slightly sinuous keel. Temporal ridges preserved bilaterally behind their point of emergence; take form of roughened ridges that arc gently backwards. Below them, tiny preserved portions of frontal show oddly smooth surfaces. Following trajectory of frontal bone, appears that the cranium would have been quite high domed. Internally, frontal crest not preserved inferiorly; its preserved superior part quite sharp and elevated. Several Pacchionian depressions visible on either side of midline. Frontal lobes extend well forward over anterior roof of orbits (seen on R). Meningeal artery grooves also visible quite far forward.

REFERENCES

Bräuer, G. 1980a. Die morphologischen Affinitäten des jungpleistozänen Stirnbeines aus dem Elbmündungsgebiet bei Hahnöfersand. *Z. Morph. Anthrop.* 71 (1): 1–42.

Bräuer, G. 1980b. Nouvelles analyses comparatives du frontal Pleistocène supérieur de Hahnöfersand, Allemagne du Nord. *L'Anthropologie* 84 (1): 71–80.

Bräuer, G. 1981. New evidence on the transitional period between Neanderthal and modern man. *J. Hum. Evol.* 10: 467–474.

Repository

Helms-Museum für Vor- und Frühgeschichte, Museumsplatz 2, Harburg, 21073 Hamburg, Germany.

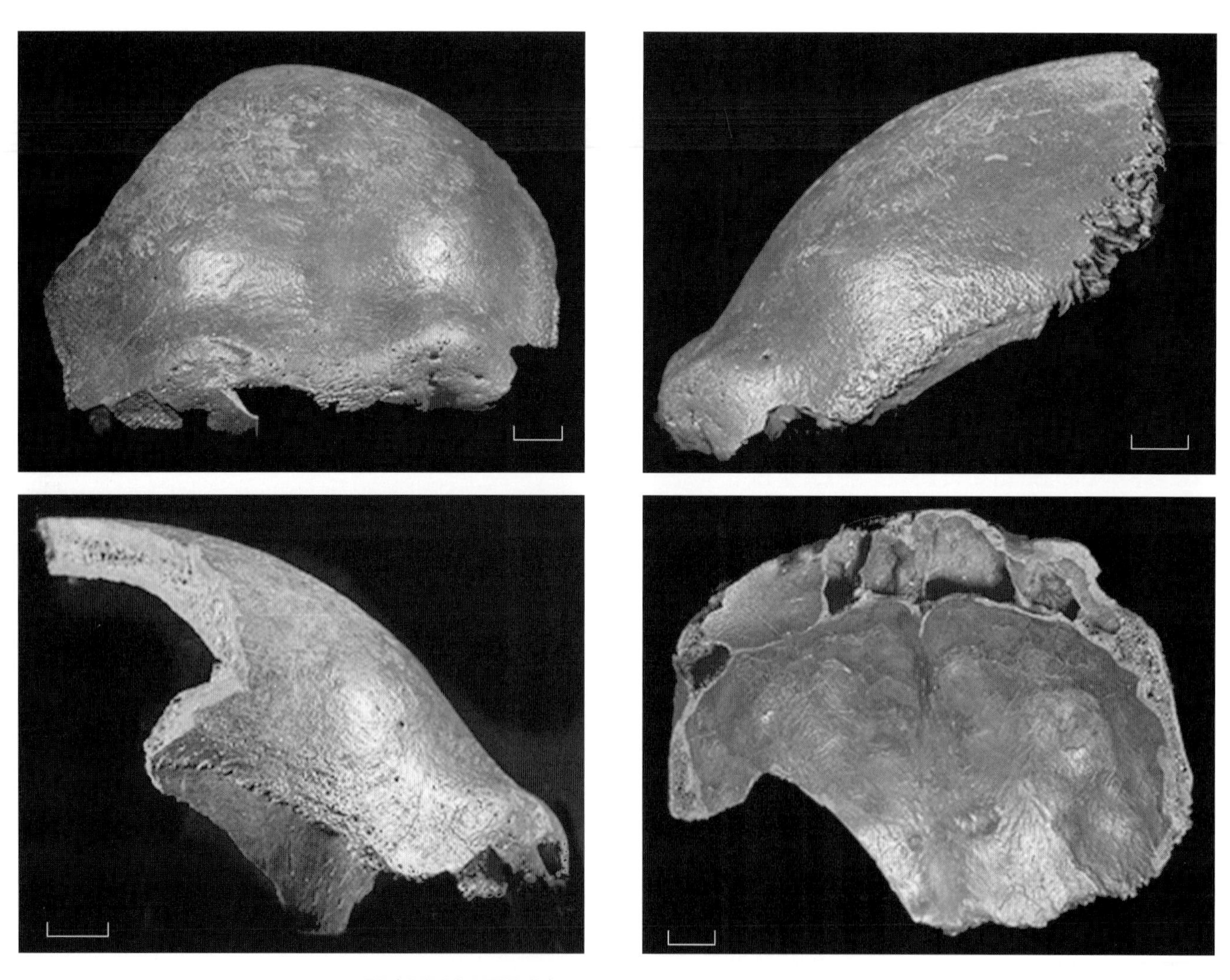

HAHNOFERSAND Figure 1. Scale = 1 cm.

Hortus

Location

Cave in the Hortus Massif in the Terrien Valley, N of Montpellier, southern France.

Discovery

Excavations of H. and M.-A. de Lumley, 1960–1964.

Material

Various fragmentary hominid fossils, most notably the front of the jaw of a child (HII) and a fragmentary adult mandible (HIV).

Dating and Stratigraphic Context

Cave deposit with frost-shattered rock fragments in clay matrix. Fauna also suggests cold conditions; a middle Würm II age has been suggested (de Lumley, 1965; de Lumley et al., 1972).

Archaeological Context

"Typical Mousterian of Levallois facies" (Piveteau et al., 1963; de Lumley, 1965; de Lumley et al., 1972), but the industry is quite sparse.

Previous Descriptions and Analyses

The principal description is that of Piveteau et al. (1963), who are clear that the remains are Neanderthal.

Morphology

Hortus II

Front of mandible of child with alveoli for I1–2s, with crowns and some root development of R and LCs and P1–2; only occlusal surfaces of premolars visible.

For a juvenile, symphyseal region very broad; swollen out on the L (and presumably the R) by unerupted tooth crowns. Low scar runs vertically from just below alveolar margin, down external midline of symphysis, to fade out just before reaching inferior border. In profile, symphysis runs slightly backward and down, curving around to inferior margin. Viewed from below, symphysis appears to be uniformly thick a/p at its inferior border, but slightly peaked profile due to central keel also visible beyond, slightly further up external symphysis. Digastric fossae very wide m/l, deeply excavated, somewhat anteriorly situated (thus where they meet in midline is slight downward projection of bone), and oriented downward and only slightly posteriorly. Postincisal plane moderately long, quite steeply inclined. Approximately midway up internal midline is shallow depression with deep pit in center. From pit, slight midline scar runs between digastric fossae to meet peak between the latter.

I1 alveolae bear low, vertical keels along mesial and distal walls; much larger I2 alveolae (as seen on the L) bear more pronounced distal keels. P1 crowns appear to be longer m/d than P2 crowns. P1 protoconids lie just mesial to midline (= mesial edge slightly shorter m/d than distal edge); from both buccal corners thick crests run lingually to become confluent with low but distinct lingual swelling connected to internal base of protoconid by low crest (thus P1s have small but distinct anterior and larger and even more distinct posterior foveae). P2s have large, somewhat mesially placed protoconids, at bases of which lie small metaconids. Stout crest runs between protoconid and metaconid, enclosing fairly large anterior fovea. Very thick crest runs from distal side of protoconid, distending distolingual corner of the tooth before turning up and forward to meet metaconid; this crest encloses deep, b/l wide but m/d thin posterior fovea. P2 enamel somewhat wrinkled.

Hortus IV

Fragmentary mandible with part of L ramus and partial R and L corpora. Contains RC–P1, R and L M1–2; associated isolated LC_1; alveolae for RI2 and RP2. Corpora not very thick boned.

Much of symphysis missing, but mandible seems to have been tallest s/i in this area, getting shorter posteriorly. Symphyseal region broad between Cs, arcuate from side to side; corpora slightly divergent posteriorly. On the R, enough bone left to reconstruct symphyseal profile; seems to have been vertical for most of way down from alveolar margin, then curves back toward inferior border. Region from RC root to mesial side of RI2 alveolus shallowly excavated from alveolar margin to level of tooth root tips. Seen from below, inferior margin of entire broad symphyseal region uniformly thick a/p. Digastric fossae very wide m/l, shallowly excavated, point mostly down, although there is a posterior component. Above R digastric fossa, internally, is swelling (appears pathological). Medium-sized mental foramina lie under region between P2 and M1. On the R, bone below mental foramen thickened slightly externally; this thickening continues both fore and aft although quickly fading. On the L, anterior root of ramus takes origin level with mesial wall of preserved M3 alveolus; judging by the angle of its shallowly concave preserved anterior border, ramus would have been tilted posteriorly; no retromolar space. Internally on the L, faint, shallowly oblique, anteriorly displaced mylohyoid line; below it, a fairly well excavated fossa. On fragment of preserved interior surface of ramus, appears to be evidence of very strong continuation of alveolar crest as pillar running up internal surface below where coronoid process was.

RC short crowned, well excavated on lingual surface, with modest vertical lingual pillar lying just medial to midline of crown. P1 very long m/d on buccal side. Apex of protoconid lies at midline of crown (= mesial and distal edges of equal length). Low but moderately swollen lingual tubercle lies at base of protoconid; stout crest from lingual swelling runs mesially to terminate below and in front of apex of protoconid. Stout crest from mesiobuccal corner runs down to side of lingual swelling from which separated by deep crease (this crest also encircles well-defined anterior fovea); less stout crest runs directly forward from distobuccal corner to become confluent with lingual swelling, enclosing m/d long, large, well-defined posterior fovea. Buccal side of P1 has steep slope (caused by apex of protoconid being somewhat centrally shifted?). Preserved R and L molars elongate, somewhat rounded in occlusal outline; buccal cusps only minimally more centrally placed than very peripherally placed lingual cusps (thus long talonid basins also quite broad and open). On all molars, apex of very wide hypoconulid lies along midline of tooth and is subdivided by crease into smaller lingual wedge and larger central-to-buccal wedge. Crestlike extension runs from base of hypoconid mesially and lingually to base of metaconid; thick paracristid courses between protoconid and metaconid (these cusps enclose thin but deep trigonid basin). M2 enamel deeply but sparsely wrinkled; M1s apparently similar.

Isolated tooth identified as LC has well-excavated lingual surface but lacks pillar; like in situ RC, has steep buccal slope as well as marked lateral flare; root very long, although not stout; tooth could just as easily be an RC^1.

Cranial fragments

Two small juvenile braincase fragments, very thin.

Isolated Teeth

About 3 dozen.

REFERENCES

de Lumley, H. 1965. *Le Paléolithique inférieur et moyen dans son cadre géologique (Ligurie, Provence, Bas-Languedoc, Roussillon, Catalogne).* Thesis, Paris, pp. 924–927.

de Lumley, H. et al. 1972. *La Grotte moustérienne de Hortus. Mémoire 1, Etúdes Quaternaires.* Marseille, Université de Provence, 688 pp.

Piveteau, J. et al. 1963. Découverte de restes néanderthaliens dans la Grotte de Hortus (Valflaunès, Hérault). *C. R. Acad. Sci. Paris* 256: 40–44.

Repository

Laboratoire d'Anthropologie, Université de la Méditerranée, Faculté de Médecine, 13916 Marseille, France.

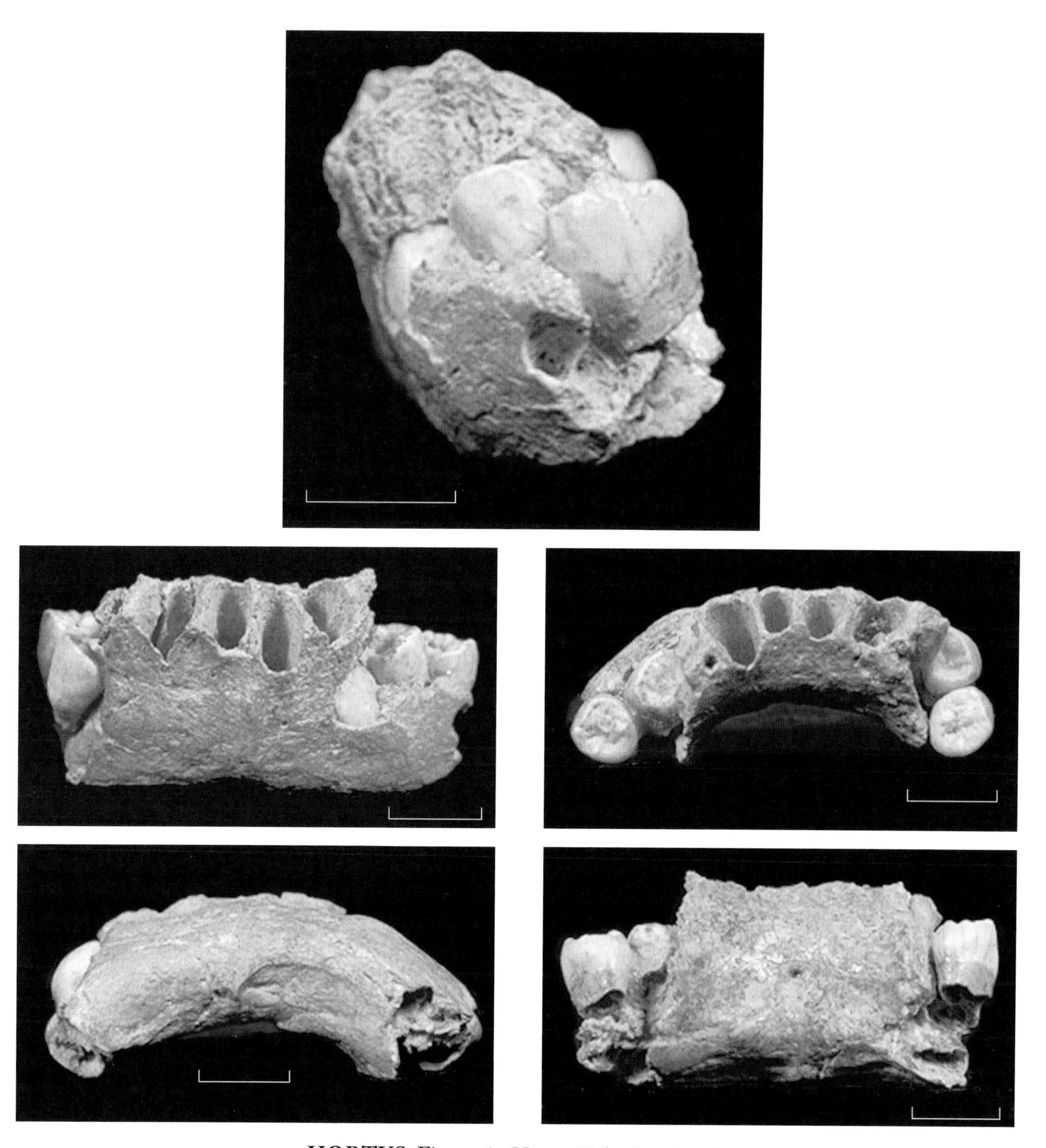

HORTUS Figure 1. Hortus II (scale = 1 cm).

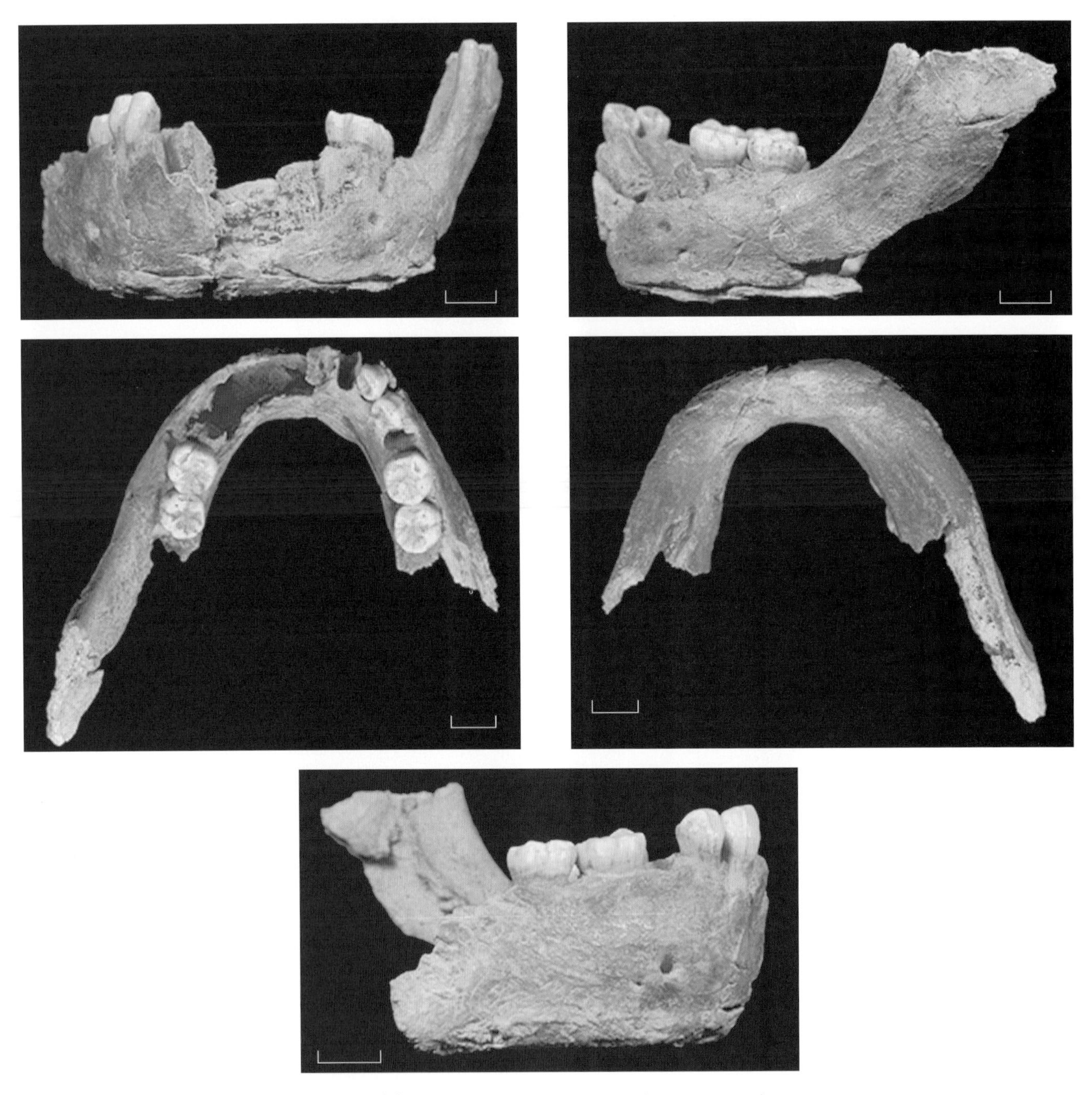

HORTUS Figure 2. Hortus IV (scale = 1 cm).

Isturitz

Location

Cave site on the Mont de Castelou, near the village of Saint-Martin-d'Aberoue, 30 km SE of Bayonne, France.

Discovery

Various discoveries, both in and independent of controlled excavation, between 1895 and 1939.

Material

Numerous hominid finds were made in various crannies of this large stratified (and decorated) cave site, mostly of mandibles or fragmentary crania, plus some postcranial elements.

Dating and Stratigraphic Context

All the Isturitz hominids appear to derive from the latter part of the last glacial and are dated by archaeological context.

Archaeological Context

All Isturitz hominids (except for some Bronze Age materials, now lost) are Upper Paleolithic, although the earliest finds (Isturitz 1–3) are of uncertain stratigraphic provenance. Isturitz 4 is Solutrean, Isturitz 5 and series 7 are Aurignacian; series 6 is Gravettian. Series 8 and 9 are late Magdalenian.

Previous Descriptions and Analyses

There has never been a systematic overall description and analysis of the Isturitz hominids; the closest thing is the works of R. and S. St-Périer (e.g., 1930, 1936, 1952). They have always been regarded as fully modern *Homo sapiens*.

Morphology

Fragments of multiple individuals, juvenile through adult, represented primarily by cranial vault fragments, partial maxillae and mandibles, isolated teeth, and a few postcranial elements. In general, cranial vault pieces characterized by thin bone, highly denticulate and segmented sutures, and prominent parietal eminences; mandibles bear well-developed central keels and mental trigons. Upper Ms decrease in size from M1 to M3, e.g., in specimen 71, in which only M1 has hypocone (all Ms, although worn flat in this specimen, appear devoid of basins).

Cranial fragments

Many show traces of cut marks (defleshed?); at least one is engraved.

Parietals and Occipital (IE alpha 1914 77152 O; Ist II 1930–22)

Vertical vault side; strong curve to occipital plane in both directions; segmented coronal and sagittal sutures; strongly interdigitated lambdoid suture rises sharply and peaks at lambda; bow-shaped superior nuchal line with prominently raised, ridgelike central part.

Frontal (Ist II 1930 77160F)

Fragmentary frontal preserving part of L supraorbital region with L part of glabellar "butterfly" terminating just lateral to large supraorbital notch. Lateral supraorbital area narrow, platelike, oriented backward. Frontal quite broad; arcs moderately in both directions. Preserved part of coronal suture segmented. Frontal sinus limited to region of "butterfly." Moderate temporal ridge rises high up on zygomatic process of frontal and fades out before reaching coronal suture. R frontal fragment colored differently from L; both sides heavily incised.

Adult Maxillae (Ist II 1934 no. 72, Ist II 1934 no. 73, Ist 71)
Broadly parabolic with anterior slope becoming posteriorly deep with vertical walls below molars; maxillary sinuses not expanded into nasal cavity; floor of nasal cavity perfectly flat; crisp-to-blunted lateral (nasal aperture) crests merge with anterior nasal spines; no internal nasal margin (= no prenasal fossa). May bear maxillary torus. Ms decrease in size from M1 to M3 with concomitant decrease in size of hypocone.

Juvenile Mandibles (Ist II 1931–66, Ist II 1932–68, Ist II 1934–65, Ist II 1943)
Front of jaw tightly arced; corpora strongly divergent posteriorly. Partially fused symphyseal region bears distinct midline keel, with distinct fossae on either side, that fans out inferiorly into mental trigon with lateral corners absent or blunt to pointed on inferior margin. Mental foramina under region of dm1–2. From below, symphyseal region peaked to straight across and moderately to quite thick a/p; inferior margin may be pronounced. Essentially no postincisal plane (inner aspect of symphysis more or less vertical). Small, shallow digastric fossae face back. Mylohyoid line faint. Submandibular fossae moderate to deep. Ramus fairly vertical. Gonial region reflected inward with faint muscle scars low on internal side; gonial angle tightly curved. Sigmoid notch probably broad, shallow, and deepest at midpoint; sigmoid notch crest terminates at lateral margin of condyle. Large mandibular foramen somewhat compressed, partially incomplete inferiorly, points back and slightly up; lingula not developed. Dm2s elongate and rounded distally and bear good-sized hypoconulid just buccal to midline; trigonid basin not delineated.

Adult Mandibles (Ist II 1932:69, No. 70)
Symphyseal region with very variably indistinct to low, blunt, midline keel, and well-developed, pointed mental trigon that may bear lateral corners inferiorly. Broad fossae bilaterally delineate mental trigon and, when present, the central keel. Large mental foramen under P2–M1. In inferior view, inferior margin quite narrow to very broad a/p, pointed; thickest at midline. Internal surface of symphysis more or less vertical. Inferior margin may be curved (rocker jaw). Small, well-excavated, digastric fossae point backward. Faint to well-developed mylohyoid line; very shallow to pronounced submandibular fossa. Tall, compressed mandibular foramen points back and slightly up.

REFERENCES

St-Périer, R. 1930. La grotte d'Isturits I, Le Magdalénien de la Salle de St- Martin. *Arch. Inst. Paléontol. Hum.* 7: 1–124.

St-Périer, R. 1936. La grotte d'Isturits II, Le Magdalénien de la Grande Salle. *Arch. Inst. Paléontol. Hum.* 17: 1–139.

St-Périer, R. and S. St-Périer. 1952. La grotte d'Isturits III, Les Solutréens, les Aurignaciens et les Moustériens. *Arch. Inst. Paléontol. Hum.* 25: 1–310.

Repository
The whereabouts of some Isturitz specimens, notably individuals 1–3, are unknown. The others are scattered among various institutions, among them the Institut de Paléontologie Humaine, 1 rue René Panhard, 75013 Paris, France, and the Musée des Antiquités Nationales, 78103 St-Germain-en-Laye, France.

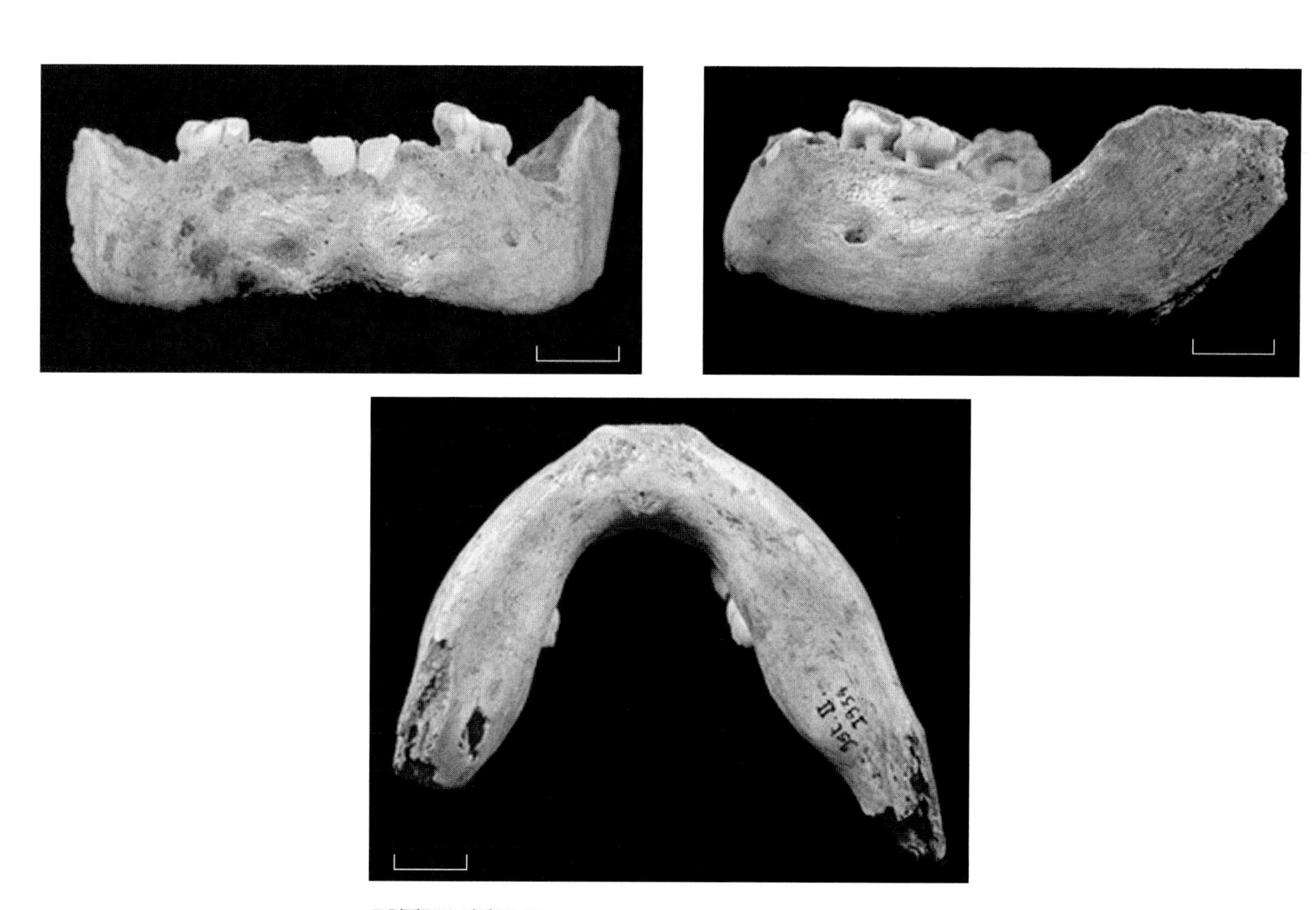

ISTURITZ Figure 1. Ist II 1936-61 (scale = 1 cm).

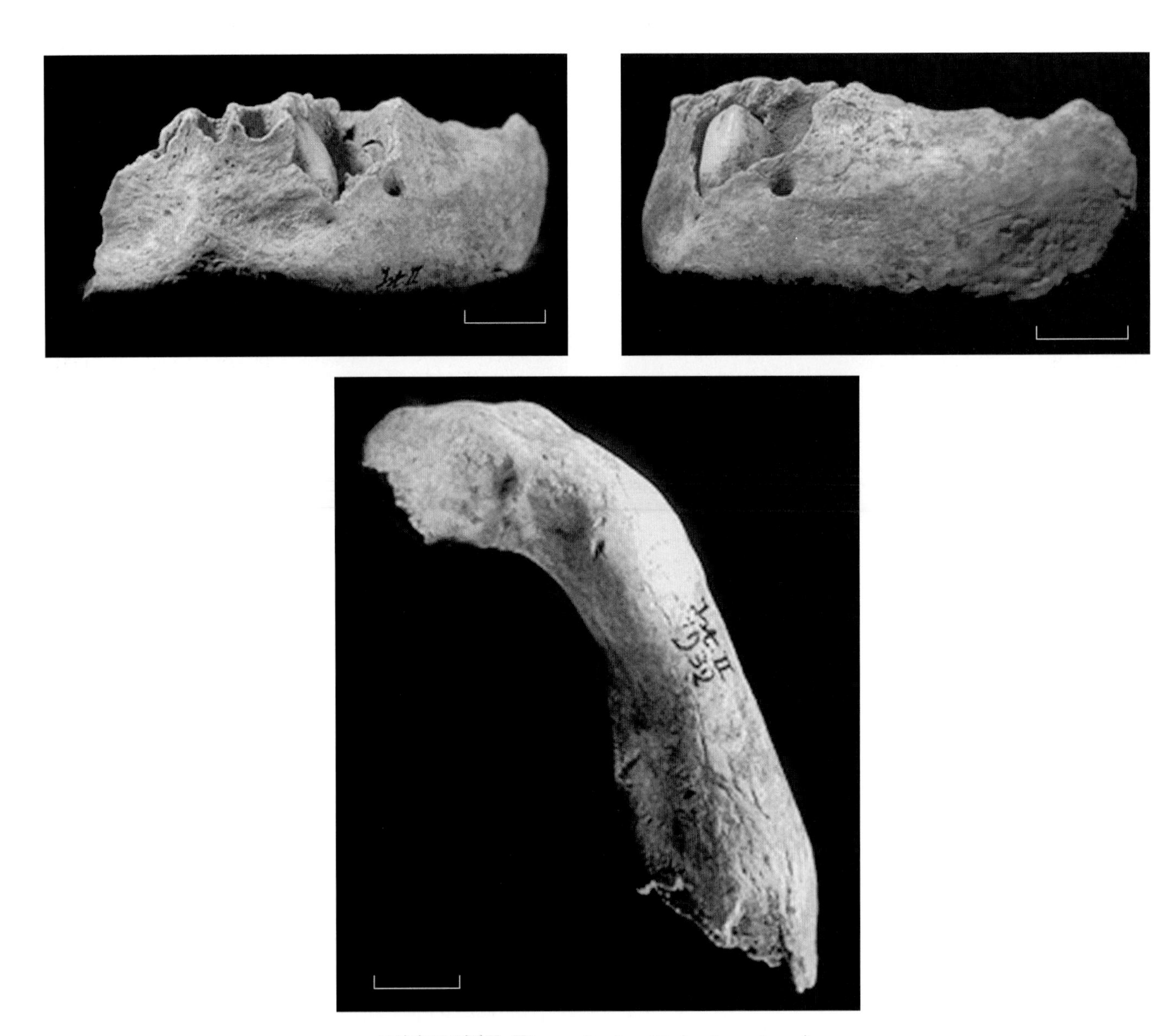

ISTURITZ Figure 2. Ist 68 (scale = 1 cm).

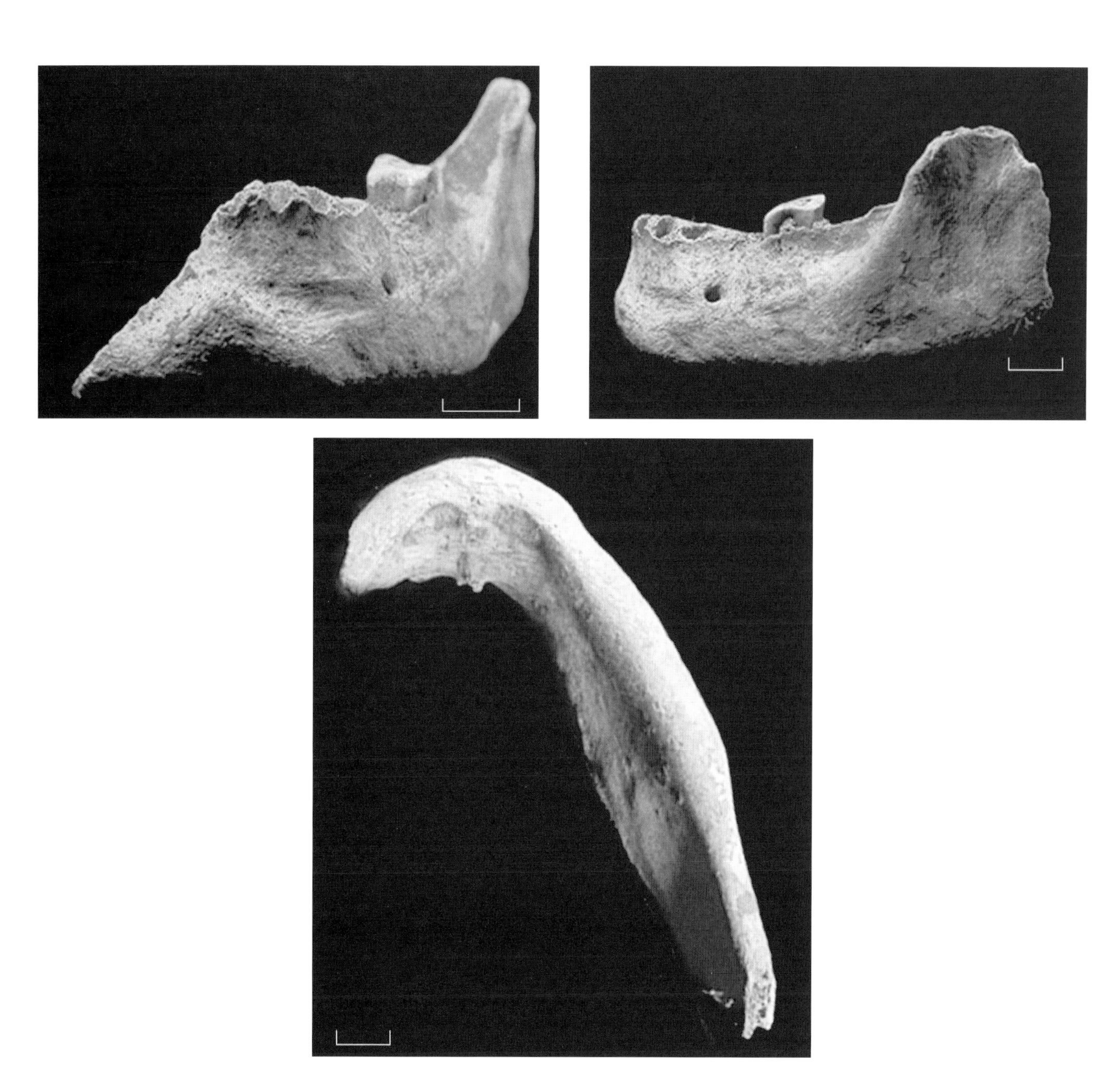

ISTURITZ Figure 3. Ist 69 (scale = 1 cm).

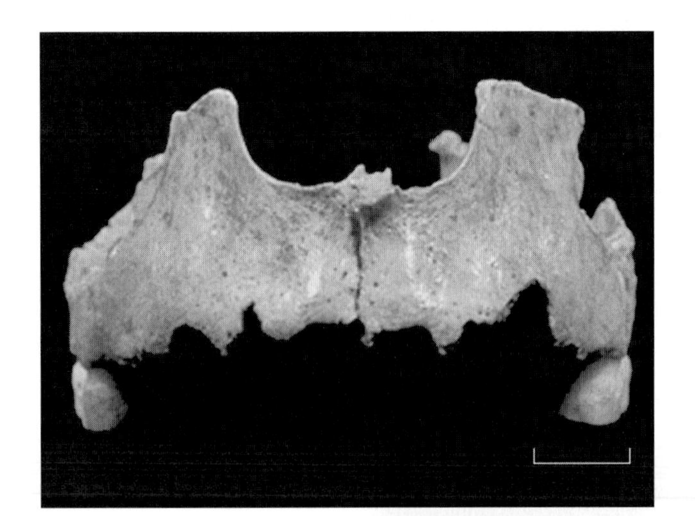

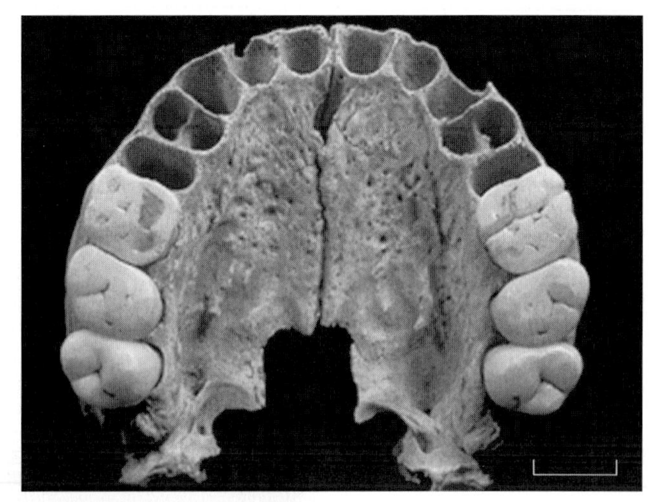

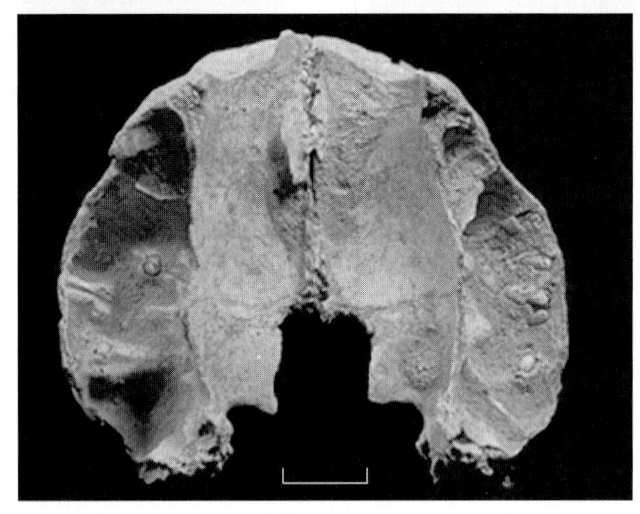

ISTURITZ Figure 4. Ist 71 (scale = 1 cm).

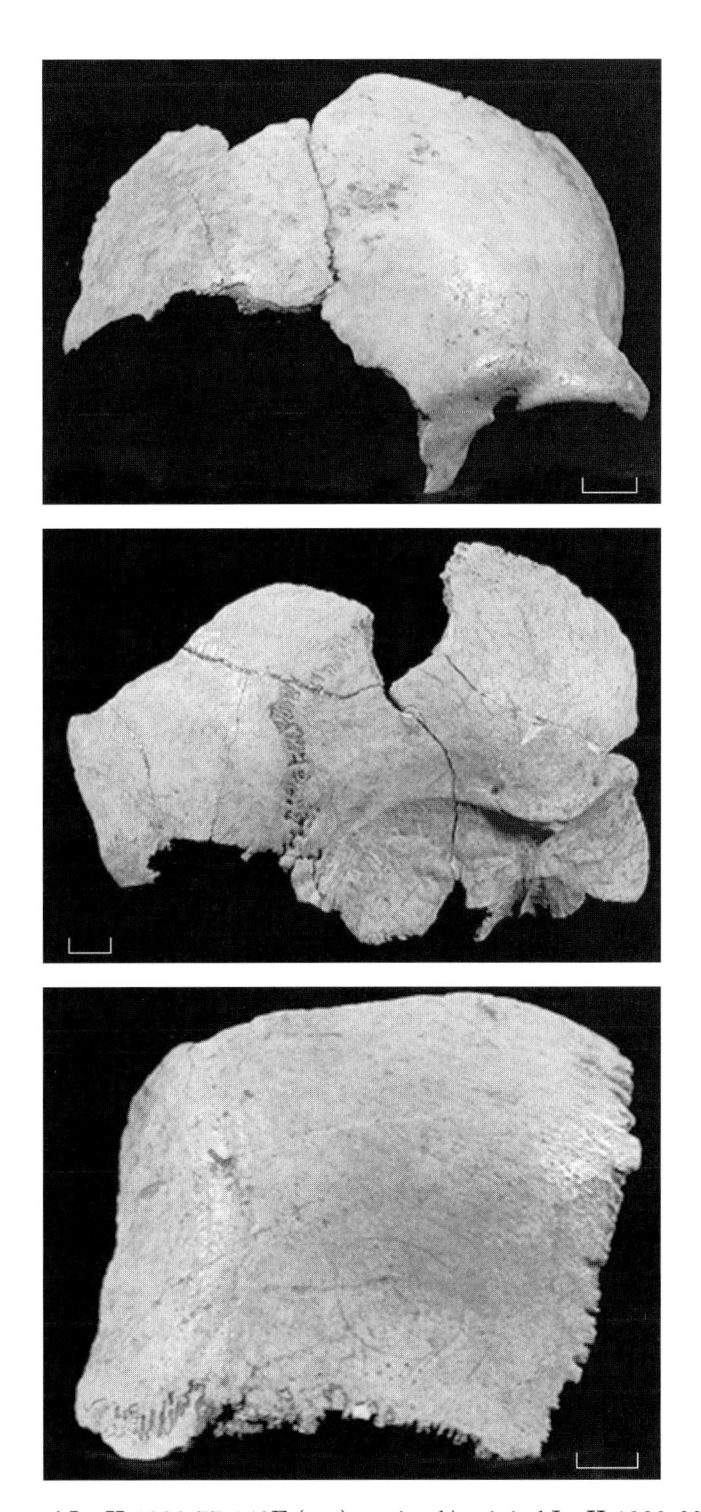

ISTURITZ Figure 5. Frontal Ist II 1930 77 160F (top); parietal/occipital Ist II 1930-22 (middle); parietal IE alpha 1914 77152 0 (bottom), (scale = 1 cm).

Krapina

Location

Rockshelter on Husnjakovo Hill, in the town of Krapina, 40 km NW of Zagreb, Croatia.

Discovery

Excavations of D. Gorjanovic-Kramberger, 1899–1905.

Material

Almost 400 skull and postcranial fragments and isolated teeth, representing perhaps as many as 30 individuals.

Dating and Stratigraphic Context

Gorjanovic-Kramberger (1899, 1900) identified 9 major levels in the 11 m of rockshelter infill, which he nonetheless believed had accumulated quite quickly. On faunal evidence of temperate conditions he eventually decided that this event was during the last interglacial (Gorjanovic-Kramberger, 1913); but recent ESR dating (Rink et al., 1995) shows that although the entire infill was indeed rapidly deposited, it was laid down during the previous interglacial at around 130 ka (end of isotope stage 6 or beginning of stage 5e). All of the hominids from the site (levels 3–4, the "Hominid Zone," and the later levels 5–7) are thus penecontemporaneous.

Archaeological Context

The lithic assemblages from Krapina have recently been analyzed by Simek and Smith (1997), who find them to represent a scraper-rich Mousterian of Charentian type throughout the section, although greater selectiveness in raw materials and a preference for cobble-wedge over Levallois technology is evident in later levels.

Previous Descriptions and Analyses

Gorjanovic-Kramberger (e.g., 1899, 1900) began by referring to the Krapina hominids simply as "diluvial" humans but by 1902 had come to regard them as Neanderthals, which from 1904 on he placed in Schwalbe's species *Homo primigenius*. He resisted the remarkably tenacious notion promulgated by Klaatsch and Hauser (e.g., 1910) that modern humans were present in the assemblage. Since Minugh-Purvis (1997) replaced her view that the Krapina 1 cranium was "considerably more modern than the rest of the Krapina skulls" (1988; p. 388) with the conclusion that this specimen simply stands at the extreme of Neanderthal variation, paleoanthropologists have been unanimous in viewing the Krapina hominids as fully Neanderthal, at least to the extent permitted by their favored evolutionary models. The fullest descriptions in English of the Krapina Neanderthals are the monograph by Smith (1976) and the catalog by Radovcic et al. (1988). Estimated cranial capacity of Krapina 2 is 1450 cc (Holloway, 1985).

Morphology

The extremely extensive Krapina assemblage consists of almost 400 cranial and mandibular fragments and isolated teeth. We begin with the more complete specimens.

Krapina 1 (= Cranium A)

Immature (open metopic suture), thin boned, very large. Preserves most of frontal region, R parietal, and L temporal, plus half of L parietal.

Forehead high. Bregmatic bulge with depression behind. Broad across parietal eminences. L supraorbital region shows slight toral development medially, with lateral thinning. Frontal sinuses not yet expanded. Squamosal suture was probably low, arcuate; parietomastoid suture long. Mandibular fossa deep, with small medial articular tubercle. Mastoid process minimally protruding, but well pneumaticized with large air cells; points down. Zygomatic process of temporal originates anterior to modest auditory meatus; low vaginal process arises at medial border of carotid foramen, fades laterally, and flares up again to wrap around large (broken) styloid process. Stylomastoid foramen aligned with, but clearly separated from, styloid process. Mastoid notch narrow and deep. Huge superior petrous sinus extends along most of length of broad petrosal. Modest, sharply ridged arcuate eminence more domed than in Krapina 3. Subarcuate fossa completely closed off.

Krapina 2 (= Cranium B)

Subadult. Part of the R, most of L parietal; part of occipital. Bone thin, muscle markings strong; some porotic hyperostosis. Cranium quite low, occipital somewhat "bunned," and region of lambda flat. Lambdoid suture arced; it and sagittal suture not segmented. Straight occipital torus delimited bilaterally by two inferior depressions; it runs completely across the bone. Suprainiac depression shallow, coarsely pitted, and not delimited by a raised margin.

Krapina 3 (= Cranium C)

Adult. Upper face, R portion of frontal, part of R parietal, most of R temporal, part of R occipital. Relatively small, gracile, thin boned.

Frontal moderately receding; lacks supra- or posttoral sulci. Gracile supraorbital tori smoothly rolled, with slight lateral flattening above; continuous across broad, rather flat, nonprotruding glabella. Exposed frontal sinuses extend some distance beyond glabella. Viewed from above, tori appear arcuate, lower face prominent and wedgelike. R supraorbital margin swollen in area of supraorbital notch. Orbital roof formed by frontal lobe depressions above. Anterior and posterior lacrimal crests (preserved on R) sharp, continuous; form smoothly arcuate crest. Medial orbital margin not puffy; slopes out inferiorly. Medial orbital floor slightly raised although not inflated. Distinct but broken, unpneumaticized medial projections preserved on both R and L sides; as seen on the R, superior nasal crest well recessed. Maxillary frontal process not inflated. At least four infraorbital and zygomaticofacial foramina preserved on the R; most of latter above inferior orbital margin. Interorbital region broad. Nasal bones smoothly flexed ca. 1 cm below nasion; frontonasal suture peaks. Nasal bone profile on R almost horizontal.

Squamous suture short, highly arced, with tall, straight anterior border; parietomastoid suture long, horizontal with large mastoid foramen above and small one below. Demarcation of temporal from infratemporal fossa not clear cut. Mandibular fossa very deep; moderate medial articular tubercle does not close off fossa. Posterior root of zygomatic arch originates in front of small auditory meatus; slight suprameatal crest runs into more bulbous supramastoid crest. Vaginal process runs from margin of carotid foramen to rather medially placed styloid process (now missing) but not to lateral margin of auditory meatus. Stylomastoid foramen lies almost midway between bases of styloid and mastoid processes; the latter is directly in line with mastoid notch.

Deeply muscle-scarred mastoid process very tiny, narrow both at base; tip downwardly directed; lacks lateral adornment. Parietal notch lies near midline of mastoid process. Mastoid notch broad and relatively shallow, lacks digastric fossa behind. Paramastoid crest absent. Occipitomastoid suture may have borne eminence. Waldeyer's crests probably lacking. Lambdoid suture would not have risen sharply.

Petrosal extremely broad, flat, with only slight elevation above superior semicircular canal; faint trace of superior petrous sinus. Area of subarcuate "fossa" puffy. Sigmoid sinus deep, short; may be entirely in the temporal or just across lower part of parietal. An unusual ridge follows endocranial course of squamous suture.

Krapina 4 (= Cranium D)

Adult. Partial frontal and L parietal. Bone relatively thin; fairly pronounced porotic hyperostosis.

Frontal quite steep behind long posttoral plane. R supraorbital torus pronounced, smoothly rolled, quite uniform in thickness laterally, continuous across glabella. Frontal lobes extended fully above orbit. Interorbital distance would have been large and the frontal crest strong. R frontal sinus (single large chamber with only minor septum) extends laterally to midline of orbit and well back into postorbital region.

Krapina 5 (= Krapina D)
Adult. Consists of top of braincase with parts of occipital and R temporal. Bone relatively thick. Similar degree of porotic hyperostotic pitting suggests association of all fragments.

Occipital torus quite strong, broad, defined inferiorly by a sulcus. Suprainiac depression lacks raised margin but has roughened surface. Temporal bone very short a/p. Posterior root of zygomatic arch originates anterior to small auditory meatus. Suprameatal crest weak and indented but confluent with more pronounced supramastoid crest. Mandibular fossae were wide and very deep. Mastoid process moderately bulky at base, inferiorly thin and tapering; angles forward before pointing down; bears small mastoid crest, but lacks anterior tubercle. Mastoid notch deep but was probably not very broad. Vestige of vaginal process lies just medial to styloid process but fades out well medial to auditory meatus. Large stylomastoid foramen lies well behind, but only slightly medial to, styloid process. Interiorly, R transverse sinus was dominant.

Krapina 6 (= Cranium E)
Adult. Smaller, even more gracile than Krapina 3. Long, low, thin-boned partial cranium with much of L malar, most of frontal, part of L maxillary frontal process and nasal bone, much of R parietal, part of R occipital.

Frontal recedes quite sharply, with very long posttoral surface. Low-arched brow ridge smoothly rolled, continuous across glabella, slightly swollen in region of supraorbital notch. Brows relatively thin s/i but quite pronounced anteriorly. From above, glabella not pronounced; torus arcs backward fairly smoothly.

Interorbital region relatively broad. Lower face would have been quite projecting. Nasal bones arc out smoothly; nasonasal suture peaks at nasion. Posterior lacrimal crest pronounced and sharp. Frontal sinus large, lacks septa; would have extended laterally just beyond midline of orbit and up to posttoral sulcus. Frontal crest well developed. Frontal lobes extend above roofs of orbits. Eight zygomaticofacial foramina preserved on L.

Lambdoid suture appears to have been low. Occipital torus weak; suprainiac depression large (both only partially preserved).

Krapina 10
Adult. L petromastoid with part of parietal and occipital. Posterior root of zygomatic arch lies anterior to moderately large auditory meatus. Vaginal process wraps around very thin, medially placed styloid process, then fades out laterally. Fairly large stylomastoid foramen lies behind, and slightly lateral to, styloid process. Narrow mastoid process downwardly pointed; bears large mastoid foramen. Mastoid notch was moderately deep and broad. Paramastoid crest small but long; occipitomastoid crest long but low. Waldeyer's crest low but distinct. Parietomastoid and "anterior lambdoid" sutures long and horizontal; lambdoid suture turns upward well behind asterion. Sulcus lies inferior to occipital torus. Petrosal wide and flat in region of arcuate eminence. Petromastoid region bears large air cells.

Krapina 16
Fairly complete adult parietal, heavily porotic, no temporal lines visible. Preserved portion of squamosal suture very oblique and deep (implies short squamous portion). Lambdoid suture was probably low and curved; may have contained ossicles.

Krapina 18.5
Large portion of juvenile R parietal with petromastoid portion. Parietomastoid suture very long and horizontal. Parietal notch lies at posterior part of mastoid process. Huge mastoid foramen lies well up on the temporal. Low parietomastoid suture suggests broad and low occipital. Sutures regular, not segmented. Mastoid process slightly elevated, with deep and narrow notch. Paramastoid crest lower than mastoid process; occipitomastoid suture lower yet. Fairly large stylomastoid foramen situated just behind and medial to deep styloid pit, which contains base of tiny styloid process. Very heavy vaginal process limited to region of styloid process. Superior petrous sinus and subarcuate fossa absent. Region of subarcuate eminence slightly raised. Petrosal quite broad; contains large sinus and many small air cells.

Krapina 23
Adult. Central portion of frontal bone to frontomaxillary and peaked frontonasal sutures. Brows quite

thick, smoothly rolled, continuous across glabella. Posttoral region quite backwardly sloping, with frontal sinuses penetrating far into it. Frontal sinuses partially subdivided by septa, continue laterally beyond midpoint of orbits. Frontal crest long, moderately pronounced.

Other Cranial Elements

There are 10 isolated temporals. Region of arcuate eminence broad in all, with upward swelling ranging from slight to moderate. Subarcuate fossa totally flat or puffed out in all but one, which has tiny depression. Superior petrous sinus absent in all, except for faint trace posteriorly in one (unlike Krapina 1, in which sinus is large). Broken specimens reveal extensive pneumatization with well-developed air cells throughout mastoid region. Sigmoid sinus varies from deep to absent. Posterior root of zygomatic arch always in front of moderate to small auditory meatus. Suprameatal crest normally not well developed. Medial articular tubercle always present where area preserved; does not fully close off mandibular fossa. Supramastoid crest variably developed, but arcs upward. Mastoid process varies in length and thickness. Mastoid notch never very deep or gutterlike. Parietomastoid suture long and horizontal where preserved. Mastoid crest and anterior tubercle absent in all.

Numerous fragments of occipitals of varying thickness. When preserved, occipital torus extends far laterally, curves anteriorly at its extremity. Inferior margin of torus defined by sulcus bilaterally; superior margin defined only by suprainiac depression that is roughened surficially and varies in size from very tall and wide to small. Protrusion of occipital variable. Internally, superior sagittal sinus of all deeply incised, with crisp and prominent edges. Where preserved, transverse and sagittal sinuses diverge far up, at level of suprainiac depression.

Numerous frontal fragments, all showing fair degree of porotic hyperostosis. In the 10 fragments preserving the supraorbital region, torus variable in size but typically very protrusive. Some tori substantially more robust than those in the more complete crania. Where supraorbital margin preserved, there is slight notch with bulge lateral to it (as in crania). Most have long posttoral plane. Large, essentially single-chambered frontal sinuses generally extend lateral to midpoint of orbit and to rear of posttoral plane. All tori smoothly rolled and more or less uniformly thick from side to side. Three significant malar fragments preserve multiple zygomaticofacial foramina.

There are 37 small cranial roof fragments; thicker ones show porotic hyperostosis. Sutures rather uniformly denticulated. Preserved temporal lines very faint (as on more complete skulls).

Six maxillary fragments, including fairly complete palate. Maxillary sinus extends less far back over alveolar region and puffs out face less markedly in younger individuals. Anterior root of zygomatic arch flares and retreats. Nasal cavity floor slightly sunken. Nasoalveolar clivus very long and quite vertical; anterior nasal spine double. External margin of nasal aperture (lateral crest) on each side descends from lateral margin and curves down, around, and then up to meet base of anterior nasal spine. Internal margin (spinal crest) runs from anterior nasal spine laterally to parallel external margin. Palate has shallow anterior slope, but steep sides. Large single incisive foramen lies close to incisors.

Mandible

Twenty significant mandibular fragments. Where preserved, morphology of all as follows. Corpora vary from shallow to rather tall s/i. Symphyseal region smooth externally; varies from being broadly curved to flat across. In some, inferior border of this region rises; inferior mandibular tubercles (sometimes large and bulky) below the P2s. Typically large primary mental foramina lie below M1; accessory mental foramina vary in size and typically lie below M1 (but may occur below P2). Inwardly deflected, externally smooth gonial angles vary from rounded to squared to cut off; their inferior margins may be slightly lipped. Sigmoid notches shallow, deepest close to neck of condyle. Sigmoid notch crests more typically run to lateral aspect, rather than to midline, of condyles. Condyles approximately equal in height to tips of coronoid processes. Digastric fossae modestly to deeply excavated; may face backward. When present, genial tubercles may be single and moderately prominent. Internal variation noted in mylohyoid lines (faintly to strongly developed), submandibular fossae (moderately to well developed), medial pterygoid tubercles (slight to marked), and mandibular foramina (both circular to long and compressed and oriented horizontally to obliquely up and back). When present, lingulae small. Alveoli/roots of front teeth forwardly displaced, with depression below them. Retromolar spaces vary from small to marked.

Krapina Dentitions
Collection includes several mandibles and maxillae, mostly partial, plus large number of isolated teeth. Crowns of all teeth in good condition, not chipped, in high proportion of individuals; few teeth heavily worn in this (young) population. Descriptions below by tooth type.

Deciduous Dentitions

Upper di1. From the R, two maxillae and one isolated tooth; from the L, one isolated tooth. All crowns very broad, short, not shoveled, and extremely thick buccolingually, with moderate to large lingual swelling but no definite tubercle. Buccal surface bulges. Crown set at angle to root. Root broad buccally, narrow lingually, triangular in section.

Upper di2. From the R, one maxilla and one isolated tooth; from the L, one isolated tooth. Not shoveled; no lingual tubercle; slight buccal bulge. Crown narrower and roots more conical than in di1.

Upper dc. One isolated tooth from each side. Crown symmetrically triangular; some buccal swelling; hint of lingual tubercle; roots long and slender.

Upper dm1. From the R, one maxilla and two isolated teeth; from the L, two maxillae and one isolated tooth. Rather variable but paracone and protocone distinct; little swelling of hypocone region; cusps relatively tall; basins quite deep; three roots present (may/may not be connected by septum).

Upper dm2. From the R, one maxilla and four isolated teeth; from the L, two maxillae and two isolated teeth. Trigon well defined and basin deep; preprotocrista courses mesially around paracone; paracone with internal crest or conule at base. Large hypocone distended distolingually; distinct postcingulum from hypocone to metacone defines very large and deep talon basin. Protocone and hypocone apices less centrally placed and lingual slope less pronounced than in M1/2; distinct protocone fold; centrocones absent; Carabelli's pit and cusp present. Stout, long roots bifurcate close to neck and splay out. In unworn crowns, cusps quite tall and pointed.

Lower di1. One isolated L. Heavily worn, but with small buccal swelling; lateral edge not markedly flared; lingual surface may have been concave.

Lower di2. One isolated from each side. Both lingually concave, with pronounced lateral flare; no flexure at neck. Root quite long and stout.

Lower dc. One mandible from each side. Both dcs worn. Crowns quite long mesiodistally; probably had short mesial and long distal edges. Lingual surface slightly concave; slight buccal swelling above neck; small lingual margocristids.

Lower dm1. L mandible and isolated tooth; crowns worn. Lingual cusps were tall and pointed, with distinct metaconid. Enamel swells over buccal root; mesial part of tooth elevated relative to distal part; trigonid is shifted and opens lingually. Two roots, not very splayed.

Lower dm2. From the R, one mandible and one isolated tooth; from the L, five isolated teeth. Cusps tall, sharp, distinct; trigonid narrow with very deep, distinct basin; talonid broadens distally, with long, well-defined basin. In unworn crowns, extensive wrinkling or some furrowing and pitting. Two roots diverge quite close to crown.

Permanent Dentitions

Upper I1. From the R, three maxillae and five isolated teeth; from the L, two maxillae and six isolated teeth. Crowns bulky and stout; unworn crowns quite tall; roots rounded. All shovel shaped, with bulbous lingual tubercle divided by vertical furrows into two components; each component further subdivided in two by smaller furrows. Tubercles cupped by rest of crown, which rises significantly above them. In unworn teeth, both sides of crown flare laterally; lateral flare more rounded, medial more cornered. Buccal surface of crown bulbous; in side view, root angles back (when buccal surface held vertically). Variability low.

Upper I2. From the R, three maxillae and six isolated teeth; from the L, three maxillae and five isolated teeth. Same buccal swelling and root-crown angle as I1s but more variable, especially in degree of shoveling and lingual tubercle development. In unworn teeth, main part of crown rises considerably above lingual tubercle, which may not be subdivided into two. Root compressed laterally; quite long relative to crown.

Upper C. From the R, three maxillae and four isolated teeth; from the L, one maxilla and seven isolated teeth. Unworn crown (no. 102) tall and asymmetrically triangular (very long distal slope, much shorter medial slope). All have buccal bulge, angled root, and lingual tubercle (in most, tubercle delineated by small furrows on either side). Linear hypoplasias common. Roots very long, extremely compressed laterally, deep buccolingually.

Upper P1. From the R, six isolated teeth; from the L, two maxillae and four isolated teeth. Paracone and protocone closely approximated with angle between them very deep and strongly V shaped. In five, roots distinct but conjoined; in only three, roots fully bifid. In unworn crowns, some wrinkling and conule formation visible.

Upper P2. From the R, one maxilla and five isolated teeth; from the L, two maxillae and five isolated teeth. Paracone and protocone widely spaced. In most, roots distinct although conjoined; one P2 is single rooted. Hypoplasia absent to severe.

Upper M1–3. In all, preprotocrista swings mesially around paracone; crest runs down face of paracone into trigon basin.

Upper M1. From the R, one maxilla and three isolated teeth; from the L, four maxillae and one isolated tooth. Very distinct trigon constricted by long lingual slope. Paracone noticeably larger than metacone; protocone quite central, giving pronounced lingual bulge; huge hypocone swells distolingual corner. Carabelli's pit present; distinct postcingulum encloses large talon basin; well-defined pre- and postprotocristae with small conules; enamel modestly crenulated. Roots very long for crown size; only one tooth with distinct individual roots; sometimes cleft between mesial and distal buccal roots starts high above crown; in three-rooted no. 45, cleft far from crown. Little variation in size and morphology, but see below.

Upper M2 and Combined M1/M2. From the R, seven isolated teeth; from the L, two maxillae and six isolated teeth. Differences in size make definitive identification difficult; small M2s may be females' or, in combined M1/2 sample, small teeth may be M2s and the larger teeth M1s. Not enough associated teeth to resolve question. In combined M1/2 sample, protocone and hypocone apices quite centrally placed, lingual slopes long, trigon basins constricted, postcingulum thick. Talon basin deeply incised, presenting clear definition of anterior and posterior basins. Paracone larger than metacone. Postprotocrista better developed than preprotocrista. Most teeth with Carabelli's pit and some enamel wrinkling. Roots long and stout; few teeth with separate roots (when there is bifurcation, it is far from crown).

Upper M3. From the R, at least seven isolated teeth; from the L, four isolated teeth. More variable than M1/2s in hypocone development. Protocone not centrally shifted in all. Enamel somewhat wrinkled. Postprotocrista well developed. Talon and trigon basins distinct. All roots long with high bifurcation when present (no. 102 three rooted; nos. 106, 169, 173 two rooted).

Lower I1. From the R, four mandibles and three isolated teeth; from the L, four mandibles and two isolated teeth. Crown relatively narrow, tall, deep buccolingually, and buccally bulging. Lingual surface concave and lacks well-defined margocristids. Root stout, very long, angled back (relative to orthal crown).

Lower I2. From the R, three mandibles and one isolated tooth; from the L, four mandibles and two isolated teeth. Crown large, mesiodistally compressed, and buccolingually deep, with raised margocristids on both margins. Root stout, long, set at angle to crown.

Lower C. From the R, three mandibles, and four isolated teeth; from the L, three mandibles and three isolated teeth. Crown tall, mesiodistally compressed, buccolingually deep; buccal bulge near neck; signs of mesial and distal margocristids. Steep distal edge long, mesial edge shorter. In unworn crowns, mesial edge almost horizontal. Root long, stout, mesiodistally compressed, backwardly angled relative to crown. Linear hypoplasia prevalent.

Lower P1. From the R, one mandible and four isolated teeth; from the L, four mandibles and four isolated teeth. Relatively small tooth. Protoconid very tall and mesiodistally long. Large paracristid runs mesially, then around to terminate at base of lingual swelling (from which it may be separated by crease). Posterior basin defined by crest that courses from distal side of protoconid and around to lingual swelling. Buccal face quite swollen, with apex of protoconid almost in midline of crown. Roots very long and compressed mesiodistally.

Lower P2. From the R, three mandibles and four isolated teeth; from the L, six mandibles and four isolated teeth. Uniformly bicuspid, with very mesially placed protoconid and metaconid that are quite close to each other, subequal in height, and connected by crest. Metaconid juts lingually in some teeth. Thick paracristid encircles small, deep trigonid basin. Talonid basin quite large, deep; size of trigonid basin relative to huge talonid basin varies. In unworn crowns, surface somewhat wrinkled. Roots quite short, stout, compressed mesiodistally.

Lower M1/M2. From the R, four mandibles and six isolated teeth; from the L, four mandibles and six isolated teeth. M1 and M2 cannot consistently be distinguished. All M1/2s very long and narrow; some lingual slope. Trigonid basin well defined and b/l narrow. Talonid basin b/l narrow, but well developed. Hypoconulid present (in some teeth, central; in others more buccally placed). Some M1/2s have centroconids. In unworn crowns, considerable enamel wrinkling (usually also discernible in worn teeth). Root systems very long relative to crown heights; five specimens with at least two separate roots, but very low bifurcations; in a few others, roots grooved but not separated; in rest, solid single root with one canal.

Lower M3. From the R, five mandibles and three isolated teeth; from the L, two mandibles and five isolated teeth. Lower M3s vary considerably in shape and presence of centroconid; all with distinct trigonid basins and heavy wrinkling. Roots long, stout, and unseparated; a break in one root shows pulp cavity extending far below neck.

REFERENCES

Gorjanovic-Kramberger, D. 1899. Der paläolithische Mensch und seine Zeitgenossen aus dem Diluvium von Krapina in Kroatien. *Mitt. Anthropol. Ges. Wien* 30: 203.

Gorjanovic-Kramberger, D. 1900. Der diluvale Mensch aus Krapina in Kroatien. *Mitt. Anthropol. Ges. Wien* 31: 164–197.

Gorjanovic-Kramberger, D. 1904. Der paläolithische Mensch und seine Zeitgenossen aus dem Diluvium von Krapina in Kroatien (zweiter Nachtrag, als dritter Teil). *Mitt. Anthropol. Ges. Wien* 34: 187–199.

Holloway, R. L. 1985. The poor brain of *Homo sapiens neanderthalensis*; see what you please. In: E. Delson (ed), *Ancestors; The Hard Evidence.* New York, Alan R. Liss, pp. 319–324.

Klaatsch, H. and O. Hauser, 1910. *Homo aurignacensis Hauseri*, ein paläolithische Skelettfund aus dem unteren Aurignacien der Station Combe-Capelle bei Montferrand (Périgord). *Prähist. Z.* 1: 273–338.

Minugh-Purvis, N. 1988. *Patterns of Craniofacial Growth and Development in Upper Pleistocene Hominids.* PhD thesis, University of Pennsylvania.

Minugh-Purvis, N. 1997. Variation in immature hominid neurocranial morphology at 100–130 ka: a comparison of the es-Skhul and Krapina 1 juveniles. In: T. Akazawa and O. Bar-Yosef (eds), *Neanderthals and Modern Humans in West Asia.* New York, Plenum, pp 339–352.

Radovcic, J. et al. 1988. *The Krapina Hominids: An Illustrated Catalog of Skeletal Collection.* Zagreb, Mladost.

Rink, W. et al. 1995. ESR ages for Krapina hominids. *Nature* 378: 24.

Simek, J. and F. Smith. 1997. Chronological changes in stone tool assemblages from Krapina (Croatia). *J. Hum. Evol.* 32: 561–575.

Smith, F. 1976. *The Neandertal Remains from Krapina: A Descriptive and Comparative Study.* Knoxville, University of Tennessee Press.

Repository

Hrvatski Prirodoslovni Muzej, Demetrova 1, 4100 Zagreb, Croatia.

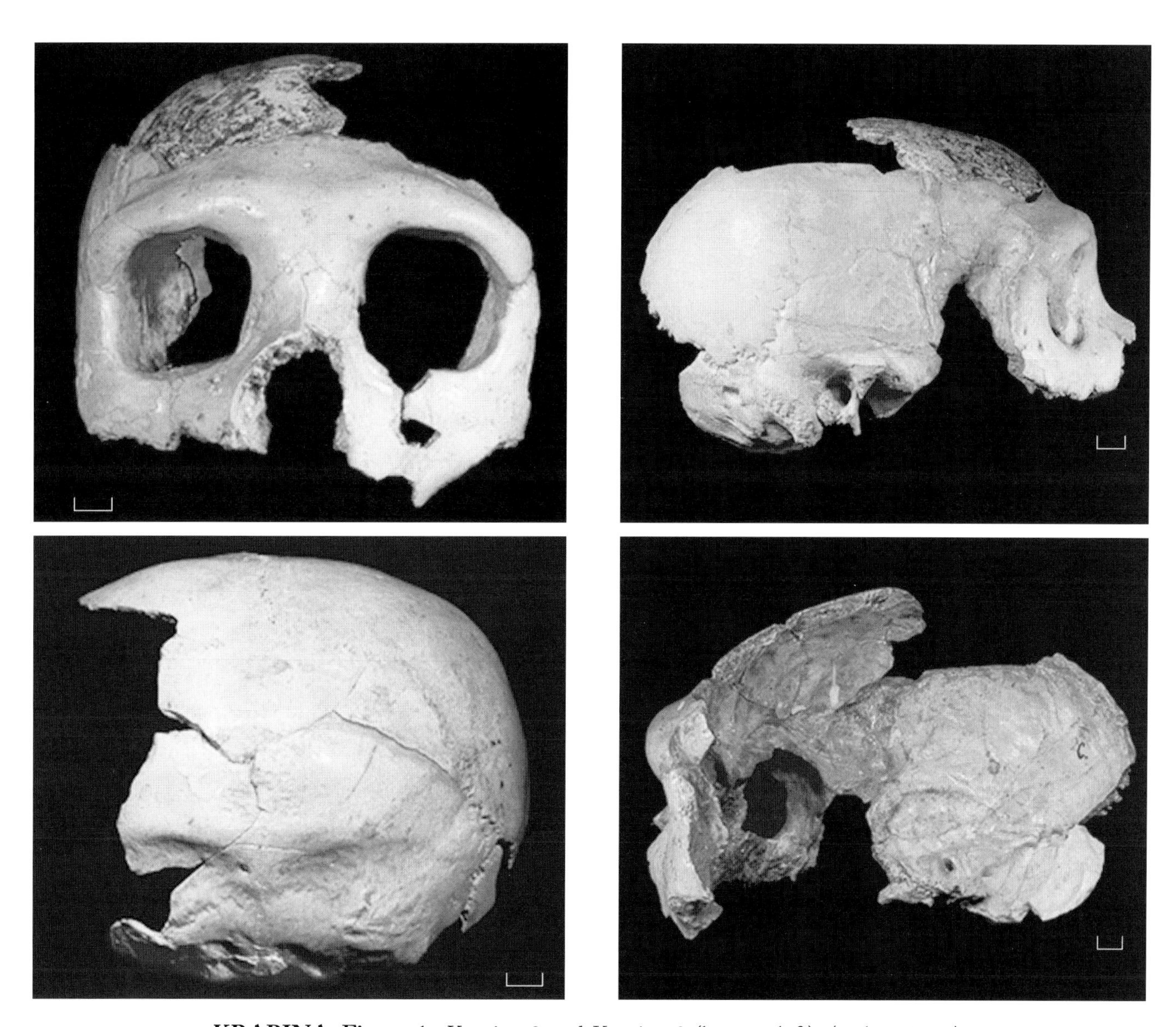

KRAPINA Figure 1. Krapina 3 and Krapina 2 (bottom left), (scale = 1 cm).

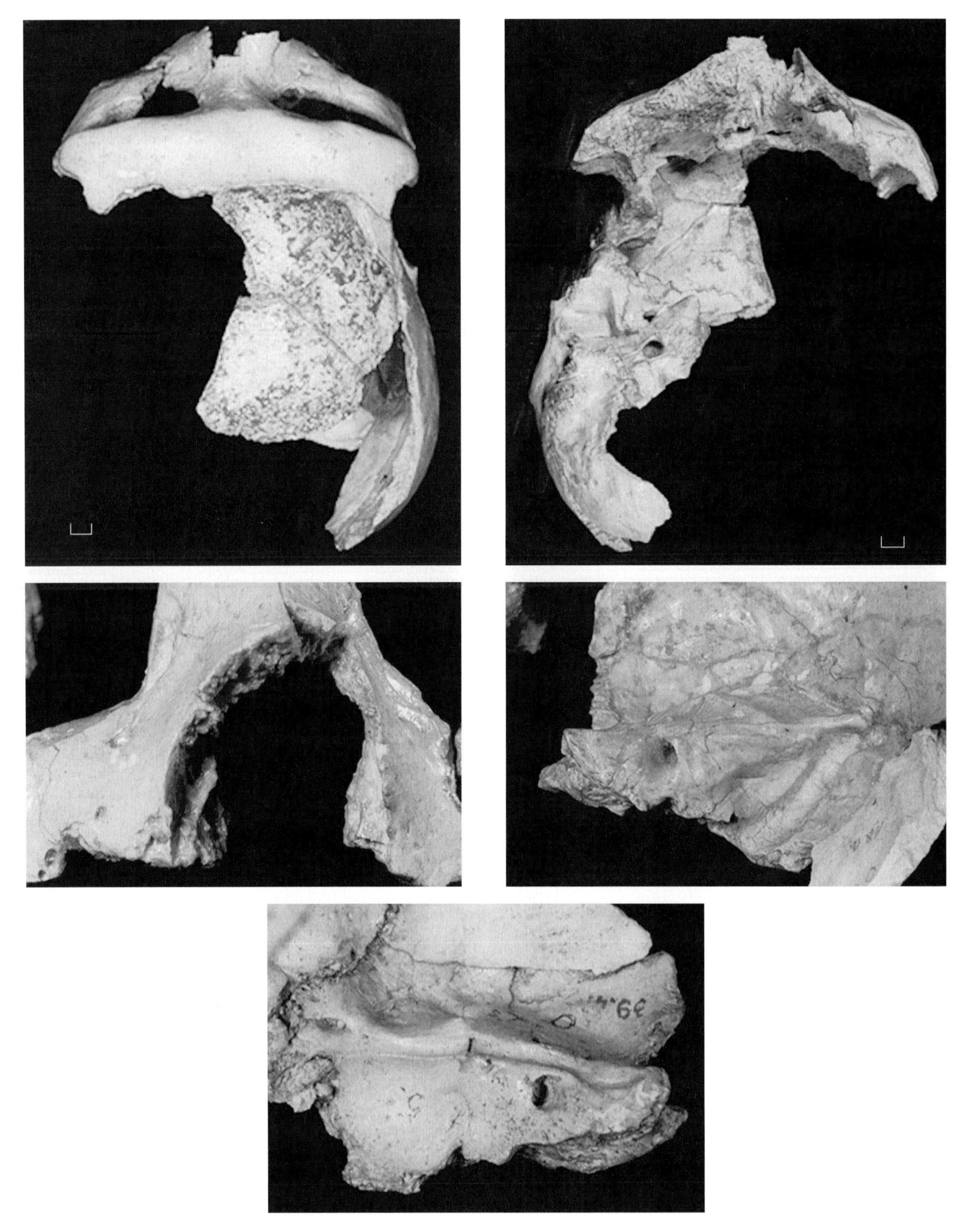

KRAPINA Figure 2. Krapina 3 (top; middle with close-ups of damaged medial projection in nasal cavity and of petrosal) and Krapina 1 petrosal (bottom) (scale = 1 cm).

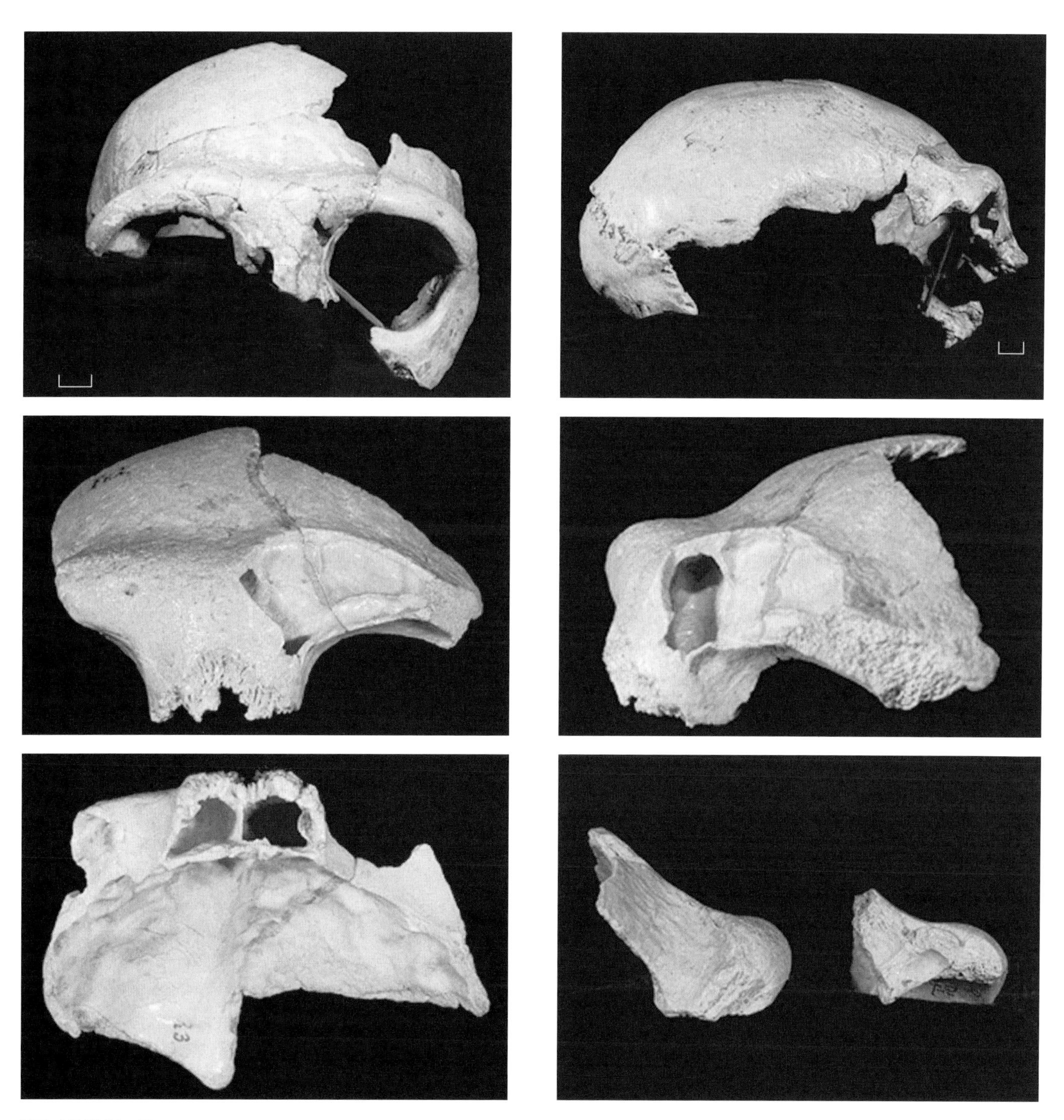

KRAPINA Figure 3. Krapina 6 (top row); Krapina 23 (middle row and bottom left); Krapina 37.3 (left) and 37.8 (right) (bottom right), (top: scale = 1 cm; others: not to scale).

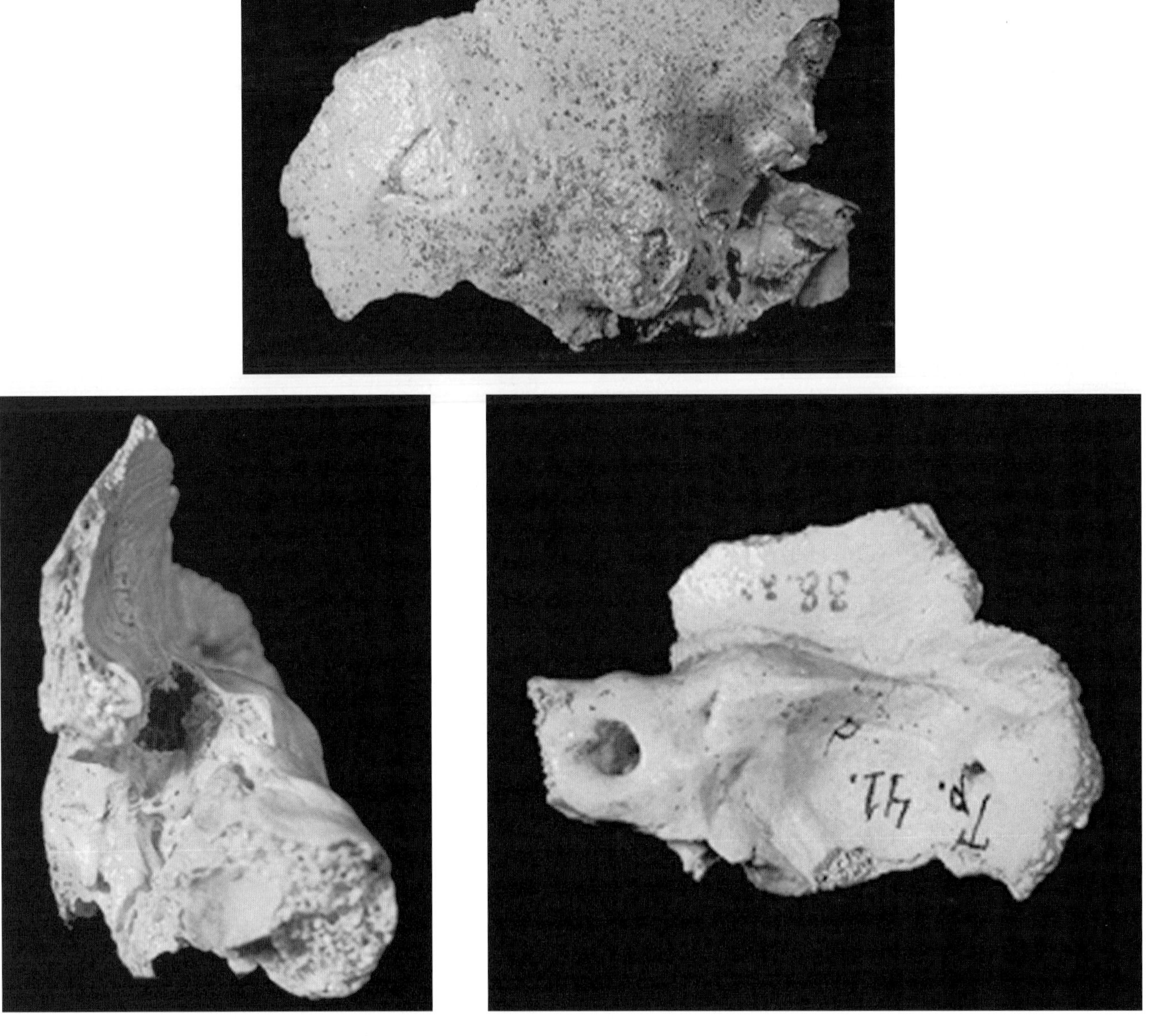

KRAPINA Figure 4. Krapina 38.22 (temporal), (not to scale).

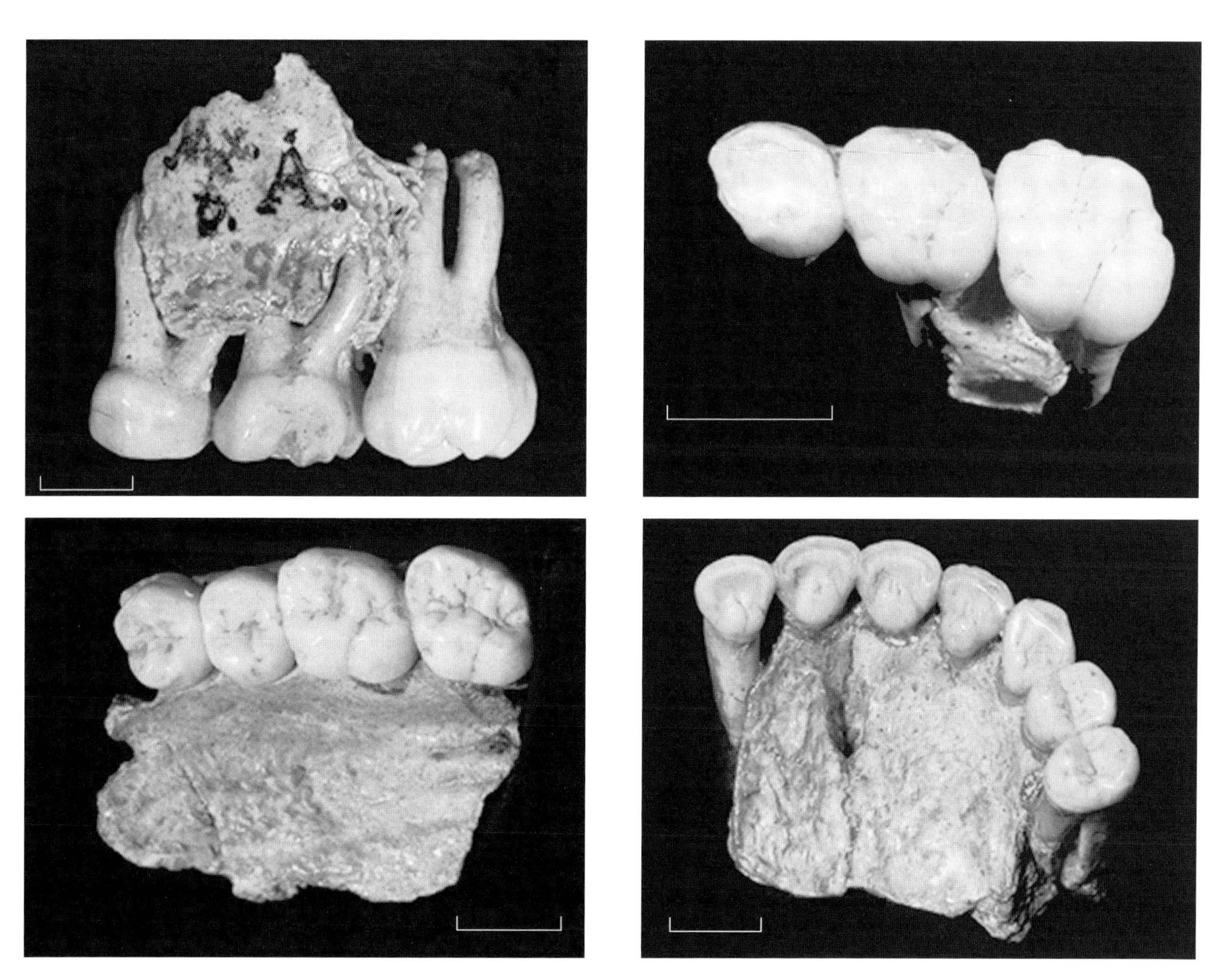

KRAPINA Figure 5. Krapina 45 (top row); Krapina 48 (bottom left); Krapina 49 (bottom right), (scale = 1 cm).

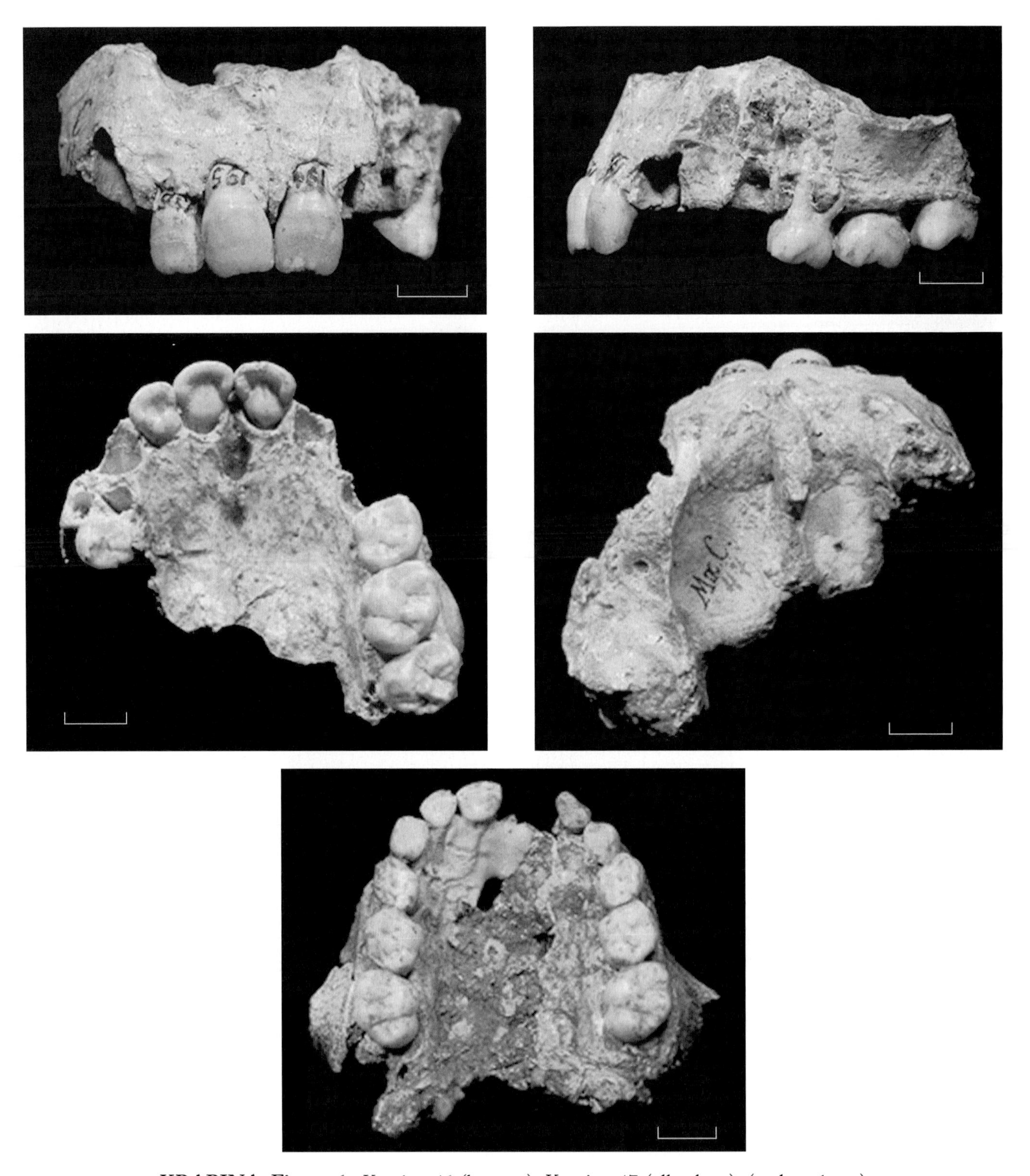

KRAPINA Figure 6. Krapina 46 (bottom); Krapina 47 (all others), (scale = 1 cm).

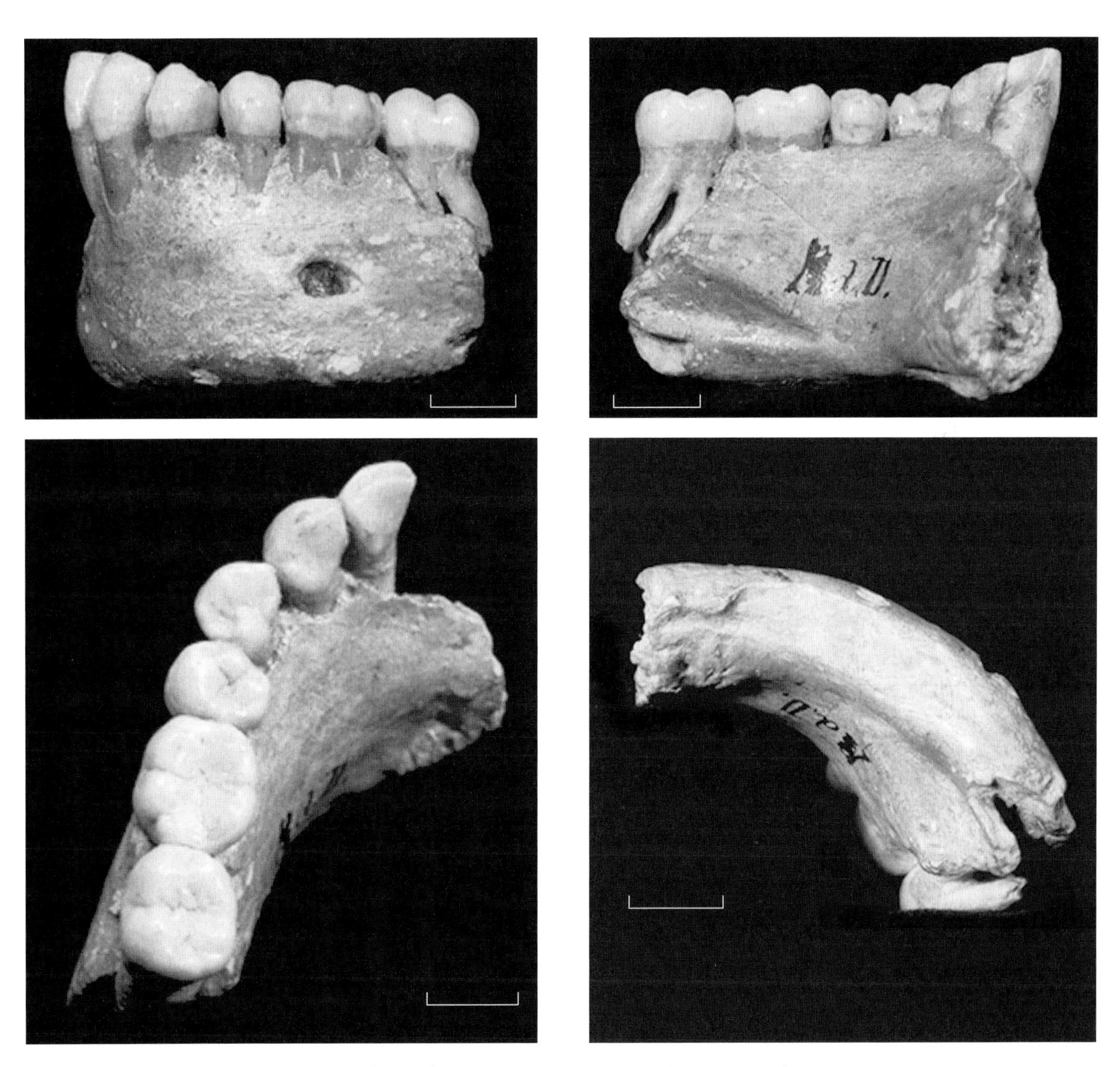

KRAPINA Figure 7. Krapina 57 (scale = 1 cm).

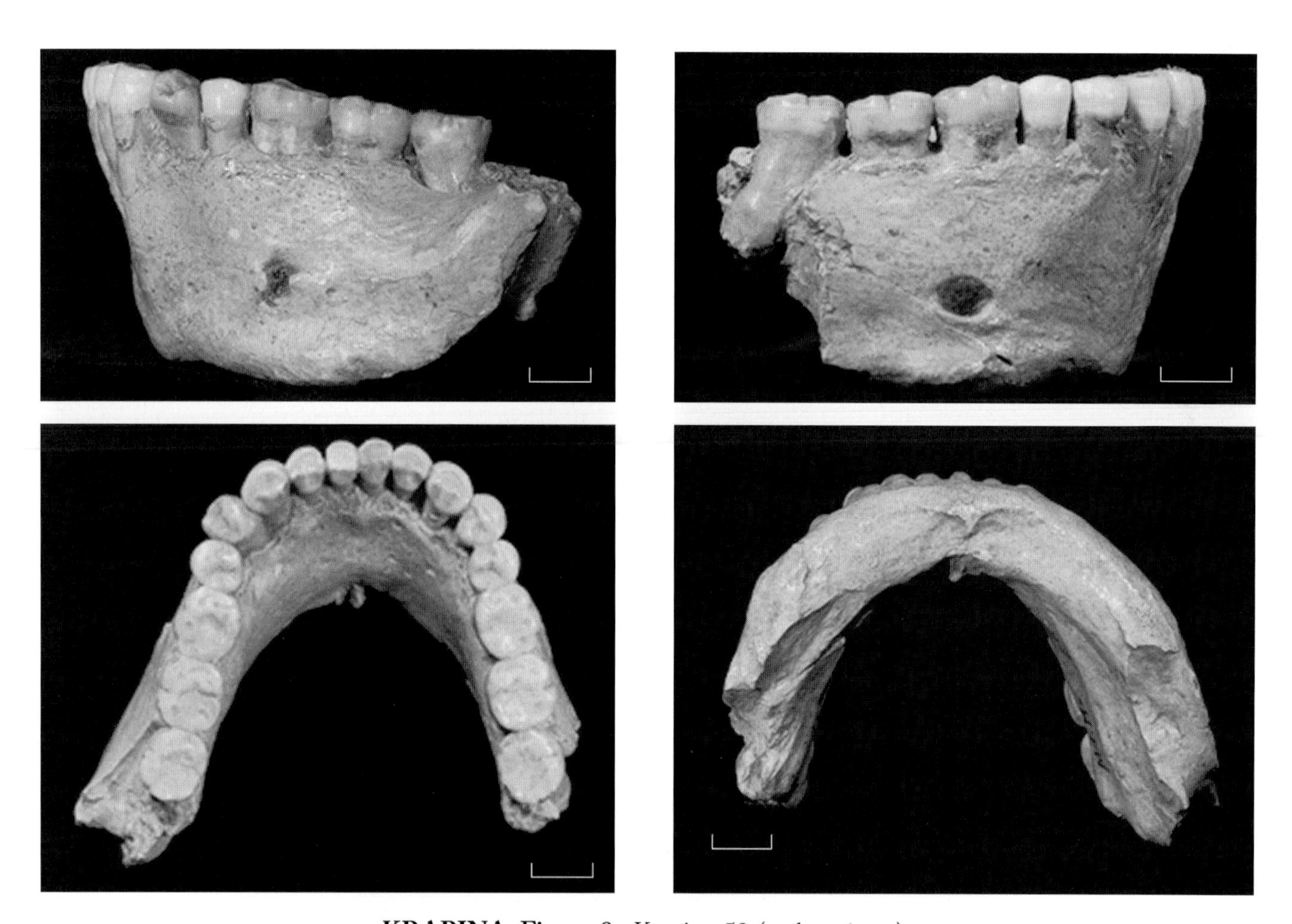

KRAPINA Figure 8. Krapina 58 (scale = 1 cm).

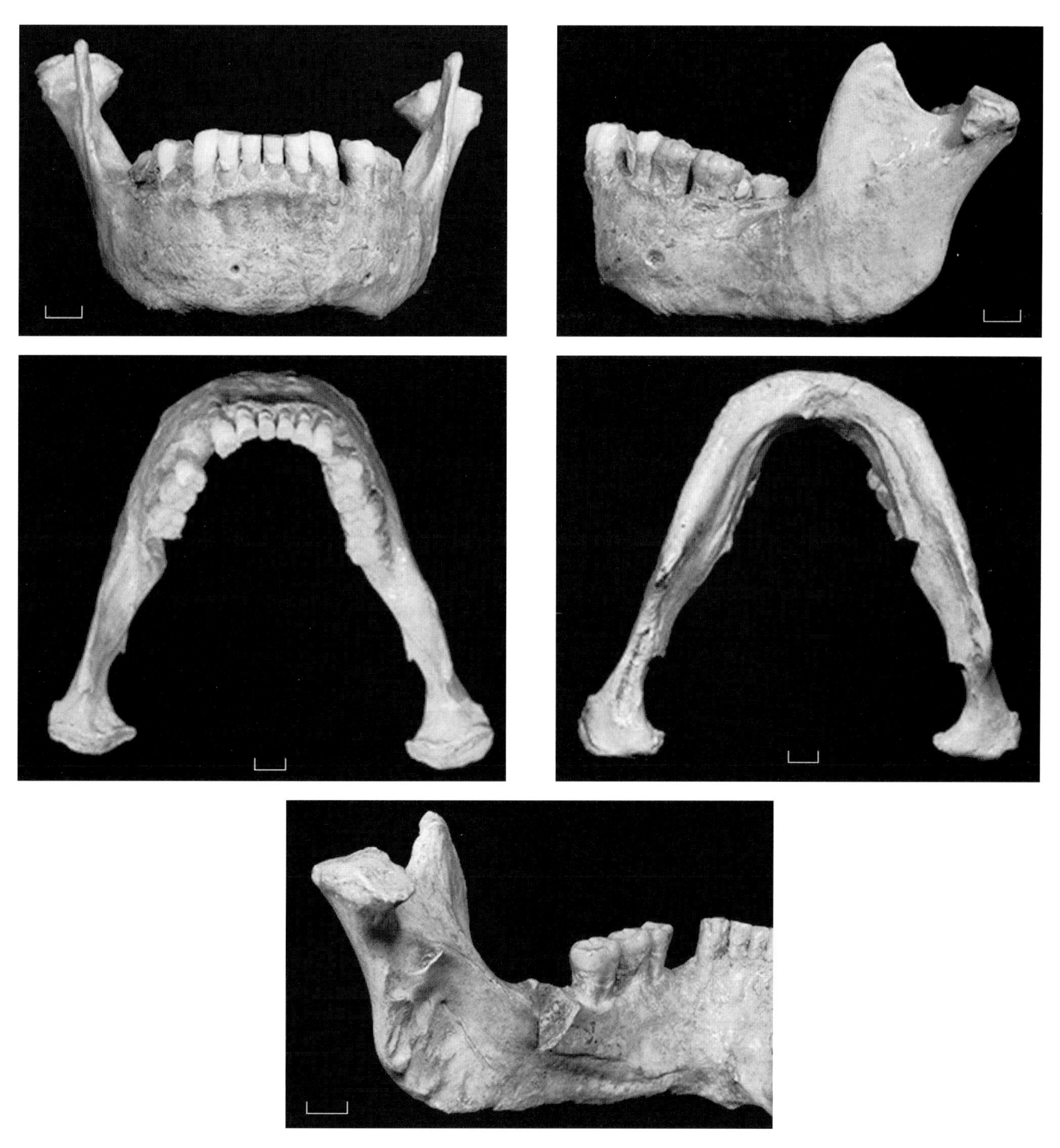

KRAPINA Figure 9. Krapina 59 (scale = 1 cm).

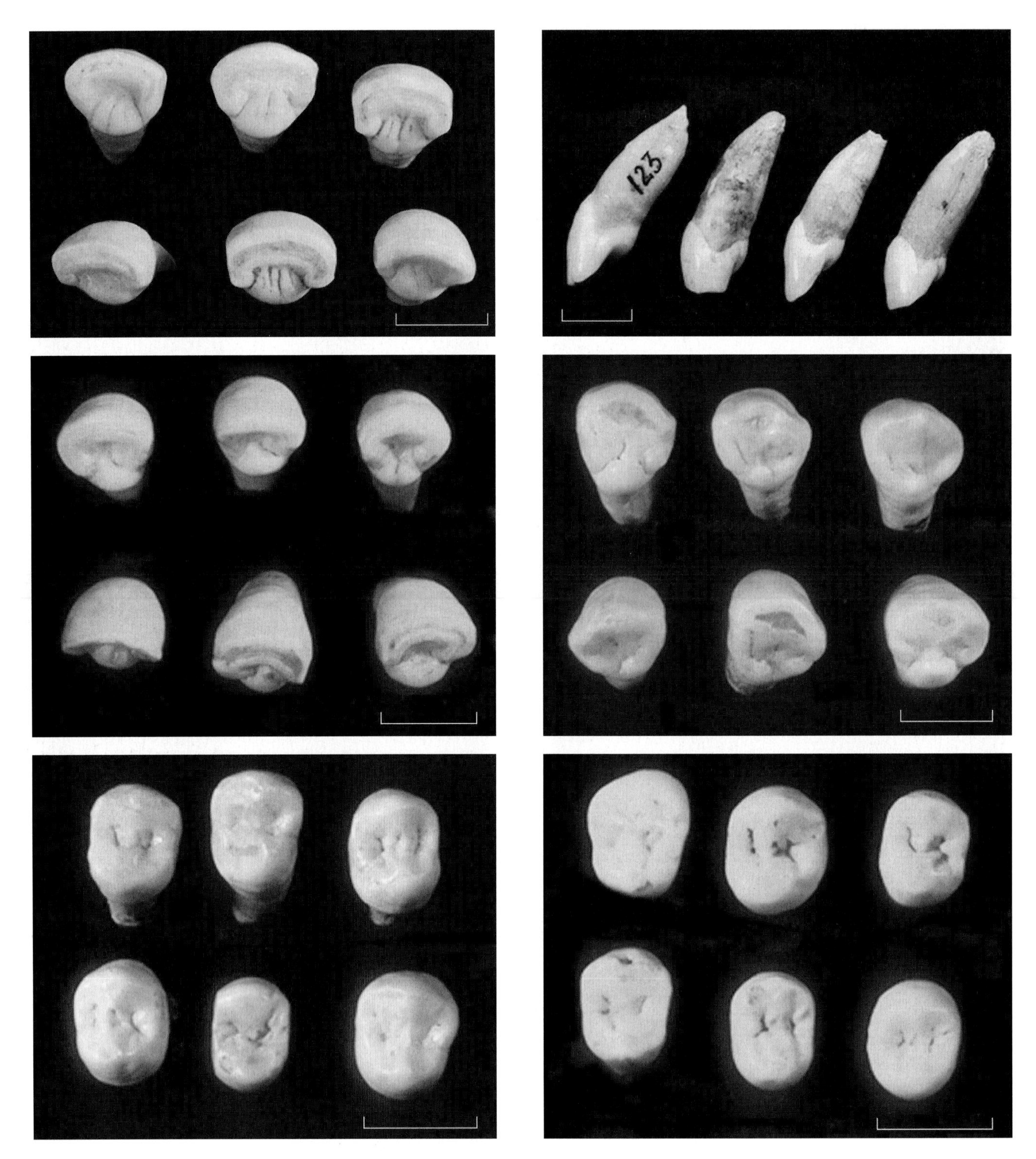

KRAPINA Figure 10. I^1s (top left; top row, KR 154, 129, 57; bottom row, 76, 147, 123); I^1s (top right, 123, 157, 126, 160); I^2s (middle left; top row, 128, 122, 160; bottom 126, 127, 125); C^1s (middle right; top row, 144, 139, 102; bottom row, 76, 147, 37); P^1s (bottom left; top row, 45, 38, 39; bottom row, 52, 47, 53); P^2s (bottom right; top row, 55, 41, 49; bottom row, 48, 42, 44), (scale = 1 cm).

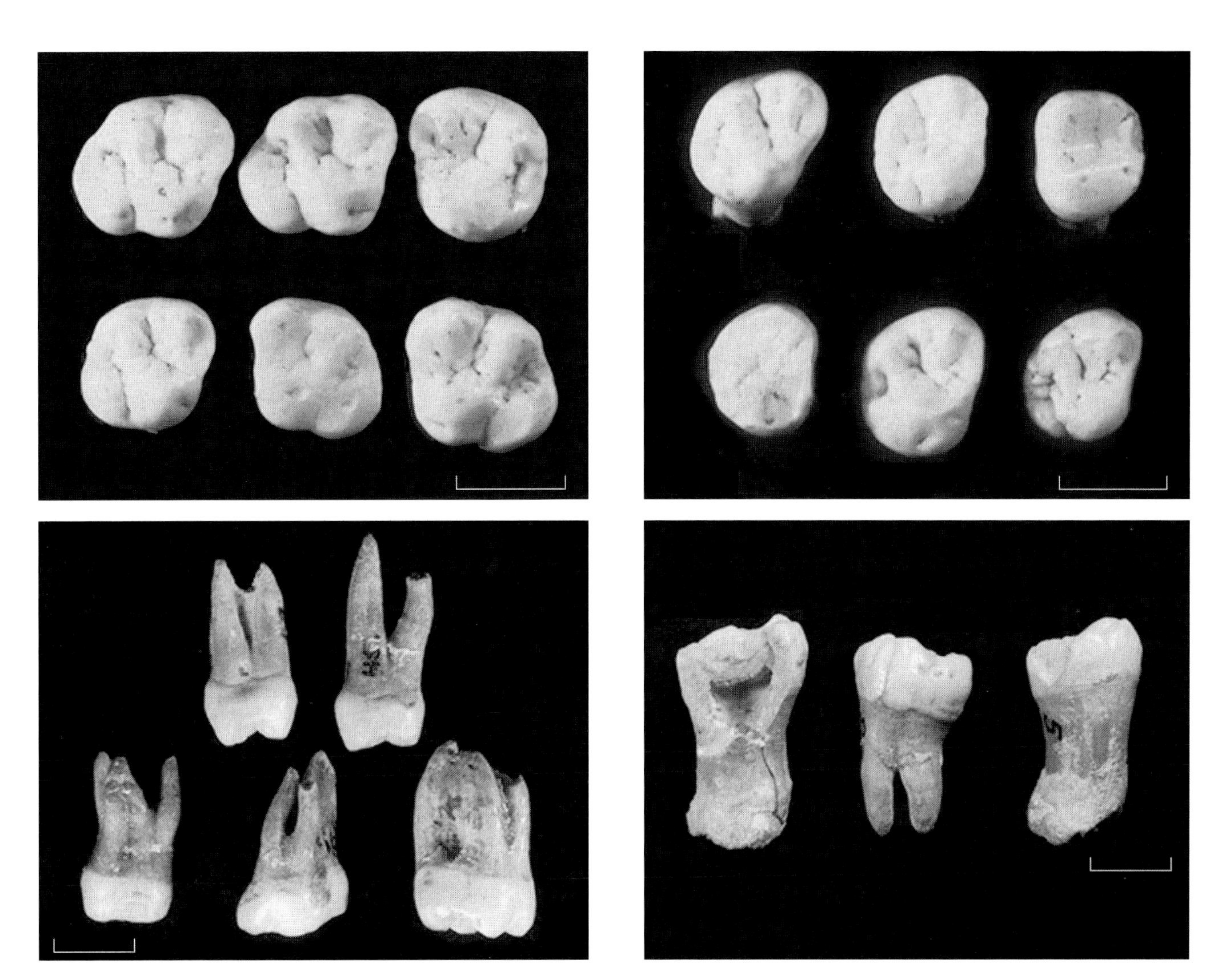

KRAPINA Figure 11. M^{1-2}s (top left; top row, KR 166, 134, 165; bottom row, 165, 164, 136); M^3s (top right; top row, 58, 163, 173; bottom row, 162, 102, 170); upper molar roots (bottom left; top row, 52, 45; bottom row, 162, 164, 177); lower molar roots (bottom right, 8, 85, 5), (scale = 1 cm).

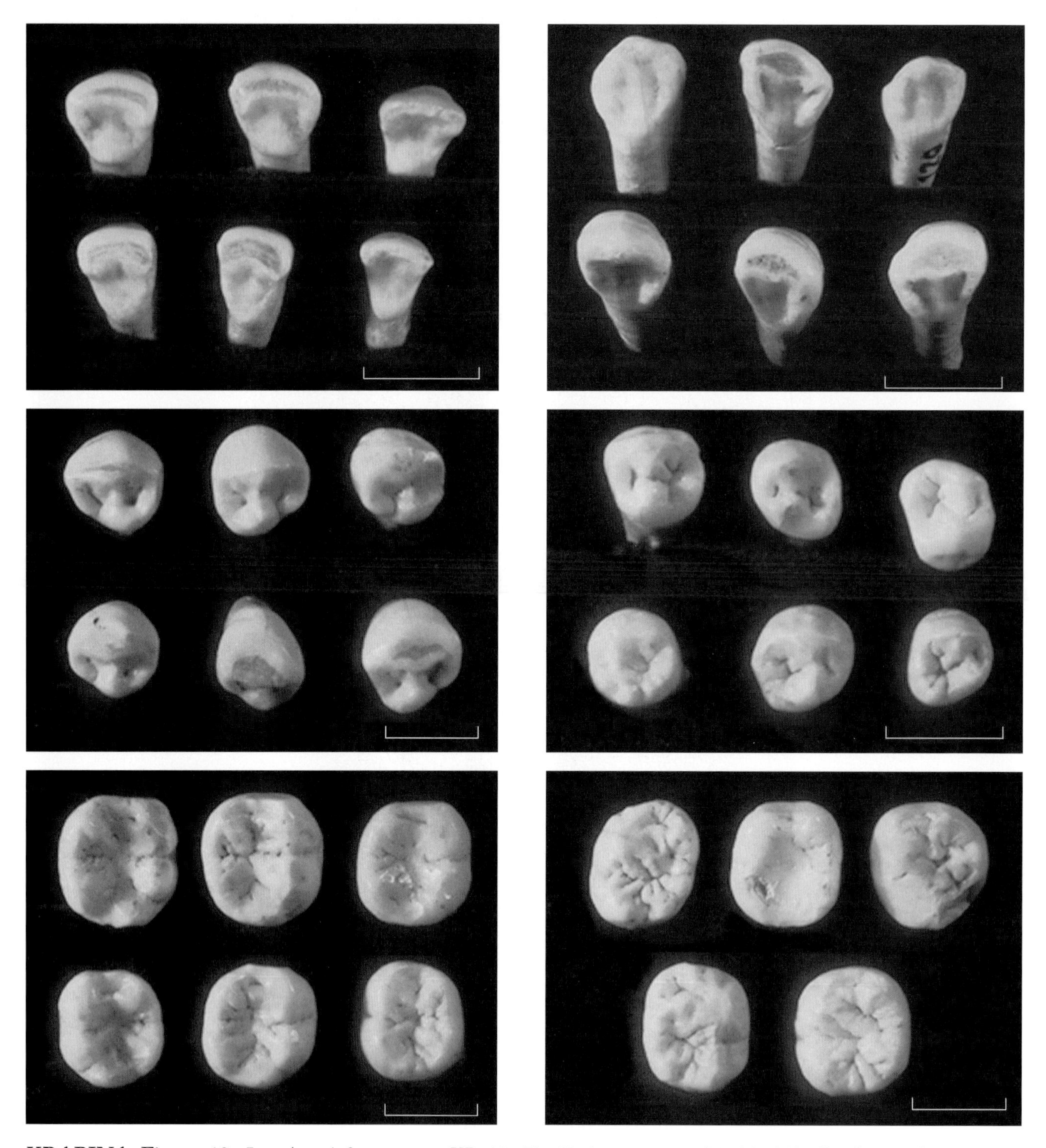

KRAPINA Figure 12. I_{1-2}s (top left; top row, KR 69, 71, 90; bottom row, 74, 70, 72); C_1s (top right; top row, 119, 143, 120; bottom row, 75, 145, 138); P_1s (middle left; top row, 25, 14, 34; bottom row, 27, 28, 29); P_2s (middle right; top row, 30, 31, 35; bottom row, 26, 32, 15); M_{1-2}s (bottom left; top row, 79, 1, 77; bottom row, 81, 80, 6); M_3s (bottom right; top row, 85, 4, 5; bottom row, 7, 106), (scale = 1 cm).

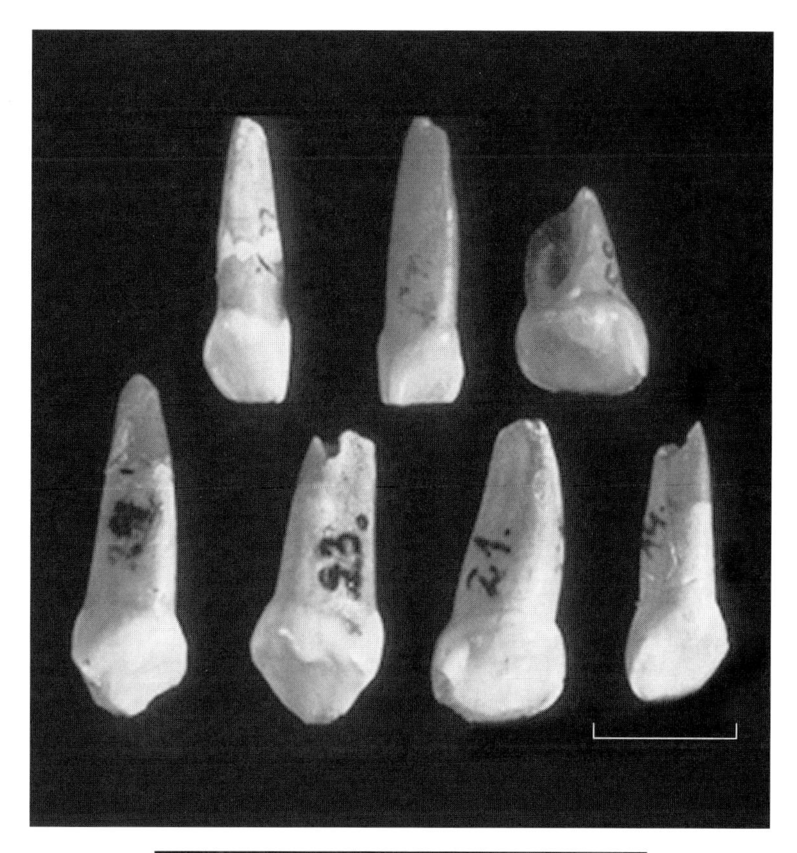

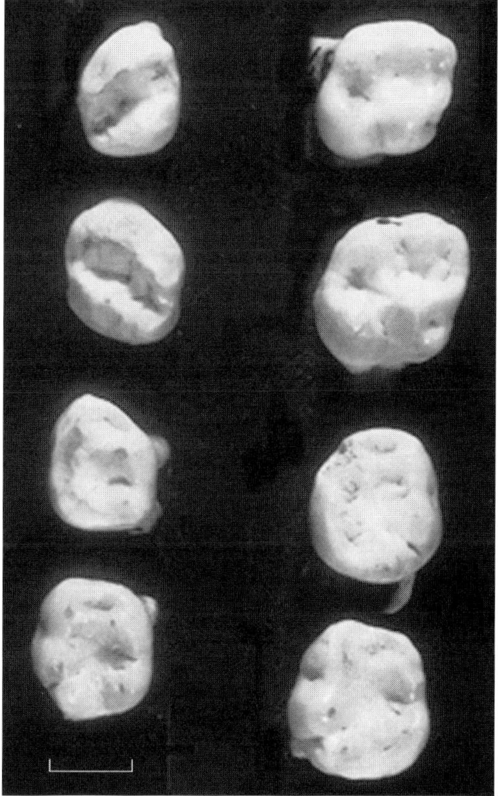

KRAPINA Figure 13. Deciduous upper incisors and canines (top; top row, KR 16, 13, 17; bottom row, 24, 23, 21, 14). (Bottom) deciduous upper molars (first row, 187, 3; second row, 3, 187); deciduous lower molars (third row, 68, 67; fourth row, 63, 62), (scale = 1 cm).

Kulna

Location

Stratified cave site, Sloup village, in the Moravian Karst some 30 km N of Brno, Czech Republic.

Discovery

Excavations directed by K. Valoch and V. Gebauer, July 1965 (Kulna 1 and 3); 1970 (Kulna 2).

Material

R maxilla with four teeth (Kulna 1), partial R parietal (Kulna 2), three deciduous teeth (Kulna 3).

Dating and Stratigraphic Context

The maxilla was found in level 7a of the Valoch and Gebauer excavation, a brownish soil layer with moderate limestone chunks. Rink et al. (1996) derived an average ESR date for large mammal teeth from this layer of 50 ± 5 ka, consistent with the macrofauna, if apparently a little young for the lithics.

Archaeological Context

The industry found in level 7a was characterized by Valoch (1967) as "Mousterian of Micoque tradition"; later he revised this diagnosis to "Micoquian of Beckstein type" because it essentially lacked Levallois technique (Valoch, 1988). Used fragments of bone, tooth, and antler were also present in this level.

Previous Descriptions and Analyses

The hominid fragments have been described by Jelinek (e.g., 1966, 1988), who finds them to be fairly typically Neanderthal.

Morphology

Kulna 1

Adult (A 17 092). R maxilla containing C–M1, I alveoli; anteriorly, midline preserved, with inferolateral corner of nasal aperture showing double margin (prenasal fossa); anterior nasal spines, lateral nasal margin missing. In area of P2–M1, face begins to swell out just above alveolar margin. Maxillary sinus extends forward to frontal process; intrudes into nasal cavity (would have "ballooned" slightly). Nasal cavity floor depressed, with gradual slope. Area at base of frontal process thick medially (suggesting medial projection). Nasoalveolar clivus, long, vertical. Incisor alveoli slightly procumbent and very deep. C root was long.

Upper teeth not remarkably large. C has large lingual tubercle, strong margocristae and central sulcus. In both Ps, thickened crest comes down inside protocone to posterior fovea (more pronounced in P2, in which cusps also more approximated); both also bear distinct anterior foveae. M1 has large, lingually placed protocone with well-developed preprotocrista that arcs around to parastylar region. M1 also with very sharp postprotocrista that runs to metacone, well-defined but constricted trigon basin, distinct Carabelli's pit, huge hypocone that swells out distolingual corner, stout postcingulum that arcs to meet metacone, low and deep talon basin, and enamel wrinkling. Exposed roots show no indication of any bifurcation.

Kulna 2

Adult (A 17 093, 1970). R parietal fragment with part of sagittal suture. Temporal lines relatively well marked and quite broad. Sagittal suture quite uniform and minimally denticulate. Bone quite thick in region of suture, with thick diploic region (not thick cortical bone).

Kulna 3

Upper dm2 (no number). L; worn; large hypocone; thick postcingulum around talon basin, and up face of metacone, which is lower than paracone.

Lower dm1 (no number). R; heavily worn; preserves deep trigonid basin opening lingually (paracristid incomplete); large enamel swelling below protoconid.

Lower dm2 (A 17 092, 1965). Isolated. Heavily worn; rather buccally placed, large hypoconulid; trigonid basin very deep; talonid basin well developed; roots bifurcate low.

References

Jelinek, J. 1966. Jaw of an intermediate type of Neanderthal Man from Czechoslovakia. *Nature* 212: 701–702.

Jelinek, J. 1988. Anthropologische Funde aus der Kulna-Höhle. In: K. Valoch (ed), *Die Erforschung der Kulna-Höhle, 1961–1976.* Brno, Moravske Muzeum.

Rink, W. et al. 1996. ESR dating of Micoquian Industry and Neanderthal remains at Kulna Cave, Czech Republic. *J. Archeol. Sci.* 23: 889–901.

Valoch, K. 1967. Die Steinindustrie von der Fundstelle des menschlichen Skelettrestes aus der Höhle Kulna bei Sloup (Mähren). *Anthropologie* 5: 21–32.

Valoch, K. 1988. *Die Erforschung der Kulna-Höhle, 1961–1976.* Brno, Moravske Muzeum.

Repository

Anthropos Institute, Moravske Muzeum, Brno, Czech Republic.

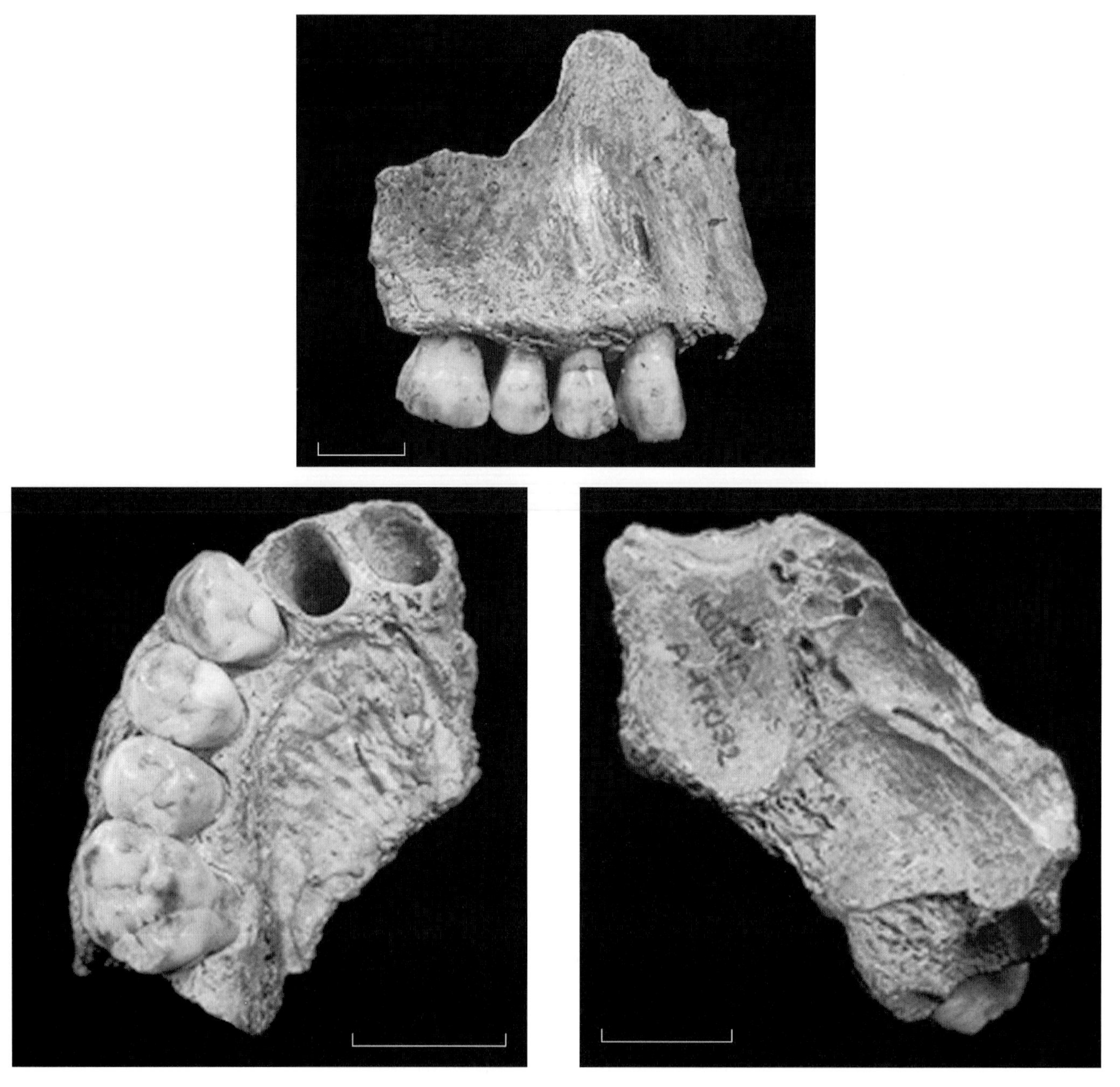

KULNA Figure 1. Kulna 1 (scale = 1 cm).

La Chapelle-aux-Saints

Location

Bouffia Bonneval Cave, in the Sourdoire River valley near the village of La Chapelle-aux-Saints, some 40 km SE of Brive-la-Gaillarde, Corrèze, France.

Discovery

Bouyssonie brothers and L. Bardon, August 1908.

Material

Single skeleton of aged and somewhat arthritic adult, presumed male. Quite well-preserved edentulous cranium; upper body skeleton fairly complete; lower body less so.

Dating and Stratigraphic Context

The skeleton was found in a flexed position within a shallow depression apparently scooped into the marly sediments of Bed 1 of the cave (Bardon, Bouyssonie, and Bouyssonie, 1908). It almost certainly represents a simple burial, though early claims of grave goods are highly problematic. Sediments above the skeleton yielded stone tools and a fauna indicative of cold conditions, suggesting that the skeleton dated from some time in the last glacial, probably Würm II, and if so, around 50 ka. Recent ESR age estimates on mammal teeth from Bed 1 (Grün and Stringer, 1991) average around 47 ± 3 and 56 ± 4 ka, depending on the uptake model; opinion appears to favor the older end of the range.

Archaeological Context

Charentian Mousterian, La Quina variant (Mellars, 1988).

Previous Descriptions and Analyses

First described in a brief note by Boule in 1908. Boule's later (1909–11) three-part monograph established the La Chapelle fossil as the archetype of the late "classic" Neanderthals of western Europe (a role it still occupies) and also entrenched the erroneous stereotype of the Neanderthals as shuffling brutes. This error was rectified by Arambourg (1955) and Cave and Straus (1957). More recently, Heim (1989) has reconstructed the skull, slightly reducing its width and cranial volume, increasing basicranial flexion, and correcting numerous details. Estimated cranial capacity is 1625 cc (Holloway, 1985).

Morphology

Most of aged adult skull (presumably male) (MNHN/MH 1908–37; 24,493–1960.3), lacking nasal bones, orbital cones, almost all of sphenoid region, part of L parietal, regions of jugular foramina. Bone not very thick.

Skull long, low, and very wide, especially posteriorly. Forehead low, with distinct rise originating well behind orbital margins (creates posttoral plane, not sulcus). Viewed from the front, face broad above, tapers inferiorly. Smoothly rolled, tall s/i, slightly laterally tapering, vermiculate-boned supraorbital tori confluent with swollen glabella; seen from above, tori angle slightly back from glabella. Temporal lines originate in ridge that rises behind zygomatic process of frontal, just at lateral edge of orbits; they proceed backward fairly low on cranial wall in weak but highly visible bandlike muscle scar that curves back from behind mastoid process to run toward asterion. Interorbital space very wide. Lacrimal fossa a/p narrow, relatively shallow; anterior and posterior crests surround upper part of fossa in smooth curve. On the L, infraorbital groove very long, with short infraorbital canal. Infraorbital foramen large, opens inferiorly; two accessory infraorbital foramina on the R, one on the L. Orbits "aviator glasses" shaped, but medioinferior portion of orbital floor not extremely puffed out (= less inflation of maxillary sinus in this area).

Nasals missing; would have flexed straight forward below nasion (indicated by long, horizontal nasomaxillary suture). Frontal process of maxilla elongated anteriorly; upper part of nasal region projects; lower part of aperture unremarkable. Lateral crest of nasal margin quite sharp, inferiorly delineates narrow prenasal fossa that is defined from behind by a short spinal crest. Lateral crests become confluent with distinct R and L anterior nasal spines; although somewhat damaged, spines still large, horizontal, and projecting. Posteriorly, anterior nasal spines confluent with preserved base of vomer that persists along length of nasal cavity; there is a groove in posterior surface of R anterior nasal spine (for cartilaginous nasal septum?). Slight depression below anterior nasal spines. Nasoalveolar clivus relatively vertical; originally was quite long.

From point well behind spinal crest, posteriorly sunken floor of nasal cavity descends sharply. Well within nasal aperture, bilaterally, two fairly large areas of roughened, spiculate bone project medially into nasal cavity; on the R, rugose and elevated area is delineated inferiorly by raised, roughened, more or less horizontal edge (short conchal crest?); no equivalent structure present on the L (on both sides, dominant orientation of rugosities is verticality). No evidence of any bone-to-bone contact or fusion along area of horizontal roughening or along anterior lacrimal crest. On both sides, faint evidence of a low "turbinal crest" coursing down to spinal crest and contributing to poorly developed internal nasal margin. On both sides, paper-thin anterior lacrimal crest continues down into nasal cavity; does not extend significantly over lacrimal groove posteriorly.

Maxillary sinus expands quite far anteriorly and superiorly into both maxillary frontal processes as well as into anterior root of zygomatic arch but not greatly into body of zygoma. Superiorly, maxillary sinus extends up beneath thin plate of orbital floor. As especially well preserved on the L, maxillary sinus also expands quite aggressively medially, deflecting its medial wall into the nasal cavity (posterior "balloon" swelling into region where ethmoturbinals normally found). Superior part of ethmoid preserved; posterior portion of perpendicular plate still in place. Extension of anterior ethmoidal air cell into frontal region seen on both sides; on the L, only second air cell preserved posterior to this (no evidence of third or more ethmoidal air cells behind this one). Sphenoid sinus penetrates to region above pterygoid plates.

Viewed from the front, lower margins of anterior root of zygomatic arch straight and laterally oblique; diverge posteriorly from their origin close to alveolar margin in region where M1 had been. Zygomas oriented quite posterolaterally. Zygomatic arches remarkably thin and horizontal; run straight back with no flare, limiting the rather narrow temporal fossa, which (on preserved L) is divided by crease into temporal and infratemporal portions. Cranial wall in this area marked by sclerotic pathology.

Temporal region better preserved on L. Squamous portion small and very short s/i, its anterior suture almost vertical and ridged (delineates anterior and posterior portions of temporal fossa). Mastoid process downwardly pointing, moderately prominent, and m/l compressed from its moderately broad base. Suprameatal crest less prominent than upwardly curving supramastoid crest. Small anterior mastoid tubercle lies high up on mastoid process, opposite external auditory meatus. Posterior root of zygomatic arch originates in front of auditory meatus. Mastoid notch well excavated and wedge shaped, with no evidence of digastric fossa posteriorly. Parietal notch lies above posterior limit of mastoid process. Long, horizontal parietomastoid suture continues into short, horizontal anterior lambdoid suture.

Mandibular fossa wide and quite deep, with distinct, low articular eminence in front. Articular surface extends over eminence (on the L, anterior area marked by erosive lesions). Fossa bounded medially by medial articular tubercle (extension of temporal bone) that is separate from the short ectotympanic tube. Vaginal process runs along midline of ectotympanic tube;

peaks significantly around thin styloid process; descends dramatically to fade out before reaching bottom of external auditory meatus. Both sides with meatal tori; the L has two (almost completely occluded by them). Relatively large stylomastoid foramen lies lateral to styloid process, with downward-pointing carotid foramen medial to it.

Occipital plane wider than tall; in profile, no distinct chignon. "Torus" poorly defined, straight, horizontal; its inferior border well delineated by infratoral sulcus. "Torus" relatively narrow from side to side; its lowest point coincides with opisthocranion. Superior border of "torus" defined only by suprainiac depression that is quite wide, not very high, and only slightly more pitted than surrounding bone. Nuchal plane undercuts occiput. At its lateral extremity, as seen on better preserved L, faint superior nuchal line continues laterally and forward, circumscribing shallow area of muscle attachment. Lambdoid suture rises gently from short anterior lambdoid suture, arcs across region of lambda.

Basiocciput broad, thin, delicate, quite flat. Slightly raised occipitomastoid crest along occipitomastoid suture lower than mastoid process. Medially and posteriorly, very large Waldeyer's crest descends slightly below tip of mastoid process. If orientation of basiocciput accurately reconstructed (but may not be), there is some basicranial flexion anterior to foramen magnum. Foramen magnum extremely elongated and ovoid. L occipital condyle remodeled (occludes postcondylar canal); postcondylar depression not visible.

Palate was long, narrow, probably deep, with marked anterior slope and some sloping of sides; bone of this region has clearly undergone pathological remodeling. Incisive foramen large, sunken, and deep; it lies at level of Cs. Medial and lateral pterygoid plates converge superiorly and inferiorly.

Lambdoid suture essentially uniform, although lobate (not conventionally denticulate). Coronal suture uniform and barely denticulated. Sagittal suture mostly uniform and conventionally denticulated.

Internally, petrosals very wide and lack evidence of superior petrous sinus posteriorly. On the L is a superior petrous sinus anteriorly. On the R, region of arcuate eminence bears low, broad dome; on the L, this region marked only by crest of superior semicircular canal. On both sides, subarcuate fossa closed off and filled in.

Mandible complete except for tips of both coronoid processes and L condyle; alveolar bone resorbed, particularly posteriorly; apart from LP2, edentulous. Front of mandible not excessively broad; slightly arced from side to side. Originally, may have been shallow subalveolar depression; otherwise surface of symphyseal region smooth, devoid of morphology. Presence of inferior marginal tubercles debatable. Anterior symphyseal region elevated. Broad, shallow digastric fossae present bilaterally. On the L, mylohyoid line quite marked. On both sides, low, broad pillar runs down inside of coronoid process to become confluent with mylohyoid torus (posterior continuation of mylohyoid line). Submandibular fossa well defined. Mandibular foramina tall, compressed, oriented obliquely upward. L lingula broken; would have been quite large. Medial pterygoid tubercle moderately big on the L, slightly smaller on the R. As preserved on the R, deepest part of sigmoid notch lies close and sigmoid notch crest terminates lateral to midline of highly remodeled condyle. Cannot tell whether preangular sulcus present. Medially deflected, externally smooth gonial angle "cut off." Mental foramina huge (especially on the L), would have lain beneath M1; large accessory foramen on the R.

References

Arambourg, C. 1955. Sur l'attitude en station verticale des Néanderthaliens. *C. R. Acad. Sci. Paris* 240: 804–806.

Bardon, L, J. Bouyssonie and A. Bouyssonie. 1908. Découverte d'un squelette humain mousterien à La Chapelle-aux-Saints (Corrèze*). C. R. Hebd. Séanc. Acad. Sci. Paris.* 147: 1414–1415.

Boule, M. 1908. L'homme fossile de La Chapelle-aux-Saints. *C. R. Acad. Sci. Paris* 147: 1349–1352.

Boule, M. 1911–13. L'homme fossile de La Chapelle-aux-Saints. *Ann. Paléont.* 6: 1–64 (1911); 7: 65–208 (1912); 8: 209–279 (1913).

Cave, A. J. E. and W. L. Straus, Jr. 1957. Pathology and posture of Neanderthal man. *Quart. Rev. Biol.* 32: 348–363.

Grün, R. and C. Stringer. 1991. Electron spin resonance dating and the evolution of modern humans. *Archaeometry* 33: 153–199.

Heim, J.-L. 1989. Une nouvelle réconstitution du crâne néandertalien de La Chapelle-aux-Saints. *C. R. Acad. Sci. Paris* 308 (Sér. II): 1187–1192.

Holloway, R. L. 1985. The poor brain of *Homo sapiens neanderthalensis*; see what you please. In: E. Delson (ed), *Ancestors; The Hard Evidence*. New York, Alan R. Liss, pp. 319–324.

Mellars, P. 1988. The chronology of the south-west French Mousterian: A review of the current debate. In: M. Otte (ed.) *L'Homme de Néanderthal, Vol. 4: La Technique*. Liège, ERAUL, pp. 97–119.

Repository

Laboratoire d'Anthropologie Biologique, Musée de l'Homme, Place du Trocadéro, 75116 Paris, France.

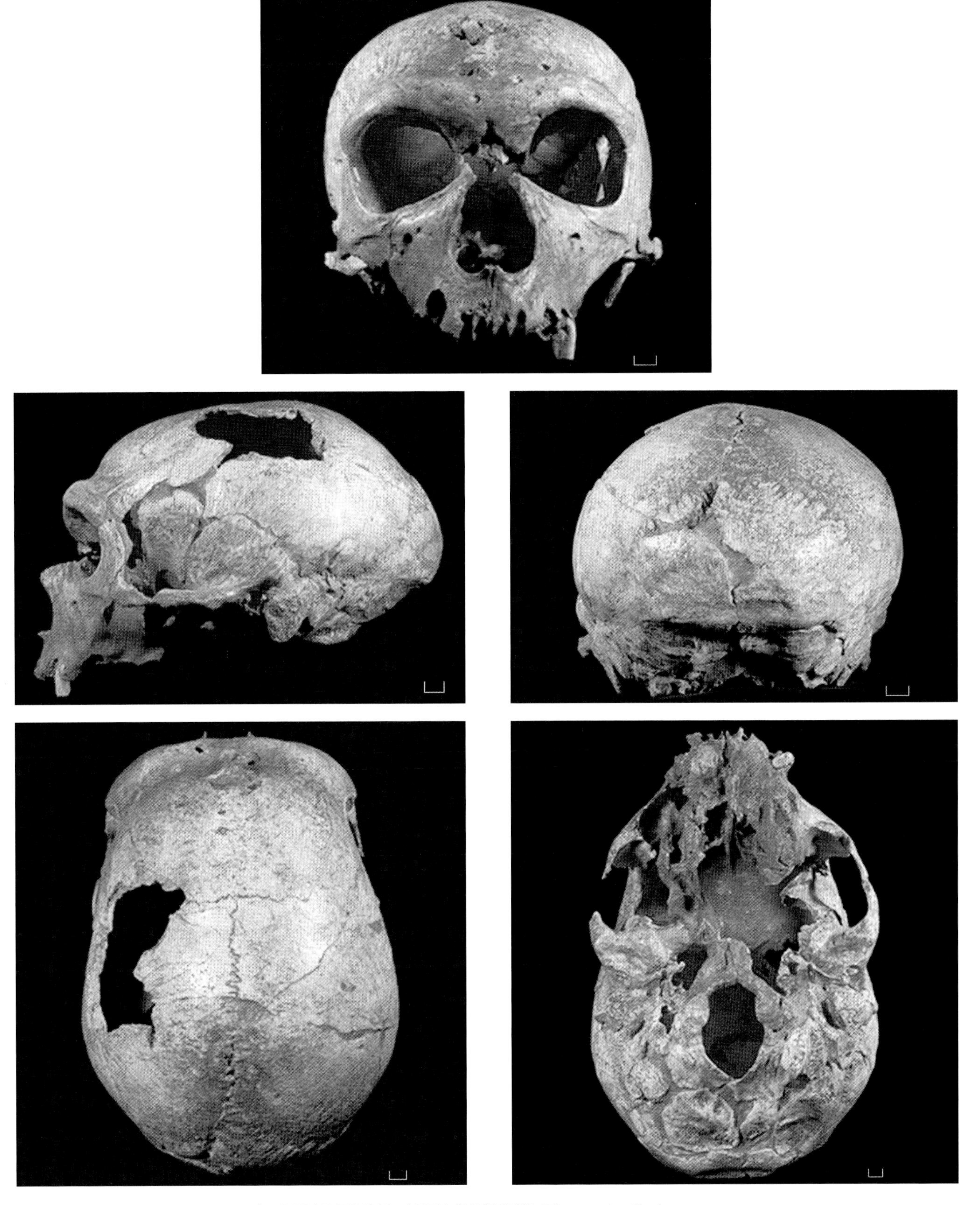

LA CHAPELLE-AUX-SAINTS Figure 1. Scale = 1 cm.

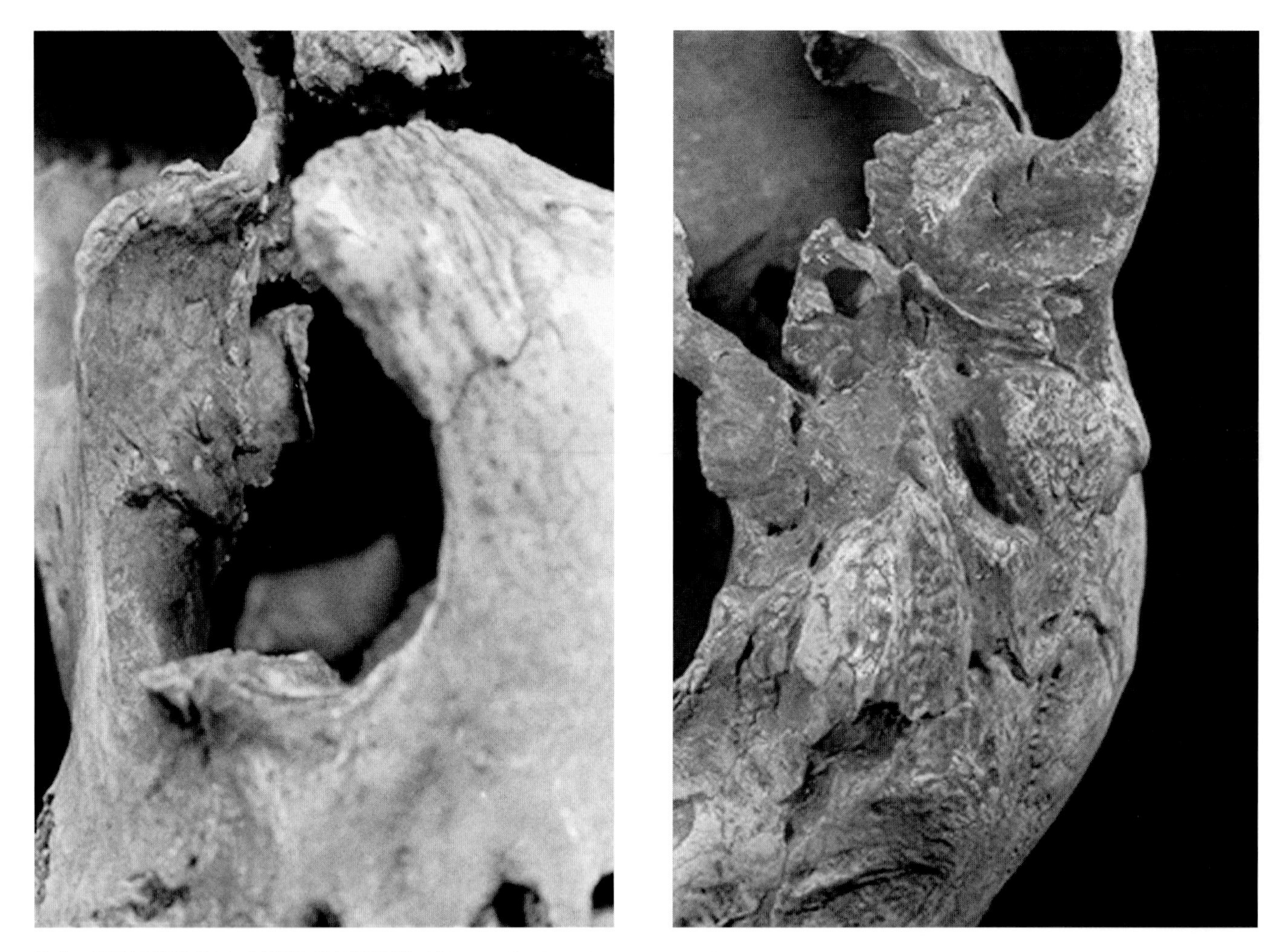

LA CHAPELLE-AUX-SAINTS Figure 2. Nasal cavity medial projection with conchal crest below; cranial base, (not to scale).

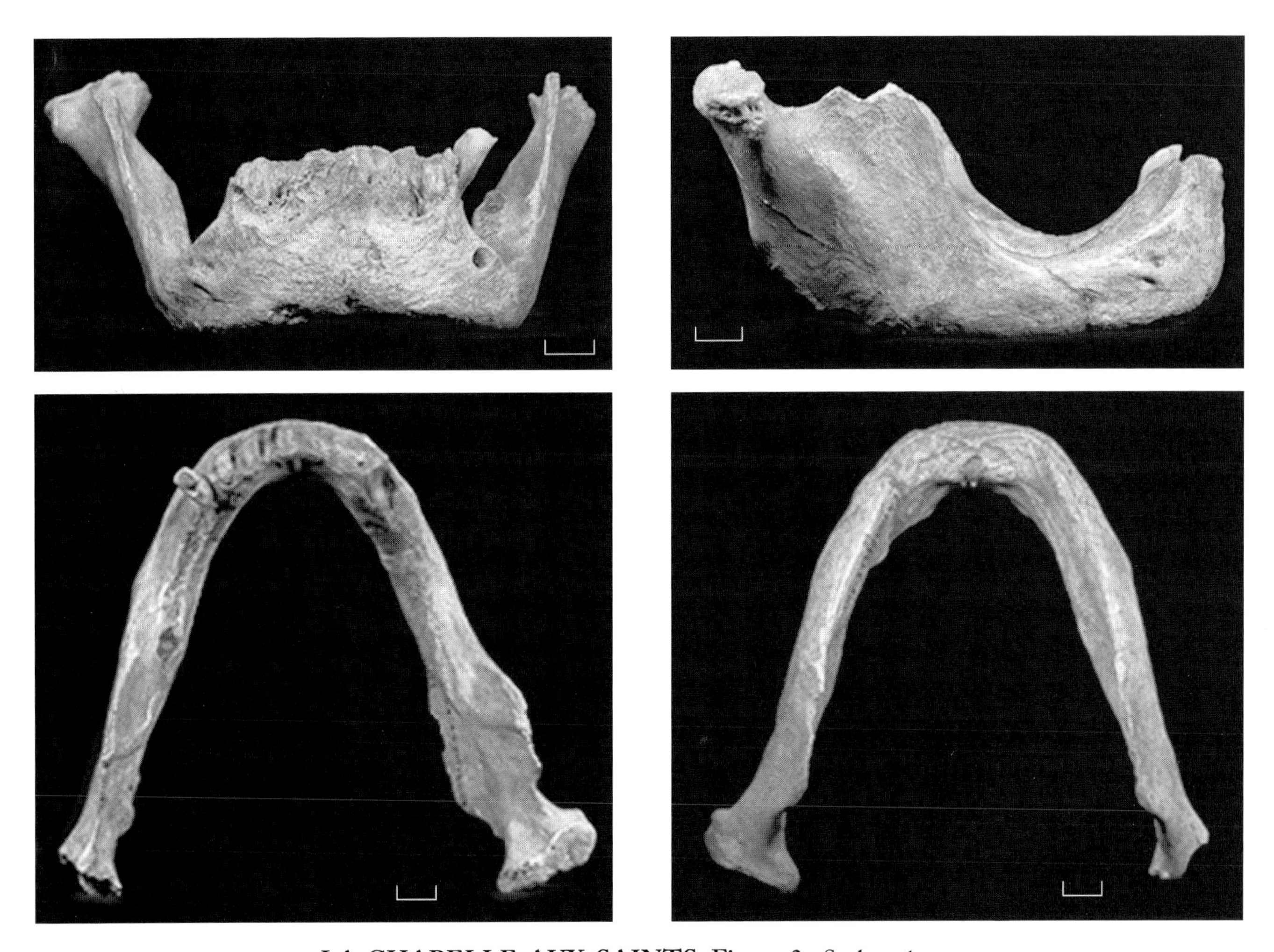

LA CHAPELLE-AUX-SAINTS Figure 3. Scale = 1 cm.

La Ferrassie

Location
Rock shelter near Savignac du Bugue, about 5 km NW of Les Eyzies de Tayac, Dordogne, France.

Discovery
Excavations of D. Peyrony and L. Capitan between 1909 (La Ferrassie 1) and 1921 (La Ferrassie 6). Excavations of H. Delporte, 1973 (La Ferrassie 8).

Material
Remains of two adult (La Ferrassie 1 and 2) and several highly incomplete and poorly preserved immature individuals (La Ferrassie 4a, 4b, 5, neonates or fetuses; 3 and 7, now believed to represent the same individual, ca. 10 years; 6, ca. 3 years; and 8, ca. 2 years). Whereas the two adult skeletons are rather more complete, and the skull of the presumed male La Ferrassie 1 is particularly well preserved, the presumed female cranium La Ferrassie 2 consists only of right maxillary, petromastoid and parietal fragments.

Dating and Stratigraphic Context
Stratified rock shelter site with occupation levels from Acheulean to Gravettian. The human fossils came from Ferrassie Mousterian deposits (levels C/D) that underlay Châtelperronian and Mousterian of Acheulean Tradition strata and may all represent burials as part of a cemetery complex. The associated fauna suggests cold conditions and is typical of the last glacial. Closer dating depends on archaeological association; Mellars (1996) places the Ferrassie Mousterian at around 70 ka at other sites, although the La Ferrassie fossils themselves may be somewhat younger than this (see below).

Archaeological Context
Bourgon (1957) regarded the industry from this site as Charentian Mousterian, but more recent authors (e.g., Heim, 1976; see review by Mellars, 1996) view the Ferrassie Mousterian as a distinct variant, also seen at other sites in southwestern France. According to Peyrony (1934) the skeletons represented crouched burials, and it is possible although far from certain that some of these interments, at least, may have been intrusive into level C from higher levels.

Previous Descriptions and Analyses
The earlier discoveries were briefly noted in field reports by Capitan and Peyrony (e.g., 1909, 1911, 1912) but were principally described by Boule (1911–13) as reference materials in his multi-part study of the La Chapelle-aux-Saints specimen. Not until much later did these specimens receive monographic treatment, notably by Heim (e.g., 1976, 1982a,b; see also 1974). La Ferrassie 1 has served as the exemplar for the hominid population at the site and since its discovery has consistently been seen as an archetypal example of the "classic" western European Neanderthals of the last glacial. Its estimated cranial capacity is 1640 cc (Holloway, 1985).

Morphology

La Ferrassie 1

Adult. Large, quite complete cranium and mandible; lacks nasal bones, medial orbital walls, sphenoid and alisphenoid regions, and parts of R and L petrosals. At least $RI^1$and LI^{1-2} placed back to front in this reconstruction, giving erroneous appearance of wear sloping upward and outward; RP^2 is a L; various other teeth (e.g., R I^2–C^1) probably also misplaced. Upper and lower cheek teeth occlude properly only when jaw displaced anteriorly, onto what would be articular eminence if it was not missing. All upper and lower teeth excessively worn, in some cases down to the pulp cavities.

Cranium long, low, broadest posteriorly. Frontal rises fairly steeply some distance behind orbital margins (thus posttoral plane, not sulcus). From the front, face broadest superiorly, tapers inferiorly. From above, face wedge shaped. From rear, neurocranium somewhat horizontal and ovoid, with slightly curved sides that flow smoothly around to top of skull.

"Double-arched," smoothly rolled, and vermiculate-boned supraorbital tori confluent with wide, prominent, and smooth glabella. Tori thin laterally only marginally; from above, they angle back slightly from glabella. L and R frontal sinuses extend both quite far back into frontal and laterally, right over lateral edges of orbits. Interorbital region quite broad. Orbits have "aviator-glasses" shape, with swollen inferomedial corner. Orbital floors flow smoothly onto puffy face below. Infraorbital foramina large and open forward and down. Temporal lines continuous with posterior margin of frontal process of zygoma; they arise close to lateral extremities of and slightly behind orbits, then fan out into bandlike scar with upper surface arcing backward to fade out before reaching anterior lambdoid suture and lower border recurving toward asterion.

Nasal aperture huge; prominent superior margins suggest that missing nasals were projecting. Preserved nasomaxillary suture shows sharp anterior projection below nasion; anteriorly, suture makes almost right angle with virtually vertical nasal aperture margin. Inferior nasal margin single (no "turbinal crest" or prenasal fossa); area of anterior nasal spine missing. Nasal cavity floor not completely preserved; was evidently somewhat sunken, with backward slope starting at inferior portion of lateral crest. Just within external nasal aperture, prominent, enlarged medial projection preserved on the L; on the R, this structure more modestly proportioned; both projections broken inferiorly. Posterior breakage shows patent lacrimal canal on both sides, lying behind medial projection. Lateral nasal walls not well preserved posterior to this. On the L, preserved nasal floor and base of nasal cavity wall show distinct incurving of huge maxillary sinus wall (suggesting some development of posterior "balloon"). Maxillary sinus also extends into nasal region, almost as far forward as external nasal margin, as well as up to nasomaxillary suture and inferior orbital margin; sinus does not penetrate very deeply into zygoma. Large sphenoid sinus (preserved on the L) extends laterally into region above roots of pterygoid plates. Nasoalveloar clivus was probably originally quite long, may have sloped a little anteriorly.

Squamosals neither very long nor very tall. Zygomatic arches quite robust, notably around posterior root, which originates anterior to auditory meatus. Anterior root of zygomatic arch originates just above alveolar margin; in front view, it angles up and outward. Inferior margin of thin zygomatic arch angled upward in side view. Arch does not flare out; runs almost straight back; subtends relatively small temporal fossa. Anterior squamosal margin raised, creating anterior and posterior temporal fossae. Distinct temporal and infratemporal fossae; horizontal infratemporal surface lies below level of bottom of zygomatic arch.

Mandibular fossae broad, large; appear to have been remodeled. Both quite shallow and not bounded anteriorly by distinct articular eminence. L mandibular fossa enlarged anteriorly (in accordance with position of condyle required for occlusion); R fossa also extended laterally by remodeling. As preserved particularly on the R, fossa bounded medially by small, distinct medial articular tubercle that is sharply separated behind from ectotympanic tube. Postglenoid plates relatively long, closely approximated to ectotympanic tubes. External auditory meati truncated laterally (i.e., not fully ossified), small, ovoid; face almost horizontally and backward. Petrosals broad laterally, flowing into interior cranial wall. On the R, superior semicircular canal more elevated into superior surface than on the L (= not domed arcuate eminences). No superior petrous sinus on either side, only longitudinal ridge along top of petrosal internally; anterior extremity of ridge delineated by notch. No subarcuate fossa on either side.

Mastoid processes (almost fully preserved on the L, partial on the R) long a/p, broad m/l, short s/i, and oriented straight down. Processes may not have borne anterior mastoid tubercle; on the R, a small tubercle lies posterior to the auditory meatus. Broken R mastoid pneumaticized by medium-sized (not notably small) air cells. Parietal notches lie over midline of mastoid processes. Parietomastoid suture on both sides relatively long and horizontal. Occipitomastoid crests were at least as tall as mastoid processes (probably taller on the L). Waldeyer's crests on both sides, situated quite far medial to occipitomastoid crests; they are relatively weak, although the R is a little larger than the L. Digastric notch (quite complete on the L) relatively narrow and only moderately deep; it continues to just behind the mastoid process, where its posterior limit is defined by low ridge of bone. Suprameatal crests weaker than supramastoid crests that sweep slightly up behind them.

Thin, sharp vaginal process (preserved on the R) runs along midline of ectotympanic tube; tube short, incompletely ossified laterally. Vaginal process far separated from mastoid process; peaks at thin styloid process, which lies lateral to carotid foramen. Lateral to styloid process, vaginal process merely low crest. Bilateral pits may indicate location of stylomastoid foramina, which lay quite lateral to styloid processes.

Anterior lambdoid sutures obscured by large Wormian bones (preserved particularly well on the R), behind which the unsegmented, not very highly denticulated lambdoid sutures rise quite sharply to arc around region of lambda. Occipital plane wider than tall. Occipital torus straight, horizontal; it is delimited below by a broad infratoral sulcus/depression and above only by smallish (ca. 3 cm), not very tall suprainiac depression that is only very slightly pitted. Inferior margin of torus appears to correspond to superior nuchal line; only possible indication of "inferior nuchal line" is an inward deflection of the posterior ends of the Waldeyer's crests. In side view, a classic chignon is discernible; entire occipital plane bulges below lambda. Foramen magnum long and ovoid; postcondylar region missing. Occipital condyles broad, strongly arced, and very anteriorly positioned; they face directly downward, defining plane that forms distinct angle with plane of basiocciput. Basiocciput quite broad but broken anteriorly. Fair degree of basicranial flexion.

Palate deep; slopes posteriorly, but with steep sides. Incisive foramen apparently quite far forward, level with C^1s. Medial and lateral pterygoid wings close together, do not converge to peak; are confluent front and bottom; may also have converged superiorly (reconstruction seems unreliable on the L).

Mandible complete except for the R condyle, part of the L, and tiny portion of L gonial region. Corpus not very deep; thins dramatically posterior to level of M3. No midline swelling of symphysis externally (appearance of "incipient chin" due merely to subalveolar depression that is most deeply indented in region between C and I2 roots). Symphysis rather narrow from side to side, slightly arced. Low, blunt inferior marginal tubercles lie below septum between P1–2 roots. Broad, shallow digastric fossae present bilaterally. Mental foramen lies below M1, is double on the R. Retromolar space huge, accentuated further by preangular notch. Ramus long from back to front; bone of gonial region very thin. Coronoid process tall, undercut inferiorly by well-developed preangular sulcus; rises considerably higher than condyle. Gonial angle fairly smoothly curved, with hint of being "cut off"; external surface essentially smooth. L gonial angle slightly inflected inwardly, the R slightly reflected outward. Internally, strong descending pillar becomes confluent at level of mandibular foramen with strong ridge (mylohyoid torus) extending posteriorly from mylohyoid line. Mandibular foramina both wedge shaped, opening upward; were probably not overlapped by lingulae. Sigmoid notch deepest close to condyle. Sigmoid notch crest runs quite far toward lateral side of condyle. Partially preserved L condyle remarkably broad; shows ridge across front (osteoarthritis?). Medial pterygoid tubercle highly developed on the L, poorly on the R.

Upper and lower M1–2 roots bifurcate relatively close to neck; a little farther away in M3.

La Ferrassie 2

Adult. R maxillary and petromastoid fragments; tiny parietal fragment.

R maxilla contains I1–M1, all heavily worn and flat. Small inferior portion of nasal aperture preserved, showing sloping anterior nasal cavity floor. Anterior nasal margin "double" (part of narrow prenasal fossa preserved; delineated posteriorly only by spinal, not turbinal crest). Nasoalveloar clivus was not short (original length cannot be determined). Appears that anterior root of zygomatic arch originated quite close to alveolar margin in region of M1 and flared up and out. Maxillary sinus expands forward quite close to margin of nasal aperture; appears to have encroached quite strongly into nasal cavity (indicated by orientation of broken lateral nasal wall relative to plane of midline). Palate was deep with straight sides.

R temporal bone lacks most of squamous portion. Mandibular fossa broad, moderately deep. Articular eminence better developed than La Ferrassie 1. Anterior part of fossa bears erosive lesions; articular surface seems to have extended over it. Fossa closed medially by moderately developed medial articular tubercle that is separated from ectotympanic tube. Low postglenoid plate also separated from tube.

Posterior root of zygomatic arch originates anterior to moderately sized, subcircular auditory meatus. Ectotympanic tube incompletely ossified laterally. Poorly developed vaginal process (merely crease of bone) runs along midline of tube; peaks around what would have been thin styloid process. Vaginal process stays well separated from mastoid process. Stylomastoid foramen lies posterolateral to and far away from styloid process; both foramen and process in line with mastoid notch; notch relatively broad, quite deep, delineated posteriorly (as in La Ferrassie 1). Downwardly pointing mastoid process moderately sized; is moderately long a/p, compressed m/l, and moderately thick at base, with low "mastoid crest" (not anterior mastoid tubercle). Parietal notch at posterior margin of mastoid process. Preserved parietomastoid suture long, horizontal. Occipitomastoid crest taller than mastoid process. Laterally broad petrosal lacks medial portion; region of arcuate eminence low but expanded laterally; crest in place of superior petrous sinus; subarcuate fossa lacking. Carotid foramen faces downward.

Two isolated teeth, possibly lower I and P; both very heavily worn.

La Ferrassie Immatures

La Ferrassie 3. Partial cranium (and skeleton) of child, possibly less than 5 years old, with R and part of L petrosal, L sphenoid with part of middle cranial fossa, and part of L parietal/occipital region.

Foramen ovale contained well within sphenoid; fairly large foramen spinosum behind may not be in sphenoid. Smallish foramen rotundum lies directly in front of foramen ovale, its axis skewed somewhat laterally. Nonconfluent superior roots of pterygoid plates preserved; may have been parallel to one another. Auditory meatus small, as is carotid canal, which was apparently fully ossified to tip of petrosal. Small carotid foramen points directly down, lies well anterior to peak of thick vaginal process. Vaginal process descends laterally to fade out at lateral extremity of ossified portion of ectotympanic tube; apparently bears vertical groove on posterior surface for styloid process (represented by styloid pit well posterior to carotid foramen). Stylomastoid foramen lies lateral and somewhat posterior to region of styloid process. Floor of ectotympanic tube very incompletely ossified laterally; moderately sized foramen of Huschke just on anterior side of peak of vaginal process. Mastoid process has just begun to form; it is well separated from the most lateral extent of the vaginal process. On damaged R petromastoid, emergent mastoid process less well defined than on the L. Parietomastoid suture was long and horizontal, as is preserved anterior lambdoid suture, from which the shallowly denticulate lambdoid suture rises quite sharply. In parietal/occipital fragment, occipital portion bears broad shallow sulcus that seems inconsistent with its reconstructed position. Preserved sagittal suture on parietal modestly denticulate and unsegmented, as is lambdoid suture.

Internally, both petrosals preserve region of arcuate eminence; on the R, eminence moderately globular, flows into superior surface; on the L, eminence truncated m/l and delineated from rest of superior surface by crease in bone (petrosals from two individuals?). L petrosal fused to squamosal; no sign of superior petrous sinus groove; subarcuate fossa depressed but totally closed over; cochlear canaliculus apparently large. Groove for sagittal sinus broad, shallow. Transverse sinus runs just below asterion to become sigmoid sinus.

Potential parietal fragment bears muscle scar.

La Ferrassie 4 (part of La Ferrassie Bloc A). R femoral and humeral diaphyses of neonate.

La Ferrassie 4 bis (part of La Ferrassie Bloc A). Small fragments from an entire skeleton, probably not more than 1–2 years old. Also, L petrosal with almost fully closed off subarcuate fossa; superior semicircular canal well elevated above superior surface.

La Ferrassie 5. Probably a little younger than 4 bis. Many fragments, mostly cranial. Subarcuate fossa of R petrosal almost entirely closed off; superior semicircular canal protrudes above superior surface; apparently, tympanic ring was fused to adjacent bone, but with no tube development. Small carotid foramen points straight down; small carotid canal apparently well ossified.

La Ferrassie 8(?). Probably from 1973 excavations; fragments probably all one individual. Many tiny postcranial fragments, mostly axial skeleton, plus cranial fragments, isolated teeth, and tooth crowns, of ca. 1-year-old individual.

Largest specimen crushed and reconstructed partial occiput. Occiput fairly wide and low, with broad, pitted, but not sunken or depressed, suprainiac region (no well-defined inferior border to area of suprainiac depression). In profile, nuchal region undercuts occipital plane; angle between the two planes rather wide. Nonsegmented but quite deeply and jaggedly denticulated lambdoid suture curves across lambda. Bone broken laterally; horizontal anterior lambdoid suture had been long. Internally, impression for transverse sinus is more marked on the L than the R.

Several preserved teeth. Apparent Rdi^1, moderately worn with very short root; crown well excavated lingually and bounded by fairly straight medial margocrista and laterally kinked lateral margocrista. Rdi^2, more extensively worn, with relatively thin, short neck; crown moderately tall with shallowly concave lingual surface that was bounded by apparently thin, slightly curved mesial margocrista and somewhat laterally kinked lateral margocrista; the latter separated by groove from small lingual swelling. Unworn Rdi^2 has thin, short root; short crown distended at mesial occlusal margin and flared at distal margin near neck, with very shallow lingual surface bounded by thin mesial margocrista and thicker lateral margocrista that is separated from small lingual swelling by groove. Rdc^1 root tip incompletely formed; crown equilateral triangle in buccal outline; buccally, crown angles slightly in on neck; internally, lower half of crown slightly swollen, surmounted by much more compressed mesial and distal edges. Rdm^1 with three roots, probably not fully formed. Protocone somewhat internally placed; large mesial protostyle connected by crista to swollen hypocone region; well-developed preprotocrista swings far forward before arcing back to apex of compressed paracone that lies opposite protocone. Thicker postcingulum curves gently down from paracone to hypocone region; large paraconule lies internal to preprotocrista; short postprotocrista terminates at base of paracone. Enamel thinly grooved; buccal side angles in from neck; lingual side swollen. R and Ldm^2 necks ca. one-third formed; possibly three roots would have developed. Crowns very long m/d with very internally placed cusps; large cingulum-like protostyle extends fully around protocone. Crest extends from protocone markedly posteriorly before curving back strongly to terminate as small cuspule on side of metacone enclosing quite large basin. Preprotocrista swings up to mesially placed paracone; stout postprotocrista goes straight to metacone. Trigon basin deep but not expansive. Enamel deeply grooved, especially in basins.

R and Ldc_1 roots ca. 80% formed; crowns angle slightly on neck; lingual surfaces steep and shallowly concave; mesial edge much shorter than distal edge. R and Ldm_1 roots not fully formed; marked exodaenodonty over mesial root; wedge shaped in occlusal outline. Compressed, ridgelike buccal cusps very internally placed. Protoconid much taller but thinner than closely appressed metaconid lying slightly distal to it. Paracristid descends sharply, then curves down to base of metaconid. Notch between entoconid and hypoconid opens the long, cristid-ringed talonid basin. Ldm_2 root ca. half formed; crown long, ovoid; lingual and internally placed buccal cusps compressed. Trigonid basin short but deep; long, b/l constricted talonid basin entirely enclosed. Enamel deeply wrinkled.

R and LM^1 crowns ca. one-third formed; stout postcingulum runs from very enlarged hypocone almost directly up to metacone, enclosing large basin. Preprotocrista with centrally placed large conule runs out from protocone to swing around to buccal side of tooth; postprotocrista runs directly to metacone. Enamel thickly crenulated.

Half-formed crown of possible LI_1 shallow on lingual surface; lacks distinct margocristids. RM_1 crown more than one-third formed; slightly centrally shifted buccal cusps (buccal more bulbous than lingual side); stout paracristid runs mesially before coming back to metaconid, enclosing well-developed basin; hypoconulid buccally placed; deep talonid basin fully enclosed.

REFERENCES

Boule, M. 1911–13. L'homme fossile de La Chapelle-aux-Saints. *Ann. Paléontol.* 6: 1–64 (1911); 7: 65–208 (1912); 8: 209–279 (1913).

Bourgon, M. 1957. Les industries moustériennes et pré-moustériennes du Périgord. *Arch. Inst. Paléontol. Humaine* 27: 1–141.

Capitan, L. and D. Peyrony. 1909. Deux squelettes humains au milieu de foyers de l'époque moustérienne. *C. R. Acad. Inscrip. Belles-Lettres,* Paris: 797–806.

Capitan , L. and D. Peyrony. 1911. Un nouveau squelette humain fossile. *Rev. Anthropol.* 21: 148–150.

Capitan, L. and D. Peyrony. 1912. Trois nouveaux squelettes humains fossiles. *Rev. Anthropol.* 22: 439–442.

Heim, J.-L. 1974. Les hommes fossiles de La Ferrassie (Dordogne) et le problème de la définition des Néandertaliens classiques. *L'Anthropologie* 78: 81–112, 321–378.

Heim, J.-L. 1976. Les hommes fossiles de La Ferrassie. Tome 1: Le gisement, les squelettes adultes (crâne et squelette du tronc). *Arch. Inst. Paléontol. Hum.* 35: 1–330.

Heim, J.-L. 1982a. *Les Enfants Néandertaliens de La Ferrassie.* Paris, Masson.

Heim, J.-L. 1982b. Les hommes fossiles de La Ferrassie II. Les squelettes d'adultes: Squelette des membres. *Arch. Inst. Paléontol. Hum.* 38: 1–272.

Holloway, R. L. 1985. The poor brain of *Homo sapiens neanderthalensis*; see what you please. In: E. Delson (ed), *Ancestors; The Hard Evidence.* New York, Alan R. Liss, pp. 319–324

Mellars, P. 1996. *The Neanderthal Legacy.* Princeton, NJ: Princeton University Press.

Peyrony, D. 1934. La Ferrassie. *Préhistoire* 3: 1–92.

Repository

Laboratoirc d'Anthropologic Biologiquc, Muséc dc l'Homme, Place du Trocadéro, 75116 Paris, France.

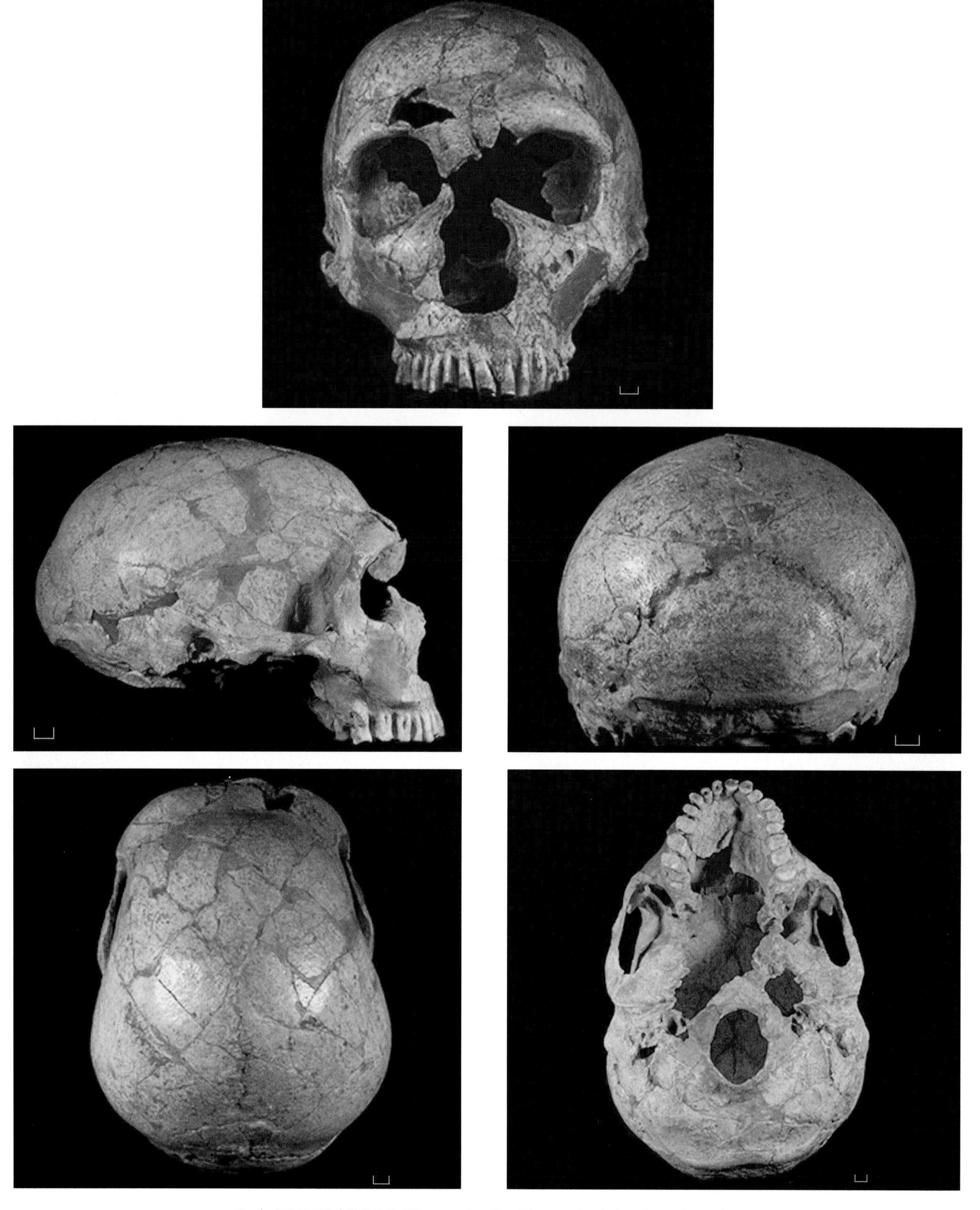

LA FERRASSIE Figure 1. La Ferrassie 1 (scale = 1 cm).

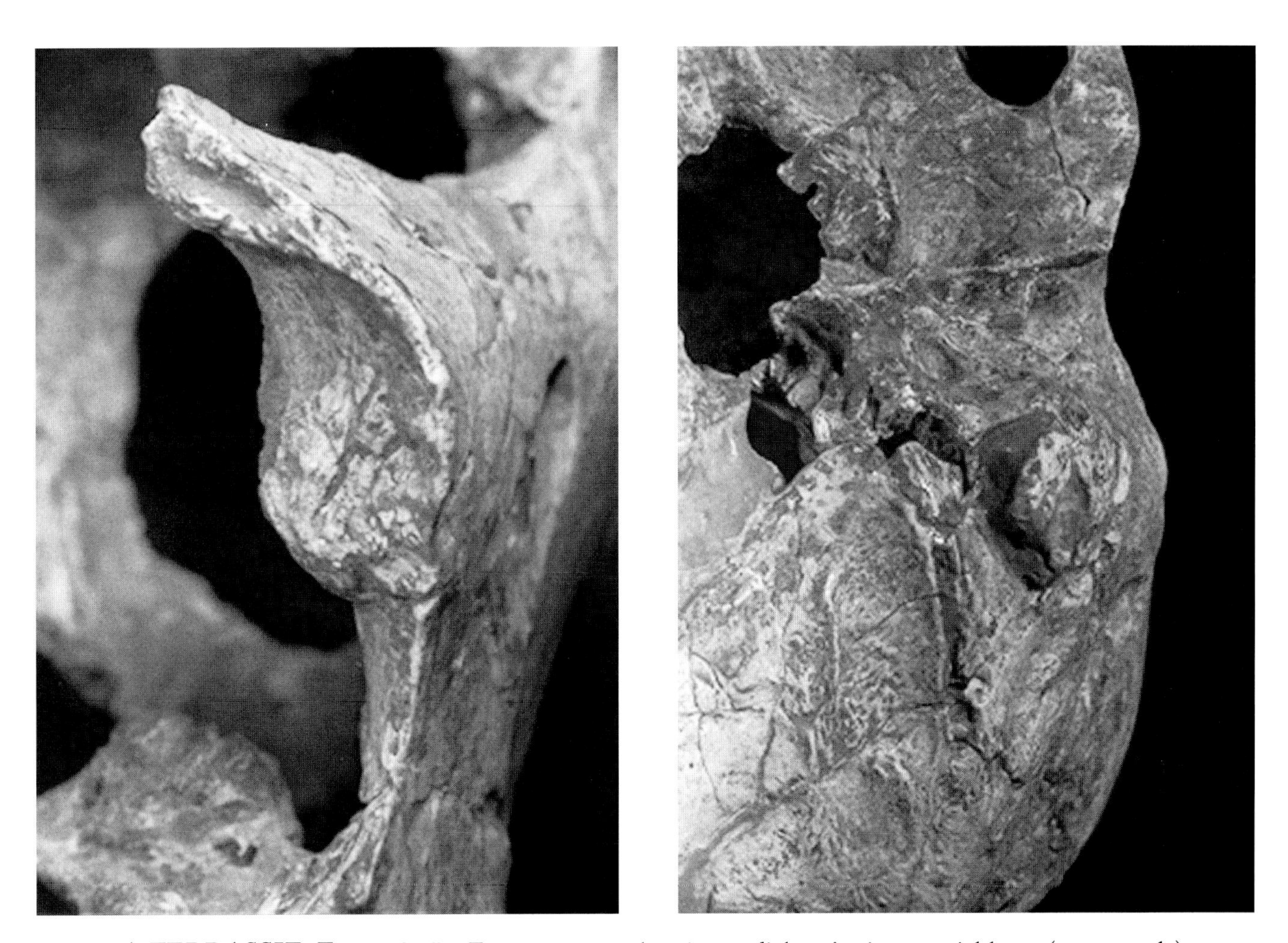

LA FERRASSIE Figure 2. La Ferrassie 1: nasal cavity medial projection; cranial base, (not to scale).

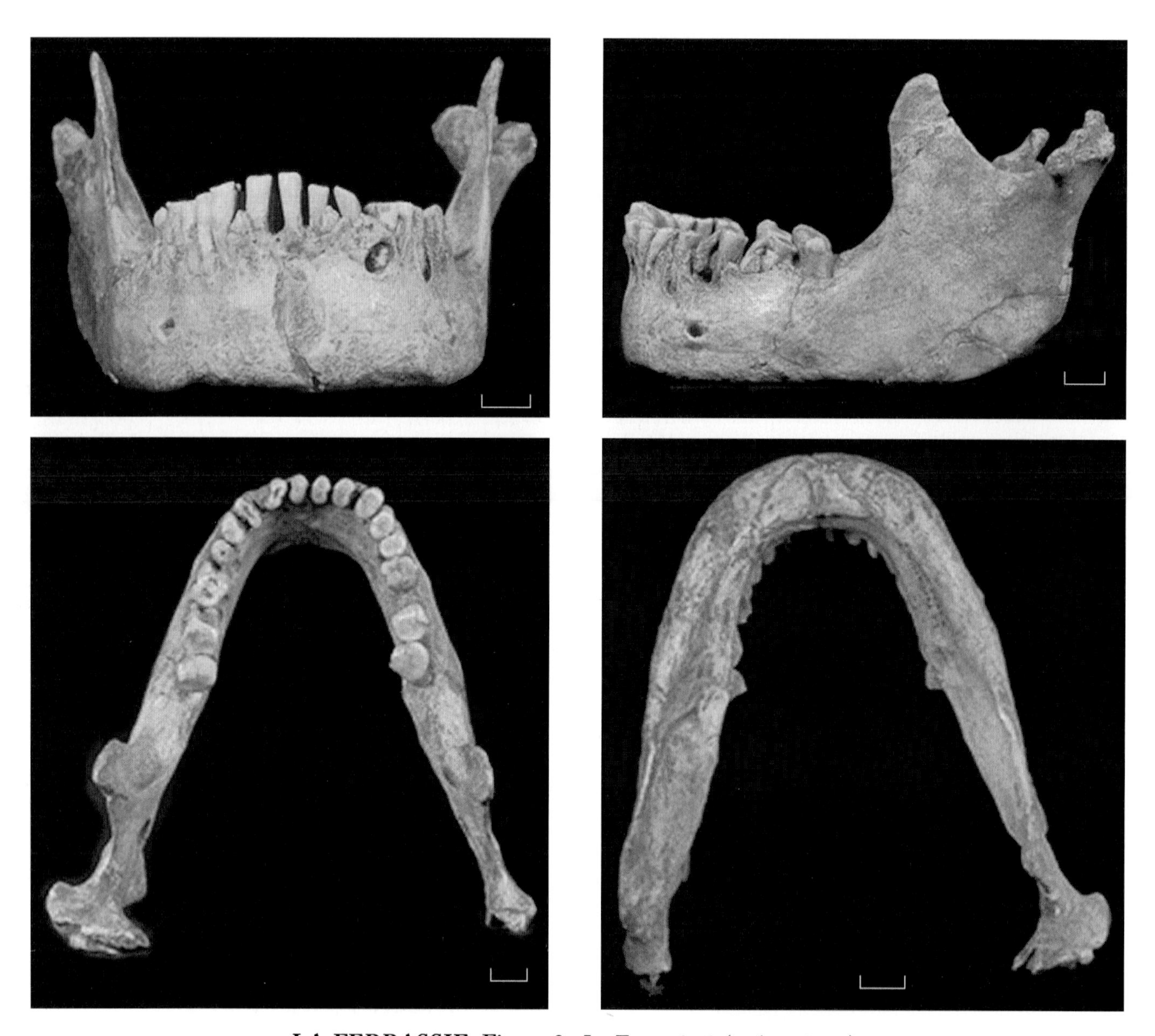

LA FERRASSIE Figure 3. La Ferrassie 1 (scale = 1 cm).

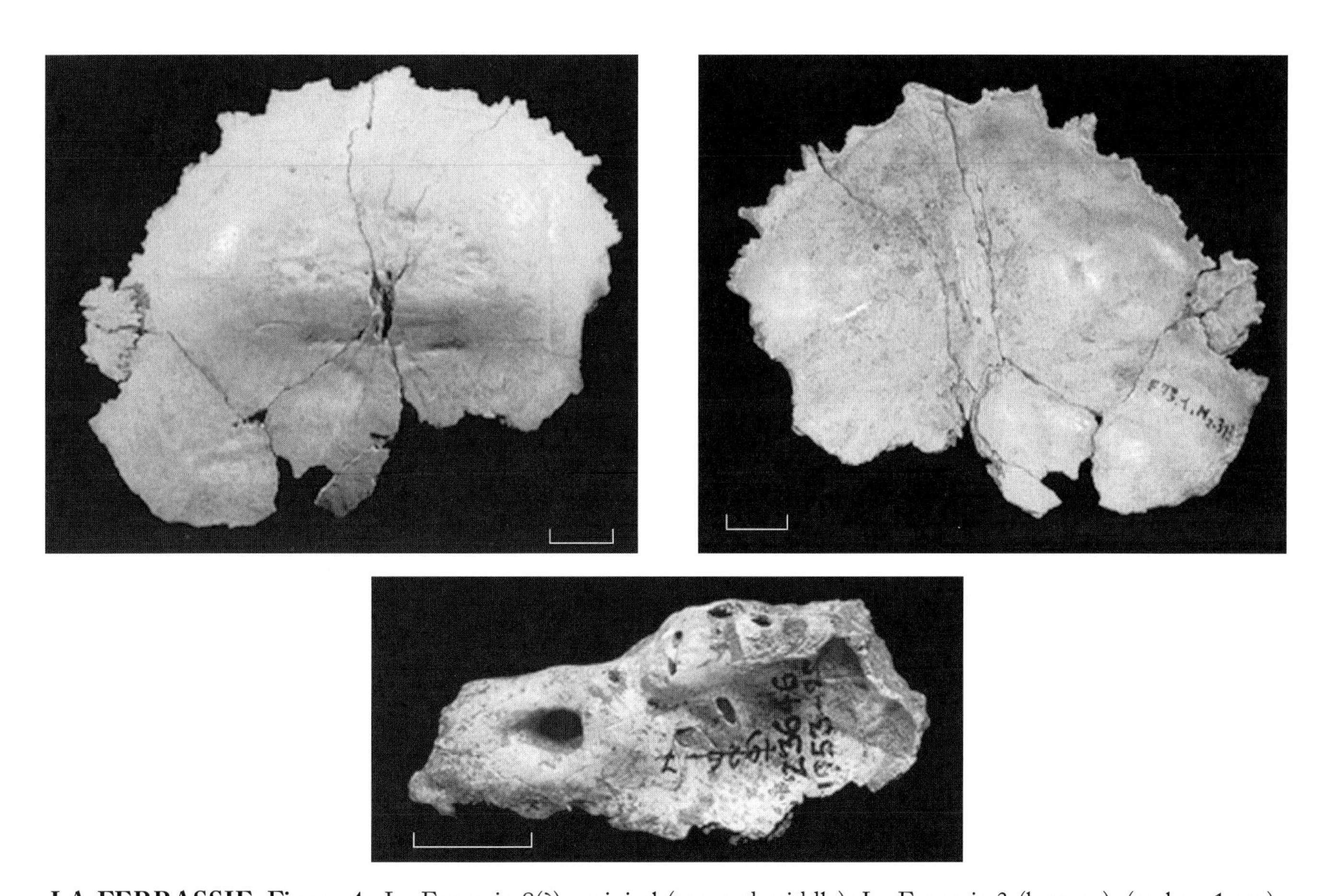

LA FERRASSIE Figure 4. La Ferrassie 8(?) occipital (top and middle); La Ferrassie 3 (bottom), (scale = 1 cm).

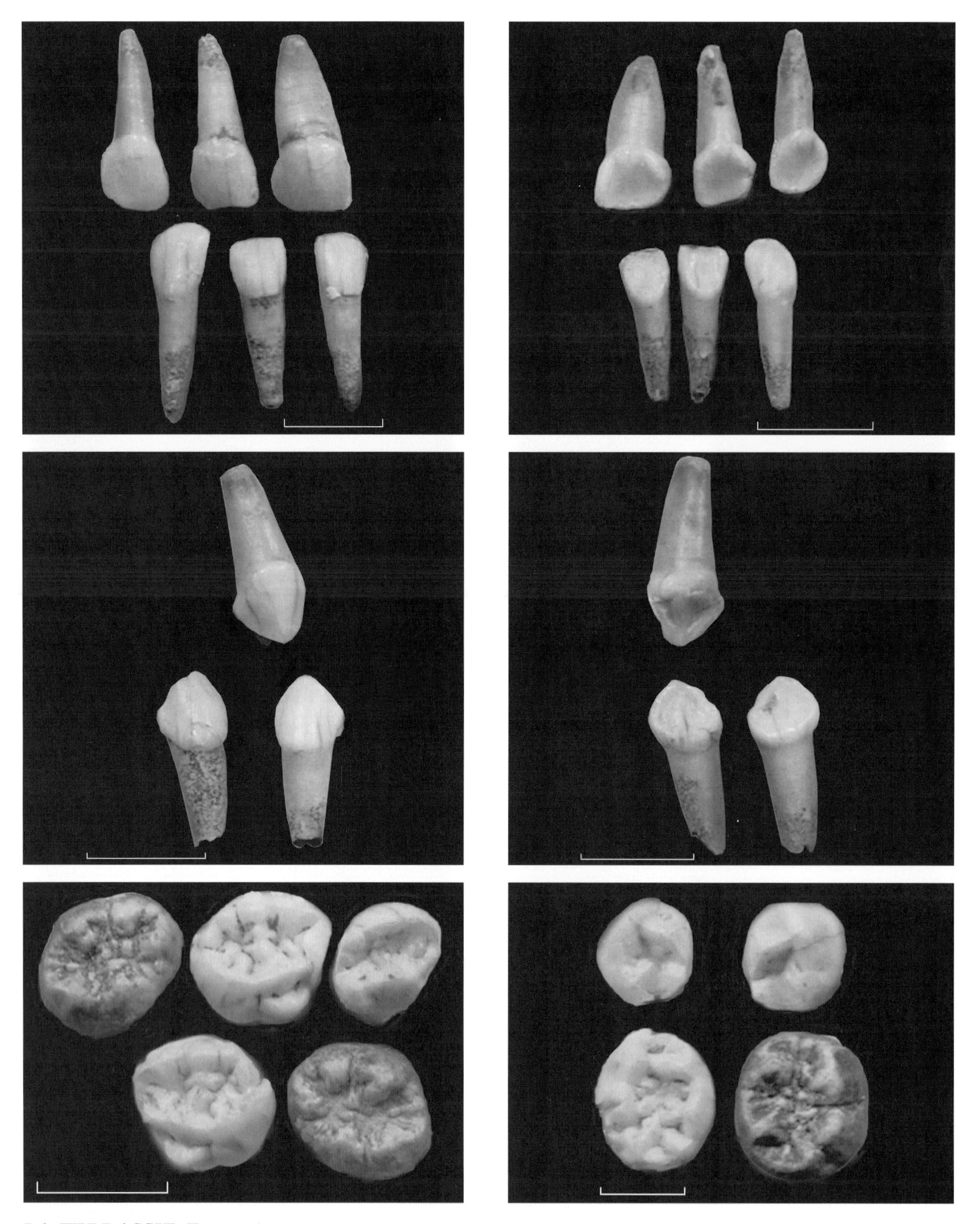

LA FERRASSIE Figure 5. La Ferrassie 8(?): upper and lower deciduous incisors, buccal and lingual views (top row); upper and lower deciduous canines, buccal and lingual views (middle row); upper deciduous molars and M1s (bottom left); lower deciduous molars and M1 (bottom right). (scale = 1 cm)

La Naulette

Location
Limestone cave (Trou de la Naulette) on the left bank of the Lesse River, just downstream of the hamlet of Chaleux, SE of Dinant, Belgium.

Discovery
E. Dupont, 1866.

Material
Edentulous left mandibular corpus, complete round symphysis to the alveolus of P2. Also a partial ulna and metacarpal, plus a canine, now lost.

Dating and Stratigraphic Context
Level 3 of Dupont's excavations (Dupont, 1866). The Pleistocene age of the hominid was demonstrated by its depth in the undisturbed deposit and by the associated fauna, which included reindeer and "giant elk." Exact age is uncertain.

Archaeological Context
None.

Previous Descriptions and Analyses
This hominid has considerable historical importance as the first fossil hominid found after the Feldhofer discovery. Its retreating symphysis was remarked upon immediately and led to a lively controversy well summarized by Trinkaus and Shipman (1992). The most recent analysis is that of Leguèbe and Toussaint (1988), who found a general morphometric resemblance to Neanderthals despite the specimen's lack of typical Neanderthal features.

Morphology
Specimen consists of L mandibular corpus complete to RP2 alveolus. Edentulous, but contains alveolae of all teeth from RP2 to LM3.

Corpus shallow, although taller at symphysis than posteriorly. Inferior border does not slope up in front of area of inferior marginal tubercle; tubercle only hinted at by slight inferior marginal thickening below P1. Alveolae of anterior teeth inclined forward, creating subalveolar depression. Externally, symphysis unadorned; arcs across moderately broadly. Viewed from below, symphysis not much thicker than corpus. Digastric fossae not long but deeply excavated; face more down than back. Weak indentation, almost at inferior margin, lies in midline of symphysis. Postincisal plane steep; plane below (with deep genial pits) more vertical. Pair of smooth swellings reminiscent of mandibular tori lie below and lateral to genial pits. Weak mylohyoid line originates from internal alveolar margin around M3; runs forward and down; shallow submandibular fossa lies below it. Moderately large mental foramen situated below P2. Rise of ramus suggests absence of retromolar space.

Roots of lower Is, Cs, and P1s appear to have been large; their alveolae bear vertical pillars down their sides suggesting that roots were grooved. P2 alveolus obliquely oriented, indicating that tooth was skewed. Judging by alveolae, Ms were not very big;

probably increased in size posteriorly. M1–2 had two roots; M3 had four.

REFERENCES

Dupont, E. 1866. Etude sur les fouilles scientifiques exécutées pendant l'hiver de 1865–1866 dans les cavernes des bords de la Lesse. *Sess. Extr. Soc. Belg. Géol., Acad. R. Belg.*: 196–207.

Leguèbe, A. and M. Toussaint. 1988. Le mandibule et le cubitus de La Naulette: Morphologie et Morphométrie. Paris, CNRS: *Cahiers de Paléoanthropol.*

Trinkaus, E. and P. Shipman. 1992. *The Neandertals: Changing the Image of Mankind.* New York, Knopf.

Repository

Institut Royal des Sciences Naturelles de Belgique, Laboratoire d'Anthropologie et de Préhistoire, rue Vautier 29, 1040 Brussels, Belgium.

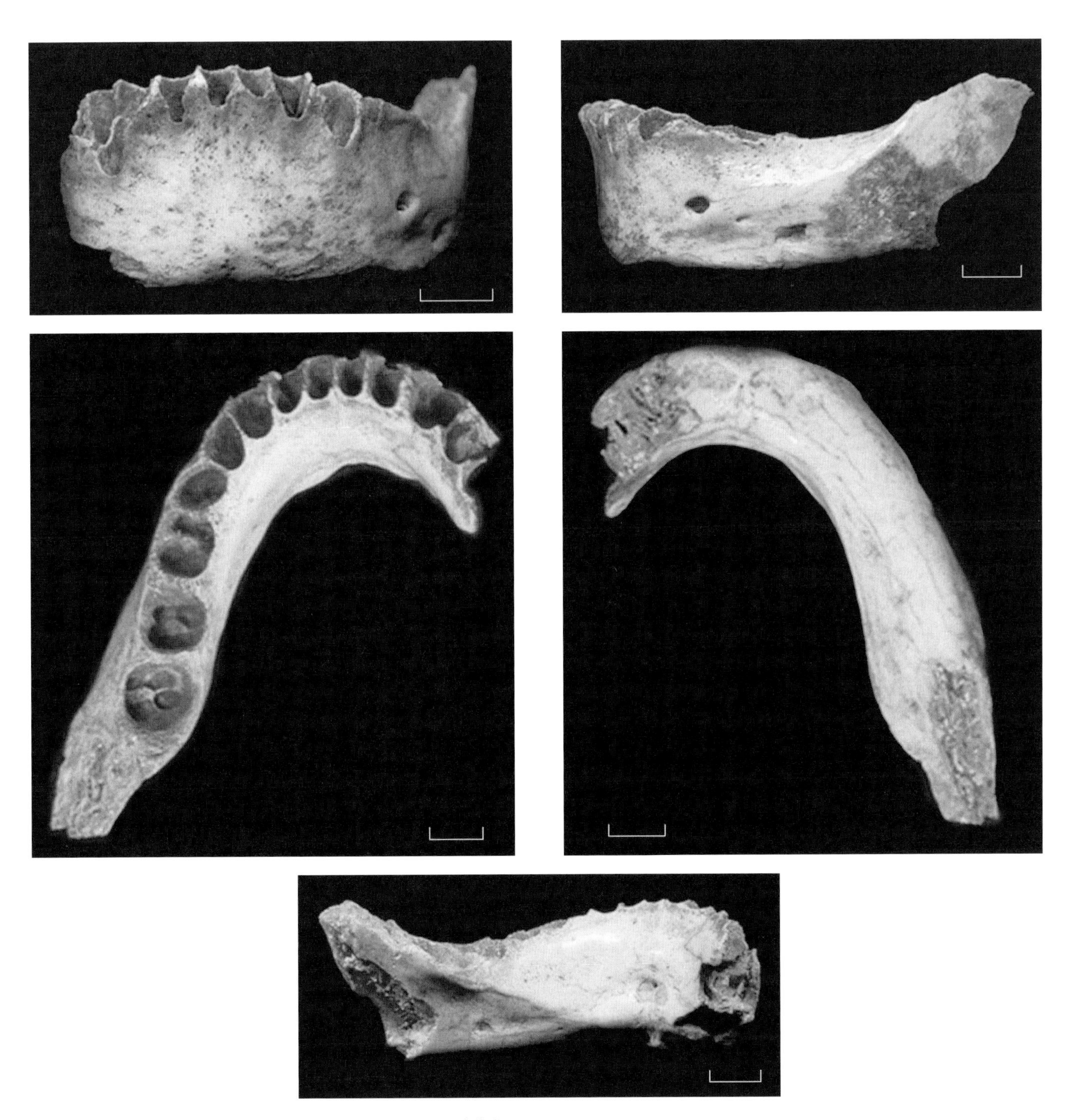

LA NAULETTE Figure 1. Scale = 1 cm.

La Quina

Location

Rock shelter complex near Villebois-Lavalette, 25 km S of Angoulème, Charente, France.

Discovery

The site was discovered in 1872, but the first hominid finds (La Quina H1, H2, and H14) were made in 1908 by L. Henri-Martin, who subsequently made numerous such discoveries and whose son G. Henri-Martin made the last find (H27) in 1965.

Material

Remains of a total of 27 hominid individuals have been reported from the site, most of them fragmentary. Notable are a partial adult skeleton (H5), an adult mandible (H9), and a child's skull (H18).

Dating and Stratigraphic Context

The "lower shelter" at La Quina has produced evidence of early Upper Paleolithic occupation, plus some underlying Mousterian and Châtelperronian. However, all of the hominids come from Mousterian occupation levels (Beds 1–4) in the "upper shelter." Burned bone from the latest of these, Bed 1, was radiocarbon dated at 35.2 ka (G. Henri-Martin, 1964), but the bulk of the hominid remains come from earlier levels and are generally reckoned by faunal and archaeological association to date from isotope stage 4, or around 65 ka (see discussion by Mellars, 1996).

Archaeological Context

As noted, all the hominids from La Quina come from the Mousterian deposits of the upper shelter. Most of the hominids are associated with the "Quina variant" of the Mousterian, which is characterized by large sidescrapers with steep retouch, and dated as above. The exact stratigraphy of the upper shelter is unclear. L. Henri-Martin divided the Mousterian industries into lower (Bed 4), middle (Bed 3), and upper (Beds 1 and 2) variants. Later excavators (see Debénath and Tournepiche, 1993), however, equated the "middle Mousterian" with the Quina variant and Henri-Martin's "upper Mousterian" with the Mousterian of Acheulean tradition and also uncovered evidence of the denticulate Mousterian, an industry apparently unrepresented in Henri-Martin's collections. The "lower" Mousterian levels appear to have been reworked, and the industries contained seem to be of uncertain affinity although with some "La Quina variant" features. The lower levels in the sequence appear to represent short episodes of butchery, the overlying MTA and denticulate levels provide evidence of longer-term occupation. Traces of fire are found throughout the deposits.

Previous Descriptions and Analyses

The hominid remains have been described piecemeal over the years (e.g., L. Henri-Martin, 1911, 1913, 1923; Vallois, 1969. They have never been fully monographed (L. Henri-Martin, 1923 comes closest), but all commentators have regarded them as "classic"

Würm Neanderthals. Estimated cranial capacity is 1350 cc for La Quina 5 (Holloway, 1985) and 1200 cc for La Quina 18 (Tillier, 1984).

Morphology

H4

LM_3, partially worn. Elliptical crown lacking cusp morphology, but surface very crenulated, trigonid basin distinct. Roots fused buccally; third lingual root.

H5

Adult. Very small partial skull, missing region around foramen magnum, most of sphenoid, most of maxilla, circumnasal region. Most of dental arcade including alveolar bone, part of nasal cavity, and maxillary sinus floor preserved on the L; very little bone preserved on the R; C–M3 preserved on both sides. Isolated I inserted on the R in reconstruction appears to be a LI (indicates palatal arcade much too broad, although articulates with badly reconstructed mandible, in which symphysis only represented internally). If L side mirror imaged, palate would be much narrower. Mandible missing much of front, but with well-preserved L and R posterior corpora; on the L, inferior border preserved to midline. Both rami present; some damage to L coronoid process and condyle; LP1–M3, RC–M3 preserved.

Cranial bone thin; face small, narrow. Viewed from rear, neurocranium en bombe. Occipital with chignon. Frontal quite domed; in midline, frontal rises smoothly from supraglabellar region; dome set back from supraorbital margins, creating sulci laterally. As seen on the R, arcing, vermiculate-boned supraorbital torus smoothly rolled with slight indication of margin at lateral juncture with orbital roof; tori confluent with prominent glabella. Slight depressions on both sides in region of supraorbital foramina; more centrally placed on the L, more medially on the R. Viewed from above, tori angle back quite sharply from projecting glabella. Frontal sinuses enormous; extend high into frontal and, with big laterally extending chamber, just beyond midpoint of L orbit, to midpoint of R orbit. Vermiculate patterning extends back toward bregma. Interorbital region fairly broad.

Portion of L maxilla from near midline to M1/2, plus few isolated bits incorporated into reconstruction of upper tooth rows from R to LM3 but lacking RI2 and LI1. Nasoalveolar clivus more or less vertical; would have been broadly curved across. No evidence of prenasal fossa (i.e., double margin). Nasal cavity, internally, shows smooth transition from anterior border to sunken floor behind. Maxillary sinus came forward probably to level of I2 root, farther back, sinus expansive both medially and laterally. Base of medial wall preserved, indicating that originally sinus had swelled medially into nasal cavity. Tiny bit of lateral nasal crest preserved on the L.

L zygoma almost entirely complete; its frontal processes gracile. On the L, no malar tubercle; R side damaged. Maxillary sinuses do not penetrate deeply into zygomas. Posterior surface of zygoma slightly concave; bone faces directly backward. Anterior root of zygomatic arch originates quite low above P1–M1 region; from the front, root appears to have been straight and angled outward; runs essentially straight back with no outward flare.

Temporal line rugose as comes off frontal process of zygoma; arcs just behind lateral extremity of supraorbital margin and courses back and up along frontal bone, becoming flatter and less ridgelike posteriorly; lies rather low on parietal and arcs down and forward well in front of asterion to terminate at parietal notch. Parietal bone not very arced in a/p plane; more arced in m/l plane. Temporal fossae would have been small and narrow from side to side. On the R, anterior squamosal suture angles in. Squamosal portion of temporal short a/p and not very tall s/i.

Mandibular fossae broad, relatively shallow; walled off medially by medial articular eminence (preserved on the L). On the L, erosive lesions occur in region of articular eminence. On both sides, articular eminence barely elevated and short a/p (posterior root of zygomatic arch does not project far laterally). Ectotympanic tube terminates at midline of mandibular fossa. External auditory meatus small and subcircular. As preserved better on the R (but also on the L), low, thin vaginal process peaks at styloid process, fades out laterally before reaching end of very short ectotympanic tube. Vaginal process well separated from mastoid process. On the L, stylomastoid foramen situated close to base of thin styloid process; more laterally on the R. Carotid foramen (preserved partially on both sides) lies medial to styloid processes. On the R, foramen ovale and relatively large foramen spinosum lie completely within alisphenoid; foramen spinosum lies almost directly lateral to foramen ovale.

Mastoid processes stout, low, oriented down; have large anterior tubercles with tall sulcus above separating them from prominent, superoposteriorly arcing supramastoid crests that are confluent with faint suprameatal crests (thus presenting an indentation between posterior root of zygomatic arch and very well-developed supramastoid crest). On the L, occipitomastoid suture partly preserved on occipital side; apparently no elevated occipitomastoid crest, but long, low, ridgelike muscle scar runs back from midpoint of suture on occipital side to fade out about 1 cm from anterior lambdoid suture. About 1 cm medial to this muscle scar, and running parallel with it, is shorter, more elevated Waldeyer's crest. Mastoid notch (partially preserved on the L) was not deep; probably quite wide. L parietal notch lies posterior to midline of mastoid process; on the R, at midline.

Occipital plane much broader than tall. Occipital torus continuous, straight across (not depressed at inion), and horizontal, being only defined as bar by long (better defined) infratoral depression below it. Infratoral depression in line with long horizontal anterior lambdoid suture. Suprainiac depression tall, very wide, a little more pitted than surrounding area; is not defined by distinct ridge of bone. On the L, nuchal line delineates bilaterally scalloped infratoral sulcus below. Nuchal plane below infratoral sulcus/depression long, steeply angled inward. Lambdoid suture rises from anterior lambdoid suture quite sharply, arcs across at lambda.

Coronal suture barely denticulate (almost smooth), especially centrally. Sagittal suture uniformly and shallowly denticulate. Lambdoid suture uniformly denticulated, more deeply so than the sagittal (which, in turn, is more deeply denticulated than the coronal suture).

As preserved on the L, palate appears to have been deep; slopes sharply anteriorly; is almost vertical at sides. Preserved anterior teeth very worn.

Internally, petrosal preserved posteriorly; very broad on both sides. Region of arcuate eminence slightly elevated on the L; on the R is an elevation (= the thin bony roof of the superior semicircular canal). Essentially no superior petrous sinus or subarcuate fossa. Large air cell internally between region of semicircular canal and squamosal (epitympanic recess?). R transverse sinus better developed than the L, but shows little definition as it approaches very short sigmoid sinus.

Mandibular corpus preserved to symphysis on the L, but only internally; on the R, corpus preserved only as far forward as M1. Rami essentially complete. L mandibular condyle with modified/resorptive lesions. Mandible was fairly tightly curved, not broad, at front. Tooth rows diverge only slightly. Symphysis would not have been thicker than bone lateral to it; gonial region quite thin, gently inflected. R mental foramen appears to have lain below P2/M1. Mandible would have been narrower at symphysis than currently reconstructed. Corpus not notably deep; ramus quite long from front to back, only slightly inclined posteriorly. Bone of corpus thins markedly inferiorly posterior to M3 and into gonial region. Long, distinct retromolar spce on both sides, prolonged on R by slight preangular notch.

Postincisal plane apparently vertical but not very tall. Preserved L digastric fossa faces more or less down; well defined by low crest at front. Mylohyoid line very indistinct; submandibular fossa long but not very tall or excavated. Gonial margin smoothly curved (not angular); inflects slightly medially. Externally, gonial region relatively smooth, with only minor muscle scarring; internally, there is an s/i tall, but somewhat low, medial pterygoid tubercle. In profile, gonial angle smoothly rounded; posterior border of ramus inclined slightly back. On the R, a/p long coronoid process rises significantly above condyle. R sigmoid notch crest terminates lateral to midline of condyle; may have been more central on the L. On the R, deepest part of sigmoid notch lies close to condyle. Mandibular foramina small, compressed, point up and back; lingulae broken but apparently extended all along opening. Internally, low pillar runs down middle of coronoid process and becomes confluent with very posteriorly extended, internal alveolar crest.

Upper teeth, especially Is, extensively worn. I1 has swollen lingual surface; on basis of preserved grooves, probably had well-developed margocristae; was probably high crowned. I2 apparently swollen basally on lingual side with margocristae. I, C, P1 buccal surfaces make distinct inward angle with their roots. Cs also worn flatly; were relatively compressed mesiodistally. Crown apex was probably quite medial, with long distal slope; base probably quite swollen lingually; enamel swells out buccally above root P1s have buccal angle between crown and root (crown swollen buccally near neck). P1 crown longer m/d buccally than lingually; these sides subequal in length on P2. P1s broader b/l and longer m/d than P2s, especially buccally. On both P1–2, quite mesially situated paracone and protocone lie opposite one another; evidence of distinct anterior, but no posterior, fovea in both. M2–3 wider b/l than M1; M1 longer m/d than M2–3, especially across lingual cusps. M3s shorter m/d than M1–2s. On all Ms, distinct, small trigon basin lies buccal to midline; lingual roots separate. M crown surfaces worn flat; on all, paracone swollen buccally, distends crown in that direction. M1 hypocone distolingually very swollen; stout postcingulum runs from cusp to slightly more mesially positioned metacone; metacone smaller than paracone. Lingual side of M1 longer than buccal; reversed in M2–3. Hypocone much smaller on M2; cusp not evident on M3. M1/2 buccal roots bifurcate some distance away from crown; this distance greater on M3.

Lower teeth: LC–M3; RP1–M3; all heavily worn with dentine exposed. C has curved root; apparently tall crown angles back slightly from root and has slightly excavated lingual surface bounded by relatively stout margocristids that converge on minimally swollen lingual base. LC and P1 show buccal angulation between crown and root. P1s longer m/d buccally than P2s. P1 crowns angle back; in occlusal outline, crowns are wedge shaped (wider b/l distally because of distension of distolingual corner). P1 protoconid quite medially positioned; short mesial cristid runs from it to base of lingual swelling, from which is separated by groove; stouter cristid runs from distal side of protoconid around distolingual corner of tooth, enclosing small basin. P2s distinctly different; the L longer m/d than the R, which is wider b/l (suggesting different individuals?). Both teeth set into plaster; if one of them is unassociated, it is the R (plausibly an upper LP2). P2s very asymmetrical with long talonid and very compressed mesial part. On the L, distal part of P2 was expanded, with moderate talonid basin. M1–2 ovoid in occlusal outline; M3 subround. M1–2 longer m/d than M3, which is wider b/l. M3s large for upper molars. Although worn, M2s show hypoconid extending lingually beyond midline of tooth. As preserved best on LM2, hypoconulid was placed just buccal to midline; M1 probably similar. On M1–2, bifurcation between roots moderate distance from neck; bifurcation not exposed on M3. RM2 with fairly stout lingual third root; third root much thinner on the LM2. Both M2s bear accessory lingual root, which is large on the R, tiny on the L.

H18

Child. Cranium, fairly complete but heavily reconstructed; wax obscures many internal details. Missing sphenoid, ethmoid, most of orbital cones, nasal cavity structures, and most of occiput. Upper dm1–M1 in place, I1–2 and M2 in various stages of eruption.

Relatively long skull; posteriorly broad; forehead quite vertical. Rather projecting snout; face tapers from midline to zygomas and from zygomaticofrontal suture down toward alveolar margin. Supraorbital region just beginning to bulge, to hint at double-arched, rolled configuration of adult; supraorbital "bulge" confluent across relatively flat glabellar region. Viewed from above, supraorbital margins angle back gently from midline. Orbits reconstructed asymmetrically. Orbital floors were sunken, with crisp, not yet puffed out, inferior margins. Nasal bones very broad in general, broadening more inferiorly; they flex gently up and out and are moderately arced from side to side. As preserved on the R, frontal process long a/p. Appears nasal aperture was tall, not very wide. As preserved on the L, lateral crest crisp, runs down to become confluent with anterior nasal spine; inferior margin single (= no prenasal fossa). Separate anterior nasal spines quite thick and projecting. As visible on the L within nasal cavity, maxillary frontal process bears raised vertical area; region behind obscured by wax. Nasoalveolar clivus very vertical; was long.

Anterior root of zygomatic arch originates just above roots of dm2; viewed from the front, angles steeply out and up. Squamosal portion of temporal long, tall; tallest point lies quite anteriorly. Modest flexure of anterior squamosal suture; inferiorly, squamosal/alisphenoid region curves inward and does not form corner (infratemporal fossa not distinct). Mandibular fossa deep, not bounded anteriorly by any marked eminence. Postglenoid plate separated from incompletely ossified ectotympanic tube by crease. On the L, vaginal process represented by spit of bone in line with, and inferior to, styloid pit, to which is connected by groove. Styloid pit in line with broad, moderately deep mastoid notch; stylomastoid foramen noticeably lateral to pit. Large, downwardly pointing carotid foramina quite medial to pits. Mastoid processes minimally developed; region from mastoid notch to very distinct Waldeyer's crest lies well below tips of processes.

Most of basiocciput preserved anterior to foramen magnum; is quite flexed upward. What remains of occiput reveals little evidence of differentiation between occipital and nuchal planes. Faint suprainiac depression discernible. As revealed in detached fragment, developing L occipital condyle entirely on occiput. Foramen magnum long. Internally, both petrosals quite broad and bear well-rounded, elevated arcuate eminence, lateral to which is another swelling (epitympanic recess/sinus?). Subarcuate fossa entirely filled in, flat on both sides. Faint groove for superior petrous sinus only present just above internal auditory meatus.

Large segments of sagittal and coronal sutures preserved. In contrast to adult, anterior part of sagittal and medial part of coronal sutures less denticulated than rest.

I1–2s deeply shoveled. M1 enamel markedly crenulated; trigon basin deeply defined.

H27

Adult. Almost complete R temporal, missing anterior part of petrosal. Externally, squamosal portion short a/p, not tall s/i. Zygomatic process of temporal rather thin. Posterior root of zygomatic arch originates in front of relatively small, ovoid, horizontally oriented auditory meatus; posterior root does not extend far out from cranial wall. Suprameatal crest faint and confluent with more marked, upwardly arcing supramastoid crest that is separated from mastoid process below by tall sulcus. Moderately a/p long mastoid process points down and slightly forward; does not project very far. Externally, surface very rough; superiorly, bears mastoid crest at level of auditory meatus. Mastoid notch quite deep and wedge shaped; notch broadens somewhat inferiorly; expands posteriorly into small digastric fossa. Mastoid notch bounded medially by paramastoid crest that descends as far as mastoid process; is separated from what was evidently an even further downwardly protruding occipitomastoid crest by groove for occipital artery. Very large, deep parietal notch lies above posterior margin of mastoid process; moderately long, horizontal parietomastoid suture runs posteriorly from it. Mandibular fossa wide, long, bound medially by large medial articular tubercle. Region of articular eminence low, does not truly confine fossa in front. Postglenoid plate wide, pressed against (but not fused to) short ectotympanic tube. Vaginal process not preserved medially; peaks around styloid process and descends laterally to fade out almost immediately. Large stylomastoid foramen lay close and posterolateral to styloid process. Internally, posterior part of petrosal very wide. Region of arcuate eminence barely protrudes; more laterally, low, broad swelling lies above large, vacuous sinus. No sign of superior petrous sinus; subarcuate fossa pitted, not closed off. Sigmoid sinus short and deep.

REFERENCES

Debénath, A. and J.-F. Tournepiche. 1993. *Préhistoire de la Charente.* Angoulème, Musée d'Angoulème.

Henri-Martin, G. 1964. La dernière occupation moustérienne de La Quina (Charente). Datation par radiocarbone. *C. R. Acad. Sci. Paris* 258: 3533–3535.

Henri-Martin, L. 1911. Sur un squelette humain trouvé en Charente. *C. R. Acad. Sci. Paris.* 153: 728–730.

Henri-Martin, L. 1913. Nouvelle série de débris humains disseminés trouvés en 1913 dans le gisement moustérien de La Quina. *Bull. Soc. Préhist. Fr.* 10: 540–543.

Henri-Martin, L. 1923. L'homme fossile de La Quina. *Arch. Morph. Gén. Exp.* 15: 1–253.

Holloway, R. L. 1985. The poor brain of *Homo sapiens neanderthalensis*; see what you please. In: E. Delson (ed), *Ancestors; The Hard Evidence.* New York, Alan R. Liss, pp. 319–324.

Mellars, P. 1996. *The Neanderthal Legacy.* Princeton, NJ, Princeton University Press.

Tillier, A.M. 1984. L'enfant *Homo* II de Quafzeh (Israel) et son apport à la compréhension des modalitiés de la croissance de squelettes Moustériens. *Paleorient* 10: 7–48.

Vallois, H. 1969. Le temporal néandertalien H27, de La Quina. Etude anthropologique. *L'Anthropologie* 73: 365–400.

Repository

Over the years La Quina hominid fossils have found their way into a variety of different collections, and no comprehensive list exists. Principal holders include Musée des Antiquités Nationales, 78103 St Germain-en-Laye, France (H1–4, H6–8, H10–20); Laboratoire d'Anthropologie, Musée de l'Homme, Place Trocadéro, 75116 Paris, France (H5); Institut de Paléontologie Humaine, I rue René Panhard, 75013 Paris, France (H27).

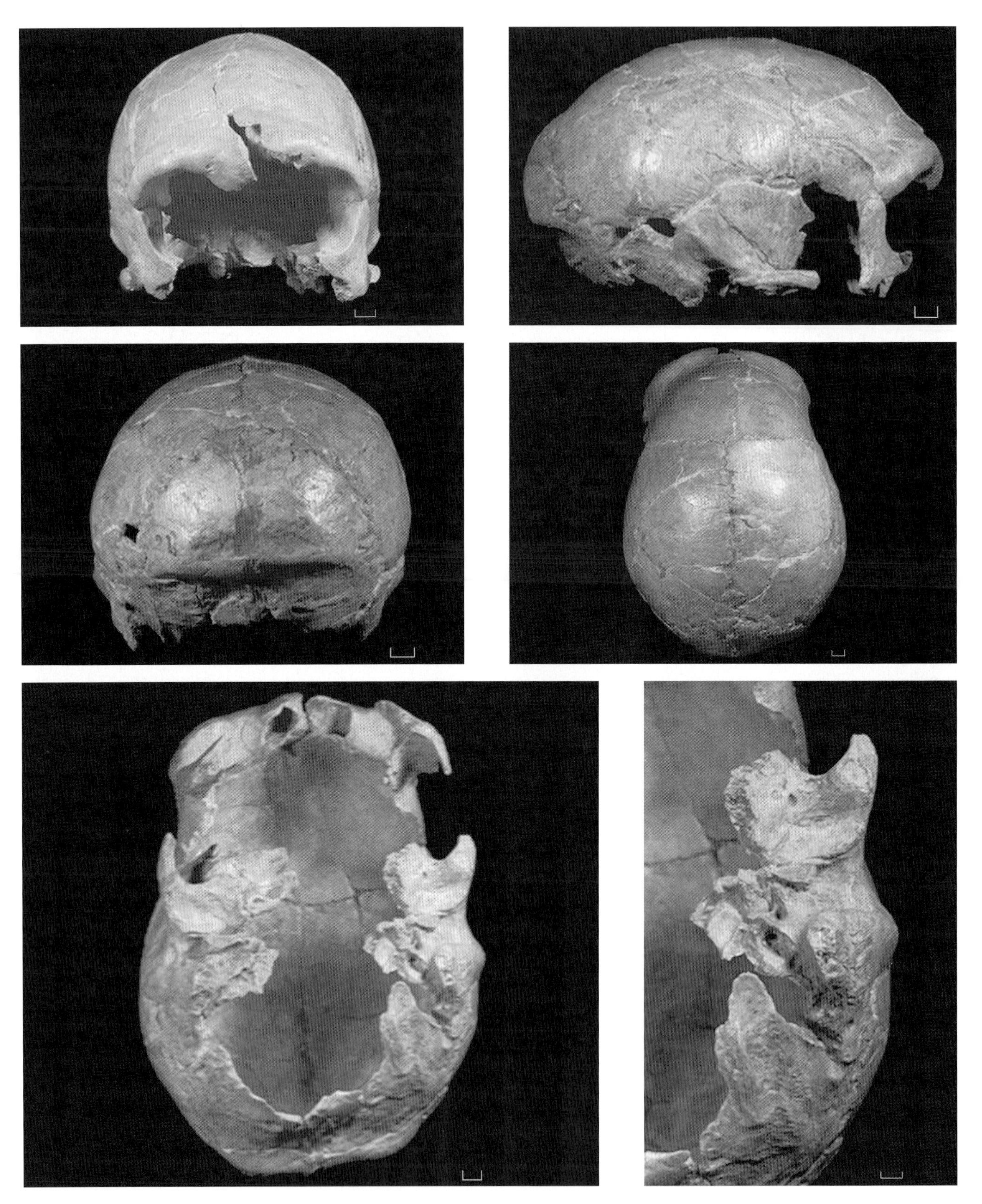

LA QUINA Figure 1. H5 (scale = 1 cm).

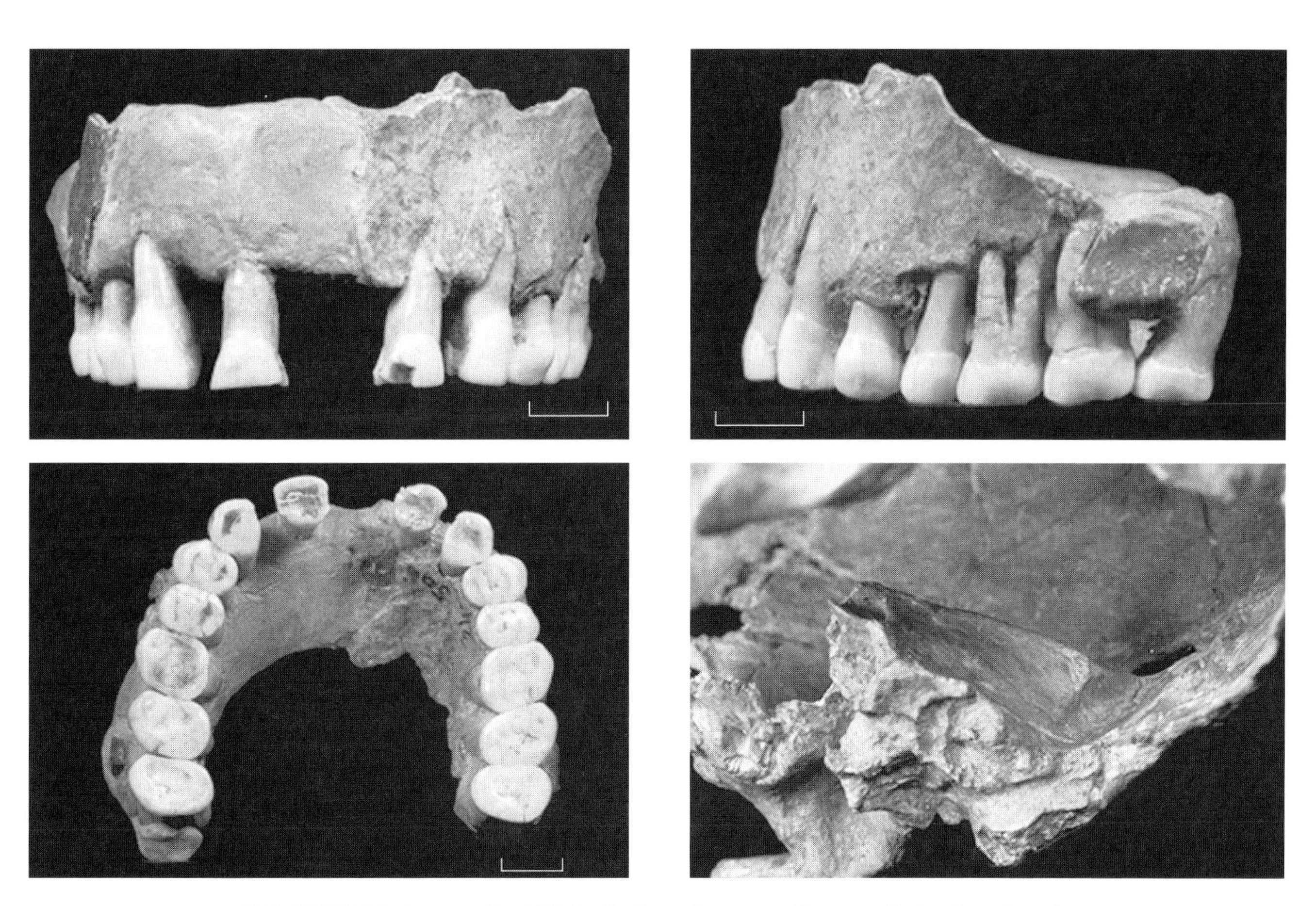

LA QUINA Figure 2. H5 (including close-up of petrosal), (scale = 1 cm).

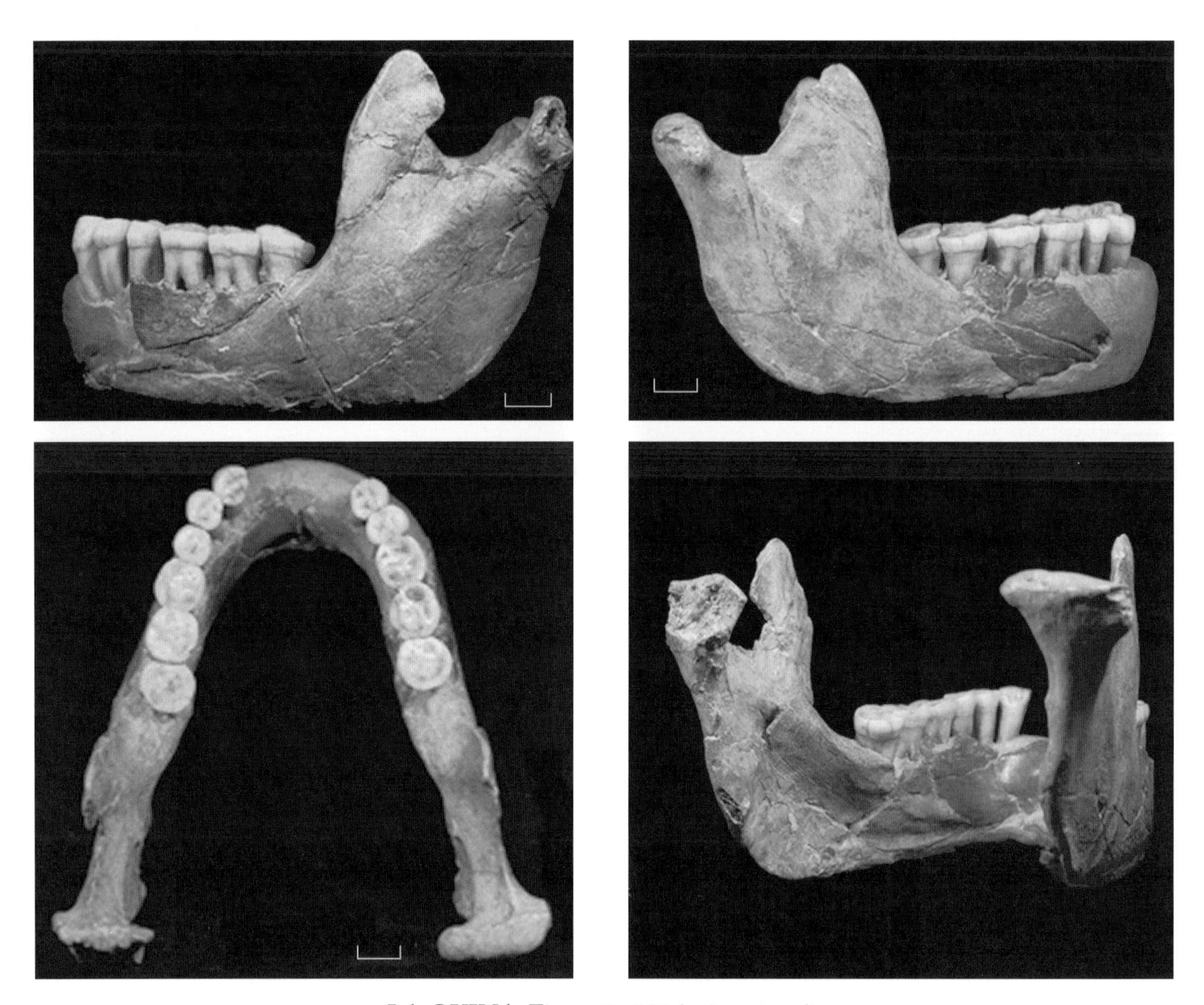

LA QUINA Figure 3. H5 (scale = 1 cm).

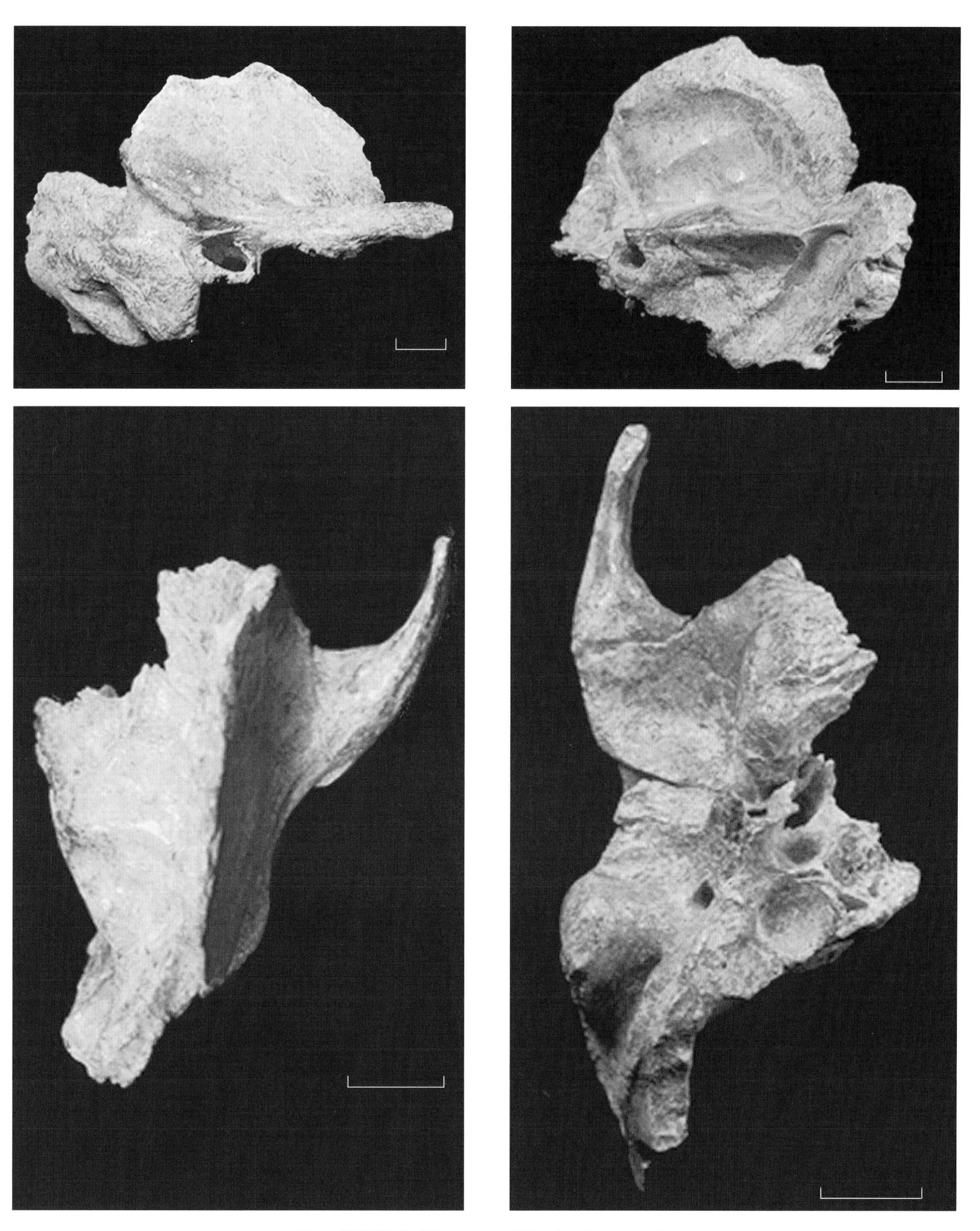

LA QUINA Figure 4. H27 (scale = 1 cm).

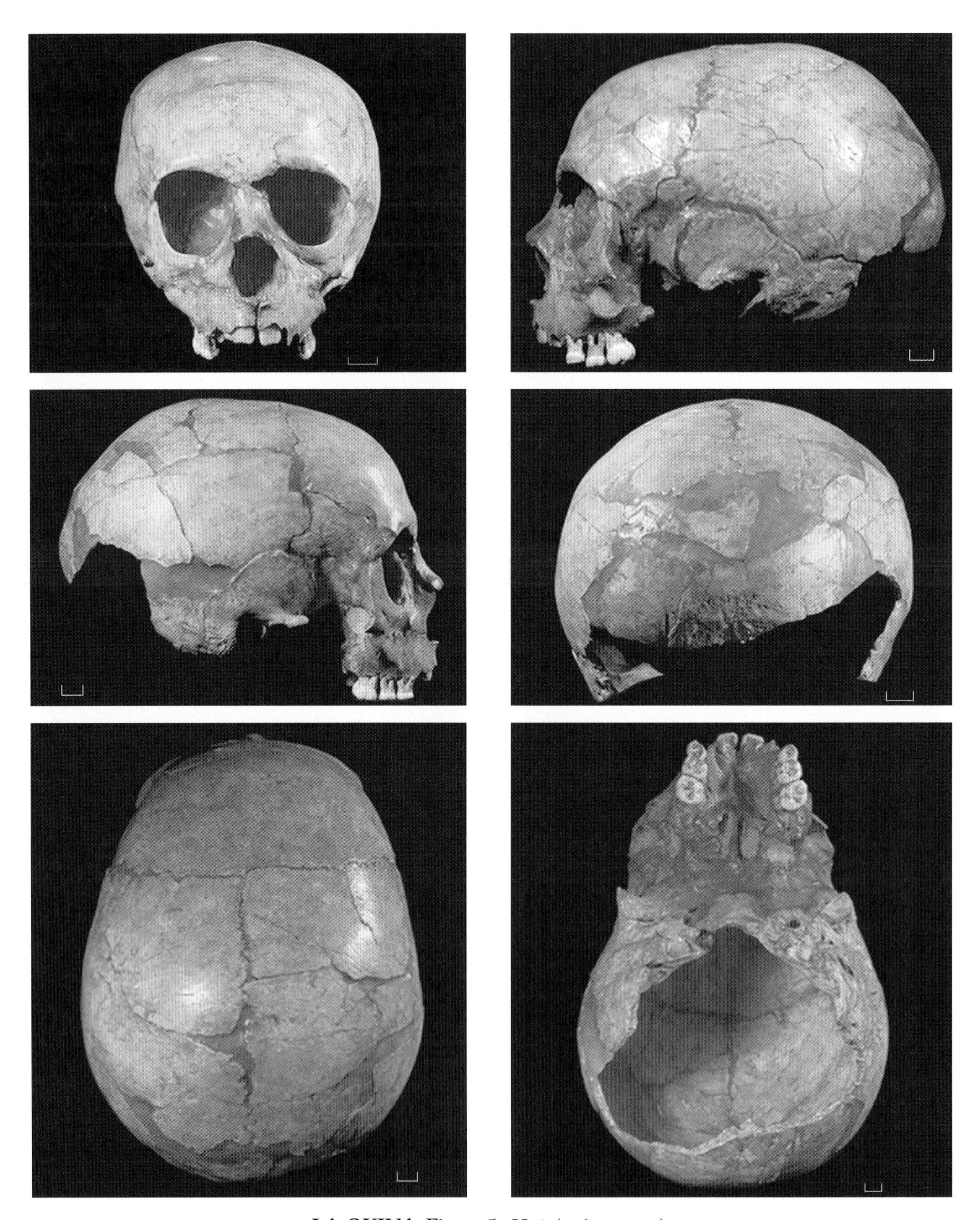

LA QUINA Figure 5. H18 (scale = 1 cm).

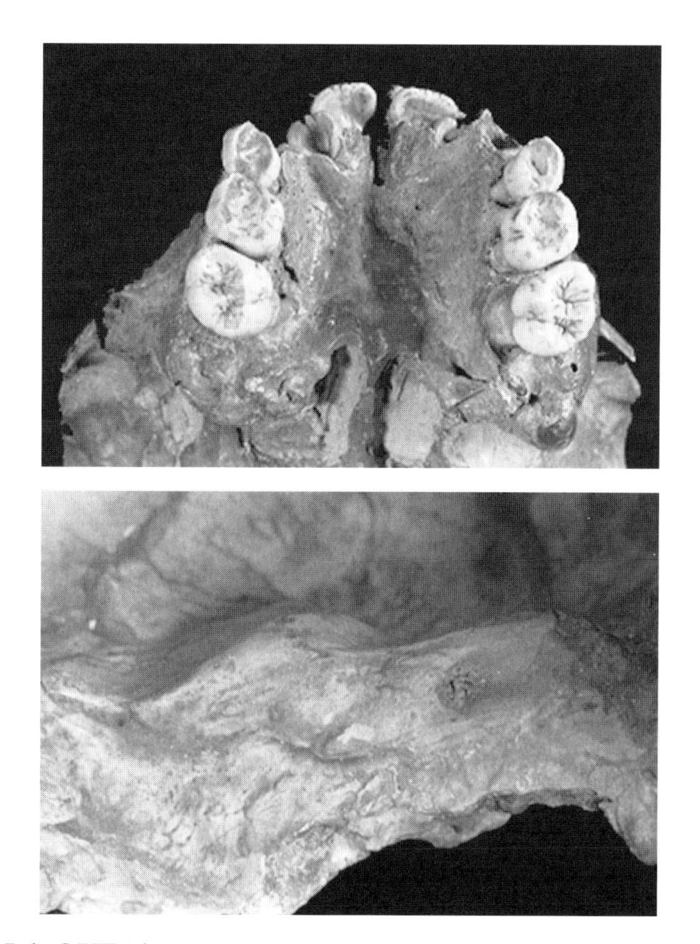

LA QUINA Figure 6. H18 (petrosal below), (not to scale).

Le Moustier

Location

Village of Le Moustier, 10 km NE of Les Eyzies de Tayac, Dordogne, France.

Discovery

Otto Hauser, August 1908.

Material

Skeleton of late adolescent (Le Moustier 1); infant skeleton (Le Moustier 2). The latter was lost before being described; Le Moustier 1 was partly destroyed in WWII. Today, all that survives is the disarticulated and incomplete skull of Le Moustier 1, plus some small fragments of the postcranial skeleton.

Dating and Stratigraphic Context

Stratified rock shelter site. Le Moustier 1 was probably recovered from Bed J of the Lower Shelter (Bordes, 1959), Le Moustier 2 from a grave cut into the underlying Beds H–I. No more will be said of the latter. A relatively late date for Bed J has long been suspected. This has now been confirmed by parallel series of TL dates on burned flint (Valladas et al., 1986) and ESR dates on mammal teeth (Mellars and Grün, 1991) obtained in a fresh excavation. The various Mousterian levels at the Lower Shelter seriate nicely by both methods, TL dates for Bed H coming in at around 42.5 ± 2.0 ka, for Bed J at 40.9 ± 5.0 ka, and for the overlying Bed J at about 40.3 ± 2.6 ka. ESR dates for Bed H of around 41 ka on the linear uptake model corroborate those by TL. Pollens in Bed J indicate fairly cold conditions, with arboreal species at low frequency (Laville et al., 1980).

Archaeological Context

Bed H is characterized by Denticulate Mousterian implements, and Bed J by Typical Mousterian (Bourgon, 1957; see discussion in Mellars, 1996). The overlying Bed K contains a Châtelperronian assemblage. According to its discoverer Otto Hauser, Le Moustier 1 was intentionally buried, but many regard this account as unreliable.

Previous Descriptions and Analyses

Le Moustier 1 was first described by Klaatsch and Hauser (1909) as the type specimen of the species *Homo mousteriensis.* It was later described in greater detail by Klaatsch (1909) and Weinert (1925). Further study was inhibited by the wartime loss of the specimen, but the cranium and a few postcranial fragments were rediscovered in 1966 (see Hesse and Ullrich, 1966). Estimated cranial capacity is 1565 cc (Holloway, 1985). The Neanderthal affinity of the specimen has never been disputed.

Morphology

Adolescent craniodental material now consists of largely complete frontal; most of L temporal with adjacent pieces of parietal, occipital, and alisphenoid; most of R temporal with large adjacent pieces of occipital; R partial parietal to which large piece of L parietal is glued; fragment of basiocciput and adjacent

R occipital; fragment of sphenoid (side uncertain); partial palate missing LI1; and most of mandible, in two pieces, missing L coronoid process. All teeth present. Certain other bone fragments in same box do not appear to be associated; they are much less mineralized.

Cranium lightly built; bone quite thin. Basiocciput shows spicules indicating beginning of closure of sphenooccipital synchondrosis; cranial vault sutures closed endocranially; lateral vault sutures closed inside and out. M3s appear to be in process of eruption (minimal development of roots visible on M^3s); M_3s, at least, look impacted (LC_1 certainly is, lying as it does beneath root of I_2); all Ms would have breached gum line. Series of cutmarks on frontal, most notably two long, parallel incisions that run obliquely and are cut at an opposite angle by shorter parallel incisions; cutmarks same color as surrounding bone.

Frontal fragment quasi-articulates with parietals behind. General cranial shape not very long in profile; moderate slope along frontal levels out well before bregma (top of braincase quite flat). Well above lambda, back of skull descends gently to reach posteriormost extent just above suprainiac depression. Viewed from above, top of skull broad; in coronal section, braincase squarish, with relatively straight side walls complementing flattish top. Greatest neurocranial width occurs far back, just behind mastoid processes, and quite far down side of braincase.

Low, double-arched supraorbital tori quite short s/i; curve smoothly backward from superior orbital margins. Orbital roofs smoothly angle into anterior surfaces of tori. Tori thickest as they emerge from glabellar region, thinning s/i toward midline of orbits; then maintain low, constant height laterally. Tori confluent across very broad, swollen glabellar region. Viewed from above, axes of tori angle back from midline, although anterior profiles concave. Superior orbital margins rounded (suggesting that orbits would originally have been ovoid and not wide). Frontal sinus exposed on the L; extended posteriorly above region of glabella; terminates laterally well before reaching midline of orbit. In its natural orientation, frontal shows slight depression in frontal profile above and behind tori, especially laterally; bone rises obliquely behind this; about midway to bregma, this plane changes to become more horizontal. Moderate lateral curvature across frontal plane seen from the front. Postorbital constriction minimal. Only slight scar (not temporal ridge) rises from behind apparently short zygomatic process of frontal and courses gently upward. Nasion quite high up on glabella; peaked frontonasal sutural surface was quite tall s/i. Appears that coronal suture, preserved in three places, was uniformly denticulate. Internally, frontal crest low, well defined, but not very long. Frontal lobes protruded quite far forward above orbital cones.

As preserved on the L, squamosal portion of temporal long but not high rising; apparently reached highest point quite anteriorly. Anterior squamosal suture makes distinct corner inward. Inferior part of alisphenoid curves (not angles) in toward basicranium (= no distinct infratemporal fossa). Parietal notch preserved on the L indistinct. Parietomastoid suture relatively long, somewhat horizontal. Posterior root of zygomatic arch originates well in front of very small, compressed auditory meatus; flares laterally only minimally, constricting temporal fossa. Low, upwardly curving suprameatal crest confluent with supramastoid crest that continues curve upwards. Mandibular fossae quite wide, relatively deep, moderately long a/p; bounded anteriorly by low articular eminence, medially by rather small, distinct medial articular tubercle. Postglenoid plate m/l wide, closely appressed to ectotympanic tube. As preserved on the R, low, barely distinct vaginal process lies along midline of relatively short ectotympanic tube; peaks slightly around apparently thin styloid process (as impressed on bone) that lay quite medial to meatus; fades out well medial to meatus. Relatively small stylomastoid foramen lies posteriorly, at base of preserved styloid pit. Ectoympanic tube separated by broad notch from base of mastoid process.

Mastoid process more complete on the L (its downward-pointing tip broken); not very long a/p; does not appear to have projected very far. Broken R mastoid process reveals many small air cells, a few large ones. Small, oblique mastoid crest level with top of auditory meatus. Mastoid notch fairly vertically oriented; appears to have been quite shallow, moderately wide; may have been bounded medially by paramastoid crest. On the R, indication of low occipitomastoid crest; on the L, bone protrudes beyond level of occipitomastoid suture on each side. On the R, seems probable that mastoid process would have protruded beyond occipitomastoid crest.

No anterior lambdoid suture. Lambdoid suture proper rises steeply from region of asterion and arcs smoothly across lambda; it and sagittal suture minimally denticulate, neither segmented. Small ossicle in parietal lies just to R of lambda. Superior part of occipital plane (like parietals) fairly flat across from side to side; superoinferiorly, begins to arc down and forward just above suprainiac depression, which lies high on occipital surface. Most of suprainiac depression preserved; is wide and tall, with generally elliptical shape and shallowly concave, coarsely pitted surface; is only defined by very shallow infratoral sulcus. Occipital plane relatively small, somewhat wider than tall s/i. Superior nuchal line (i.e., occipital "torus") poorly defined, more or less horizontal.

Preserved fragment of cranial base includes most of basiocciput and portion of R lateral part of occipital. Sphenooccipital synchondrosis was not fused. Basiocciput broad from side to side, relatively thin; external surface quite flat. Damaged occipital condyle was situated quite anteriorly on what appears to have been very wide, possibly long, ovoid foramen magnum. Postcondylar canal is not patent, but condyloid canal is.

Palate heavily reconstructed, with reconstruction material obscuring upper surface. Underneath, appears that palate was shallow; its preserved walls gently slope on all sides.

Internally, petrosals very wide, with well-developed arcuate eminence; more laterally, is another elevation of superior surface (better preserved on the R). On the R, evidence of superior petrous sinus. Subarcuate fossa represented by tiny depression. Sigmoid sinuses short bilaterally; transverse sinuses quite long; superior sagittal sinus quite indistinct.

Mandibular corpora neither very deep nor thick. Front of jaw moderately broad, moderately arced from side to side. Symphysis essentially vertical anteriorly, curving only fractionally back to inferior border. Viewed from below, symphyseal region uniformly thick a/p. Missing bone exposes roots of all anterior teeth; inferior surface of symphyseal region preserved. Symphyseal region completely smooth. Wide, well-excavated digastric fossae face down; on the L, thick, blunt inferior marginal tubercle lies under P1–2. Anterior to this tubercle, and around preserved symphyseal region, inferior margin of jaw is elevated. On both sides, two small mental foramina; one under P1, the other below region between P1–2. Internally, virtually no postincisal plane; internal symphyseal surface essentially vertical, with no indication of genial tubercles or pits. Mylohyoid line very oblique, moderately marked; submandibular fossa very shallow. On the L, broken mandibular foramen appears to have been compressed and oriented obliquely up and back. On the R, very low pillar runs down midline of coronoid process, which overhangs well-defined preangular notch. R gonial region better preserved; its straight (reflected neither internally nor externally) angle is broadly open arc that curves smoothly into posteriorly leaning posterior margin. Gonial region relatively smooth on both medial and lateral surfaces. No medial pterygoid tubercle but faint muscle scar in that region. Ramus short s/i and relatively long a/p. Very long, shallow sigmoid notch levels out posterior to a/p long, superiorly broadly blunt, coronoid process; runs horizontally to base of much lower condyle; deepest close to condyle. Sigmoid notch crest runs lateral to midline of condyle; terminates well below articular surface.

Upper dentition complete except for LI1; no teeth greatly worn. Enamel on all teeth wrinkled to some extent. Preserved RI1 tall, very wide crowned; wrinkling manifested in longitudinal grooves on lingual surface that run from shallowly concave inferior portion to tall, broad lingual swelling at base. I2s narrow, very tall; barrel shaped on lingual surfaces, which bear distinct internal pits in front of tall lingual swellings; longitudinal grooves run from lingual margin to pit; buccal surface arcs inward. Cs tall and rather narrow with very long, with steep distal slopes; internally, Cs bear almost freestanding, spikelike lingual tubercle; stout keel with thin grooves on either side runs down to tubercle from mesially offset apex. P1s slightly smaller m/d than P2s. Each has well-developed paracone and shorter, slightly mesially positioned protocone; stout mesial crest between these two cusps encloses small anterior fovea; thicker distal crest encloses slightly larger fovea (crest on P2 swells distal part of tooth somewhat). On P2, notch between internal surfaces of protocone and paracone wider than on P1.

On the L, upper Ms increase in size from M1 to M3; on the R, M2 and M3 subequal. On M1–2s, postprotocrista stouter than preprotocrista; the latter runs mesially around face of paracone. Especially on M1–2, base of protocone expansive buccally; intrudes on deep, truncated trigon basin. All Ms bear distinct hypocone (most swollen distolingually on M1), which decreases in size from M1 to M3. Enamel wrinkling increases from M1 to M3. Both M1s bear distinct pit on mesial surface of protocone; metacones, which are close to larger paracones, decrease in size from M1 to M3.

Except for the retained Ldc, lower teeth not very worn. As in upper teeth, enamel of all lower teeth wrinkled to some extent, with Is and RC grooved lingually. All Is tall crowned; bear vertical keel along midline that terminates in small lingual swelling at base. RC tall crowned with short, steep distal slope, bears broader lingual pillar than the Is. LC impacted at angle below LI2 crown, overlapping its root. P1 smaller than P2; both have somewhat swollen buccal sides. On P1, stout crest runs from moderately tall protoconid to much shorter, peaked lingual swelling; thick crest runs mesially from apex of protoconid to base of lingual swelling from which (better seen on the L) is separated by groove. Mesial crest encircles thin, deep fovea; stouter distally coursing crest runs from protoconid to become confluent with lingual swelling and enclose the larger posterior fovea. P2s narrower mesially than distally. Protoconid and almost subequal metaconid lie close together; mesially arcing, short, stout crest between these cusps encloses small anterior fovea. Distal part of P2 much larger, with strong, relatively long crest that runs distally from protoconid and then turns back to meet metaconid. In distolingual corner of P2 this crest distended into entoconid; centrally this crest expands into cusplike structure. Distal portion of P2 more talonid basin- than posterior fovea-like.

Lower Ms essentially subequal in size; all bear distinct, b/l broad trigonid basin that lies anterior to widely separated protoconid and metaconid. All Ms have distinct hypoconulid lying just buccal to midline of tooth and long, relatively wide talonid basin; stout crest runs mesially from hypoconid into center of talonid basin. In occlusal view, crowns quite rounded, especially distally, giving generally ovoid shape. Enamel wrinkling increases in complexity from M1 to M3.

References

Bordes, F. 1959. Le contexte stratigraphique des Hommes du Moustier et de Spy. *L'Anthropologie* 63: 154–157.

Bourgon, M. 1957. Les industries moustériennes et pré-moustériennes du Périgord. *Arch. Inst. Paléontol. Humaine* 27: 1–141.

Hesse, H. and H. Ullrich. 1966. Schädel der "Homo mousteriensis Hauseri" wiederfunden. *Biol. Rundschau* 4: 158–160.

Holloway, R. L. 1985. The poor brain of *Homo sapiens neanderthalensis*; see what you please. In: E. Delson (ed), *Ancestors; The Hard Evidence*. New York, Alan R. Liss, pp. 319–324.

Klaatsch, H. 1909. Die neueste Ergebnisse der Paläontologie der menschen und ihre Bedeutung für das Abstammungs-problem. *Z. Ethnol.* 41: 537–584.

Klaatsch, H. and O. Hauser. 1909. *Homo mousteriensis Hauseri. Arch. Anthropol.* 35: 287–297.

Laville, H., J.-P. Rigaud and J. Sackett. 1980. *Rock Shelters of the Périgord: Geological Stratigraphy and Archaeological Succession.* New York: Academic Press.

Mellars, P. 1996. *The Neanderthal Legacy.* Princeton, NJ: Princeton University Press.

Mellars, P. and R. Grün. 1991. A comparison of the electron spin resonance and thermoluminescence dating methods: the results of ESR dating at Le Moustier (France). *Cambridge Archaeol. J.* 1: 269–276.

Valladas, H., J.-M. Geneste, J.-L. Joron and J.-P. Chadelle. 1986. Thermoluminescence dating of Le Moustier (Dordogne, France). *Nature:* 322: 452–454.

Weinert, H. 1925. *Der Schädel der eiszeitlicher Menschen von Le Moustier in neuer Zusammenserzung.* Berlin, Springer.

Repository

Museum für Vor- und Frühgeschichte, Schloss Charlottenburg, D-14059 Berlin, Germany.

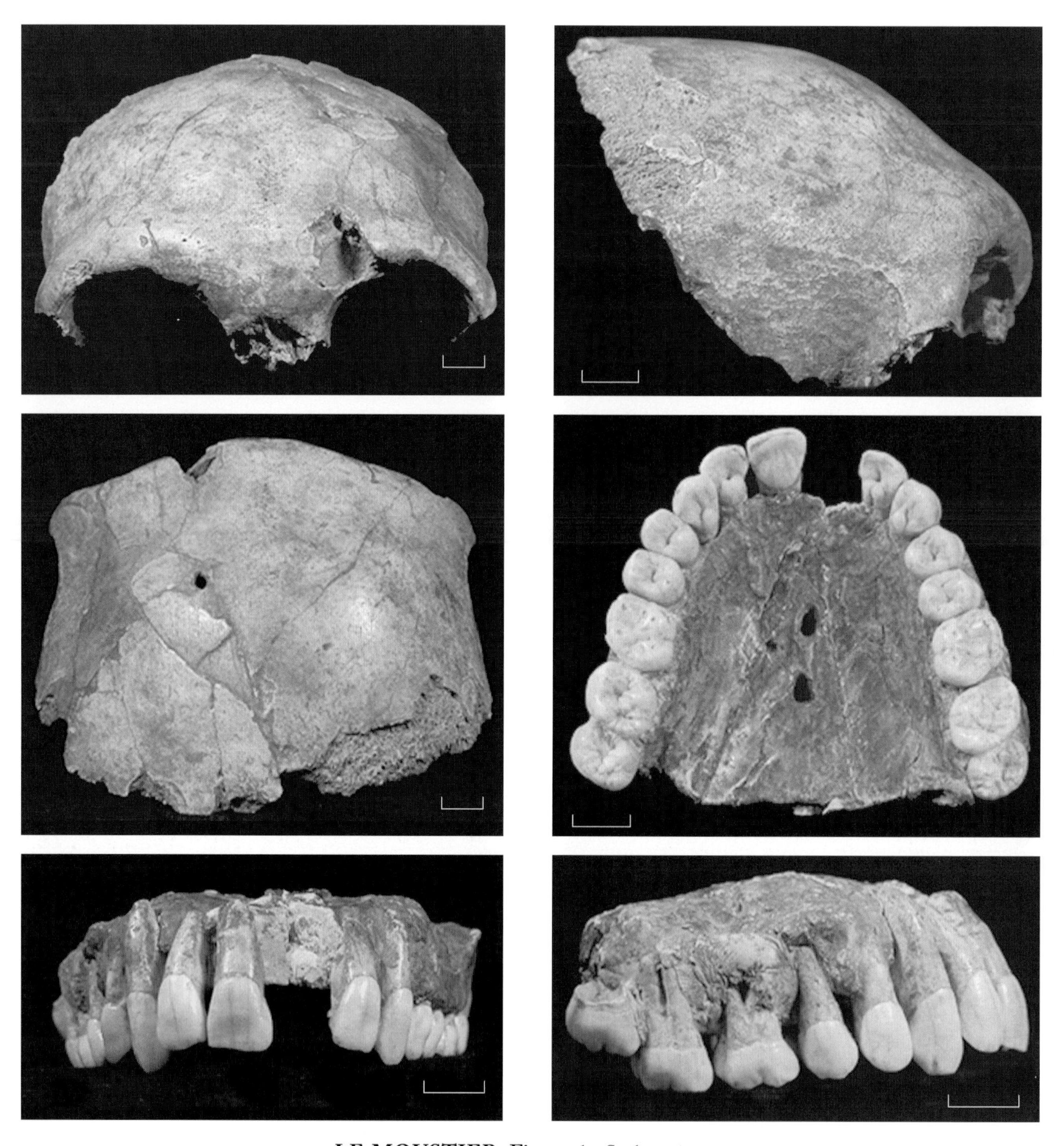

LE MOUSTIER Figure 1. Scale = 1 cm.

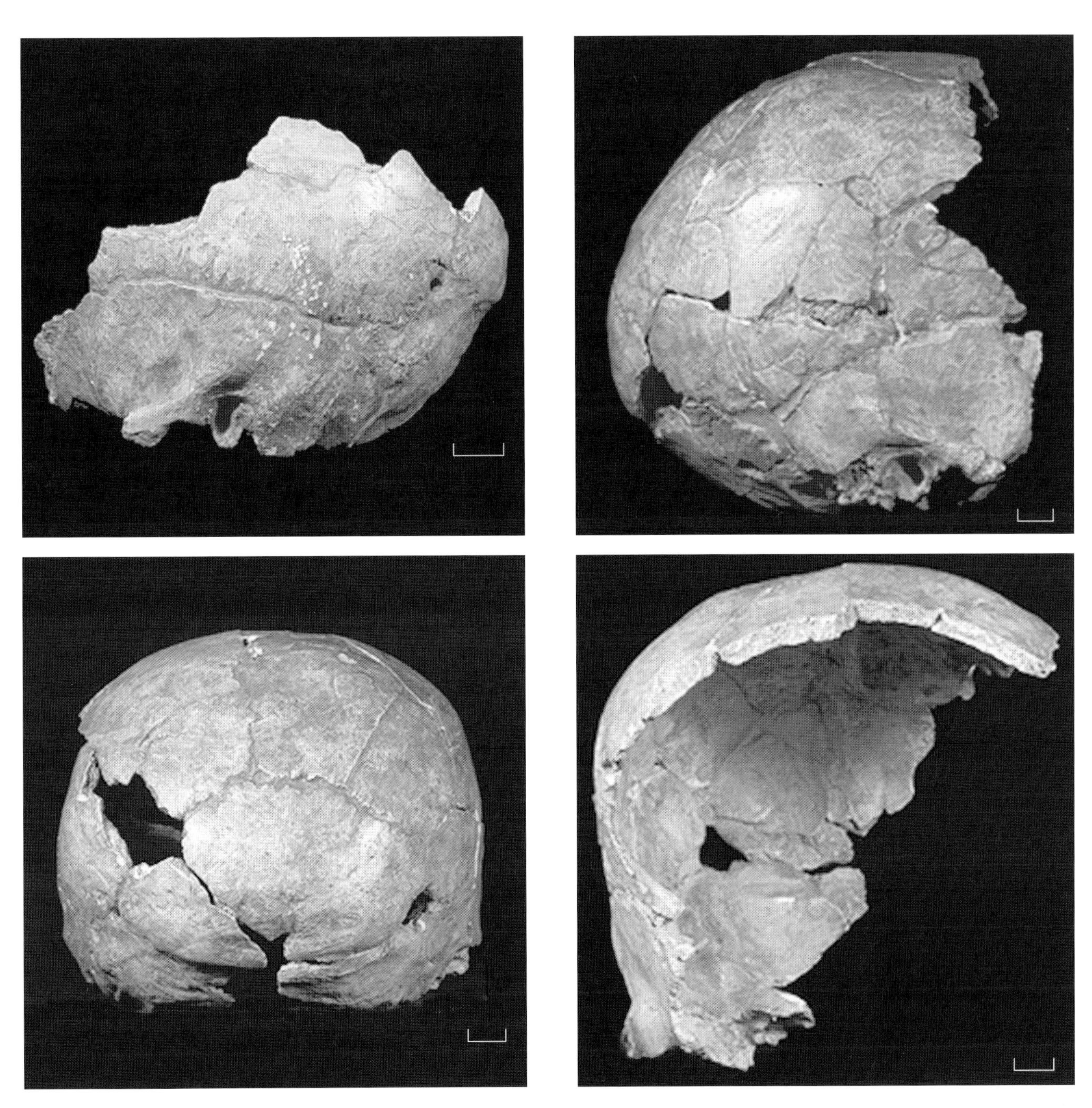

LE MOUSTIER Figure 2. Scale = 1 cm.

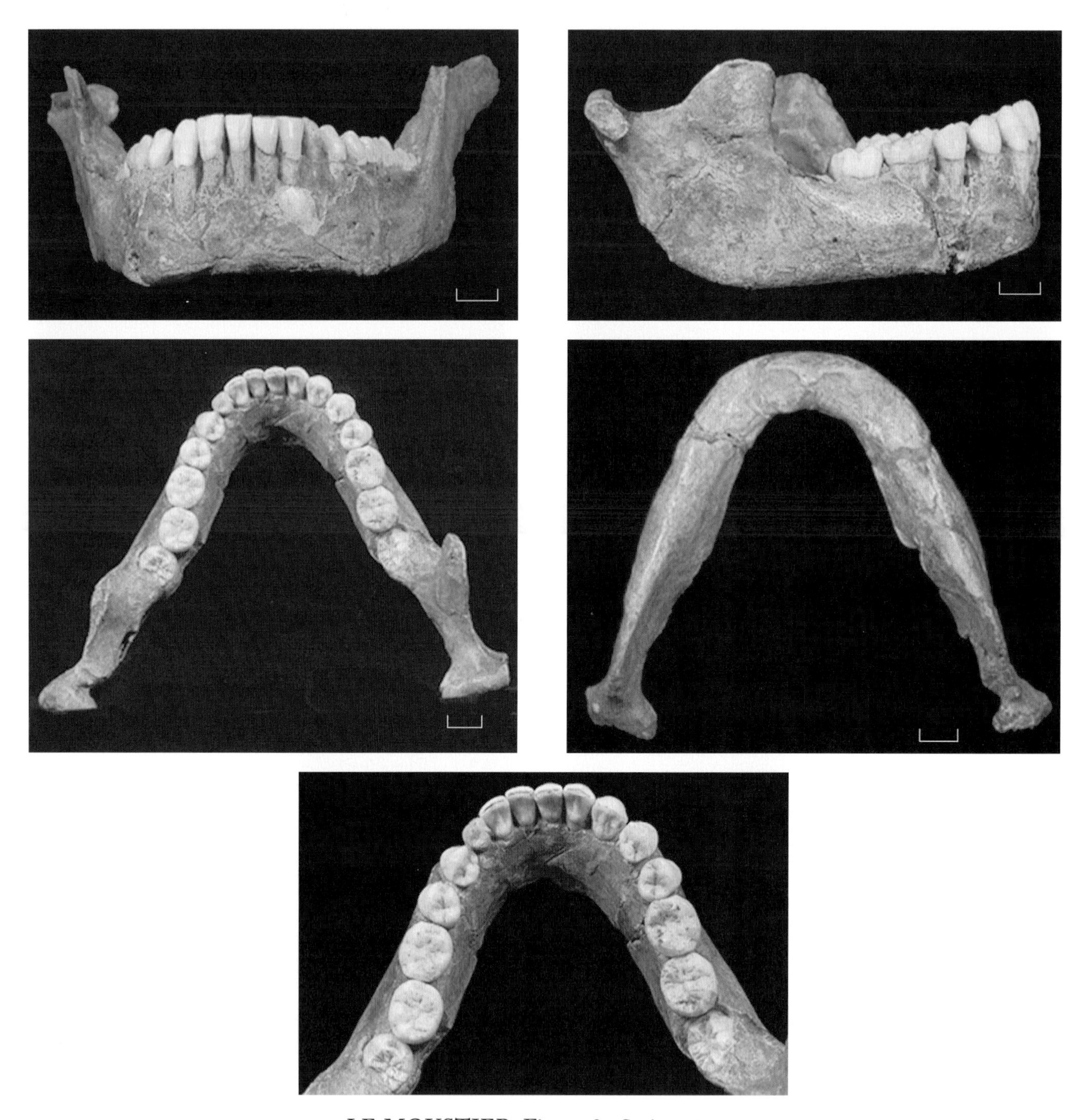

LE MOUSTIER Figure 3. Scale = 1 cm.

Mauer (Heidelberg)

Location

The Rösch sand quarry, about 1 km N of the village of Mauer, 16 km SE of Heidelberg, Germany.

Discovery

D. Hartmann, October 1907.

Material

Mandible lacking LPs.

Dating and Stratigraphic Context

The fossil was found at 25 m depth in fluviatile "Lower Sands" of the Neckar river that are overlain by gravel, silt, and loess layers. A well-preserved associated fauna suggests an age somewhere in the middle Pleistocene, between the Cromerian (Cook et al., 1982) and Holsteinian interglacials. An age of around 500 ka thus may not be too far off.

Archaeological Context

Reports of stone and bone tools from Mauer have all been contested, on the basis of either age or identification.

Previous Descriptions and Analyses

Schoetensack (1908) described the Mauer jaw as the holotype of a new hominid species, *Homo heidelbergensis,* and this specimen has subsequently served as the basis for several new genera, none of which has advanced the science of paleoanthropology. In the second half of the twentieth century discussion mostly centered around whether this specimen should be included in *Homo erectus* (e.g., Wolpoff, 1980) or with the group of "archaic *Homo sapiens*" fossils that includes Petralona and Arago (e.g., Stringer, Howell, and Melentis, 1979). Latterly, however, it has become evident to many workers that the Petralona-Arago-Kabwe group forms a distinctive morph that deserves recognition as a species in its own right rather than simply as a stage or "grade" somehow intermediate between *Homo erectus* and *Homo sapiens.* Furthermore, it has become fashionable to co-opt the Mauer jaw into this group, for which Schoetensack's name—*Homo heidelbergensis*—therefore has priority (Tattersall, 1986). It is thus especially regrettable that no cranial elements are associated with the jaw.

Morphology

Virtually complete mandible (GPIH Mauer 1) missing part of external surface of symphysis, much of both coronoid processes (entirely broken off, but smoothly, just above level of sigmoid notch, giving misleading impression of low, long processes), and crowns of LP1–M2. On the R, sigmoid notch crest also slightly damaged.

Very robust, thick corpora; a/p long rami. Symphyseal region rather broad across; arcs strongly from side to side. In profile, symphysis curves gently down and back to elevated inferior margin. Corpora do not diverge strongly; very thick bone of corpora yields tighter symphyseal contour internally than externally. Anterior to point of greatest thickening, inferior margin of corpus rises on both sides in arc that follows contour of very m/l wide, anteriorly placed, downwardly oriented digastric fossae. In midline, where digastric fossae approximate to each other, inferior margin of symphysis forms broad, low, downward peak. Viewed from front, lower margin of symphysis describes shallow bow-shaped course, with symphysis itself considerably raised above level of inferior margin of posterior corpus. Central portion of external symphyseal region remains smooth, lacking any surface features. On both sides, a large and, above it, a tiny mental foramen lie under region between P2 and M1; below the larger foramen, inferior margin of mandible thickened outward. As seen more clearly on the R, marginal thickening below mental foramen continues medially only for about two-thirds length of digastric fossa. Marginal thickening diminishes posteriorly, fading out by level of M3; particularly on the R, this marginal thickening is defined above and throughout its length by a shallow sulcus. Moderately inclined postincisal plane of modest length bounded inferiorly by fairly large, shallow, subcircular fossa, below which are small, apparently twinned (a break runs between them) genial tubercles.

Mylohyoid lines faint; submandibular fossae below very shallow. Right behind M3s, bone of corpus thins drastically into ramus. Gonial regions slightly reflected outward, bear slight external muscle scars; internally, scarring more pronounced though still modest, except for (particularly on the L) low, pronounced medial pterygoid tubercle. Gonial angle itself somewhat "cut off." Posterior margin of ramus shows shallow concavity of profile above angle and below the moderately wide m/l, thick a/p, and coronally somewhat peaked condyles. Coronoid processes would have been quite long a/p and risen to unknown extent above level of condyles. Sigmoid notch would have been very constricted a/p. Sigmoid notch crest runs (on both sides) to lateralmost extremity of condyle. Internally, coronoid processes (on each side) bear very thick, vertical buttress of bone that flexes inferiorly to become confluent with more ridgelike internal alveolar crest. Large, superoposteriorly oriented mandibular foramen incomplete inferiorly; its inferior margin forms "V shape," from which extends a short, pronounced mylohyoid groove that courses forward and down from its apex. On both sides, distal edge of each M3 visible just in front of modest preangular sulcus that indents anterior margin of ramus.

Is and Cs worn well into dentine; remaining teeth less severely worn. In spite of wear, Is and Cs seem bulkier, especially buccolingually, than the cheek teeth. Incisor lingual surfaces rather convex from side to side, bear modest lingual swellings. Cs seem to have been somewhat concave lingually. RP1 bears small, well-defined lingual tubercle at base of worn protoconid, from either side of which relatively thick crests course down to base of tubercle, confining small, relatively deep anterior and posterior foveae (no crease separating anterior crest from tubercle). P2 bears anteriorly placed, subequal metaconid and protoconid connected mesially and distally by crests that course between their apices. Anterior fovea pitlike, somewhat more developed than posterior fovea. Preserved RM1–3 quite long; would have been rounded (not squared off) in occlusal outline. M2 longer m/d and wider b/l than M1; M1 slightly longer and wider than M3. All Ms show some enamel wrinkling; their hypoconulid lies fractionally to buccal side of crown's midline. As seen especially on LM3, trigon basin very distinct (traces not seen on M1–2). Talonid basins were probably not very distinct (being encircled by cusps that are internally expansive at their bases). On M1, base of hypoconid extends lingually beyond midline of tooth (not so on other molars). Talonid of preserved LM3 considerably less developed m/d than on RM3 (thus creates subcircular occlusal outline).

REFERENCES

Cook, J. et al. 1982. A review of the chronology of the European middle Pleistocene hominid record. *Yrbk. Phys. Anthropol.* 25: 19–65.

Schoetensack, O. 1908. *Der Unterkiefer des* Homo heidelbergensis *aus den Sanden von Mauer bei Heidelberg.* Leipzig, Wilhelm Engelmann.

Stringer, C., C. Howell and J. Melentis. 1979. The significance of the fossil hominid skull from Petralona, Greece. *J. Archaeol. Sci.* 6: 235–253.

Tattersall, I. 1986. Species recognition in human palaeontology. *J. Hum. Evol.* 15: 165–175.

Wolpoff, M. 1980. Cranial remains of Middle Pleistocene European hominids. *J. Hum. Evol.* 9: 339–358.

Repository

Geologisch-Paläontologisch Institut der Ruprecht-Karls-Universität Heidelberg, 69120 Heidelberg, Germany.

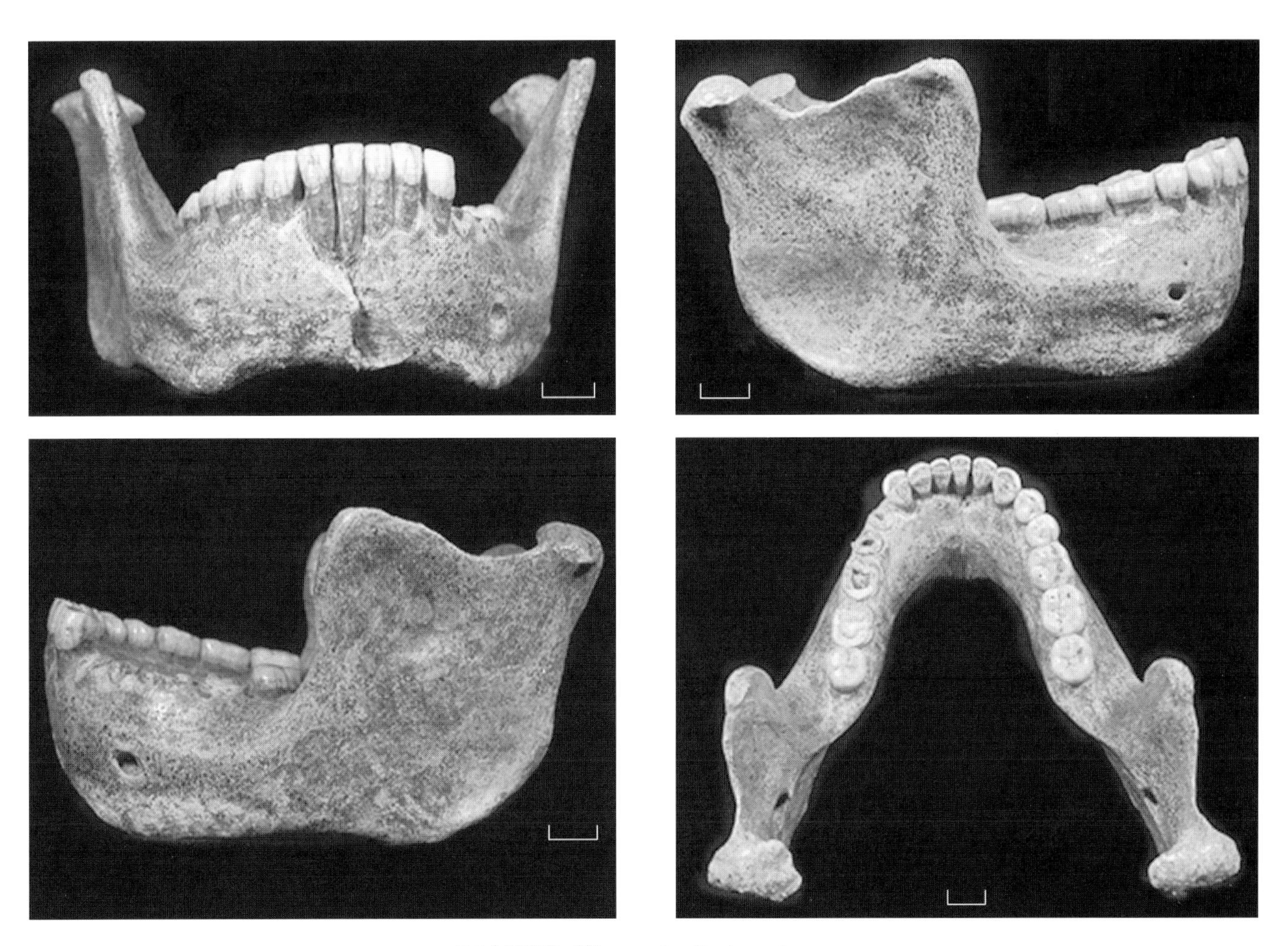

MAUER Figure 1. Scale = 1 cm.

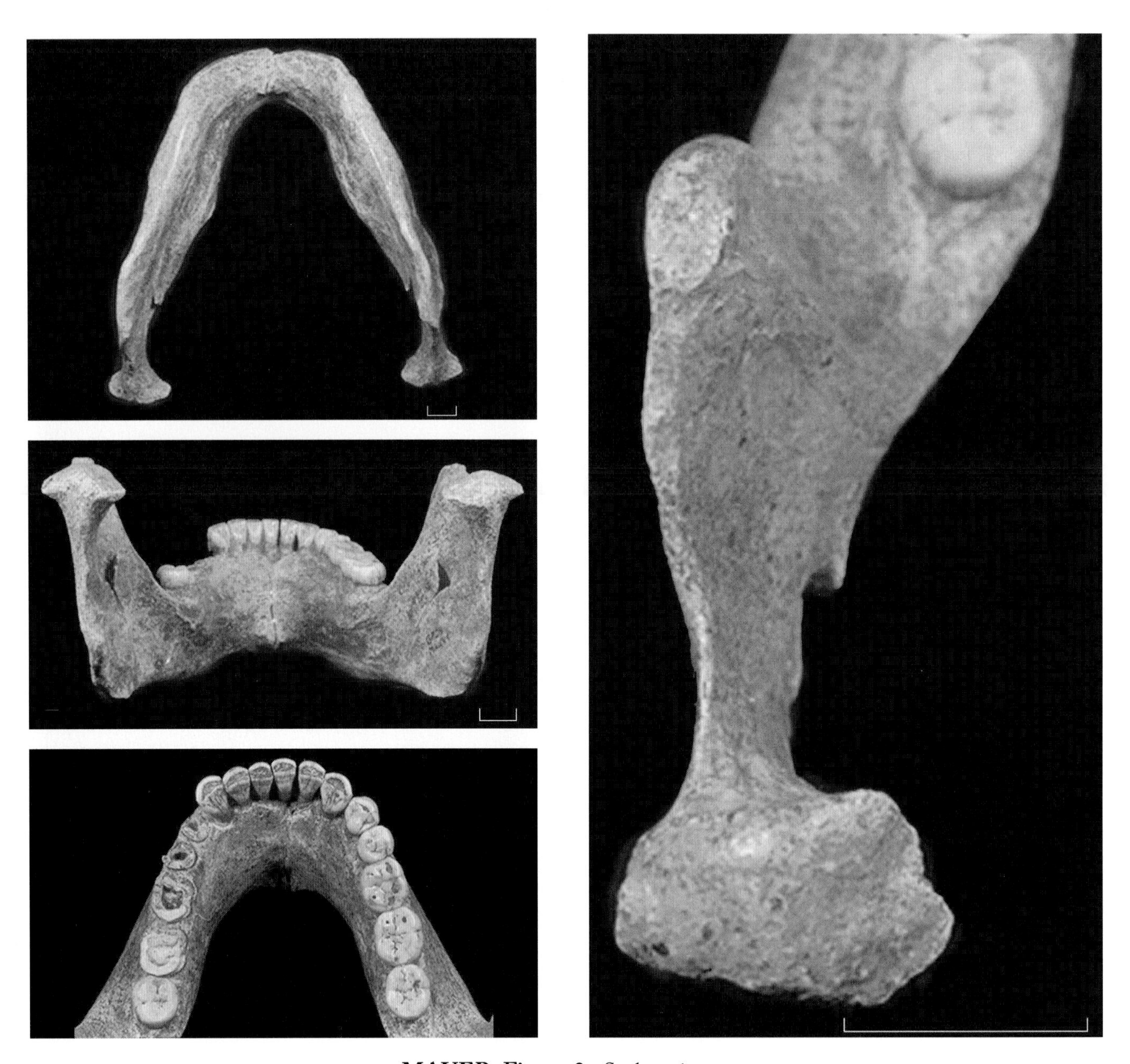

MAUER Figure 2. Scale = 1 cm.

Mladec (Lautscherhöhle)

Location

Bocek's Cave or Fürst-Johanns-Höhle, just W of Mladec (Lautsch) village, 4.5 km W of Litovel, northern Moravia, Czech Republic.

Discovery

Excavations of J. Szombathy, June 1881 (Mladec 1–3), June 1882; J. Knies and others, 1903–1922 (Mladec 5–10).

Material

Several skeletons in varying stages of completeness. After the burning of Mikulov Castle in 1945, only the Mladec 5 partial cranium, plus the materials stored in Vienna (the Mladec 1 and 2 crania and various cranial fragments and assorted postcranial elements) remain in the original.

Dating and Stratigraphic Context

The Mladec human remains in both the main cave (Mladec 1–4 and 8–10, plus some unnumbered specimens) and an adjacent smaller cave (Mladec 5–7, plus some unnumbered specimens) were associated with an Aurignacian stone and bone industry and a Würm II (i.e., earlier than ca. 32 ka) fauna (Szombathy, 1925). Because in both cases the fossils and stone tools were found in talus cone deposits, their contemporaneity cannot be conclusively demonstrated, but the association seems reasonably secure (Smith, 1984).

Archaeological Context

Apparently Aurignacian (see above).

Previous Descriptions and Analyses

Szombathy (e.g., 1900, 1925) provided the principal accounts of these fossils and found no difficulty in attributing them to *Homo sapiens* of the "Cro-Magnon type." Later commentators (e.g., Smith, 1984) have pointed to some minor and superficial similarities in supraciliary morphology with Neanderthals, especially in Mladec 5 and other presumed males, and attention has also been drawn to a certain degree of occipital protrusion in Mladec 1 and also Mladec 4–6 (Wolpoff, 1996). Smith (1984) has emphasized that such "bunning" differs in significant respects from that observed among Neanderthals, although Wolpoff (1996) would apparently not agree. Nobody has argued that these fossils are not modern *Homo sapiens*, but Wolpoff has emphasized variability in the sample and has made much of supposed Neanderthal similarities, arguing that the Mladec population "establish[es] unique anatomical links to the preceding Neandertals" (1996, p. 755). Wolpoff (1999) provides estimated cranial capacities of 1540 cc for Mladec 1, 1390 cc for Mladec 2, and 1650 cc for Mladec 5.

Morphology

Mladec 1 (Lautscherhöhle 1)
Adult. Very weathered but largely intact cranium, missing parts of R frontal, parietal and squamosal, and all teeth except R and LM1–2.

Skull long, narrow, very dolichocephalic. Frontal vertical. Orbits low; rectangular, inferior margin pronounced. Supraorbital region bipartite; medial supraorbital swellings confluent across glabella (= glabellar "butterfly") but obliquely divided from platelike plane lateral to supraorbital foramen, which is quite medially placed and high above orbital rim. Slight medial keeling present on forehead (artifact of weathering?). Nasals pinched in midline; quite strongly flexed 0.5 cm below nasion. Lacrimal fossae not visible; on the L, anterior lacrimal crest very oblique, fades out superiorly. Moderately large infraorbital foramen close to orbital margin bilaterally; groove descends from foramen on both sides to canine fossa. Lateral nasal margin quite salient, forward projecting; inferior part of lateral nasal margin somewhat scooped out. Inferior margin double (= prenasal fossa), with lateral crest coming down below margin of anterior nasal spine; flows back into nasal aperture. Anterior nasal spine fused (single), quite projecting. Nasoalveolar clivus short, anteriorly inclined. Nasal floor flat (not depressed); quite capacious maxillary sinus visible on the R. Otherwise, internal nasal structures not preserved.

Anterior roots of zygomatic arches flat, anteriorly facing, sharply angled backward laterally at maxillary tuberosity (creating very small temporal fossae). No sharp division between temporal and infratemporal fossae (because of absence of angle inferiorly along squamous and alisphenoid). Zygomas and zygomatic processes of temporals quite gracile. Auditory meatus relatively small; on the R, suprameatal crest confluent with supramastoid crest. Moderately deep mandibular fossae slightly angled forward m/l; are not closed off medially by tubercle. Well-defined articular eminence lacking. Mastoid processes low and downwardly pointed, with longer posterior slope. On the L, possibly anterior mastoid tubercle below auditory meatus. Damaged R mastoid process shows honeycomb of small air spaces. Narrow mastoid notch broadens out somewhat behind process. Parietomastoid suture short. Lambdoid suture emerges obliquely from asterion; forms sharp peak at lambda. Vaginal process present on the R but damaged. Also on the R, possible partially preserved base of styloid process situated lateral to region of carotid/jugular foramina.

Occipital region swollen posteriorly, rounded. Area of occipitomastoid suture low and slightly swollen but not crestlike. Medial to this suture (on both sides) are slight, straight, and a/p oriented Waldeyer's crests that fade out posteriorly. Foramen magnum narrow.

Palate small; its shape between "V" and "U"; anterior slope shallow, deepest posteriorly. Incisive foramen close to alveolar margin; possible slight maxillary torus, especially on the R. Region of posterior nasal spine very broad, curving.

Inside braincase, petrosals can be palpated on both sides. Especially on the R, arcuate eminence distinct; R sigmoid sinus deeper than the L.

Preserved dental alveoli relatively small. LP1 root preserved, is quite compressed m/d and not double. M1s squarish with large hypocones and large postprotocristae; posterior foveae partly invade metacone, truncating it. In M1–2, paracone larger than metacone. M2s lack hypocones; posterior fovea continues farther toward apex of metacone (tooth roundedly triangular); cuspules in midline of postcingulum.

Mladec 2 (Lautscherhöhle 2)
Adult. Partial cranium lacking occipital, sphenoid, R and L malars, and parts of R and L squamosals. No mandible, but lower RM1, root of RM2, LM1–3.

Less dolichocephalic than Mladec 1; very gracile with thin, delicate bone. Frontal quite vertical. Supraorbital region bipartite, weakly developed. "Butterfly" swelling across glabella; supraorbital plane changes lateral to supraorbital notch. Frontal sinuses not well developed. Orbit (preserved on the R) probably less rectangular than Mladec 1. Infraorbital region broader than Mladec 1; infraorbital foramen smallish. Nasal bones appear to have been similarly pinched. Nasal margin preserved on the R; seems to have been more or less vertical and not prolonged forward as in Mladec 1. Anterior lacrimal crest present on the R; is crisp inferiorly but fades out mediosuperiorly. Inferior orbital margin elevated. Floor of nasal cavity flat; no medial projection on inner lateral wall. Nasoalveolar clivus would have been short, anteriorly inclined.

Anterior root of zygomatic arch flattened, anteriorly facing; is broken off just by maxillary tuberosity, but temporal fossa was probably narrow. Mandibular fossa quite deep, inclined obliquely forward, not closed off medially by tubercle; no articular eminence. Vaginal process contacts base of mastoid process; on the L, becomes higher laterally, toward margin of small auditory meatus. Suprameatal crest confluent with upwardly angled supramastoid crest. Mastoid process bulky but not prominent; almost no mastoid notch. Swelling (not paramastoid crest) probably continued across occipitomastoid suture at about the level of tip of mastoid process.

Occipital would not have had protruded posteriorly as in Mladec 1, and the parietals curve in much more steeply. Lambdoid suture rises obliquely from asterion and peaks at lambda. On both sides (notably the L), petrosals bear distinct arcuate eminence but no subarcuate fossa. Superior petrous sinuses large, deep. R and L sigmoid sinuses both deeply excavated, well defined.

Palate an intermediate U/V shape and steep sided, with sloping at the front. Upper teeth little worn; preserved alveoli and Ms small. M1s narrower than Mladec 1 but show same basic pattern (large hypocone, closely positioned paracones and metacones, with paracones larger). Both M1s have postprotocristae, and LM2 a small hypocone, but otherwise comparable to Mladec 1, especially in having fovea invading metacone on M2. M3 small with some enamel wrinkling; paracone much larger than metacone. M2–3 have conule in midline of postcingulum.

Mladec 5

Adult. Calvaria, with healed wounds; slightly charred when Mikulov Castle was burned by Germans (1945); only surviving skull from this catastrophe. Very large, broad braincase; squatter looking than Viennese Mladec specimens. Skull long and quite tall, with maximum width fairly low, at supramastoid crests.

Frontal vertical, but not very high, with distinct eminences. Superciliary arch bipartite (= "butterfly" and distinct lateral plate), well developed. Frontal sinuses extend moderately toward sides but not far up into frontal. Temporal lines quite well marked. Parietal eminences very prominent. Posterior root of zygomatic arch originated at anterior edge of auditory meatus. Vaginal process broken bilaterally; did not run entire length of ectotympanic tube but did contact mastoid process. Moderately stout styloid process with large stylomastoid foramen at base lies just medial to mastoid process. Large suprameatal crest swept up and back into even larger supramastoid crest. Mastoid crests present bilaterally. Parietomastoid suture not very distinct. Mastoid processes broken off; bases were large; processes themselves (as preserved on the L) were swollen laterally, pneumaticized with large air cells. On the R, mastoid notch was narrow, with crest lateral to it (occipitomastoid suture not visible; unclear whether crest is paramastoid or occipitomastoid). On the L, a bit more posteriorly, is probable Waldeyer's crest.

Occiput broadly triangular. Occipital plane bulges flatly and broadly below suture. Inferiorly, occipital plane bordered by wavy, bow-shaped depression that corresponds to superior nuchal line. In midline, just above this depression, is small pitted surface with rim of bone separating it from an inferior depression (neither constitutes a suprainiac depression in position nor morphology). Lambdoid suture rises from asterion, peaks at lambda.

Cranial sutures segmented. Pacchionian depressions along posterior part of superior sagittal sinus. R petrosal, especially, shows distinctly raised arcuate eminence; subarcuate fossa not closed off on the L, but is on the R, where it is concave. Superior petrous sinus well marked on both sides.

Mladec 6 (cast)

Adult. Incomplete calotte; apparently very thick boned. Massive skull reminiscent of Mladec 5 but not quite as broad. Some trauma marks on frontal.

Frontal quite vertical. Supraciliary region quite well developed but not protruding, with glabellar "butterfly." Temporal lines broad, faint. Parietomastoid suture small. Occiput barely protruding; presents as broad-based triangle, with broad and shallow depression along suture just above lambda. Lambdoid suture curves concavely up from asterion to peak at lambda; weak muscle markings inferiorly. Sagittal suture segmented.

Mladec Fragments

Cranial fragments #5456. Various juvenile cranial fragments, including much of frontal, part of occipital, and portion of R temporal. Frontal lacks any supraorbital development; glabellar region essentially flat. Mastoid process not very distinct. Foramen of Huschke closed (individual was over 5 years old). Petrosal has well-defined arcuate eminence but no discernible arcuate fossa; displays quite vacuous pneumatization. Occipital lacks both torus and suprainiac depression.

Palate #5487 (Wolpoff #8). Large, very shallow palate, with possible low torus in midline. Floor of nasal cavity totally flat. Maxillary sinuses did not expand anteriorly. Lower nasal aperture margin (lateral crest) meets small anterior nasal spine; horizontal conchal crest present. Nasoalveolar clivus broad, flat across, and very short. Incisive foramen very far forward. Alveoli and teeth much larger than Mladec 1 and 2. Teeth very heavily worn. C very large, short rooted. LM1–2 present and worn, rectangular and broad b/l, with large hypocones; on M2, paracone larger than metacone.

Teeth

L9. Associated RC^1 (short root, good lingual tubercle, quite deep fovea) and unworn LP^1 (paracone slightly larger than protocone, both cusps situated slightly mesial of midline, creaselike anterior and posterior fovea).

L10. RM^3. Triangular with some enamel wrinkling, paracone larger than metacone, deep posterior fovea; buccal roots closely approximated.

References

Smith, F. 1984. Fossil hominids from the Upper Pleistocene of Europe and the origin of modern Europeans. In: *The Origins of Modern Humans*. New York, Alan R. Liss: 137–209.

Szombathy, J. 1900. Un crâne de la race Cro-Magnon trouvé en Moravie. *12th Int. Cong. Anthropol. Préhist. Archaeol.* 1900: 133–140.

Szombathy, J. 1925. Die diluvialen Menschenreste aus der Fürst-Johannes-Höhle be Lautsch in Mähren. *Eiszeit* 2: 1–34.

Wolpoff, M. 1996. *Human Evolution*. New York, McGraw-Hill.

Wolpoff, M.H. 1999. *Paleoanthropology*. New York, McGraw-Hill.

Repository

Naturhistorisches Museum, Burgring 7, 1010 Wien, Austria (Mladec 1–3, isolated teeth, fragments, and unnumbered pieces). Anthropos Institute, Moravske Muzeum, Zelny Trh, Brno, Czech Republic (Mladec 5; cast of Mladec 6).

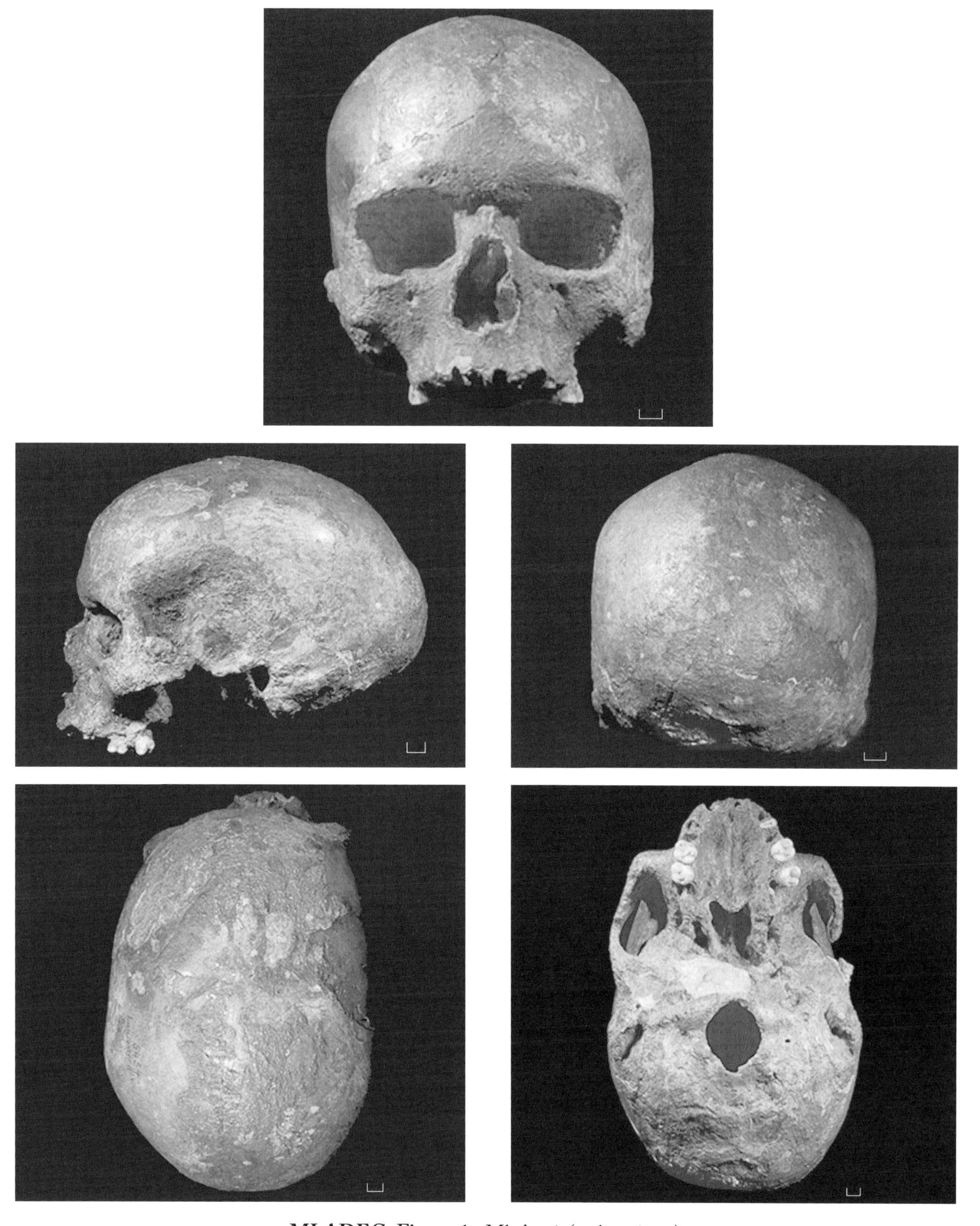

MLADEC Figure 1. Mladec 1 (scale = 1 cm).

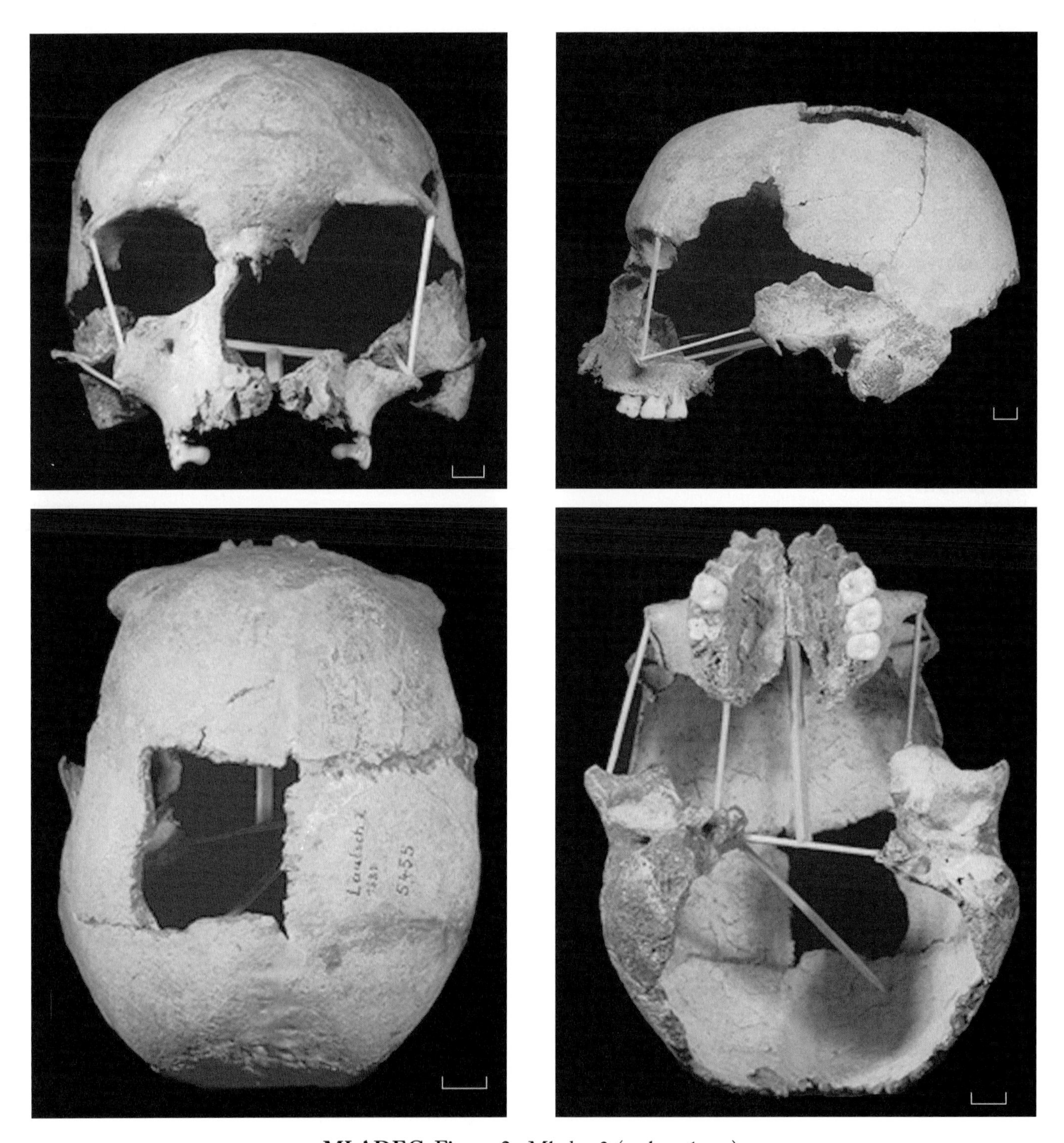

MLADEC Figure 2. Mladec 2 (scale = 1 cm).

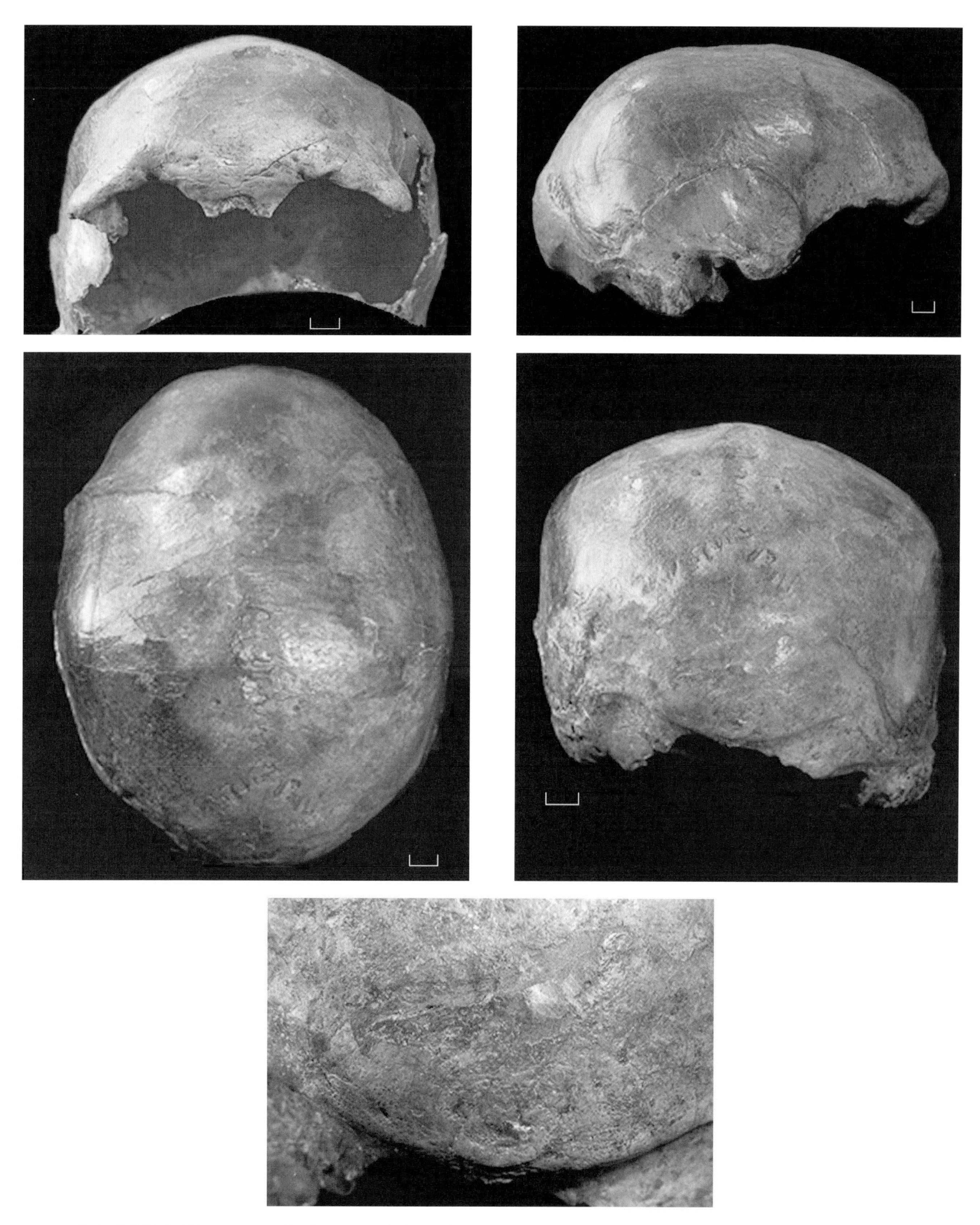

MLADEC Figure 3. Mladec 5 (including close-up of region of superior nuchal line). (scale = 1 cm)

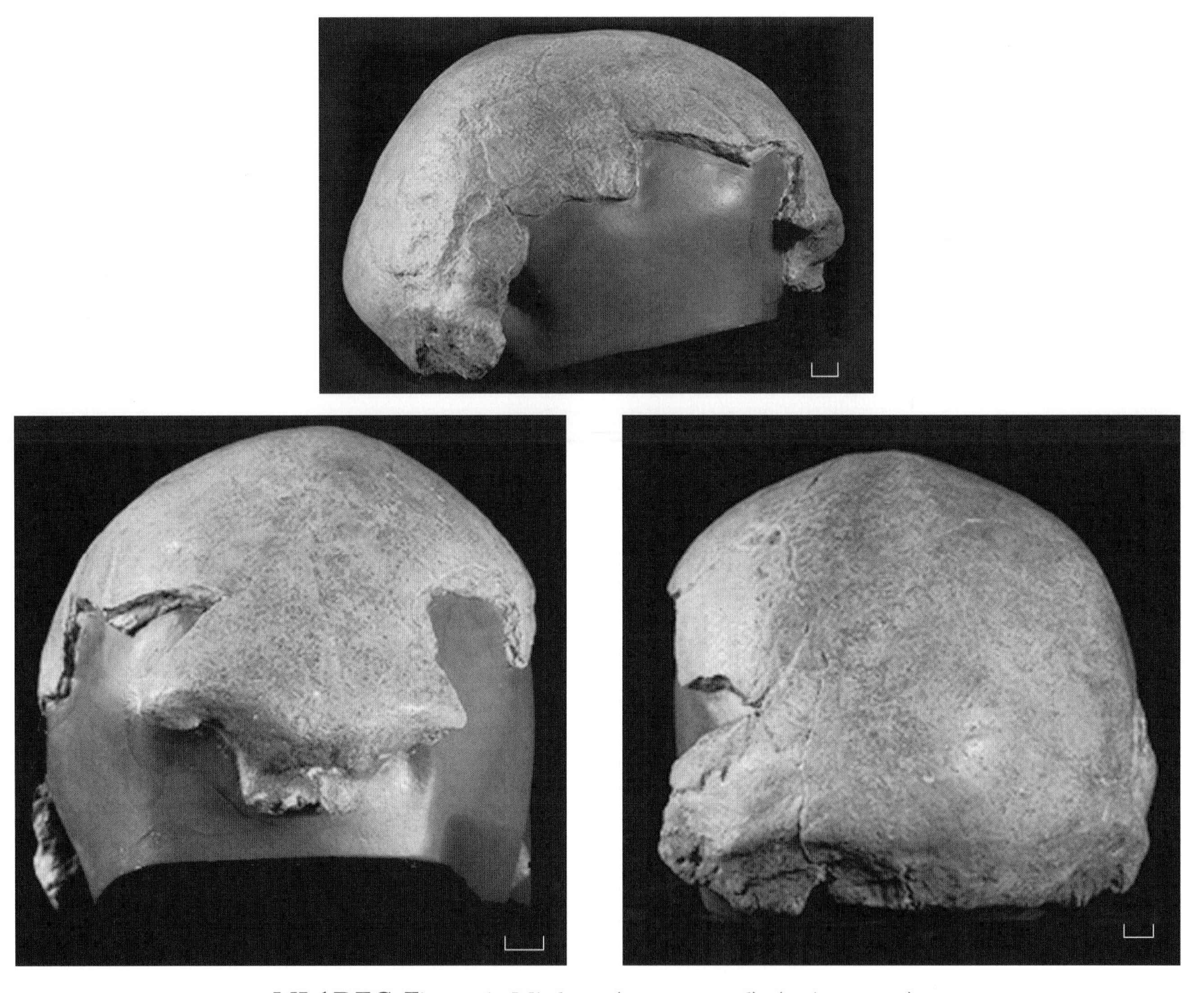

MLADEC Figure 4. Mladec 6 (cast on stand), (scale = 1 cm).

Montmaurin

Location

Deposits filling a karst chimney (La Niche) in the valley of the Seygouade river at Montmaurin, N of St-Gaudens in the Pyrenees foothills of southern France.

Discovery

Excavations of R. Cammas and colleagues, 1949 (Baylac et al., 1950).

Material

Fairly complete adult mandible missing anterior teeth, plus some isolated teeth and postcranial fragments.

Dating and Stratigraphic Context

Difficult to assess because of poor stratification, but from the beginning the associated fauna was seen to suggest a warm climate and a relatively early date. Méroc (1963) claimed that the hominid came from low in the section, such as it is. Vallois (1956) dated the human mandible as either Mindel-Riss or Riss-Würm, whereas Kurtén (1962) placed the remains approximately contemporary with Swanscombe and Steinheim. In today's terms, this would place the fossil somewhere in isotope stage 7 or earlier, perhaps 300–200 ka ago. Tavoso et al. (1990) disputed Méroc's geological model but still inclined to a late penultimate glacial or last interglacial date for the episode of fissure filling, in which case the hominid would be a little younger than Kurtén's estimate.

Archaeological Context

"Pre-Mousterian" (Vallois, 1956).

Previous Descriptions and Analyses

Basic description was done by Vallois (1955, 1956), who characterized the mandible as that of a "Pre-Neanderthal," an opinion to which he still adhered a decade later (Billy and Vallois, 1977), although he did note resemblances to the Mauer jaw as well as a strong Neanderthal character complex. Howell (1960) also believed that Montmaurin fit into a lineage leading from Mauer to the Neanderthals.

Morphology

Adult. Mandible with little-worn M1–3 bilaterally.

Corpus relatively shallow. Symphyseal region moderately broad but arcuate across; in profile, slightly retreating, with no subalveolar depression. Symphyseal surface largely featureless and lacking inferior tubercles; bottom of symphyseal region somewhat raised at front. Digastric fossae deep. Bilaterally, large mental foramen under P2. Small retromolar space; M3 barely exposed from side; no preangular sulcus.

Ramus slightly inclined posteriorly. Region of gonial angle not very sloping, forms smooth curve (not an angle). Mylohyoid line and submandibular fossa below it well marked. Gonial regions slightly inwardly inflected and externally smooth. Very large medial pterygoid tubercles high up bilaterally. Coronoid processes higher than very broad condyles. Sigmoid notch crest terminates rather laterally on the R condyle; runs more to midpoint of condyle on the L. Low pillar comes down inside of coronoid process to meet internal alveolar crest (mylohyoid torus not developed). Mandibular foramina large, vertically ovoid, compressed, slightly incomplete inferiorly, and oriented obliquely upward; both lingula broken.

Preserved Ms large, elongate, and somewhat ovoid in outline. Each bears distinct trigonid basin and narrow, well-defined talonid basin; metaconid base extends into center of talonid basin ("deflecting wrinkle"). R and LM1s bear distinct, buccally shifted hypoconulids. Enamel shows some wrinkling in talonid basin of M1; more wrinkling seen in less worn M2, most in unworn M3. In M2 and especially M3, noticeable buccal slope.

REFERENCES

Baylac, P. et al. 1950. Decouvertes recentes dans les grottes de Montmaurin (Haute-Garonne). *L'Anthropologie* 54: 262–271.

Billy, G. and H. Vallois. 1977. La mandibule pré-Rissienne de Montmaurin. *L'Anthropologie* 81: 273–312.

Howell, F. C. 1960. European and northwest African Middle Pleistocene Hominids. *Curr. Anthropol.* 1: 195–232.

Kurtén, B. 1962. The relative ages of the australopithecines of Transvaal and the pithecanthropines of Java. In: G. Kurth (ed), *Evolution und Hominisation*. Stuttgart, Fischer, pp. 74–80.

Méroc, L. 1963. Les éléments de datation de la mandibule humaine de Montmaurin (Haute-Garonne). *Bull. Soc. Geol. Fr.* 7: 508–515.

Tavoso, A. et al. 1990. La Grotte de la Niche à Montmaurin (Haute-Garonne, France). Nouvelles données biostratigraphiques et approche taphonomique. *C. R. Acad. Sci. Paris* 310 (III): 95–100.

Vallois, H. 1955. La mandibule humaine pre-mousterienne de Montmaurin. *C. R. Acad. Sci. Paris* 240: 1577–1579.

Vallois, H. 1956. The pre-Mousterian human mandible from Montmaurin. *Am. J. Phys. Anthropol.* 14: 319–323.

Repository

Laboratoire d'Anthropologie Biologique, Musée de l'Homme, Place du Trocadéro, 75116 Paris, France.

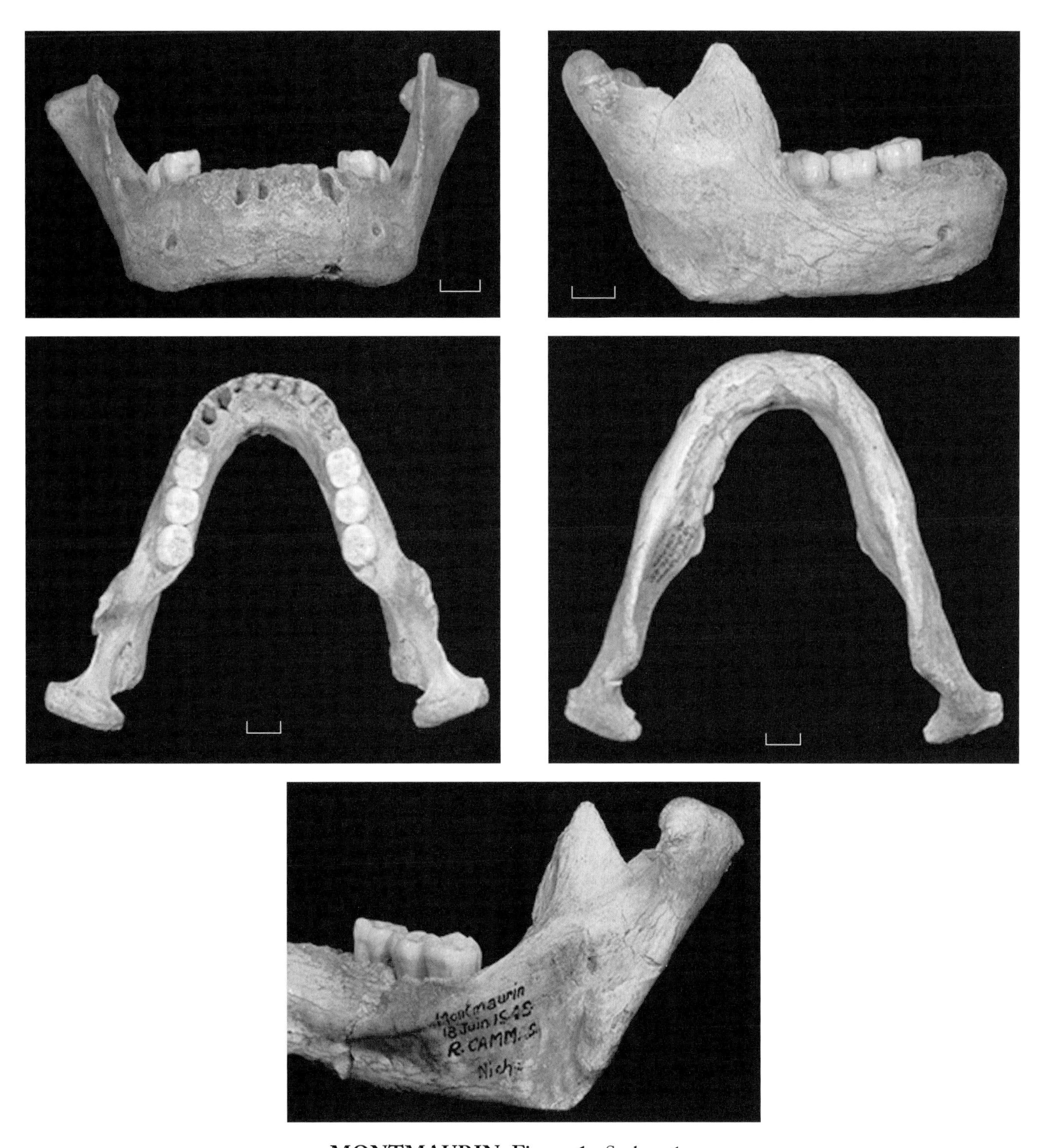

MONTMAURIN Figure 1. Scale = 1 cm.

Ochoz

Location

Svedove Stolu (Schwedentischgrotte), cave site SW of Ochoz village, near Brno, Moravia, Czech Republic.

Discovery

K. Kubasek, 1905 (Ochoz 1); Z. Vanurova and J. Vanura, 1964 (Ochoz 2).

Material

Partial mandibular corpus with LM3–RM2 (Ochoz 1); dental and cranial fragments (Ochoz 2).

Dating and Stratigraphic Context

The stratigraphic position of the original find in the layered cave deposits are unclear: Ochoz 1 came from one of two very similar reddish clay strata, the upper corresponding to the last interstadial and the lower to the last interglacial (J. Jelinek, personal communication). Both the earlier and later excavations produced "early Würm" faunas (Rzehak, 1905; Klima et al., 1962; Vanura, 1965).

Archaeological Context

No stone tools were found in the original excavation (Rzehak, 1905), but subsequent excavations have found "Eastern Mousterian" in both red clays (Klima et al., 1962).

Previous Descriptions and Analyses

Rezhak (1905) described the Ochoz 1 mandible as *Homo primigenius* (i.e., Neanderthal). Bayer (1925) thought it more like modern *Homo sapiens*, but opinion since then has reconverged on the Neanderthal status of all the Ochoz material (e.g., Jelinek, 1962; Klima et al., 1962; Vanura, 1965).

Morphology

Ochoz 1

Adult. Mandibular corpus (AP501/A 17 089) missing inferior part; preserving LM3–RM2; all teeth heavily worn.

Symphysis appears to have been broad, quite flat across front; internally, large depression in midline of symphysis. Anterior tooth roots exposed; all long, stout; slight procumbency produces subalveolar depression (when crowns orthal). Mental foramen lay below P2–M1. Seems to have been retromolar space. Cs had distinct margocristids. P1s had large lingual swellings bearing evidence of crest descending from protoconid. P2 metaconids small, low, smaller than P1s. Molars rounded in outline, subequal in size. Judging by M3, Ms had distinct trigonid basins. RM1–2 have centrally placed centroconid. M3 hypoconulid region expanded. Talonid basins deep; were probably wide. Long M3 roots bifurcate very low (1 cm) below neck.

References

Bayer, J. 1925. Das jungpaläolithischer Alter des Ochozkiefers. *Eiszeit* 2: 35–40.

Jelinek, J. 1962. Der Unterkiefer von Ochoz. Ein Beitrag zu seiner phylogenetischen Stellung. *Anthropos* 13: 261–284.

Klima, B. et al. 1962. Die Erforschung der Höhle Sveduv stul 1953–1955. *Anthropos* 13: 1–297.

Rzehak, A. 1905. Der Unterkiefer von Ochoz. Ein Beitrag zur Kenntnis des altdiluvialen Menschen. *Verh. Naturforsch. Ver. Brünn* 44: 91–114.

Vanura, J. 1965. *Neue Funde von Resten des Neanderthalmenschen aus der Höhle Sveduv stul (Schwedentisch Grotte) in Mährischen Karste.* Brno, Prírodovedecky.

Repository

Moravske Muzej, Zelny Trh 6, Brno, Czech Republic.

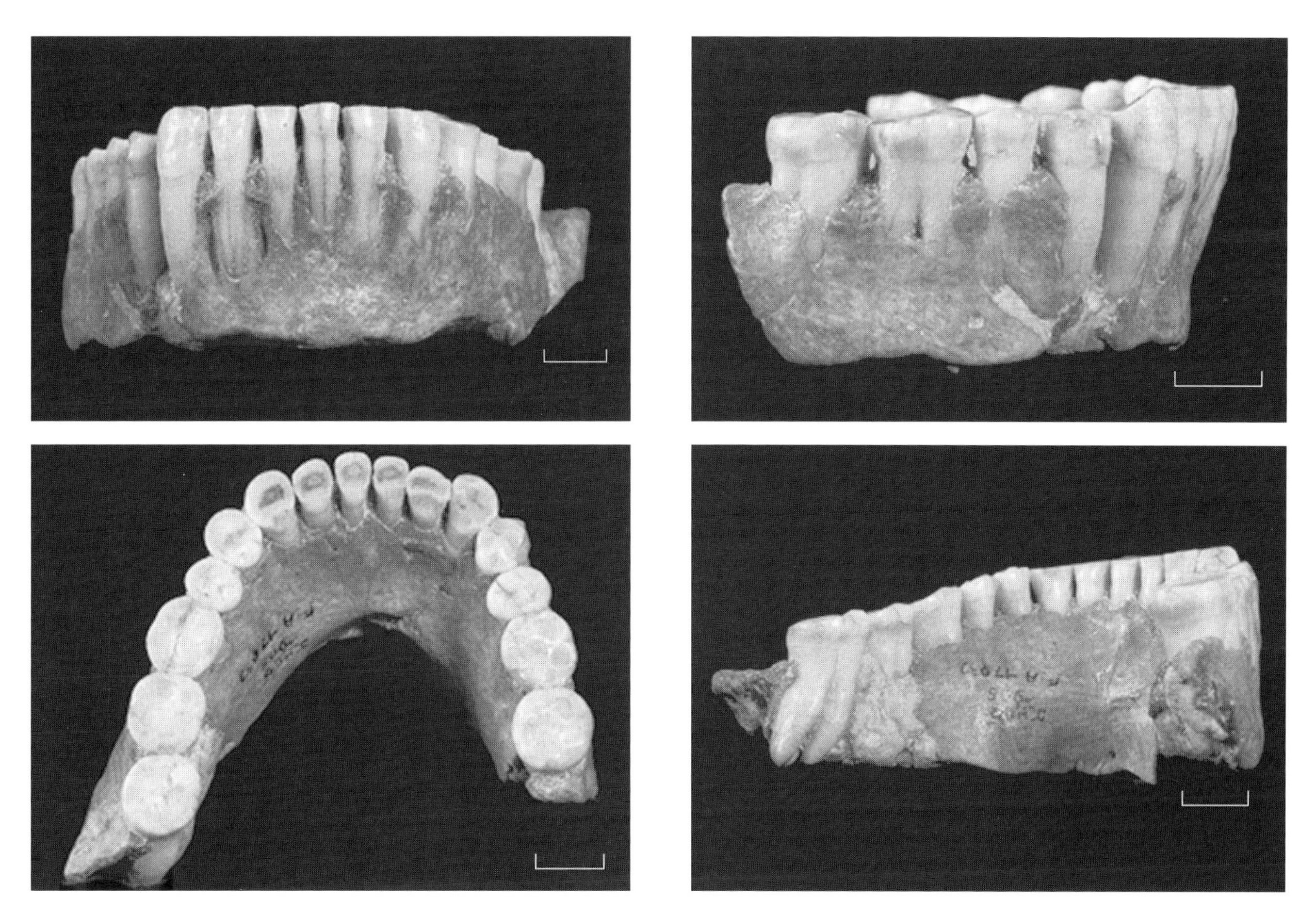

OCHOZ Figure 1. Ochoz 1 (scale = 1 cm).

Pavlov

Location

Occupation site near Pavlov village, at the base of the northern slope of the Pavlovske Hills, southern Moravia, Czech Republic.

Discovery

Excavations of B. Klima, 1954–1957.

Material

Most of the skeleton of adult, probably male, including a partial cranium (Pavlov 1); maxilla (Pavlov 2) and mandible (Pavlov 3), plus isolated teeth.

Dating and Stratigraphic Context

Open-air site in soliflucted loess, which appears to be part of a large complex of sites including those at nearby Dolni Vestonice. Radiocarbon dates on charcoal of around 26 ka (Klima and Kukla, 1963) agree with those from elsewhere in the complex.

Archaeological Context

Pavlovian, otherwise known as Eastern Gravettian (Klima, 1959). Pavlov 1 represents a contracted burial below a mammoth scapula and other bones.

Previous Descriptions and Analyses

The Pavlov remains have been most comprehensively described by Vlcek (e.g., 1961a,b), who found them to be early modern *Homo sapiens* with their closest affinity to Engis 1 and Brno 2. Wolpoff (1996) remarks on the robusticity of Pavlov 1 and considers it an "archaic link" (p. 709). Vlcek (1993) estimated the cranial capacity of Pavlov 1 at 1472 cc.

Morphology

Pavlov 1

Adult. Large, incomplete skull lacking face and cranial base but with associated maxilla and teeth. In profile, skull had been long, with somewhat projecting brows and occipital region.

Frontal rises quite vertically. Sides of cranial vault quite parallel, with greatest width above squamosals. Supraorbital area, including lateral part of supraciliary arch, moderately well developed but clearly bipartite; glabellar "butterfly" bears slight vertical glabellar crease. Frontal sinuses small. Well-marked temporal lines low on cranium. Maxilla with inferior portion of nasal aperture; no double margin (no prenasal fossa); nasal floor seems flat. Nasoalveolar clivus short, quite vertical; palate moderately deep. Maxillary sinus does not intrude into nasal cavity. Zygomas quite gracile; zygomatic arches curve back close to the cranial wall, do not "corner" anteriorly. Mandibular fossa not closed off medially by tubercle. Posterior zygomatic root originates just at anterior margin of auditory meatus. Suprameatal crest low; supramastoid crest quite prominent; no mastoid crest. Vaginal process runs entire length of ectotympanic tube; probably peaked around the stout styloid process. On the L, vaginal process contacts a/p very long, broad-based, and strongly projecting mastoid process. Parietomastoid suture short.

Occipital plane triangular. Lambdoid suture angles up from asterion to peak at lambda; bone bulges below lambdoid suture. Superior nuchal line bow shaped and prominent, especially in midline. External occipital protuberance marked by sulcus above. Cranial sutures segmented. Internally, no Pacchionian depressions. Petrosal bears low arcuate eminence, subarcuate fossa depressed but closed off; no groove for superior petrous sinus.

Teeth highly worn; all roots short.

REFERENCES

Klima, B. 1959. Zur Problematik des Aurignacian und Gravettien in Mittel-Europa. *Archeol. Austriaca* 26: 35–51.

Klima, B. and G. Kukla. 1963. Absolute chronological data of Czechoslovakian Pleistocene. *Conf. Int. Assoc. Quat. Res. Warsaw 1961* 1: 171–174.

Vlcek, E. 1961a. Posustatky mladopleistocenniho cloveka z Pavlova. *Pamatky Archeol.* I52: 46–56.

Vlcek, E. 1961b. Nouvelles trouvailles de l'homme du Pléistocène récent de Pavlov (CSR). *Anthropos* 14: 141–145.

Vlcek, E. 1993. *Fossile menschenfunde von Weimer-Ehringsdorf.* Stuttgart, Konrad Theiss Verlag.

Wolpoff, M. 1996. *Human Evolution.* New York, McGraw-Hill.

Repository

Institute of Archaeology, Dolni Vestonice, Czech Republic.

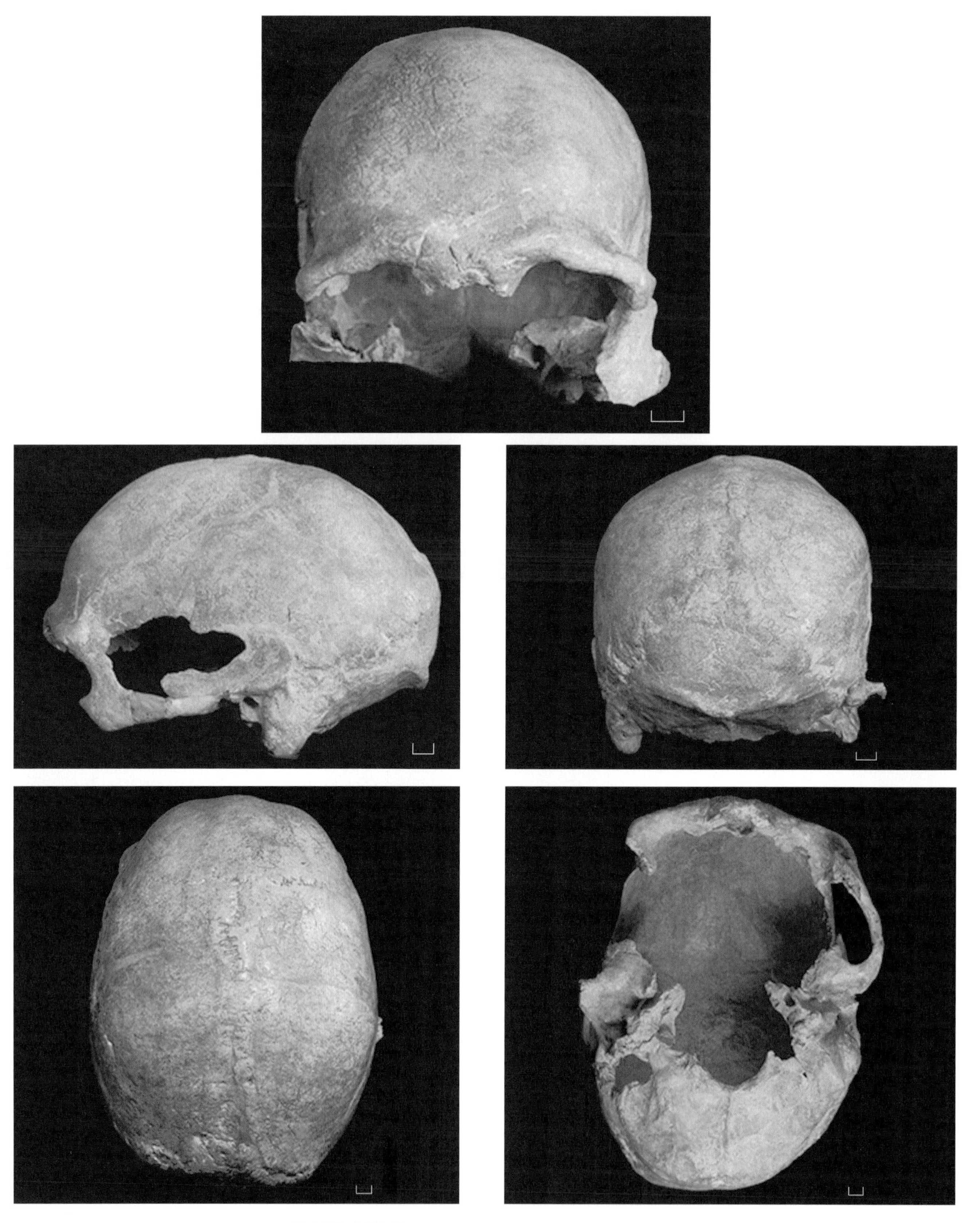

PAVLOV Figure 1. Pavlov 1 (scale = 1 cm).

Pech de l'Azé

Location
One of a series of solution cavities near Carsac, 5 km S of Sarlat, Dordogne, France.

Discovery
Excavations of L. Capitan and D. Peyrony, 1909.

Material
Partial skull of juvenile, ca. 4 years old, missing parts of parietal, occipital, and basicranium, all of sphenoid, internal facial structures, and most of the left half of the mandible.

Dating and Stratigraphic Context
Pech de l'Azé I (PA I) was the subject of numerous investigations from very early on in the history of archaeology, and the original context of the cranium is less clear than it might be. In this part of the Pech de l'Azé site there seem to have been two major sequences of deposits, the lower one sterile. The cranium apparently came from Bed 4 of the upper deposit, which contained 12 layers in 6 m of sediment (see Bordes, 1972). The whole sequence contains lithics belonging to the Mousterian of Acheulean Tradition and was attributed to the Würm II by Bordes (1955). The sequence apparently postdates the deposits in the adjacent PA II site, which contains lithics of earlier Mousterian industries. A calcite sample from Layer 3 of the PA II sequence was U-series (Th/U) dated by Schwarcz and Blackwell (1983) to 103 ka, with a very wide margin of error; at the low end this margin overlaps (just) with ESR dates of 55–73 ka on mammal teeth from this layer obtained by Grün et al. (1991). Some teeth from the top of the PA II sequence came in at around 60 ka, which gives an approximate maximum date for the later PA I sequence. Because there was apparently a significant hiatus between the washing out from PA I of PA II-age sediments and the deposition of the later deposits there, it seems reasonable to estimate an age of ca. 45–55 ka for the hominid.

Archaeological Context
As noted above, the associated lithics are Mousterian of Acheulean Tradition. There is also evidence of hearths and some polished pieces of manganese dioxide that might have been scraped to produce powder or used as crayons (see discussion in Mellars, 1996).

Previous Descriptions and Analyses
There has never been any doubt that the PA I juvenile skull is that of a Neanderthal. This has been made plain in the monograph of Patte (1957) and in numerous subsequent commentaries. Estimated cranial capacity is 1150 cc (Ferembach, 1970).

Morphology
Partial skull of child (MNHN/MH 24.378.1956–6), ca. 4 years old. Missing parts of parietal, occipital, basicranium, all of sphenoid, and internal facial structures, as well as most of L side of mandible. Upper

I1s, dcs, dm1–2s, and M1s, and lower I1–2s and dm1–2s present (although some misplaced).

Skull very long and very broad, especially posteriorly. Vertical frontal curves back sharply into very flat superior cranial surface. No supraorbital tori; glabellar region flat. Frontal (metopic) suture still patent. Very broad interorbital region. Orbits tall, ovoid, oriented slightly sideways, markedly asymmetrical (pathological?). Nasal region largely reconstructed; enough of frontal process preserved on the R to confirm presence of small, vertical, medial projection.

On the basis of what is preserved on both sides, appears that zygomatic arches did not flare out but ran straight back. Mastoid processes small but clearly defined, a/p long, very flat on lateral surfaces. Mandibular fossae very broad and quite deep, with small, laterally placed postglenoid plate well separated from ossifying ectotympanic tube. Tympanic ring may still have been quite horizontally oriented. Carotid foramen lies midway on petrosal, points down. R jugular foramen much smaller than the L, which is quite pocketed; jugular foramen points markedly forward (axes of carotid and jugular foramina cross at broad angle). Peak of vaginal process already developing; lies toward carotid foramen and away from mastoid process. Tiny styloid pit below peak of vaginal process; medial to pit lies stylomastoid foramen, in line with narrow, well-incised mastoid notch.

Occipital condyles very flat; confined essentially to lateral part of occiput, with only tiny contribution from the basiocciput. Foramen magnum missing posteriorly; was evidently large, probably long. Basiocciput very broad, flat, angled up (basicranium already somewhat flexed).

Petrosals very broad laterally. On both sides, region of arcuate eminence domed; more laterally, bone puffed out. Subarcuate fossa closed off, almost completely filled in. No groove for superior petrous sinus.

R half of mandible preserved. Medial pterygoid tubercle present, already pronounced. Broad, quite deep digastric fossa already developed. Symphyseal region very broad across, slightly arced transversely. Symphyseal surface preserves some bone in midline; shows detail only from impression of I1 roots. Below I2 and C roots, close to margin of digastric fossa, is odd horizontal crease in bone that follows curvature of fossa. Although reconstructed externally, preserved inferior border of mandible hints at presence of large inferior marginal tubercle lying below dm1. Viewed from below, it is evident that mandible was not only broad from side to side but corpus was thinnest a/p at symphyseal region and thickest posterior to dm2. Inferior margin of corpus slightly concave in region below dm2 and developing M1. Coronoid process and most of condyle missing. Hint of pillar that had descended down internal face of coronoid process to meet developing mylohyoid torus. Sigmoid notch preserved; appears to have been long and straight (not scooped out). Large mandibular foramen incomplete; is vertically ovoid, horizontally oriented (presumably because individual is a juvenile); probably bore large lingula.

Upper I1s present, shoveled. I2s missing. Both dcs present (L and R switched in reconstruction); are swollen buccally; bear lingual cingula and somewhat excavated lingual surfaces; in outline, very triangular. dm1s have well-developed posterior parts, with prominent postcingula terminating in small metacones. Stout preprotocristae run mesially and up to apices of paracones; very stout, shorter postprotocristae run more directly across to metacones; protocristae encircle fairly distinct, deep trigon basins. Small, stylelike metacones lie at bases of paracones, just distal to postprotocristae; thick precingula run from metacones to apices of hypocones, enclosing narrow, deep basins; protocone somewhat centrally placed. Buccal surfaces of upper Is, dcs, and dm1s quite distended, form sloping surface that creates angle with roots; buccal side of dm2 straighter. dm2s have ledgelike cingula around protocone (more lingually placed than on dm1). Stout preprotocrista runs mesiobuccally around to meet paracone, which is connected to only slightly smaller metacone by compressed, sharp centrocrista; stout postcingulum runs down from metacone to apex of well-developed hypocone, which is joined to protocone by short protocone fold. Trigon and talon basins distinct, with grossly pitted, wrinkled enamel. M1 crowns calcified, very crenulated; have huge, distally expanded, ledgelike hypocones that extend right across crown. Protocones very lingually placed; stout postprotocrista runs directly to metacone, which is only slightly smaller than paracone; very weak preprotocrista arcs mesially and buccally up to paracone. Root formation lacking.

Lower I1s arcuate on sides; preserved RI2 distended distally. Tooth in position of Rdc probably missing LI2; is shoveled. Rdm1 has distinct trigonid and talonid basins; cusps distinct (does not have "figure 8" configuration of cusps). Stout paracristid runs down from base of protoconid to base of metaconid;

cristids do not entirely encircle talonid basin (notch between hypoconid and entoconid). Buccal surface swollen markedly toward neck, most swollen mesiobuccally. dm2 somewhat ovoid; cusps well separated from each other and very distinct (enclose deep, very b/l wide, m/d long talonid basin). Talonid basin slightly pitted, bears distinct centroconid. Stout paracristid runs between protoconid and metaconid; encloses m/d narrow, deep, b/l wide trigonid basin. Hypoconid close to protoconid; smaller, well-defined hypoconulid lies just behind hypoconid, just buccal to midline of tooth; entoconid compressed, with long crests connecting it to metaconid and base of hypoconulid. Buccal side of crown somewhat swollen.

References

Bordes, F. 1955. Les gisements de Pech de l'Azé (Dordogne). I. Le Moustérien de Tradition Acheuléen. *L'Anthropologie* 58: 1–32.

Bordes, F. 1972. *A Tale of Two Caves*. New York, Harper & Row.

Ferembach, D. 1970. Le crâne de l'enfant du Pech-de-l'Azé. *Arch. Inst. Paleontol. Hum.* 33: 13–51.

Grün, R. et al. 1991. ESR chronology of a 100,000-year archaeological sequence at Pech de l'Azé II, France. *Antiquity* 65: 544–551.

Mellars, P. 1996. *The Neanderthal Legacy*. Princeton NJ, Princeton University Press.

Schwarcz, H. and B. Blackwell. 1983. ^{230}Th/^{234}U age of a Mousterian site in France. *Nature* 301: 236–237.

Patte, 1957. *L'enfant néandertalien du Pech de l'Azé*. Paris, Masson.

Repository

Laboratoire d'Anthropologie Biologique, Musée de l'Homme, Place Trocadéro, 75116 Paris, France.

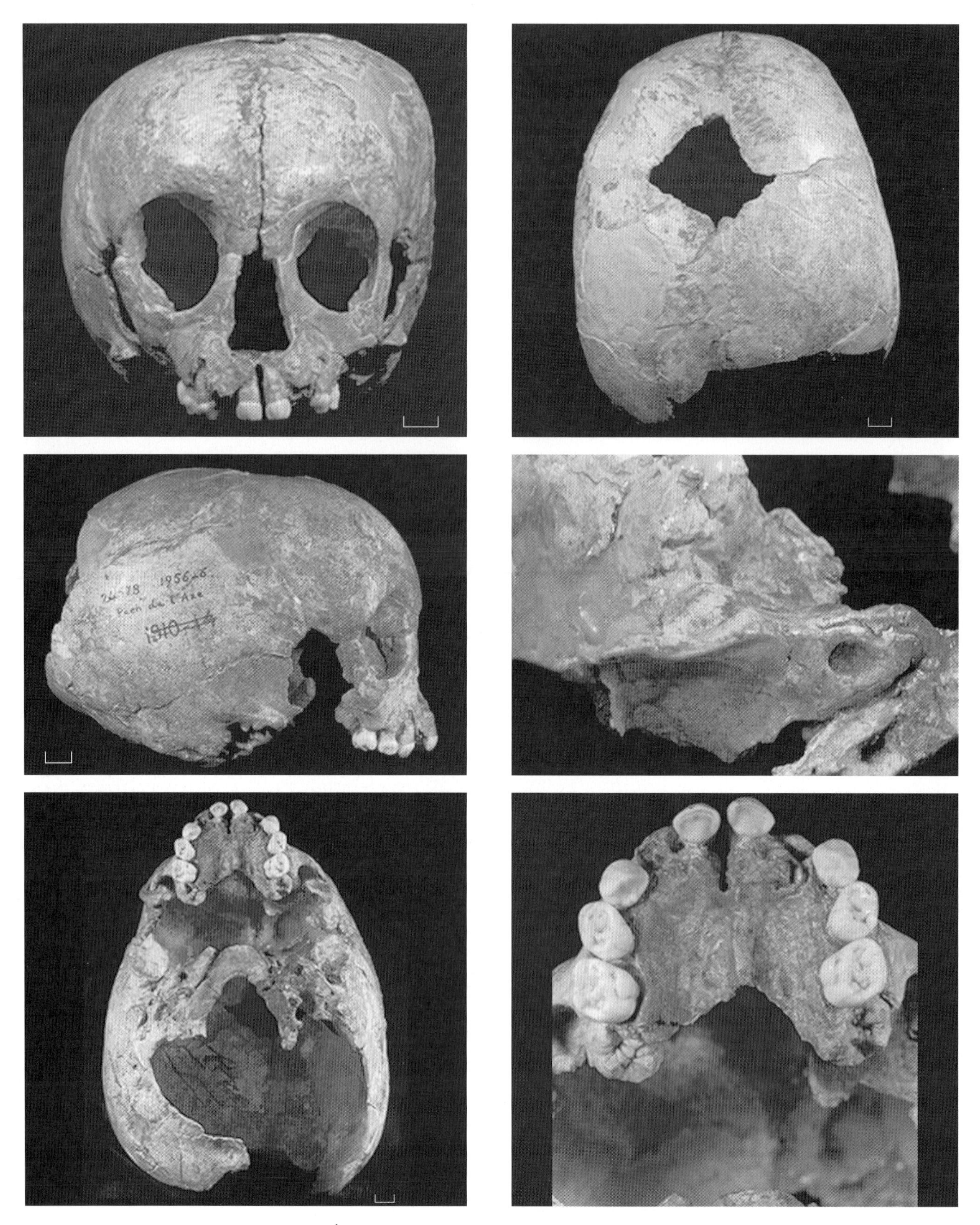

PECH DE L'AZÉ Figure 1. Including close-up of petrosal, (scale = 1 cm).

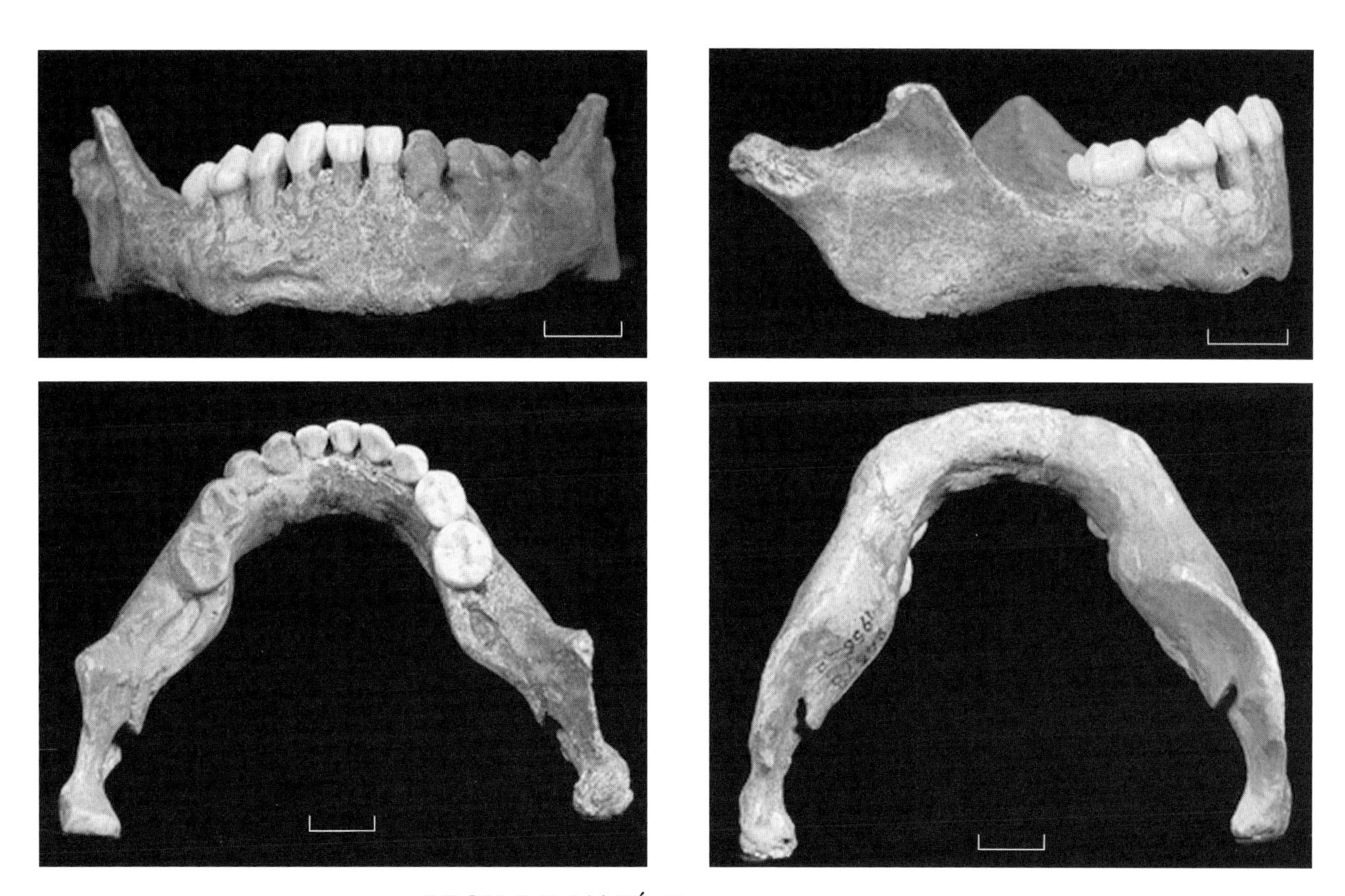

PECH DE L'AZÉ Figure 2. Scale = 1 cm.

Petralona

Location
Cave site near Petralona village, some 37 km SE of Thessaloniki, northern Greece.

Discovery
J. Malkotsis and others, September 1959.

Material
Well-preserved adult cranium, presumed male.

Dating and Stratigraphic Context
Dating of the Petralona cranium, an accidental discovery by a group of local villagers, has been controversial from the beginning. The floor of the cave chamber in which the discovery was made was covered by a deep flowstone layer dated several times by U-series. Dates from deep in the flowstone are all beyond U-series range, i.e., greater than about 300 ka, whereas those for the its topmost layer are well within this date (see review by Grün, 1996). However, it is not clear whether the specimen when found was lying on top of the flowstone or on the cave floor below (although the former seems substantially more probable; see Cook et al., 1982). Thus, although Poulianos (1980) claimed that the rest of the skeleton was found below the flowstone and later disappeared, others have strongly contested this account (e.g., Xirotiris et al., 1982). The fauna collected in the cave has also proven contentious as regards its age. Cook et al. (1982) pointed not only to intrinsic problems with earlier estimates but also to the fact that none of the faunal elements recovered from the cave could be clearly associated with the hominid. ESR dates on encrustations on the specimen and on the top levels of the flowstone have also diverged strongly and have been variously interpreted to yield ages between 200 and 700 ka (e.g., Hennig et al., 1981; Poulianos, 1984). Grün (1996) has recently made a heroic attempt to sort out the mess by reanalyzing earlier ESR dates and reexamining the U-series dates. He concludes that the balance of the evidence points to an age well within U-series range, and plausibly within about 150–250 ka.

Archaeological Context
Some stone and bone artifacts were apparently found in the cave, but they are undiagnostic in themselves and their association with the cranium is dubious.

Previous Descriptions and Analyses
The affinities of the cranium have been almost as contentious as its dating. Described in many early studies as a Neanderthal of some kind (e.g., Kanellis and Savva, 1964; Mann and Trinkaus, 1973), it was later compared more closely to *Homo erectus* (e.g., Hemmer, 1972), specimens such as Kabwe (e.g., Howells, 1967), or both (e.g., Stringer, 1974), or it was given in its own genus, species, and subspecies (e.g., Poulianos, 1980). Most recent assessments follow Hublin (1985) and Stringer (1985) in placing Petralona squarely among the Arago/Kabwe group, now increasingly placed in the species *Homo heidelbergensis.* Es-

timated cranial capacity is 1220 cc (Olivier and Tissier, 1975).

MORPHOLOGY

Adult. Externally almost complete cranium, lacking R zygomatic arch and various details of mastoid region; internally lacks much of bone around nasal cavity; retains upper LC–M3, RP2–M3 (M1 damaged), and root of RC.

Skull relatively long, not extremely low, fairly broad; greatest width across unusual squamosal bulges at low parietal notches; asymmetrical in rear view. Face swollen but not greatly protruding; maxillae filled out lateral to nasal margin by inflation of maxillary sinus. Supraorbital tori thick but not excessively protruding. Forehead recedes sharply from tori with no posttoral sulcus. Glabella protrudes markedly beyond nasion; slight midline crease. On either side of glabella, plane of torus is broad and flattish; it faces forward and down adjacent to glabella, then twists backward as it proceeds laterally, and then comes to face forward and up on reaching zygomaticofrontal suture. Plane of torus thickest at midpoint; viewed from above, it recedes from glabella at an angle (not as an arc). Behind orbit, barely any temporal muscle scars; merely sharp edge to postorbital bar. Temporal lines weak (surprisingly so for size of face and cranial vault); extend far posteriorly beyond very low-situated parietal notch. Anterior lacrimal crest (hence also medial orbital margin) set far back; lacrimal fossae defective on both sides but were probably short and ovoid. Superior orbital margin rises from top of fossa (set back within orbit), sweeps forward as it proceeds laterally to orbital midpoint, then retreats slightly again.

Interorbital region, nasal bones, and frontal processes of maxilla extremely broad. Nasal bones flare laterally as approach nasal aperture. Region above nasal aperture broad and swollen out by capacious maxillary sinuses that extend to within 1 cm of maxillofrontal suture. Originally long nasal bones (now broken anteriorly) were broad but not inflated (but carry the swollen contour across top of nose while sloping gently down in profile); flex only slightly forward about 1 cm below nasion. Upper margin of wide, but not tall s/i, nasal aperture would have projected beyond inferior margin. Floor of nasal cavity somewhat depressed; slopes gently down as it retreats from inferior nasal margin. Superior part of lateral nasal wall somewhat swelled out by sinus that is separate outpocketing of the ethmoid, most superior extension of maxillary sinus up the frontonasal process, or sinus of independent origin (from CT scan, could be the latter). To judge from straight inferior margin of damaged maxillary sinus wall, this sinus did not intrude into nasal cavity at least inferiorly. Low, distinct, beaded conchal crest runs horizontally back from just within lateral crest of nasal aperture. Lateral nasal margin crest sharp superiorly; inferiorly becomes more rounded and leads into weakly formed double inferior margin with weak prenasal fossa. Superior part of fossa runs to large anterior nasal spine (which projects more forward than up); lower part of fossa runs in front of internal margin to base of anterior nasal spine. Depression between the two margins weakly defined. Nasoalveolar clivus broad and flattish between Cs; slopes slightly forward. Roots of anterior teeth short and displaced far below nasal margin.

No maxillary incisura. Zygomas retreat gently; no sign of malar tubercle or angle. Anterior root of zygomatic arch broad, low on face; inferior border of arch sweeps up and back, but not sharply. Infraorbital foramen preserved on L large; lies well below orbital margin; faces down. One zygomaticofacial foramen lies below orbit. Zygomatic arch flares somewhat. Anterior squamosal portion of temporal presents distinct margin onto alisphenoid. Squamosal very short, tall, triangular. Temporal fossa very small, with generally gracile structures within it squeezed far forward. Bone smooth along very short and small temporal and infratemporal fossae. Low ridge posteriorly arcs up from anterior edge of posterior root of zygomatic arch. Inferior margin of posterior root of zygomatic arch narrow, lacking pronounced muscle scars; masseter markings cease at zygomaticomaxillary suture.

Mandibular fossa very broad from side to side; is not closed off by well-defined medial articular tubercle. Articular eminence very low. Posterior root of zygomatic arch originates over auditory meatus, flows directly into laterally projecting supramastoid region (corresponding to large, unusual sinus within temporal) that stands out almost 1 cm laterally beyond parietal notch, and bulges out over matrix-obscured bases of mastoid processes (processes themselves not preserved). Base of mastoid process displaced downward by supramastoid bulge. Possible supramastoid crest lies low and forward on the parietal, well above mastoid process. Parietomastoid suture very long, horizontal, also very low (at level of auditory meatus).

Biasterionic breadth extremely wide; nuchal plane very short from top to bottom. Lambdoid suture rises only gently from asterion until arcing up toward lambda, where it peaks at distinct angle (enclosing large ossicle?). Very slight bulge lies below lambda. Profile arc of occipital plane smooth down to superior nuchal line, where slightly interrupted by mild transverse depression. External occipital protuberance moderate, blunt, low; lies below and anterior to superior nuchal line, which forms continuous arc between the two asterions, and marks an almost right angle between nuchal and occipital planes. No occipital torus in any strict sense (because no depression below superior nuchal line). No suprainiac depression, but small sulcus lies above inion. Paired large cerebellar bulges occur bilaterally between external occipital protuberance and foramen magnum. Little muscle marking on nuchal plane except for Waldeyer's crest situated far posterior and medial to mastoid region. Occipitomastoid region not preserved; there probably was no occipitomastoid crest. On the L, impossible to say whether any paramastoid crest was present; on the R, hint of crest. Matrix obscures area medial to auditory meatus; seems that vaginal process peaked around very medially placed styloid process.

Palate very broad and U shaped, although tooth rows somewhat divergent posteriorly (some distortion on L side). Palate also moderately deep, with shallow anterior slope and steeper profile at sides. Incisive foramen single, small, very anteriorly placed; no palatal torus. Pterygoid wings join inferiorly in broad arc, do not taper together, but instead are more or less parallel.

Only coronal suture distinctly segmented; sagittal and lambdoid sutures more uniformly interdigitated and relatively deeply denticulated.

Internally on the R, petrosal large with domed arcuate eminence (L counterpart obscured by matrix). May have been an excavated superior petrous sinus (but matrix in this area). Subarcuate fossa closed over, depressed. Superior surface of petrosal broad, but not markedly. Foramen magnum round, not excessively large; its anterior margin lies posterior to plane of mandibular fossa. Basiocciput broad, short; slopes upward toward sphenooccipital synchondrosis, reflecting considerable basicranial flexion. Occipital condyles missing, but lay quite far forward.

Anterior teeth would probably have been rather large compared with molars. Cs stout rooted, with creases delineating margocristids on either side of a broad, low lingual pillar. All cheek teeth have rounded lingual outlines. P1 larger than P2. P1 bears large anterior and very large posterior foveae, but protocone and paracone not very asymmetrically offset. P2 narrower m/d than P1 and less symmetrical (although similar in having largish anterior and large posterior foveae). Molars decrease in size from M1 to M3. Metacone low compared to paracone in M1; former cusp becomes lower and less distinct from latter cusp in M2–3. M1 has fair-sized hypocone that does not swell out distolingual corner of crown; well-defined postcingulum encloses distinct talon basin. M1 protocone large; encroaches on trigon basin. On M3, essentially no hypocone; postcingular corner beaded; large protocone bulges into trigon basin. Molar roots large and long compared with those of anterior teeth. Roots of M3 visible on the R; buccal roots not fully demarcated, with buccolingual bifurcation high up. On M1–2, as well, root bifurcation appears to be well above neck.

References

Cook, J. et al. 1982. A review of the chronology of the European Middle Pleistocene hominid record. *Yrbk Phys. Anthropol.* 25: 19–65.

Grün, R. 1996. A re-analysis of electron spin resonance dating results associated with the Petralona hominid. *J. Hum. Evol.* 30: 227–241.

Hemmer, H. 1972. Notes sur la position phylétique de l'homme de Petralona. *L'Anthropologie* 76: 155–162.

Hennig, G. et al. 1981. ESR-dating of the fossil hominid cranium from Petralona cave, Greece. *Nature* 292: 533–536.

Howells, W. 1967. *Mankind in the Making,* new ed. New York, Penguin.

Hublin, J.-J. 1985. Human fossils from the North African middle Pleistocene and the origin of *Homo sapiens.* In: E. Delson (ed), *Ancestors: The Hard Evidence.* New York, Alan R. Liss, pp. 283–288.

Kannellis, A. and A. Savva. 1964. Craniometric study of the *Homo neanderthalensis* of Petralona. *Aristot. Panep. Thess.* 9: 65–92.

Mann, A. and E. Trinkaus. 1973. Neandertal and Neandertal-like fossils from the upper Pleistocene. *Yrbk. Phys. Anthropol.* 17: 169–193.

Olivier G. and H. Tissier. 1975. Determination of cranial capacity in fossil men. *Am. J. Phys. Anthropol.* 43: 353–362.

Poulianos, A. 1980. The postcranial skeleton of the *Archanthropus europaeus petraloniensis. Anthropos (Athens)* 7: 13–29.

Poulianos, A. 1984. Once more on the age and stratigraphy on the Petralonian man. *J. Hum. Evol.* 13: 465–467.

Stringer, C. 1974. A multivariate study of the Petralona skull. *J. Hum. Evol.* 3: 397–404.

Stringer, C. 1985. Middle Pleistocene variability and the origin of late Pleistocene humans. In: E. Delson (ed), *Ancestors: The Hard Evidence.* New York, Alan R. Liss, pp 289–295.

Xirotiris, N. et al. 1982. Die phylogenetische Stellung des Petralona Schädel auf Grund computertomographischer Analysen und der absoluten Datierung mit der ESR-methode. *Humanbiol. Budapestensis* 9: 89–94.

Repository

Department of Geology, Aristotle University of Thessaloniki, Thessaloniki, Greece.

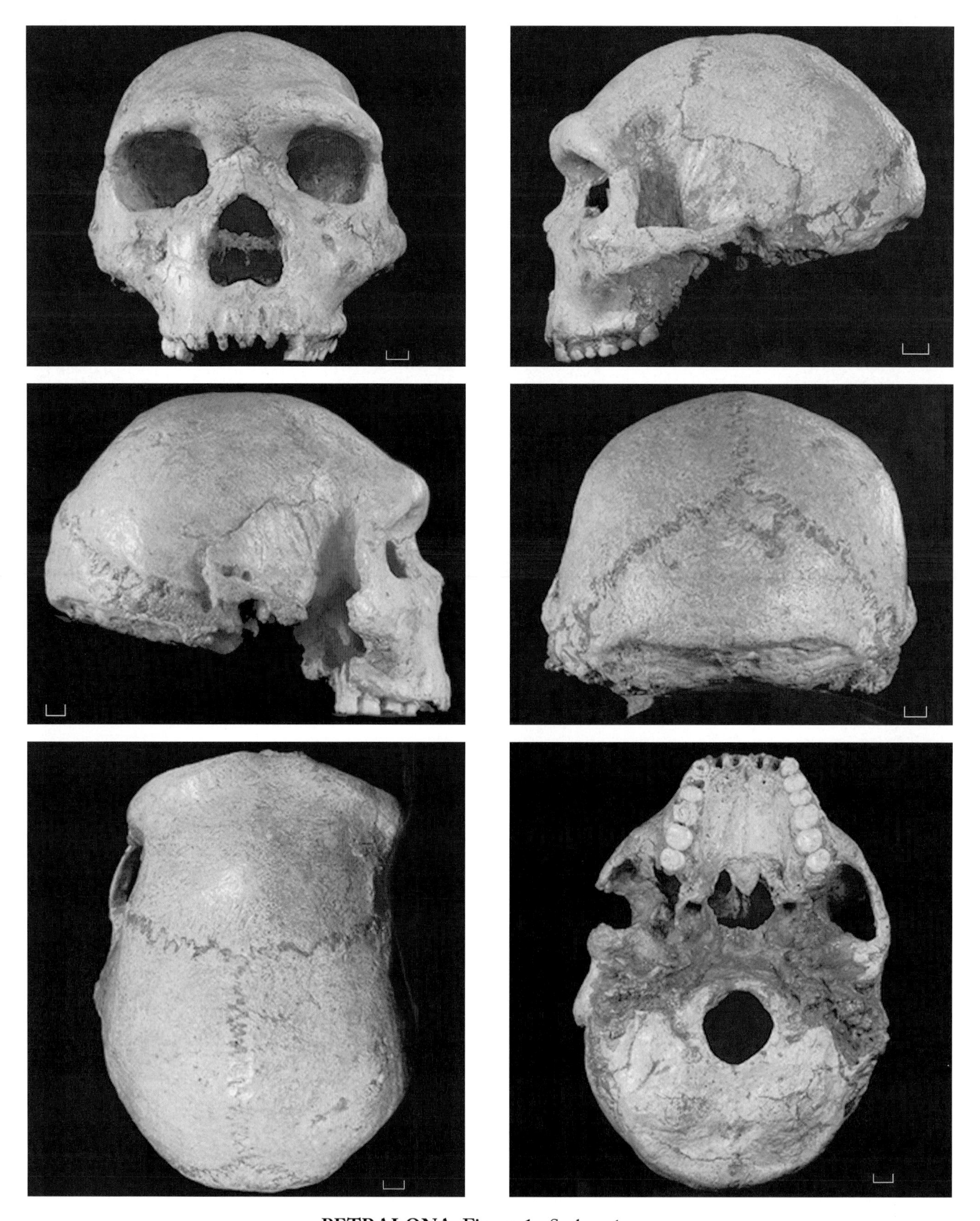

PETRALONA Figure 1. Scale = 1 cm.

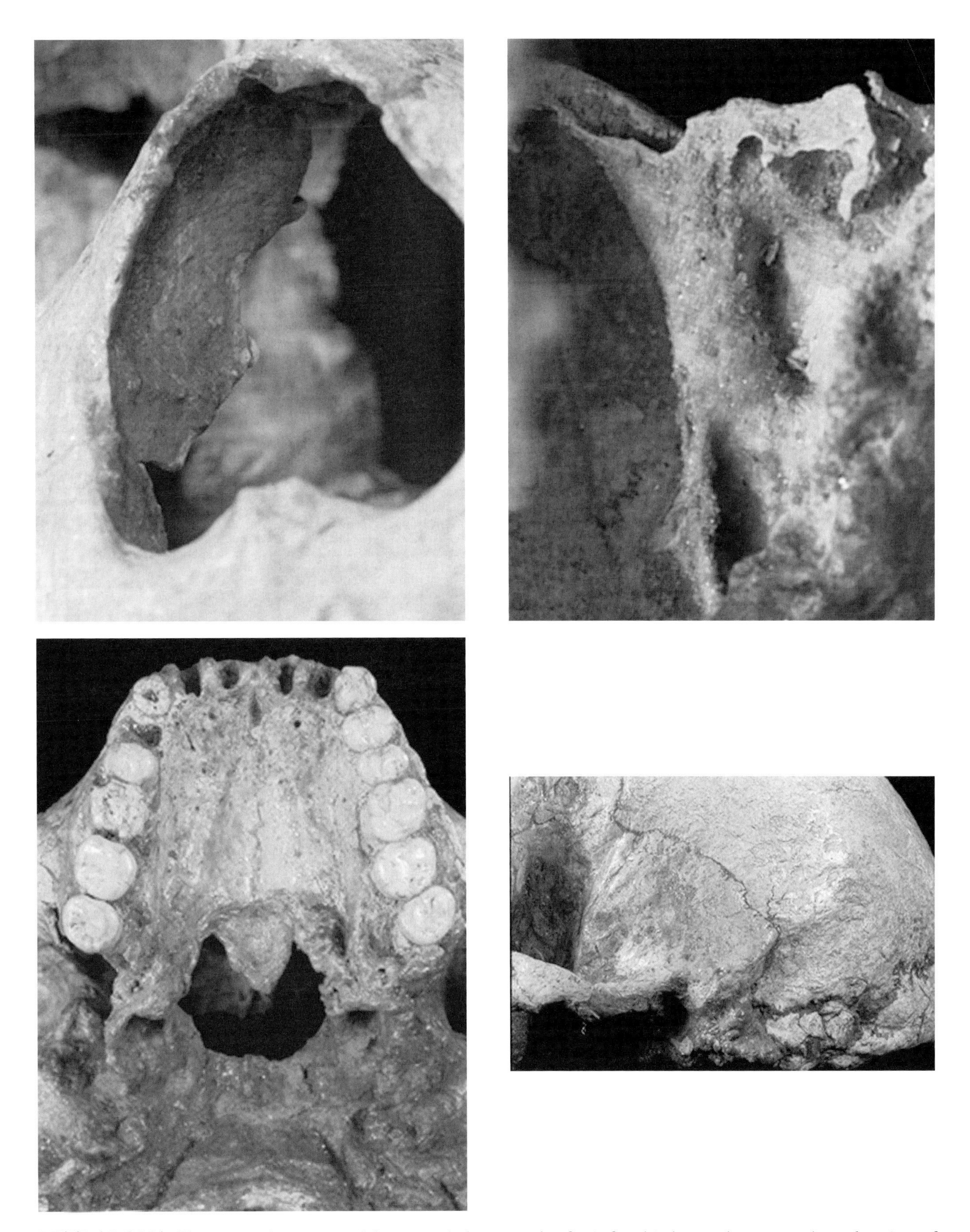

PETRALONA Figure 2. Close-ups of faint conchal crest, tube for infraorbital nerve/artery, teeth, and region of parietal notch, (not to scale).

Predmostí

Location

Huge open-air site or complex of sites in and around Chromecka's brickyard, 1 km N of Prerov, northeastern Moravia, Czech Republic.

Discovery

Various discoverers, April 1884–July 1895 (Predmostí 1–24, 28, and 29); K. Absolon, August 1928 (Predmostí 27).

Material

Remains of 29 individuals in greatly varying degrees of completeness, 20 of them (8 adults and 12 juveniles and infants) associated in an apparent mass grave. Collection totally destroyed in WWII; casts survive of two adult skulls, Predmostí 3 and 4.

Dating and Stratigraphic Context

Predmostí was dug in an era of primitive excavation techniques, and it is unclear whether the burials came from a single site or from multiple sites or if the site(s) were multilayered. Allsworth-Jones (1986) has provided a composite stratigraphy, and radiocarbon dates indicate occupation around 26 ka (Allsworth-Jones, 1986; Vlcek, 1991; Jelinek, 1991).

Archaeological Context

Abundant lithics from Predmostí are all Upper Paleolithic and have been assigned to the Aurignacian and Pavlovian (Eastern Gravettian) industries.

Previous Descriptions and Analyses

The human remains were described piecemeal but monographed by Matiegka (1934; see references therein). Although they were recognized from early on to be of general "Cro-Magnon type," they did manage to acquire a couple of rather exotic names, the most fanciful of them being *Notanthropus eurafricanus archaius* Sergi 1911. Estimated cranial capacity is 1580 cc for Predmostí 3, 1250 cc for Predmostí 4, 1555 cc for Predmostí 9, and 1452 cc for Predmostí 10 (Holloway, 2000).

Morphology

Because these skulls are indisputably those of *Homo sapiens*, their peculiarities and variations are described.

Predmostí 3

Large, quite massive skull. Although braincase tall, maximum width falls quite low, at supramastoid crests.

Supraorbital tori moderately prominent but quite thin, with glabellar "butterfly" exhibiting small vertical crease in midline (= not totally confluent across glabella). Inferior orbital margin protrudes slightly. Canine fossa broad. Nasal bones slope forward, down. Maxillary frontal process not enlarged a/p. Quite thin lateral crest bony of nasal aperture fades out inferiorly; sharp spinal crest runs out from broad-based and projecting anterior nasal spine to define inferior margin of aperture. Horizontal conchal crest just visible within nasal cavity on the R. Alveolar prognathism slight. Nasoalveolar clivus short, fairly vertical beneath anterior nasal spine. Zygoma arced back close to cranial wall. Zygomatic arch very long; its posterior origin lies above external auditory meatus. Temporal and infratemporal fossae not demarcated. Upper margin of suprameatal crest not confluent with large supramastoid crest. Parietomastoid suture short. Mastoid process extremely broad based, thick and stubby even at downwardly pointing tip. Mastoid notch narrow, lacks digastric fossa behind. May have been an occipitomastoid crest, with Waldeyer's crest fairly close to it. Styloid process was huge, with vaginal process peaking around it and running entire length of ectotympanic tube to contact mastoid process. Lambdoid suture angles up from asterion to peak at lambda, creating shallow, broad-based occipital triangle. Bone above lambda damaged; below this point, occipital bulges only slightly. Broad, stout, Y-shaped external occipital protuberance delineated above by broad, crescentic, moderately deep excavation. Nuchal markings otherwise quite minor.

Mandible has distinct mental trigon and retromolar space. Digastric fossae small, shallow. Twin genial tubercles pronounced. Mylohyoid line well marked, with deep submandibular fossa below. Gonial region reflected outward at angle, but inward posteriorly; bears distinct masseter markings externally; internally, low-lying pterygoid rugosities discernible.

Teeth heavily worn.

Predmostí 4

Skull relatively tall, long, quite gracile.

Frontal rises high, with well-marked eminences. Glabellar "butterfly" only slightly protruding with almost no development lateral to it. Nasal bones slope downward. Maxillary frontal processes do not protrude anteriorly. Nasal aperture of modest size, with sharp lateral crest margins. Nasoalveolar clivus short but quite vertical. Zygomas gracile, with backwardly curving arches that originate posteriorly over auditory meatus. No suprameatal crest; only small supramastoid crest. Preserved R mastoid crest broad based, stubby; process projects moderately downward. Parietomastoid suture short. Lambdoid suture angles gently up from asterion to peak at lambda, yielding broadly triangular occipital plane. Occipital bulges moderately below lambda; bulge marked below in midline by modest external occipital protuberance. Lateral to protuberance run fairly faint superior nuchal lines; rest of nuchal plane modestly rugose.

Mandible (cast K319) identified as Predmostí 4 but does not appear matched with cranium; is very long and gracile, with retromolar space and large mental foramen below P2. Appears to have borne mental trigon, with its bulge lying low on symphysis. Masseter scars lie low down on ramus externally; internal aspect of ramus seems to have been sculpted. Condyle and coronoid process widely spaced. Teeth highly worn; upper and lower anterior teeth slender rooted.

References

Allsworth-Jones, 1986. *The Szeletian and the Transition from Middle to Upper Palaeolithic in Central Europe.* Oxford, Clarendon Press.

Holloway, R. L. 2000. Brain. In: E. Delson et al. (eds), *Encyclopedia of Human Evolution and Prehistory.* New York, Garland Publishing, pp. 141–149.

Jelinek, J. 1991. Découvertes d'ossements de la population gravettienne de Moravie. *L'Anthropologie* 95: 137–154.

Matiegka, J. 1934. *Homo predmostensis,* Vols. 1 and 2. Prague, Anthropologica.

Vlcek, E. 1991. L'homme fossile en Europe Centrale. *L'Anthropologie* 95: 409–472.

Repository

Lost in the 1945 fire at Mikulov Castle. Casts of Predmostí 3 and 4 are in the Anthropos Institute, Moravian Museum, Brno, Czech Republic.

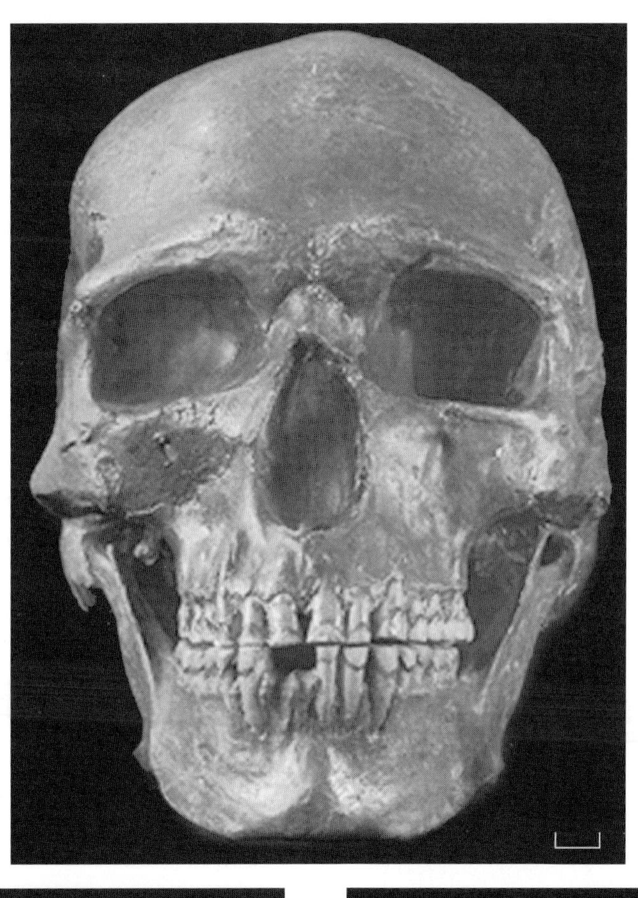

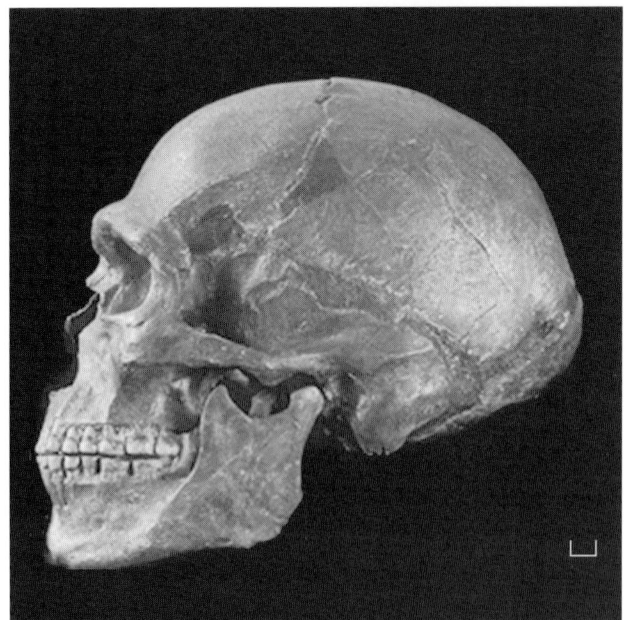

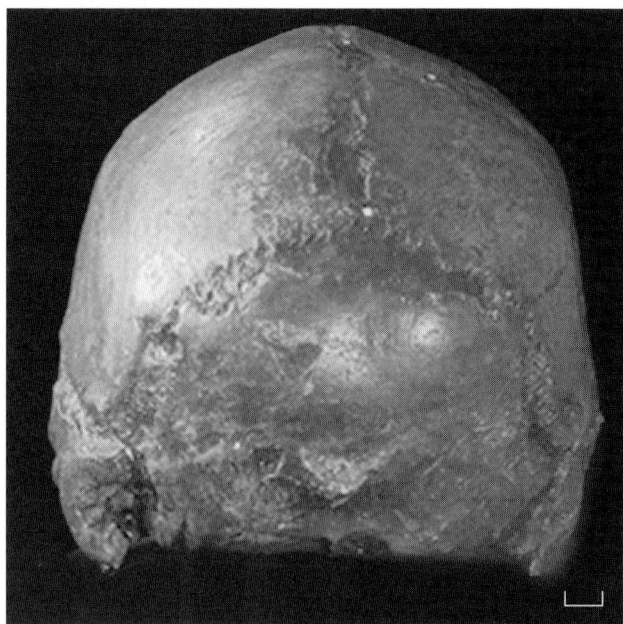

PREDMOSTÍ Figure 1. Cast of Predmostí 3 (scale = 1 cm).

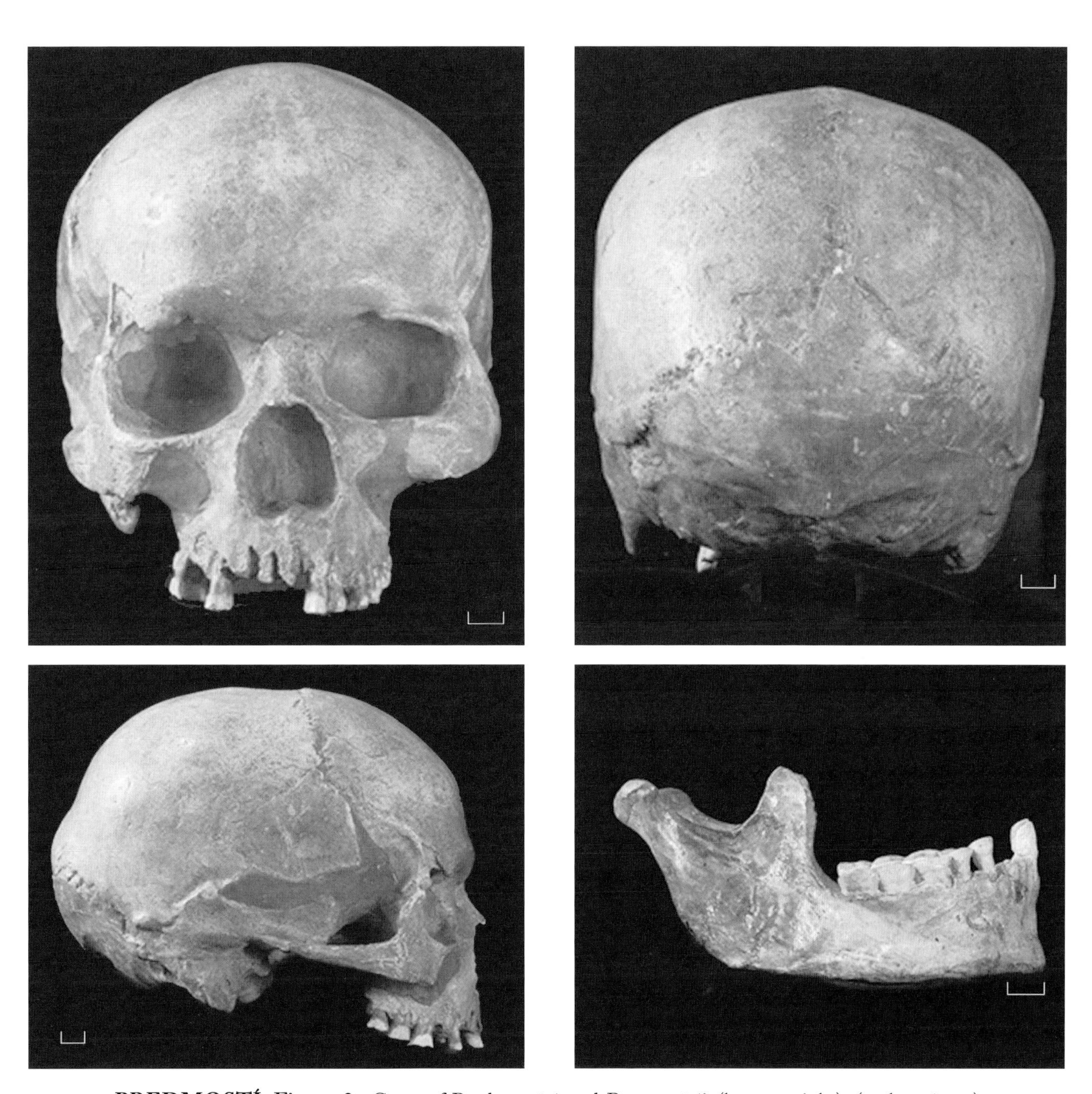

PREDMOSTÍ Figure 2. Casts of Predmostí 4 and Premostí 4? (bottom right). (scale = 1 cm)

Régourdou

Location

Collapsed limestone cave about 2 km N of Montignac, Dordogne, France; just up the hill from Lascaux.

Discovery

R. Constant, September 1957.

Material

Well-preserved mandible and partial skeleton of young adult (Régourdou 1); some pedal elements (Régourdou 2).

Dating and Stratigraphic Context

Up to the time of discovery of the skeleton the Régourdou site had been excavated in a rather haphazard fashion. Later work by E. Bonifay revealed 10 stratigraphic levels, of which the hominid was discovered in level 4, along with a scattering of lithics and charred animal bones. Levels 3–8 contain a mammal assemblage indicating temperate conditions. Level 4 was correlated by Bonifay (1964) with the early part of the last glacial (late Würm I or beginning of the Würm I/II interstadial: oxygen isotope stage 4).

Archaeological Context

Lithics in levels 3–8 are sparse but Mousterian in character (Bonifay, 1964); in the overlying level 2 they are sufficiently abundant to be allocated to the Quina variant of the Charentian Mousterian (Bonifay and Vandermeersch, 1962). Régourdou is publicized by its proprietor as the site of a "bear cult" because of the presence of cave bear bones in the deposits and the discovery of the skeleton below a pile of limestone slabs. There is no substantive evidence for this notion.

Previous Descriptions and Analyses

The mandible was described by Piveteau (e.g., 1963–65) and the postcrania by various authors, most recently Vandermeersch and Trinkaus (1995). The remains are unanimously regarded as Neanderthal.

Morphology

Fairly complete mandible, missing most of L ramus, R gonial region, and part of R condyle. Two chips out of symphyseal region. All teeth present. In general, quite light boned.

Corpora relatively tall s/i, decreasing in height slightly toward ramus. Symphyseal region quite broad from side to side and very straight across, with two distinct corners on inferior margin below M1–2; behind, corpora angle back quite strongly, diverging gently. Buccally, bone around anterior teeth overhangs symphyseal region below. Bone preserved around roots of LI1–C, well depressed around roots. On the L, inferior to incisor root tips, this depression becomes confluent into continuous fossa, which was presumably mirrored on missing L side. Below anterior teeth, plane of symphyseal region quite flat, with central area thrown into relief by adjacent shallow depression(s). At inferior margin, corners described above swell modestly outward around cornered region where corpora begin to diverge. On the L large single mental foramen and on the R three small foramina lie under M1. Very well excavated, a/p long, m/l moderately wide digastric fossae oriented downward and somewhat posteriorly. In midpoint where they converge, fossae so scalloped that distinct ledge of bone is formed. Very long postincisal plane runs obliquely down almost to inferior border, where it curves forward towards digastric fossae. In midline, this plane interrupted just above inferior border by three vertical rugosities in "genial" region.

Inferior margin of symphyseal region elevated slightly between inferior marginal tubercles. In inferior view, symphyseal region essentially uniformly thick from front to back; corpora continue to be of about same thickness back to region of M3s. Bone does not thin significantly toward inferior margin below mylohyoid line. Mylohyoid lines fairly well developed, very long; slope down from well behind M3s almost to digastric fossae in front; seem to take origin from pillar, as seen on the R, that comes down from internal tip of coronoid process and curves anteriorly, bifurcating into confluence with mylohyoid line and with moderately developed internal alveolar crest. Mandibular foramen preserved entirely on the R; is compressed, points more posteriorly than upward; is overhung by narrow shelf of bone. Inferior border of mandibular foramen creased; becomes confluent with long anteroinferiorly directed mylohyoid groove (also visible on the L). On both sides, anterior root of ramus emerges well behind M3; as seen on the R, shallow preangular sulcus adds to length of huge retromolar space. Coronoid process very long a/p, bluntly tipped (as preserved on the R); rises significantly above level of apparently very m/l wide, medioposteriorly angled, slightly coronally arcuate condyle. Sigmoid notch consists essentially of posterior margin of coronoid process, which descends almost directly to base of condyle (thus deepest point of crest well posterior to midpoint). Sigmoid notch crest terminates just medial to what appears to be midline of condyle. Gonial region on the R broken off; preserved internally, low series of rugosities in general area of medial pterygoid tubercle.

Teeth relatively small in proportion to size of mandible; not disproportional relative to each other. I1s very slender, somewhat worn occlusally. Equally worn I2s only a little longer m/d and a bit more concave on their lingual surfaces. Cs relatively slender, with thick posterior margocristids delineated by vertical grooves on their medial sides. Roots of all anterior teeth long, gently outwardly curved (to slightly overhang symphyseal region below); crowns angled back somewhat relative to roots. P1s longer m/d on buccal side than P2s. P1s bear moderately developed lingual swellings; as seen on the R, swellings connected to internal face of worn protoconids by thick crest. On either side of this crest, small but deep fovea; each fovea bounded on its outside by thick crest running from buccal corner of tooth that becomes confluent with lingual swelling. From I1s through P1s, buccal slope and swelling of crowns increasingly pronounced. P2s buccally more straight sided; distended in distolingual corners (thus posterior foveae wider b/l than very small, pitlike anterior foveae). Protoconid and large metaconid opposite it very mesially placed. B/l wide posterior crest around posterior fovea lies lower relative to protoconid height than crest around anterior fovea.

Ms somewhat squatly ovoid; not very long m/d or narrow b/l. On M2–3, and presumably on somewhat worn M1, buccal cusps quite centrally placed compared with quite peripherally placed lingual cusps. Moderately thick paracristid encloses tiny trigonid basin on all Ms. As seen on M2–3, moderately deep, quite m/d long talonid basin truncated buccally by buccal cusps. On all Ms, enamel apparently was only modestly crenulated. All Ms with large hypoconulid; this cusp quite buccally placed on M1, more centrally placed on M2, and central on M3. As seen on M2–3 and presumably also was on M1, sides of tooth quite buccally swollen. The four cusps all meet at lingually displaced midline of talonid basin.

References

Bonifay, E. 1964. La grotte de Régourdou (Montignac, Dordogne). Stratigraphie et industrie lithique moustérienne. *L'Anthropologie* 68: 49–64.

Bonifay, E. and B. Vandermeersch. 1962. Dépots rituels d'ossements d'ours dans le gisement moustérien de Régourdou (Montignac, Dordogne). *C. R. Acad. Sci. Paris* 255D: 1635–1636.

Piveteau, J. 1963–1965. La grotte de Régourdou (Dordogne), paléontologie humaine. *Ann. Paléontol. (Vert.)* 49: 285–305; 50: 155–194; 52: 163–194.

Vandermeersch, B. and E. Trinkaus. 1995. The postcranial remains of the Régourdou 1 Neandertal: the shoulder and arm remains. *J. Hum. Evol.* 28: 439–476.

Repository

Musée du Périgord, Cours Tourny, 24000 Périgueux, France.

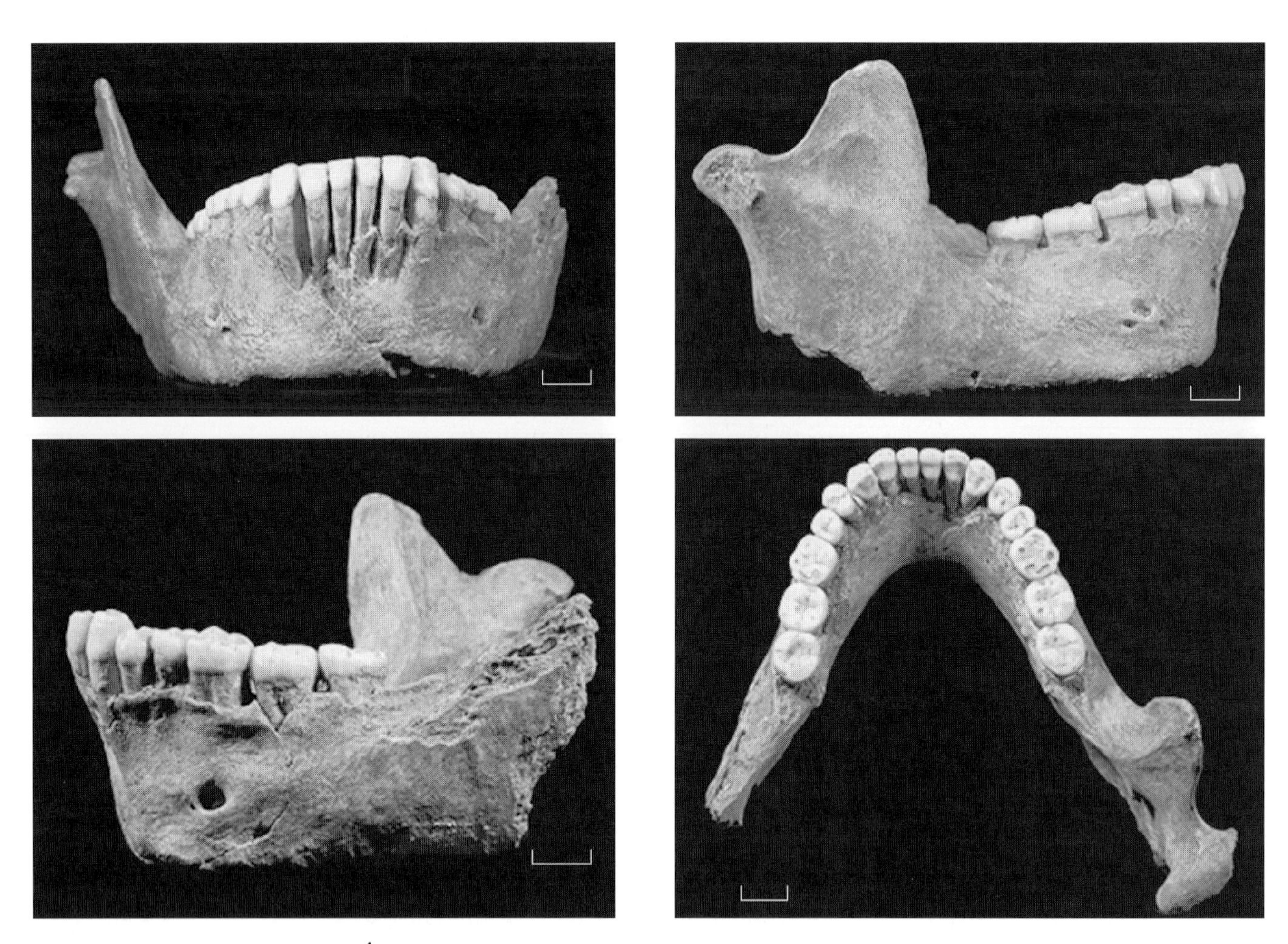

RÉGOURDOU Figure 1. Régourdou 1 (scale = 1 cm).

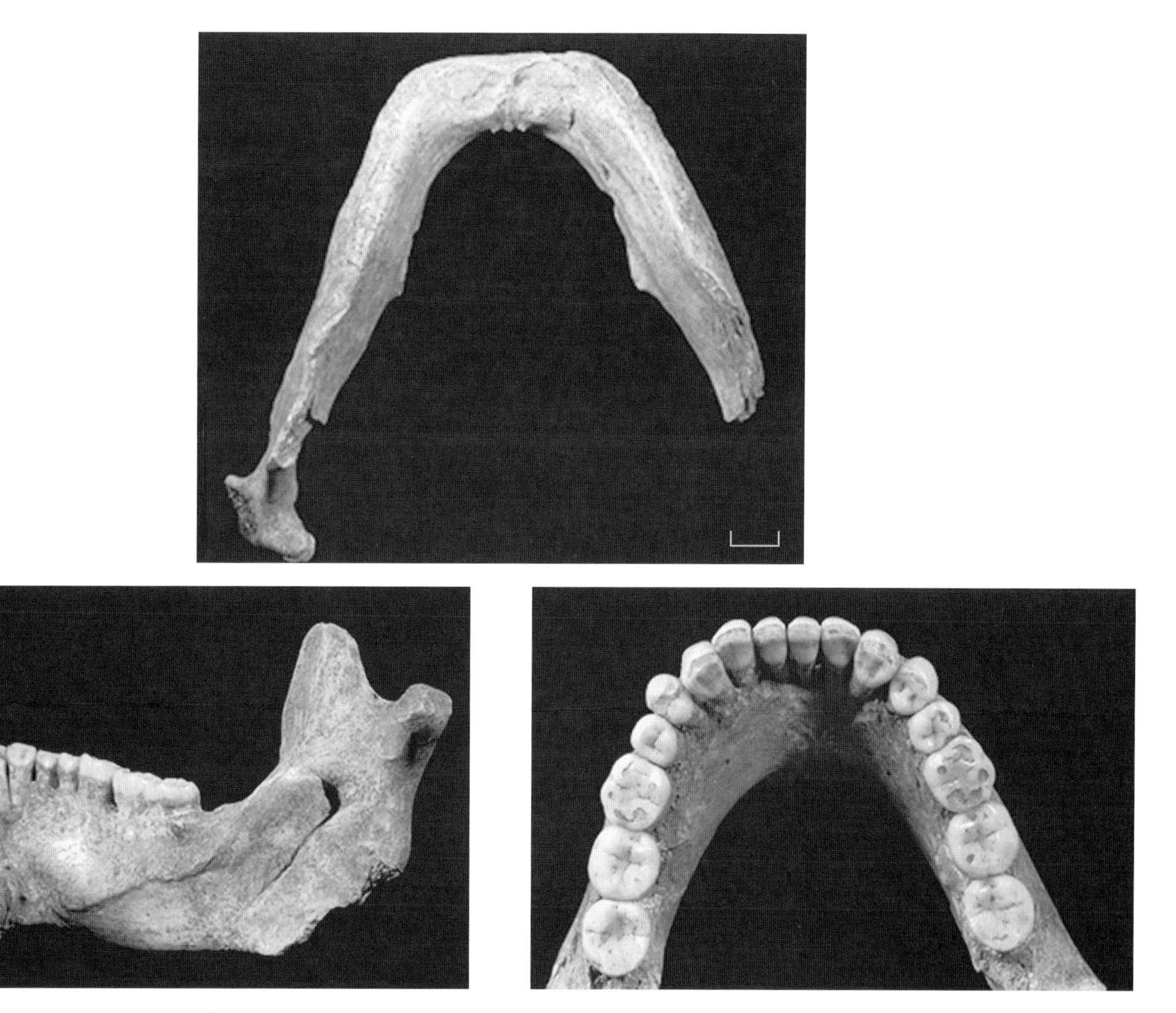

RÉGOURDOU Figure 2. Régourdou 1 (scale = 1 cm).

Reilingen

Location

Walther AG gravel pit NW of the town of Reilingen, Spies subdistrict, Baden-Württemberg, Germany. About 25 km S of Mannheim.

Discovery

Gravel pit workers, spring 1978; identified by E. Schmid.

Material

Partial adult calvaria, with fused parietals, occipital and right temporal.

Dating and Stratigraphic Context

Reilingen is situated in the Upper Rheingraben, which contains substantial deposits of fluviatile sands and gravels of Quaternary age. These deposits are commercially mined, and the Reilingen hominid, along with many other mammal fossils, was found by mining workers after being dredged from the bottom of a flooded gravel pit. Fauna from various stratigraphic levels became mixed during the dredging operation, but because the Reilingen pit is only 30 m deep (into a thickness of about 180 m of Quaternary sediments) it would not ordinarily be expected to sample sediments earlier than the last (Eem) interglacial, ca. 125 ka (Ziegler and Dean, 1998). However, the fauna recovered does include the extinct beaver *Trogontherium cuvieri*, which at other sites in the region is a marker for the preceding Holstein interglacial (Ziegler and Dean, 1998). Some elements of the Reilingen fauna are thus some 250 ka old, the hominid possibly among them because it was recovered during dredging of the lowest level of the pit (it shows few signs of transport, although it might have rolled down from a higher level). In any event, its faunal associations bracket the Reilingen hominid in the period between 250 and 125 ka; Ziegler and Dean incline toward the earlier age.

Archaeological Context

None.

Previous Descriptions and Analyses

In an early analysis Czarnetzki (1989) attributed the Reilingen hominid to *Homo erectus reilingensis*. Adam (1989) and Schott (1990) preferred an allocation to "archaic *Homo sapiens*," whereas Waddle (1993) and Dean (1993) emphasized Neanderthal resemblances. In the most complete account of the specimen, Dean et al. (1998) consider it most similar to "pre-Neandertal" specimens such as Steinheim, Swanscombe, and Atapuerca SH5. These authors accommodate the Reilingen hominid within a "lineage running from earliest European 'archaic' *Homo sapiens* to later early-, and finally, classic-Neandertals." Condemi (1996) considered the Reilingen specimen evidence of "very ancient presence of the Neanderthal lineage," a conclusion she reiterated in 1997. Estimated cranial capacity is 1430 cc (Holloway, 2000).

Morphology

Partial adult calvaria consists of both parietals, fused, with all sutures intact; also a R temporal missing part of squamous and anterior part of petrosal and an occipital missing part of R side (all sutures preserved on the L).

Parietals broad, low, "en bombe" (not "roofed"). Temporal lines weak. As seen on R temporal, posterior root of zygomatic arch originates anterior to auditory meatus. Suprameatal crest hardly developed. Supramastoid crest large; arcs upward and backward. Parietal notch lies anterior to mastoid tubercle and crest. Squamosal suture tall, oblique. Mandibular fossa broad a/p but deep, bounded anteriorly by thin articular wall and medially by medial articular tubercle (more a projection than a swelling). Vaginal crest peaked along midline of ectotympanic tube; lies medially, close to carotid foramen, which may have pointed downward; does not contact mastoid process; posteriorly, bears impression of very thin styloid process. Stylomastoid foramen large, lying near base of styloid process; is in line with moderately a/p long, relatively narrow, deep mastoid notch. Jugular fossa large, moderately deep. Inferior margin of ectotympanic tube short (interior part of margin preserved intact, although external part slightly broken); evidently was incompletely ossified.

Mastoid process quite long a/p at base, relatively elongate and quite pointed at (slightly broken) tip; points downward. Small hole in posterosuperior margin of mastoid process reveals large sinus inside: cavernous, not as pneumaticized as further distally. Small part of occipitomastoid suture appears preserved posteriorly. Lateral to it, bounding the mastoid notch, is sharp, quite prominent ridge (but lower than mastoid process) that deviates laterally from occipitomastoid suture, defining medial margin of mastoid notch (paramastoid crest?). On occipital, on the L, is preserved rear part of long, low, Waldeyer's crest that lies well lateral to occipitomastoid suture. Appears to be no occipitomastoid crest, unless very slightly raised area along L occipitomastoid suture corresponds to this structure.

Anterior lambdoid suture was quite long but not very straight (though straighter on the L than on the R). Lambdoid suture broad, arcuate from side to side; does not peak at lambda. Occipital torus strong, very broad, horizontal; delineated by infratoral depression that is divided longitudinally in midline by low external occipital crest. Midpoint of occipital torus (inion) not its highest point. Suprainiac depression distinct, quite broad; represents area of maximal roughness within field of generally pitted bone; depression forms only definable upper boundary of occipital "torus."

Internally, petrosal very broad laterally; forms essentially flat surface right across to cranial wall. Arcuate eminence low; area lateral to it swelled out with large vacuous sinus (visible through two holes). Area around carotid canal also swollen with sinusial inflation. Cochlear canaliculus very large, deep. Tiny trace of superior petrous sinus; equally tiny trace of subarcuate fossa. Sigmoid sinus short; curves slightly anteriorly (jugular foramen pointed anteriorly?). L transverse sinus only faintly impressed. Superior sagittal sinus also minimally impressed; disappears anteriorly.

REFERENCES

Adam, K. D. 1989. Alte und neue Urmenschenfunde in sudwest-Deutchland-eine kritische Wardigung. *Quarter* 39/40: 177–190.

Condemi, S. 1996. Does the human fossil specimen from Reilingen (Germany) belong to the *Homo erectus* or to the Neanderthal lineage? *Anthropologie* 34: 69–77.

Condemi, S. 1977. Le statut phylogénétique du fossile de Reilingen (Baden-Württemberg, Allemagne). *Anthropol. Préhist.* 108: 135–146.

Czarnetzki, A. 1989. Ein archaischer Hominidencalvariarest aus einer Kiesgrübe in Reilingen, Rhein-Necker-Kreis. *Quartär* 39/40: 191–201.

Dean, D. 1993. *The Middle Pleistocene* Homo erectus/ Homo sapiens *transition: New evidence from space curve statistics.* PhD thesis, City University of New York.

Dean, D. et al. 1998. On the phylogenetic position of the pre-Neandertal specimen from Reilingen, Germany. *J. Hum. Evol.* 34: 485–508.

Holloway, R. L. 2000. Brain. In: E. Delson et al. (eds), *Encyclopedia of Human Evolution and Prehistory.* New York, Garland Publishing, Inc., pp. 141–149.

Schott, L. 1990. "*Homo erectus reilingensis*"—Anspruch und Wirklichkeit eines Schädelfundes. *Biol. Rundsch.* 28: 231–235.

Waddle, D. 1993. Affinities of the Reilingen partial cranium. *Am. J. Phys. Anthropol.* 16 (Suppl.): 201.

Ziegler, R. and D. Dean. 1998. Mammalian fauna and biostratigraphy of the pre-Neandertal site of Reilingen, Germany. *J. Hum. Evol.* 469–484.

Repository

Staatliches Museum für Naturkunde, Rosenstein 1, 7091 Stuttgart, Germany.

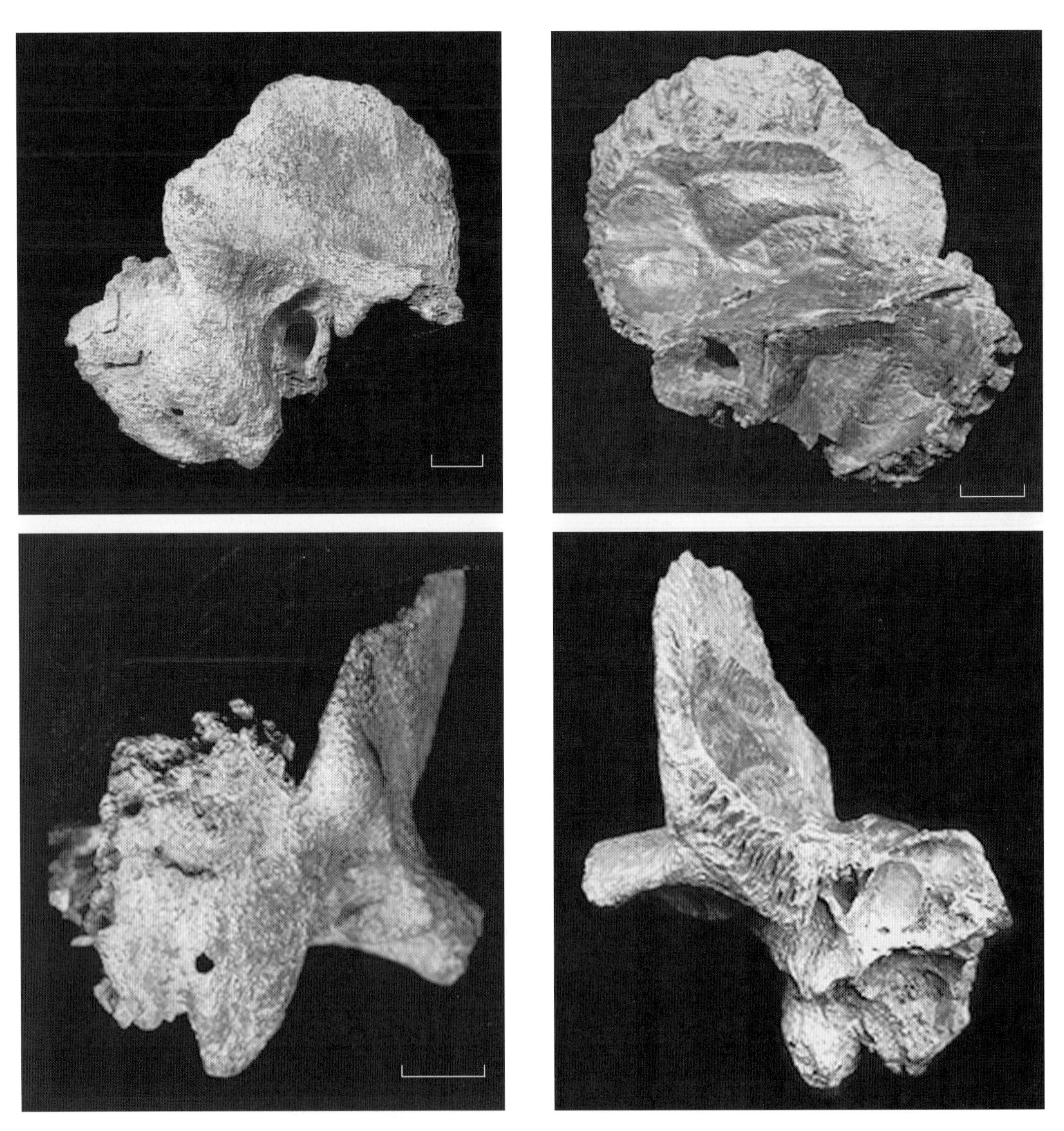

REILINGEN Figure 1. Temporal (scale = 1 cm).

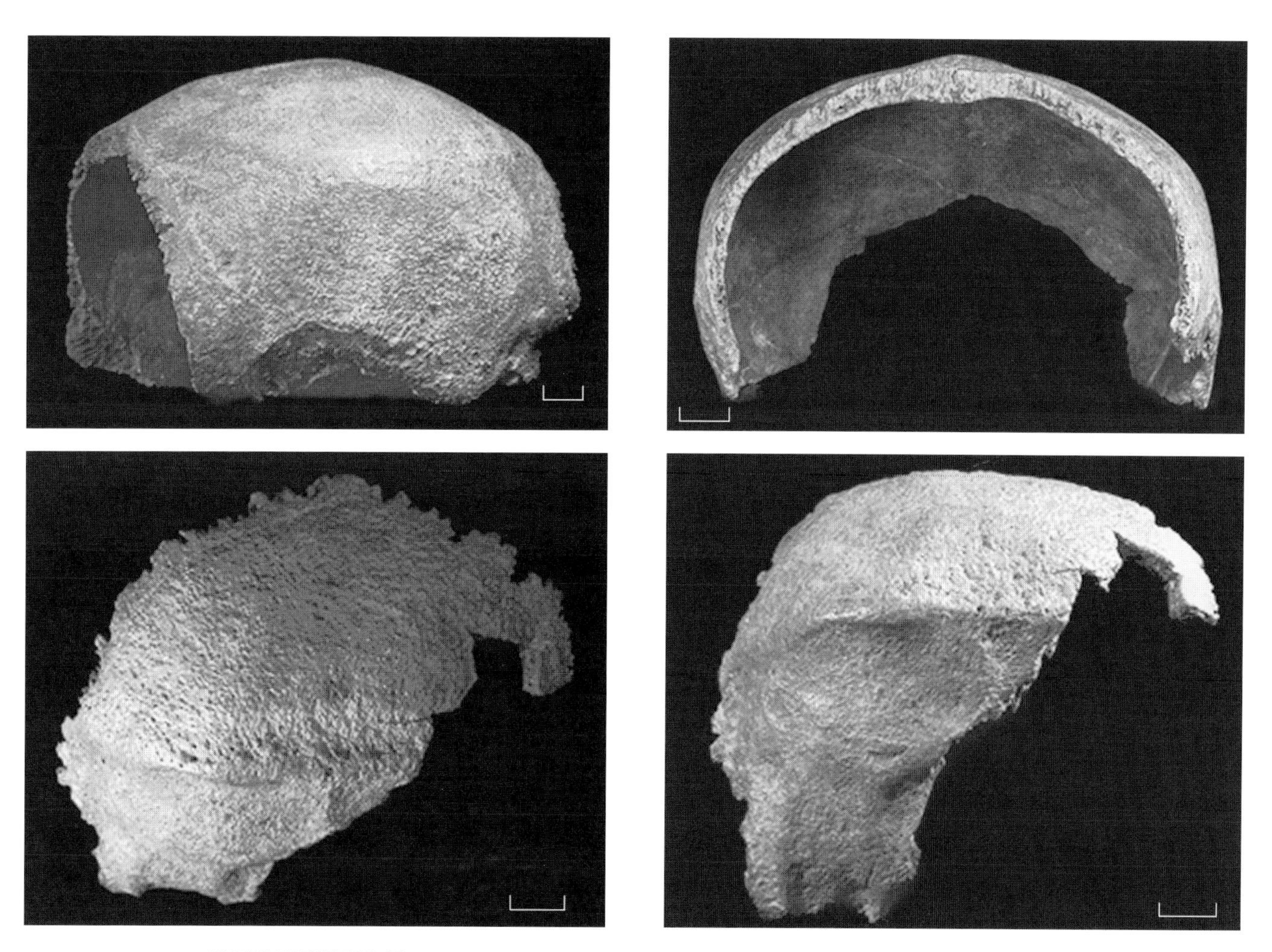

REILINGEN Figure 2. Parietal (above) and occipital (below), (scale = 1 cm).

Roc de Marsal

Location

Stratified cave entrance deposits in a side valley of the Vézère river at Campagne-du-Bugue, 5 km SW of Les Eyzies-de-Tayac, Dordogne, France.

Discovery

J. Lafille, August 1961.

Material

Partial skeleton of infant, including fragmentary skull with most of face.

Dating and Stratigraphic Context

The infant skeleton was found in a pit dug into the lowest level (Bed I) of the deposits, probably from Bed V (Bordes and Lafille, 1962). Van Campo and Bouchud (1962) deduced relatively temperate conditions from a pollen analysis. Based on this and on the fauna, Vandermeersch (1965) suggested a late Würm I/II or early Würm II age for Bed V. Turq (1979), however, reviewed the available evidence and concluded that a date within (or even toward the end of) Würm II was more likely. The lithic industry implies an age in excess of 50 ka (J.-J. Cleyet-Merle, personal communication).

Archaeological Context

Bordes and Lafille (1962) described the industry from Roc de Marsal as "typical Mousterian." Later, however, Turq (1979) characterized the Bed I–IV lithics as "Quina" Mousterian but the Bed V industry as Mousterian "rich in scrapers." The bottom of the grave in which the infant was interred was littered with Mousterian lithics (Bordes and Lafille, 1962), but the exact tradition they represent is unclear. The association of the burial with Bed V remains accepted, however.

Previous Descriptions and Analyses

In their initial notice of the Roc de Marsal discovery, Bordes and Lafille (1962) referred the skeleton to *Homo neanderthalensis*. Subsequent authors, including the skeleton's monographer Madre-Dupouy (1992), have all concurred that the fossil is Neanderthal. Opinions as to age at death are divided, however. Legoux (1965) estimated this as 30 months at most, whereas Madre-Dupouy (1992) suggests 30–48 months.

Morphology

Cranium with most of face; frontal, R parietal, bit of L parietal; part of squamous region of R temporal, parts of occiput including basiocciput. Mandible quite complete, missing part of R condyle, all of L condyle, and L gonial angle. All deciduous teeth present; crowns of upper and lower I2s and Ms visible in crypts; crypt for M_2 moderately developed. Many postcranial bones. Probably 4–5 years old on osteological and dental criteria.

Cranium broad across frontal; apparently even broader further back along parietals. Tall frontal ascends very steeply directly from supraorbital region, then flattens out after curving back strongly. Curve down to occipital plane smooth; in profile, slight posterior jutting below lambda.

Frontal quite vertical. Supraorbital region flat except for incipient swelling medially on both preserved supraorbital regions. Sharp superior orbital margins, down to which orbital roofs descend sharply. Orbits tall, somewhat ovoid in shape, with superolaterally oriented long axis; some truncation of inferomedial corners. Nasal bones and frontal process of maxilla quite protruding. Maxillary sinuses swell out floors of both orbits but do not puff out face; as judged from the R, sinuses probably did not swell into nasal cavity. Orbital floors slope down and outward toward face; infraorbital margin quite sharp, at least medially. On the R, infraorbital margin slightly overhangs face. Infraorbital plane slightly depressed just medial to zygomaticomaxillary suture. Infraorbital foramina very large, oddly constructed; seem to be oriented medially and to be unroofed as exit onto face.

Nasal bones broad. Nasal aperture very broad inferiorly, its inferolateral corners quite curved; appears not to be much taller than it is broad at its base. Lateral (marginal) crests continuous around to anterior nasal spines, which are separate, very protruding. From this juncture, crisp spinal crests run posterolaterally to meet incipient medial projection. Nasal cavity floor gently sloped downward and back but more or less flat; not excavated. On the R, anterior lacrimal crest continues into nasal cavity as short posterior shelf over large, deep lacrimal groove. Internally, keel runs along nasonasal suture. On both sides, vertical, crest-like elevation of bone runs along inside frontal processes; on the R, this vertical (superior nasal?) crest terminates in inferior thickening that probably represents base of developing medial projection. Nasoalveolar clivus tall s/i, very vertical.

Anterior root of zygomatic arch low, does not sweep back (because maxillary sinuses not yet swelled out in this region?). Anterior squamosal margin of temporal a little puffed out, providing some demarcation between longer anterior and shorter posterior temporal fossae. Alisphenoid smoothly curved, thereby providing no delineation of temporal/infratemporal fossae. Posterior root of zygomatic arch lies quite anteriorly on apparently s/i tall squamosal. Mastoid and posterior squamosal regions largely missing, but on the R, medial part of slightly raised occipitomastoid crest visible, with larger Waldeyer's crest slightly posterior and quite far medial to it and separated from it by wide, shallow sulcus.

On the L, tiny bit of shallow, pitted suprainiac depression preserved; also a bit of infratoral depression, suggesting slightly swollen, arcing superior nuchal line. Judging from the L, anterior lambdoid suture lacking. Symmetrically protruding occipital delineated by continuously arcing lambdoid suture. Curve from occipital onto parietal smooth in profile. Basiocciput short, broad, also thin. Anterior margin and part of R lateral margin suggest long, rather heart-shaped foramen magnum, broader anteriorly and more pointed posteriorly. Large, subdiamond-shaped R occipital condyle lies moderately anteriorly along margin of foramen magnum; only anterior margin of condyle derives from basiocciput (= body of condyle entirely formed from lateral part of occipital; raised anterior margin is developmental consequence of fusion of lateral part with basiocciput, in process in this specimen). Basicranial flexion clearly extreme: basiocciput articulates well with occipital.

Palate relatively deep, especially anteriorly. Incisive foramen large; points downward.

In profile, mandibular symphysis almost vertical above, then curves gently to inferior border. External surface of symphyseal region smooth and featureless. Two mental foramina on each side, below dm1. In inferior view, symphyseal region uniformly thick from front to back, broad and gently curved between dcs. Digastric fossae fairly well defined, quite wide m/l, oriented downward. Postincisal plane long, gently convex, with some degree of posterior slope; tiny foramen lies about halfway up. Very marked mylohyoid lines descend steeply, overhang fairly well-defined submandibular fossae beneath them. Condyle lies well below level of a/p long coronoid process, which bears, internally, close to its anterior margin, thickened vertical strut of bone. Almost no sigmoid notch; base of coronoid process terminates just in front of base of somewhat horizontally oriented condyle. Sigmoid notch crest runs to middle of condyle. Large, rather circular mandibular foramina point straight backward; on the R is strong, long, ledgelike, posteriorly directed lingula. On the R, gonial region curves medially. Large, thickened medial pterygoid tubercle lies quite low down on posterior internal edge of gonial angle, below lingula. Directly behind lingula (and above tubercle) is small swelling (also on posterior edge).

Mesial corners of upper dis somewhat distended; shallowly shoveled surfaces bound by thin margocristae that emanate from very small lingual swellings. dc^1s triangular in outline, somewhat concave on lingual surfaces; bounded near neck by band of enamel that is continuous with margocristae. Trigon basins of dm^{1-2}s deep but restricted buccolingually because of relatively internal position of protocone. dm^1s have small metacones lying at end of long distal slope of much taller, buccally swollen paracones. Strong preprotocrista runs from lingually swollen, slightly distally displaced protocone to mesial edge of paracone. Hypocone-like swelling on distal side of protocone appressed to metacone. General shape of dm^1s accentuates oblique axis, from mesiobuccal to distolingual corner of crown. dm^2 protocristae very stout; preprotocrista runs to front of paracone; equally stout but shorter postprotocrista runs directly to metacone. Hypocone huge, with internally displaced apex. Stout postcingulum runs from distinct metastylar region to curve around to apex of hypocone (enclosing medium-sized but quite deep talon basin). Internally placed protocone and hypocone apices give long lingual slope to crown, which bears large, pronounced protostyle. Groove between protocone and hypocone deeply incised. Enamel not wrinkled; is pitted on dm^2s.

Crown of upper M1 long m/d mesiodistally but not wide b/l; roots just beginning to develop. Enamel well wrinkled. Both protocristae well developed; postprotocrista at least as robust as preprotocrista. Preprotocrista runs around front of paracone and postprotocrista up to apex of metacone. Hypocone huge, markedly swells crown distolingually. Apices of lingual cusps not very internally placed (= M^1s do not have pronounced lingual slope). P^{1-2} crowns started to form.

Distal margins of lower di2s flared, not straight. Buccal surfaces of dc_1s swollen near neck; lingual surfaces slightly concave with cingulum inferiorly and margocristids; thin, vertical, slightly mesially displaced pillar runs to apex of crown. On dm_{1-2}s, lingual surfaces somewhat incised vertically, with distinct notches between cusps. dm_1s have distinct cusps; cristid obliqua connects hypoconid and protoconid. Subequal, large protoconid and metaconid peripherally placed but connected by stout metacristid. Well-developed paracristid runs short distance mesially from protoconid before curving lingually and up base of metaconid (enclosing well-defined, deep, slightly lingually tilted trigonid basin). Entoconid small, well separated from hypoconid opposite it. Stout hypocristid runs down from hypoconid to terminate at base behind entoconid. Talonid basin largest and deepest between four major cusps; also opens distally between hypoconid and entoconid. Crown modestly exodaenodont over anterior root. dm_2s ovoid in outline; lack angled edges. Trigonid basins deep, well defined, a/p compressed. Trigonid basin enclosed by thick, beaded paracristid that runs directly between apices of protoconid and metaconid. Metaconid and protoconid close together but not joined by crest. Cusps peripherally placed, especially on lingual side; are well defined and vertical, defining very broad, deep talonid basin. Enamel in basin bears number of tiny centroconid-like cusps. Distinct, deep posterior fovea on distolingual portion of crown. Small protostylid at cleft between protoconid and hypoconid. In talonid basin, especially visible on the R, base of hypoconid extends mesiolingually across midline of crown. M_1 morphology not visible.

REFERENCES

Bordes, F. and Lafille, J. 1962. Découverte d'un squelette d'enfant moustérien dans le gisement de Roc de Marsal, commune de Campagne-du-Bugue (Dordogne). *C. R. Acad. Sci. Paris* 254: 714–715.

Legoux, P. 1965. Détermination de l'âge dentaire de l'enfant néandertalien du "Roc de Marsal." *Rev. Fr. Odonto-Stomat.* 10: 4–24.

Madre-Dupouy, M. 1992. *L'enfant du Roc de Marsal; Etude analytique et comparative.* Paris, CNRS, pp. 1–300.

Turq, A. 1979. *L'évolution du Moustérien du type Quina au Roc de Marsal et en Périgord.* Thesis, Ecole des Hautes Etudes en Sciences Sociales.

Van Campo, R. and J. Bouchud. 1962. Flore accompagnant le squelette d'enfant moustérien découvert au Roc de Marsal, commune du Bugue (Dordogne) et première étude de la faune du gisement. *C. R. Acad. Sci. Paris* 264: 897–899.

Vandermeersch, B. 1965. Position stratigraphique et chronologique des restes humains du Paléolithique moyen du Sud-Ouest de la France. *Ann. Paléontol.* 51: 69–126.

Repository

Musée National de Préhistoire, 24620 Les Eyzies-de-Tayac, France.

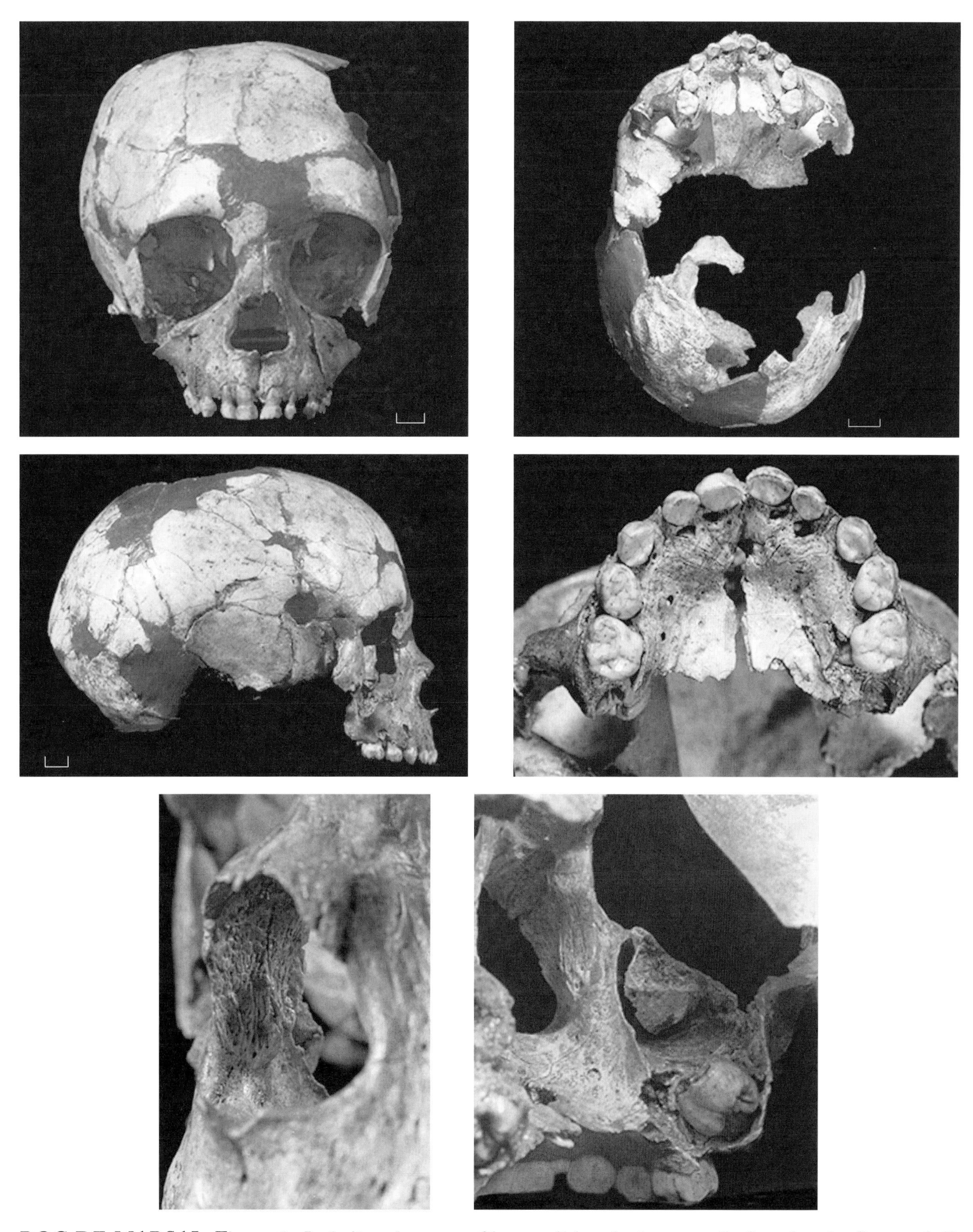

ROC DE MARSAL Figure 1. Including close-ups of low medial projection on wall of nasal cavity (bottom left) and posterior view of nasal cavity (bottom right), (scale = 1 cm).

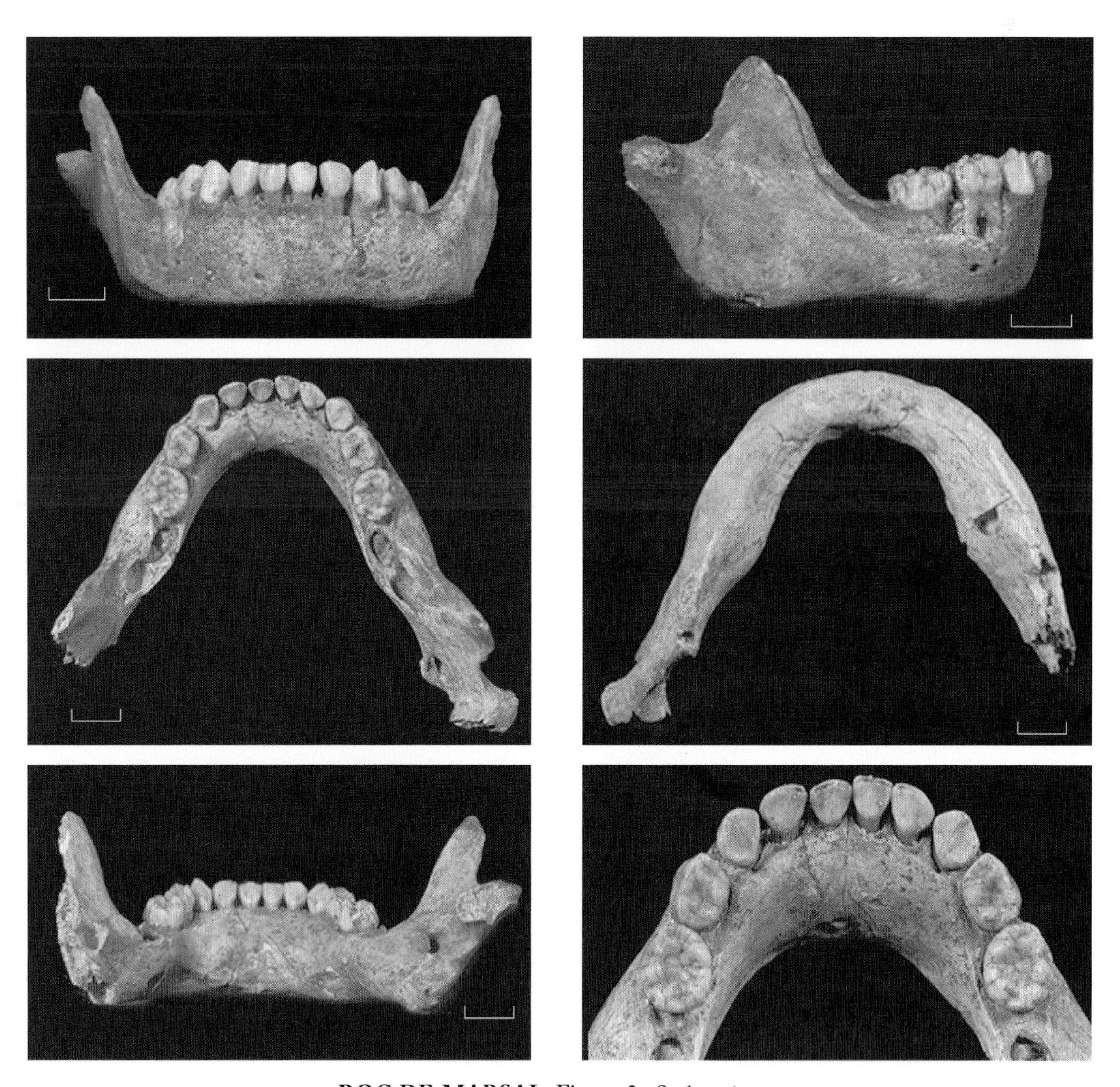

ROC DE MARSAL Figure 2. Scale = 1 cm.

Saccopastore

Location
Gravel pit, now gone, in the suburbs of Rome, Italy, in a meander of the Aniene river, a tributary of the Tiber.

Discovery
A. Giovannini, May 1929 (Saccopastore 1); A. Blanc and H. Breuil, July 1935 (Saccopastore 2).

Material
Fairly complete adult cranium lacking most of supraorbital and glabellar regions and both zygomatic arches (Saccopastore 1); partial face and cranial base of adult (Saccopastore 2).

Dating and Stratigraphic Context
Stratigraphy of the fluviatile gravels and sands at the Saccopastore site is rather poorly documented, but it was established early on that the ensemble postdated the penultimate glaciation (e.g., Köppel, 1934; Blanc, 1948; Segre, 1948). Analysis of invertebrates from level 3b of the deposits, which lies stratigraphically between the findspots of Saccopastore 1 (level 2a, slightly below) and Saccopastore 2 (level 3a, slightly above), suggested a period of quite intense cold, sometimes attributed to the beginning of the last glacial, ca. 80 ka. Segre (1948), however, emphasized evidence that these invertebrates had been secondarily redeposited, whereas the mammalian fauna in both hominid levels is typical of the Riss-Würm interglacial (e.g. Blanc, 1939). Recent authors have generally agreed on a Stage 5e date for the Saccopastore hominids, around 130–120 ka.

Archaeological Context
A small number of lithics were found in level 3a, along with Saccopastore 2. These tools were characterized by Blanc (1939) as of Pontian Mousterian type. Piperno and Segre (1982) noted close similarities to the more abundant industry at the site of Torre in Pietra, reinforcing the hominids' Stage 5e attribution.

Previous Descriptions and Analyses
From their earliest descriptions (e.g., Sergi, 1929; Breuil and Blanc, 1935) both Saccopastore hominids have been regarded as Neanderthals (Saccopastore 1 having been made the holotype of *Homo neanderthalensis aniensis* by Sergi, 1935). Condemi (1992), the most recent monographer of these specimens, has compared them with a wide range of other Neanderthals; she finds them primitive in a variety of characters compared with later European Neanderthals (and hence similar to Levantine Neanderthals). Estimated cranial capacity is 1245 cc for Saccopastore 1 and 1300 cc for Saccopastore 2 (Holloway, 2000).

Morphology

Saccopastore 1

Fairly complete cranium lacking most of supraorbital and glabellar regions and both zygomatic arches. Two holes, one in L frontal, the other mostly on the R parietal with its front edge extending into adjacent frontal bone, probably caused by pick damage. Crowns of LM1–3, RM2–3 present; other upper teeth in various states of breakage from the neck up.

Relatively small cranium, long and low, with relatively large facial structure at front. Bone appears to be relatively thin, but matrix still adhering throughout intracranial region. In rear view, very wide and en bombe. Viewed from above, skull broad posteriorly. Posterior outline asymmetrical (is plagiocephalic); thus widest point on the L lies low, at level of squamosal suture, whereas on the R, this point is much higher. In profile, short forehead curves strongly back into much shallower arc that continues almost to lambda before sharply but smoothly curving downward again; there is no distinct "chignon."

Supraorbital region missing. Very small part of glabella still exists above nasion; is placed well forward of short frontal rise. Preserved on the R is lateralmost portion of supraorbital region; is quite horizontal in orientation in contrast with central portion, which is more vertically oriented (thus slight degree of doming of frontal bone confined to area between midpoints of orbits). R and L frontal sinuses separated by thin septum; extend laterally to about midpoint of orbits. From area cleaned of matrix, appears that frontal sinuses arose from position quite far below level of nasion; matrix still present in inferior recesses. Possible (though not certain) that frontal sinuses connected with both ethmoid and maxillary sinus cavities. As best seen on the R (where matrix has been more thoroughly cleaned out), downward extension of frontal sinus descends at least to point internal to lacrimal bone (thus anterior to the ethmoid).

Difficult to assess original shape of orbits in absence of supraorbital margins. However, because frontal processes do not appear to be inflated and orbital inferior margins are essentially horizontal, would seem that orbits did not have "aviator glasses" shape. Orbital floor flows smoothly out onto face, with no distinct margin, especially medially. Both lacrimal areas damaged in medial sides of orbits; appears from the R that lacrimal fossa was short s/i and moderately deep and that anterior and posterior lacrimal crests were neither crisply defined nor convergent superiorly. Interorbital region was evidently quite wide; became wider superiorly, truncating superomedial corner of orbit.

At level of infraorbital margin, frontal process very long a/p (reflecting great anterior projection of upper part of snout). Relatively small infraorbital foramina lie well below inferior orbital margin; appear to have faced forward (although on both sides a long, well-excavated sulcus runs down maxilla from foramen). Lower face very narrow, wedge shaped when viewed from above. As judged from the L especially, anterior roots of zygomatic arches angled sharply backward to continue this wedge shape posteriorly. Inferior portion of anterior root of zygomatic arch takes origin on both sides well above region of M^2 but angles up and out quite steeply. In total, extraordinarily anteriorly projecting snout and sharply receding face.

Region around nasion obliquely sloping, quite straight in profile. Nasal bones very long, broad; widen slightly and steadily inferiorly (form trapezoidal shape); flex strongly anteriorly well below nasion, about halfway along their length; bear low keel along nasonasal suture, on either side of which the bones curve gently to the nasomaxillary suture. Nasal aperture apparently both tall and wide; postmortem damage (causing R frontal process of maxilla to partially overlap nasal bone) artificially narrows nasal aperture on that side. Inferior nasal region extensively damaged; appears that anterior nasal spines were quite large, horizontal, anteriorly projecting. No evidence preserved as to configuration of inferior nasal margin. Nasoalveolar clivus also damaged; was long, forwardly gently sloping. Interior of nasal cavity totally obscured by matrix.

As preserved especially on the R, thin temporal ridge appears to have come up from well behind zygomatic process of frontal; as seen best on the L, it becomes well-marked temporal line well anterior to area of almost obliterated coronal suture. Temporal lines lie relatively low on long parietals, recurve strongly above asterion to remain distinct right to region of parietal notch. As seen on both sides, but particularly well on the R, postorbital constriction, especially inferiorly, is relatively deep.

Squamosal rather short, almost as tall as long; its superior margin forms tight curve. As better preserved on the R, anterior squamosal suture bends obtusely into alisphenoid (thus is neither sharp corner nor smooth curve, but anterior and posterior fossae nonetheless distinguishable). In vertical plane, although no sharp flexure horizontally along alisphenoid, infratemporal fossa distinguishable. Mandibular fossae only partially preserved; were not very wide m/l or very long a/p; may have been deep; appears both were bounded medially by swollen medial articular tubercle. On the R, depth of fossa more discernible, with very little development of articular eminence (although bone may have been modified postmortem on both sides). On the L, low postglenoid plate lies lateral to and away from small, compressed, more or less vertically oriented auditory meatus. On both sides, ectotympanic tube apparently very short. On the L, scar of broken vaginal process well separated from mastoid process; how far vaginal process extended laterally is impossible to tell. Styloid process and stylomastoid foramen cannot be identified. On the L, relatively large, somewhat medially placed carotid foramen faces backward.

Posterior root of zygomatic arch takes origin anterior to auditory meatus. Above meatus, very low suprameatal crest confluent with somewhat more laterally prominent, quite swollen, upwardly arcing supramastoid crest. This latter separated by marked sulcus from a/p short, downwardly pointing, rather thin mastoid process. Mastoid notch broad, relatively deep; especially as seen on the R, its posterior extent delimited by low rim of bone. Occipitomastoid crest on both sides long, thick; bifurcates posterior to posterior margin of mastoid process; does not extend downward as far as tip of slender, a/p short, downwardly pointing mastoid process on either side. Medial to occipitomastoid crest, and originating only fractionally posterior to its anterior end, is shorter, but also relatively thick, Waldeyer's crest. Parietal notch vertical; lies over posterior edge of mastoid process; posterior to it extends short, but more or less horizontal, parietomastoid suture.

Many ossicles in region around lambda, extending into sagittal suture and down along L lambdoid suture, disrupting sutural courses; on the R, large ossicle lies immediately posterior to asterion. Occipital plane very broad between asterions; ossicles apart, occipital apparently not very tall. Superior nuchal line arises from region of asterion; becomes more prominent ("torus"-like) as it courses medially; viewed from below, is broadly, but shallowly, bow shaped, being its lowest in midline; viewed from behind, runs horizontally across occipital. Long, shallow infratoral sulcus defines superior nuchal line very distinctly inferiorly; is gently scalloped bilaterally. Superior margin of superior nuchal line "torus" only defined by broad, tall, quite asymmetrical, shallow suprainiac depression that is marked by patchwork of tiny areas of raised bone (distinct from morphology of surrounding bone). Below superior nuchal line, relatively flat nuchal plane slopes gently forward to foramen magnum with only minor undulations. Twin bilateral pits lie on either side of midline, just posterior to foramen magnum. Foramen magnum small, short and egg shaped, being broad posteriorly. A/p short, m/l wide occipital condyles lie quite anteriorly on margin of foramen. Basiocciput quite highly flexed upward and forward; is a/p short, very m/l wide posteriorly; narrows anteriorly, tapering dramatically toward sphenooccipital synchondrosis. External surface of this bone bears only very few swellings.

Palate very deep; sides virtually vertical; front very steeply sloping. Narrow opening in midline of palate at level of P^2s may be incisive foramen. As seen on the L, pterygoid plates (damaged) parallel for most of their length; from both sides, it appears that they converged at their superior roots; cannot determine shape inferiorly. Posterior pole of vomer present; bears only minuscule alae, which apparently lay in front of region of sphenooccipital synchondrosis.

Coronal suture quite finely denticulate; was apparently unsegmented. Sagittal and lambdoid sutures more heavily denticulated but also apparently not segmented. All internal anatomy obscured by matrix.

In general, judging from both preserved crowns and roots, upper dentition was small. Cs would not have been very large; I1s probably were. Both P2s have undivided roots but two distinct root canals. M1 only slightly larger than M2; M3 much smaller than both. M1–2 bear large hypocone, which on M1 is much more swollen distolingually. M3 lacks hypocone altogether. Although Ms very worn, appears that trigon basins were small and confined buccal to midline of crown and that M2s at least had well-excavated talon basins.

Saccopastore 2

Partial face with incomplete L and virtually complete R orbital region, complete R zygomatic arch, parts of R sphenoid, temporal and occipital attached. External nasal margin severely damaged; nasal cavity immediately behind this region blocked by matrix. Preserves upper RC–M3, LC, LP2–M3. Anterior teeth represented by alveoli, LP2 by roots.

In general, lower face appears less protrusive and wedge shaped, but wider, than Saccopastore 1; also faces much more forward. R supraorbital margin partially damaged; was very tall medially; tapers somewhat laterally. In profile, supraorbital margin medially bears flattish anterior plane with hint of superior border; laterally, supraorbital torus more smoothly rolled.

Piece of frontal extends backward from medial portion of orbital region; appears to rise quite steeply from directly behind orbit. Some frontal sinus preserved on R; probably penetrated frontal squama, although its matrix-filled lateral extent was probably not great. Portion of cribriform plate preserved quite far behind frontal pole and approximately halfway back along orbital cone. Upper part of snout appears to have been convex, if not very forwardly projecting. A/p axis of frontal process not very long; faces more obliquely downward than in Saccopastore 1. Entire lower face thus descends quite steeply from inferior orbital margin (more so than Saccopastore 1). On the L, preserved somewhat internal to and somewhat behind region of lacrimal fossa (not observable), is apparently fairly large, although matrix filled, concavity within frontal process (superior extension of maxillary sinus?).

Interorbital region perhaps exaggerated by postmortem compression; appears extremely wide, becomes wider superiorly and truncates mediosuperior corner of m/l wide orbit. Inferior orbital rim more crisply defined, and maxilla below more vertical, than in Saccopastore 1. Medioinferior border of orbit rises somewhat medially, truncating this corner. Moderately large infraorbital foramen on the L and two somewhat small foramina on the R; all lie well below inferior orbital margin, appear to face anteriorly. Better preserved on the R, shallow concavity lies medial and inferior to these foramina.

Nasal bones damaged; appear to have flexed forward well below nasion; were probably broad superiorly, broadening even more toward external nasal margin. On the L, inferior portion of nasal margin preserved; appears to have been continuous right across to region of damaged anterior nasal spine. Judging by proportions on the L, aperture was quite broad but probably not very tall s/i. Internally, appears that floor of nasal cavity was not depressed. Nasoalveolar clivus quite straight, inclined somewhat forward, but only moderately long; is gently arced from side to side; external surface furrowed by sulci between anterior tooth roots.

Anterior root of zygomatic arch takes origin well above level of M1; arcs somewhat outward as well as backward. Maxillary portion of zygoma faces rather anteriorly; temporal extension flexes back strongly into rather straight, horizontal, rather s/i tall but m/l thin zygomatic arch. As seen on the L, maxillary sinus did not extend far laterally into zygoma. Posterior surface of zygomatic frontal process faces straight backward; borders on deep anterior temporal fossa. This fossa delimited posteriorly by marked but blunt cornering along course of anterior squamosal suture into alisphenoid. Distinct temporal and infratemporal fossae delimited by blunt horizontal cornering of alisphenoid. Posterior root of zygomatic arch takes origin anterior to small, subcircular auditory meatus. Well-developed suprameatal crest lies above meatus; is confluent posteriorly with a/p short, s/i tall, almost rectangular supramastoid crest. Posterior portion of squamous suture runs vertically behind posterior margin of supramastoid crest; is bluntly raised.

Mastoid process moderately m/l thick, moderately projecting, externally roughened, slightly anteriorly oriented; only moderately long a/p at base (though much longer than on Saccopastore 1). Mastoid notch fairly narrow, not very deep; posterior limit marked by faint rim of bone. Medial to mastoid notch, and continuing over region of occipitomastoid suture, bone swells modestly into long, low, wide occipitomastoid crest that appears to bifurcate posteriorly around what may have been a mastoid foramen (much less marked than the bifurcation on Saccopastore 1).

Moderately sized foramen ovale lies lateral to lateral pterygoid plate. Preserved R mandibular fossa quite wide, moderately deep, longest a/p at its midline; was poorly delimited medially by low, not very pronounced medial articular tubercle; is bounded anteriorly by gentle slope (not typical articular eminence). Structure reminiscent of the latter, however, is found quite far laterally on posterior root of zygomatic arch. Postglenoid plate absent. Moderately short ectotympanic tube well separated from mastoid process. Vaginal process appears not to have extended full length of ectotympanic tube; its presence is indicated medially along petrosal. Carotid foramen relatively small, backwardly pointed; lies very medially on petrosal. Narrow styloid pit also lies very medially. Depression probably representing stylomastoid foramen lies posterior and well lateral to styloid pit.

Palate very deep, with vertical sides and a very steep anterior slope; is wide and long, especially in proportion to rather small teeth. Incisive foramen difficult to identify; appears to be matrix-filled indentation far anterior in midline, right at front of palate. Pterygoid plates (damaged) were parallel to one another; as indicated on the R, may have become confluent at superior roots.

Internally, superior surface of petrosal very wide; bears two low, keel-like elevations, one posteriorly along its internal margin and one a little more anterior and closer to vault wall. No evidence of superior petrous sinus. Subarcuate fossa filled in. Sigmoid sinus very short. Exposed sphenoidal sinus large. Preserved R side of jugum short m/l but long a/p; is shelf-like; greatly overhangs shallow but long middle cranial fossa; bears thick, blunt, very posteriorly projecting anterior clinoid process.

As in Saccopastore 1, upper teeth very worn, relatively and absolutely small. From alveoli, appears that I1s were large. C roots long; crowns apparently small; buccally, very strong inward curve to root-crown axis. RP1 was b/l wider as well as m/d longer than P2. M1 slightly larger than M2; M2 noticeably larger than M3. Hypocone on M1–2; on M1, noticeably expands distolingual corner of tooth. M3s lack hypocones. Appears on all Ms that protocone was somewhat internally placed, thus constricting trigon basin.

REFERENCES

Blanc, A. 1939. Il giacimento mousteriano di Saccopastore nel quadro del plaeostocene laziale. *Riv. Antrop.* 32: 223–231.

Blanc, A. 1948. Notizie sui trovamenti e sul giacimento di Saccopastore e sulla sua posizione nel Pleistocene laziale. *Paleontogr. Ital.* 42: 3–23.

Breuil, H. and A. Blanc. 1935. Rinvenimento in situ di un nuovo cranio di Homo neanderthalensis nel giacimento di Saccopastore (Roma). *Atti Accad. Naz. Lincei*, ser 6, 22: 166–169.

Condemi, S. 1992. *Les Hommes Fossiles de Saccopastore et leurs Relations Phylogénétiques.* Paris, CNRS.

Holloway, R. L. 2000. Brain. In: E. Delson et al. (eds), *Encyclopedia of Human Evolution and Prehistory.* New York, Garland Publishing, pp. 141–149.

Köppel, R. 1934. Stratigrafia e analisi della cava di Saccopastore e della regione circostante in riguardo alla posizione del cranio neandertaliano scoperto nel maggio. *Riv. Antrop.* 30: 475–476.

Piperno, M. and A. Segre. 1982. The transition from Lower to Middle Palaeolithic in Central Italy: An example from Latium. In: A. Ronen (ed), *International Symposium to Commemorate the 50th Anniversary of Excavations in the Mount Carmel Caves by D. A. E. Garrod.* Haifa, University of Haifa, pp. 6–14.

Segre, A. 1948. Sulla stratigrafia dell'antica cava di Saccopastore presso Roma. *Atti Accad. Naz. Lincei Rc.*, ser. 8, 4: 743–751.

Sergi, S. 1929. La scoperta di un cranio del tipo di Neandertal presso Roma. *Riv. Antrop.* 28: 457–462.

Sergi, S. 1935. Die Entdeckung eines Weiteren Schädels des *Homo neandertalensis var. aniensis* in der Grübe von Saccopastore (Rom). *Anthrop. Anz.* 12: 281–284.

Repository

Università di Roma "La Sapienza," Dipartimento di Biologia Animale e dell'Uomo, 00198 Roma, Italy.

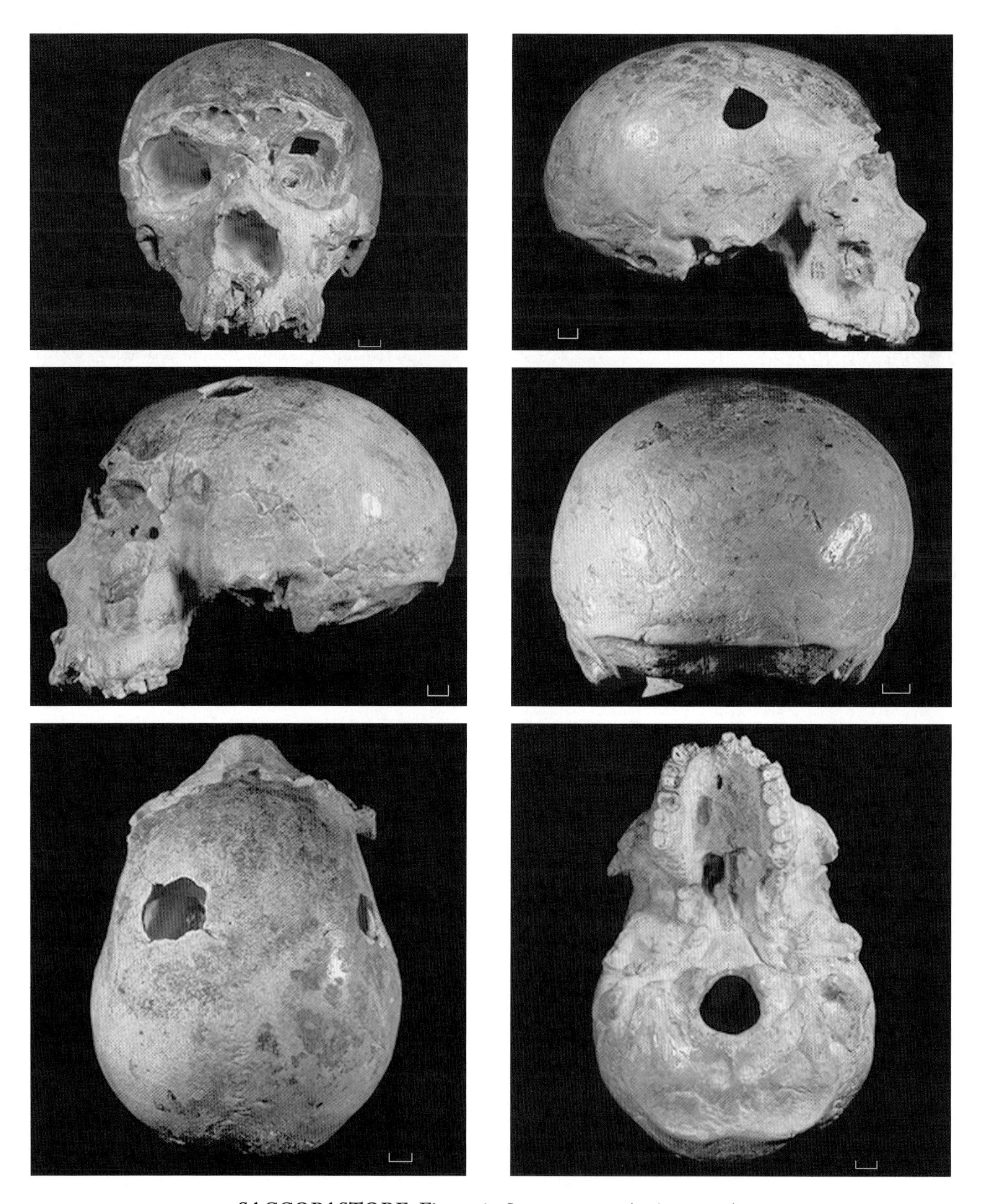

SACCOPASTORE Figure 1. Saccopastore 1 (scale = 1 cm).

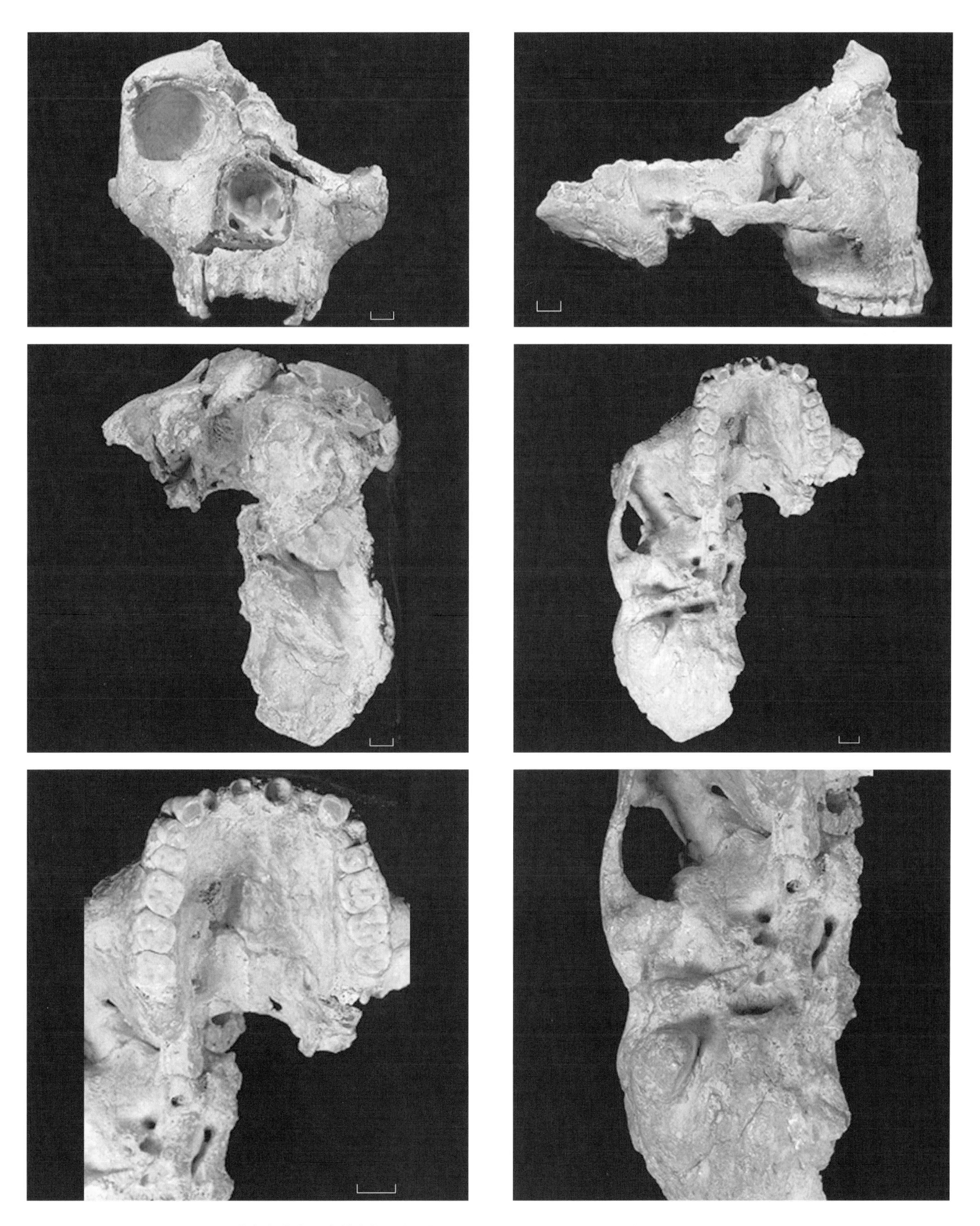

SACCOPASTORE Figure 2. Saccopastore 2 (scale = 1 cm).

Saint-Césaire

Location
La Roche à Pierrot, small collapsed rock shelter beside the tiny river Coran in the village of Saint-Césaire, 12 km E of Saintes, Charente-Maritime, France.

Discovery
Excavations of F. Léveque, July 1979.

Material
Part (mostly the right side) of the front of an adult cranium, with the right half of the mandible, plus some postcranial fragments.

Dating and Stratigraphic Context
Seventeen levels of sands and clays were identified below the rock shelter floor. The lower "gray unit" of the 2.5-m-thick deposit contained eight Mousterian levels, whereas above it the "yellow unit" produced Châtelperronian artifacts from levels 9 and 8 and "early" to "evolved" Aurignacian from levels 6 to 3 (Léveque and Vandermeersch, 1980; Mercier et al., 1991). Mercier et al. (1991) reported TL dates on burned flints from layers containing each of these industries. A dozen Mousterian dates from levels 10–12 scattered around a mean of about 40 ka; those for two Aurignacian flints (level 6) came in at 31 and 34 ka. Six dates from the Châtelperronian level 8, whence the human fossils came, averaged 36.5 ± 2.7 ka. Mercier et al. suggest that the longish apparent time gap between the Mousterian and Châtelperronian levels is due to an erosional discontinuity between them.

Archaeological Context
As noted, the hominid fossils came from the Châtelperronian level 8, which produced an abundant lithic industry with bone points, numerous backed blades, and perforated teeth (Léveque and Vandermeersch, 1980). The human remains, though sparse, were found apparently tightly folded into a small, shallow pit, and probably represent a burial. They represented the first secure association of a specific kind of hominid with the Châtelperronian industry, which is now seen as the end point of the Middle Paleolithic. Exactly how the Châtelperronian comes to show Upper Paleolithic influences is still debated, although most suggestions center on the "acculturation" of Neanderthals (see Mercier et al., 1991) by one means or another.

Previous Descriptions and Analyses
From their first description (Léveque and Vandermeersch, 1980) it has been recognized that the Saint-Césaire human fossils are unquestionably Neanderthal. This identification has never been disputed although the (by now well-established) association of the burial with the Châtelperronian has been disputed (see Bordes, 1981). The late date of the Saint-Césaire Neanderthal is now accepted as good evidence of several thousand years of coexistence in Europe of Neanderthals and moderns (see Stringer and Grün, 1991). It should be noted that although the Saint-Césaire remains are the best direct evidence for the late survival of Neanderthals in western Europe, they are not the latest evidence (see Figueira Brava, Columbeira, and Zafarraya, this volume).

Morphology

Adult. Incomplete R side of skull, lacking occipital, petrosal, part of parietal, squamous part of temporal, internal part of orbit, and all internal structures apart from portions of the R maxillary sinus; upper LI2–RM3 preserved. Appears to be healed cut just to side of sagittal suture, extending a bit forward of bregma (where bone is broken). Mandible lacks coronoid process, part of condyle, and part of gonial angle; has part of L symphysis; preserves LI2–RM3.

Frontal rises quite steeply behind a/p narrow supratoral sulcus. More than three supraorbital foramina on the R. Torus has vermiculate bone; is smoothly rolled; tapers slightly laterally; retreats from glabella, which is broad. Interorbital space would have been very broad. Frontal sinus at least partially preserved on R and L; has three components: a teardrop-shaped medial section behind glabella separated bilaterally by thin septum from lateral sinuses that extend to midpoint of supraorbital torus. Frontal process of maxilla broken inferiorly and along nasal suture (original degree of prominence not ascertainable). Maxillary sinus extends into orbit and frontal process of maxilla; extends a/p from C root to M3. Three moderate-sized zygomaticofacial foramina preserved, from below to above infraorbital margin. No temporal lines visible. Floor of nasal cavity somewhat depressed; anterior nasal spines missing. Small medial projection within margin of nasal cavity.

Squamous portion of temporal was probably not very long. Posterior root of zygomatic arch originates anterior to auditory meatus. Suprameatal crest tiny. Supramastoid crest large, but not very pronounced; angles slightly upward from front to back. Mandibular fossa was probably short from front to back. Articular eminence missing, but sinus cavity lies internal to where it would have been.

Mandibular symphysis devoid of morphology but for slight depression below I1–2s. Symphyseal region wide, broadly arcuate between canines; viewed from below, uniformly thick from front to back. Preserved R digastric fossa quite long, wide; is separated from L counterpart by raised ridge. Large mental foramen lies beneath P2–M1; smaller one beneath mesial root of M1. Slight inferior marginal tubercle lies beneath foramina. Retromolar space large. Gonial angle would have curved smoothly rather than being angled; has slight inward tilt and modest external muscle markings. Condyle compressed a/p in its preserved part. Internally, strong medial pterygoid tubercle. Mandibular foramen high, near sigmoid notch; is oriented obliquely up and back. Mylohyoid line marked but not crestlike; trends downward and forward, forming triangle with inferior mandibular border.

Upper RI1–LI2 worn flat on occlusal surface; lingual surfaces of RI2 and RC worn obliquely. I1s were shoveled; would have had vertical grooves in lingual surfaces. I2s were barrel shape. P1 paracone and protocone quite anteriorly situated (= greater part of crown lies behind transverse axis). M1–3 trigon basin truncated by quite internally placed protocone. M1–2 have very stout postprotocristae. M1 hypocone large, swollen both lingually and distally. M2 hypocone large but less swollen. Buccal roots on M2–3 unseparated; root clefts begin high up.

Lower I1–2s and C worn down to dentine. C has distinct margocristids. P1 b/l and m/d larger than P2. P1 with subequal paraconid and protoconid and small metaconid. P2 also with metaconid and small cusp behind it. M2 b/l and m/d smaller than both M1 and M3. M1–3 with small but distinct trigonid basin (very worn in M1–2), buccally placed hypoconulid, large metaconid, large talonid basin. M3 talonid basin huge, rounded at rear; bounded from hypoconulid to metaconid by continuous ridge. Unworn M3 buccal cusps high. Enamel wrinkling on all molars; best seen in least worn M3. Roots on M1 not splayed; lie close together in M2; not separated at all in M3. Root clefts begin far down.

References

Bordes, F. 1981. Un Néandertalien encombrant. *La Recherche* 12: 644–645.

Léveque, F. and B. Vandermeersch. 1980. Découverte de restes humains dans un niveau castelperronien à Saint-Césaire (Charente-Maritime). *C. R. Acad. Sci. Paris* D291: 187–189.

Mercier, N. et al. 1991. Thermoluminescence dating of the late Neanderthal remains from Saint-Césaire. *Nature* 351: 737–739.

Stringer, C. and R. Grün. 1991. Time for the latest Neanderthals. *Nature* 351: 701–702.

Repository

Laboratoire d'Anthropologie, Université de Bordeaux 1, 33405 Talence, France.

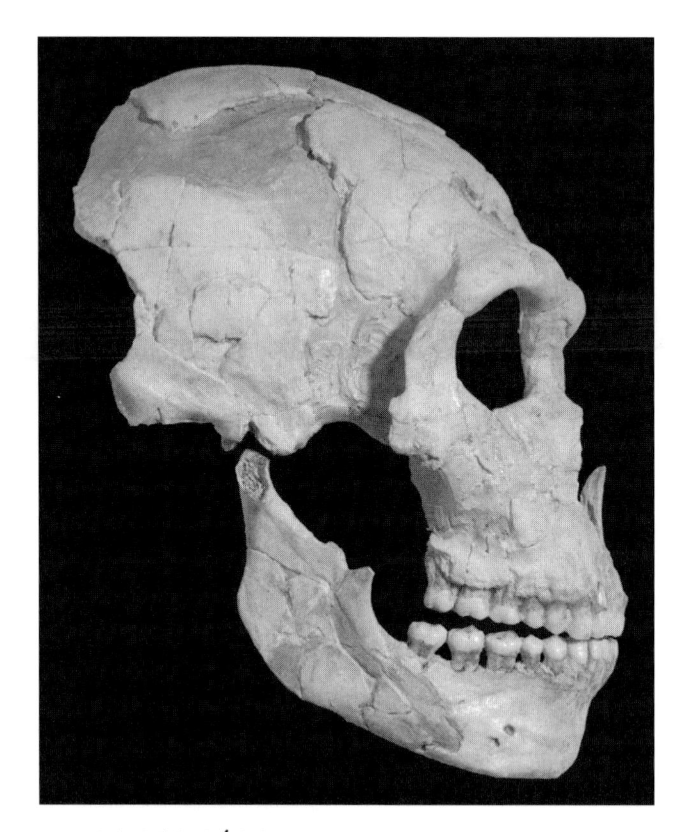

SAINT-CÉSAIRE Figure 1. Not to scale.

Sakajia

Location

Cave in the western Georgian foothills, near Tsutkhva region, E of Kutaisi, Republic of Georgia.

Discovery

Late 1980s, during archaeological prospection.

Material

Partial maxilla, plus isolated lower tooth.

Dating and Stratigraphic Context

Stratified cave site with numerous cave bears (Nioradze, 1992). Probably later Pleistocene.

Archaeological Context

Middle Paleolithic (Lordkipanidze, 1999).

Previous Descriptions and Analyses

Neanderthal (Gabunia et al., 1978; Lordkipanidze, 1999).

Morphology

Adult. Partial L maxilla with C–M1, partial alveoli for RI1–2, and a mysterious partial alveolus adjacent. Broken root of M^2. Also an isolated LM_1 or M_2. Teeth somewhat worn.

Maxilla preserves anterior face of anterior root of zygomatic arch above region of P2–M1. Also preserved is part of floor of apparently capacious maxillary sinus. Preserved prenasal fossa s/i tall, anteriorly facing and quite extensive; low, indistinct extension of lateral nasal crest courses down on to nasoalveolar clivus. Part of nasal cavity floor preserved; is flat. What appears to be a single incisive canal (seen at midline) is quite narrow and vertically oriented. Nasoalveolar clivus was probably quite long; appears to slope forward and down. Palate very deep; preserved side wall almost vertical. Canine crown was large, probably quite tall. Mesial edge flares somewhat; distal edge straighter. Buccal surface moderately arced. Lingual surface with central groove; retains evidence of distinct mesial and distal margocristae that terminated in moderate swellings at base of crown; swellings somewhat coalesced. P1–2 subequal in size; paracones centrally placed and moderately tall; protocones slightly mesially shifted and lower. P1 slightly narrower lingually than buccally; P2 more distinctly ovoid in outline. Each has distinct anterior fovea; P2 has larger posterior fovea. M1 distended lingually by hypocone and distally by thick postcingulum that runs from hypocone to side of metacone. Metacone close to and slightly smaller than paracone, which bears small pillarlike structure on its buccal surface. Buccal cusps appear to have been peripherally placed, the large protocone more centrally placed. Moderate amount of root exposed below neck; does not show root division.

The quite worn isolated lower molar is small, probably ovoid; roots undivided buccally and closely appressed lingually (giving a C shape). Appears there was a moderately large, centrally placed hypoconulid, a distinct trigonid basin, and a notch between protoconid and hypoconid.

REFERENCES

Gabunia, L., M. Nioradze and A. Vekua. 1978. O must'erskom chelovke iz Sakayia. *Voprosy antropologii* 59: 154–164.

Lordkipanidze, D. 1999. The settlement of mountainous regions: a view from The Caucasus. *Anthropologie* (Brno) 37 (1): 71–78.

Nioradze, M. 1992. *Dzeveli Kvis Khanis Mgvime-Namosakhlarebi Tskhaltsitelas Kheobashi.* Metsniereba, Tsibilisi.

Repository

Georgian State Museum, 3 Purtseladze Street, 380007 Tbilisi, Republic of Georgia.

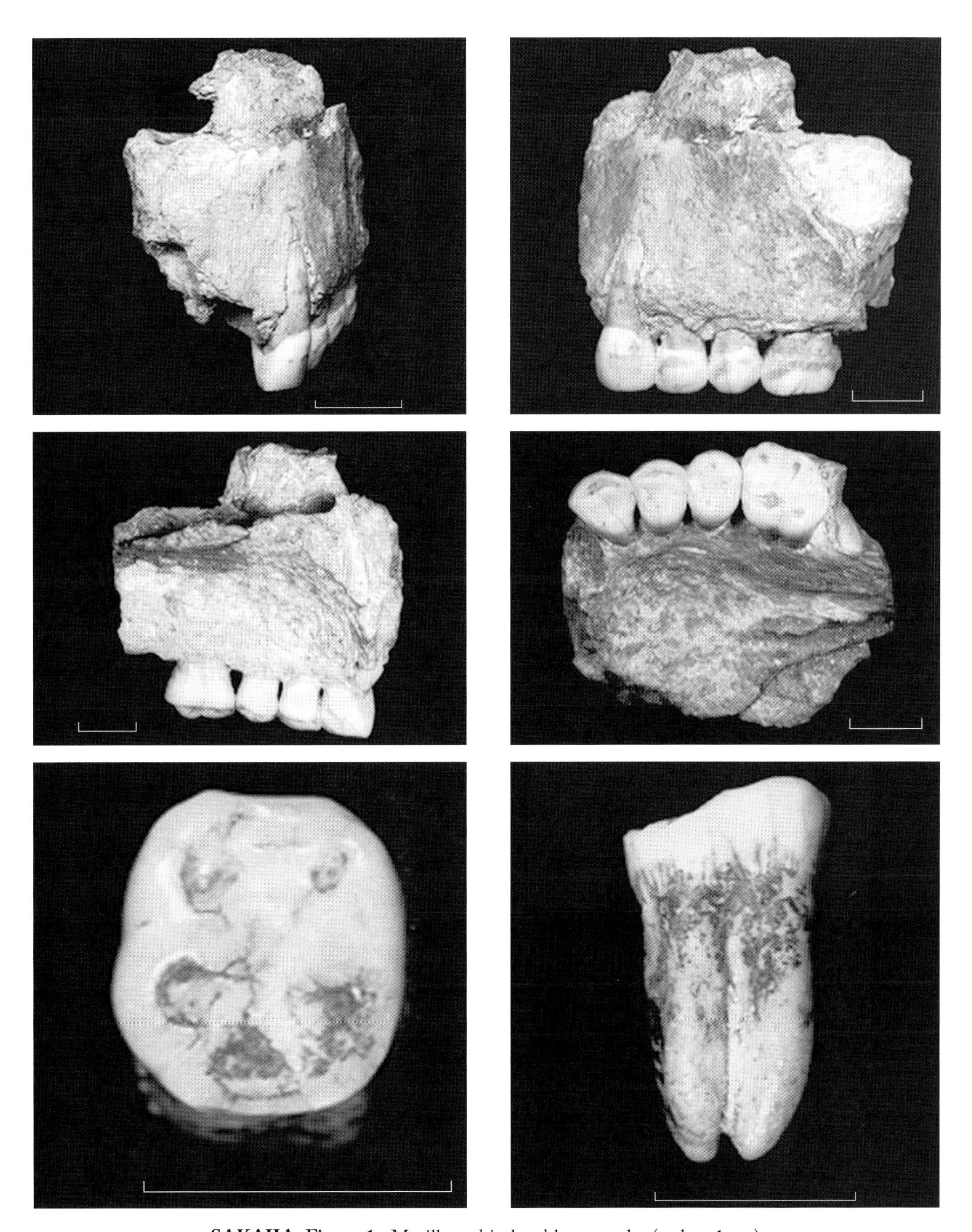

SAKAJIA Figure 1. Maxilla and isolated lower molar (scale = 1 cm).

Scladina (Sclayn)

Location
Limestone cave in the N wall of the valley of the Ri de Pontainne, an ephemeral stream that drains into the river Meuse at the village of Sclayn, near the town of Andenne in Namur Province, Belgium.

Discovery
M. Otte and colleagues, July 1993 (right hemimandible), July, 1996 (left hemimandible). Maxilla and isolated teeth recovered from materials excavated between March 1990 and October 1992.

Material
Matching L and R mandibular halves of adolescent, with L and RM1–2, and RM3 visible in crypt; right maxillary fragment with dm1–2 and M1. Several isolated teeth. All appear to belong to the same individual.

Dating and Stratigraphic Context
The hominid fossils were found quite widely scattered through level 4A of the stratified site. Palynological and micropaleontological studies have placed this level and an overlying stalagmitic floor at about 80 ka (Bastin et al., 1986; contributions in Otte, 1992). However, TL dating of stalagmitic deposit has yielded an age of about 120 ka (Debenham, cited by Bonjean et al., 1997), and Yokoyama and Falguères (cited in Bonjean et al., 1997) obtained a direct gamma spectrometry date on the right hemimandible of ca. 127 ka.

Archaeological Context
The lithic assemblage from Level 4a has been described as Mousterian (e.g., Otte et al., 1993).

Previous Descriptions and Analyses
The Sclayn hominid has not been formally described, but it has been noted that typical Neanderthal traits such as the retromolar space are lacking (e.g., Otte et al., 1993, Bonjean et al., 1997). This lack has been attributed to the young age of the individual.

Morphology
Juvenile. Fragment of R maxilla with alveoli for I1–2, C, dm1–2, and M1. Also R hemimandible behind partial C alveolus; coronoid process and condyle present. Lower M1 in place; M2 not fully erupted; M3 in crypt; one premolar in crypt. Some cutmarks visible. Probably ca. 10–11 years old.

Maxillary sinus expansive both medially and laterally; descends almost to level of nasal floor, which steps down slightly posteriorly just behind tiny preserved portion of nasal aperture. R incisive fossa in nasal cavity floor just visible; descends at angle, then turns down into vertical incisive canal that emerges just behind I1. Incisive foramen would have been very far forward. Nasoalveolar clivus was apparently relatively vertical. Above M1, anterior root of zygomatic arch begins to sweep outward. Palate not very deep; has moderately steep slope anteriorly and steeper slope at the side.

Mandible rather gracile; corpus somewhat shallow. Modestly sized mental foramen lies below P2; smaller secondary foramen below and behind. Appears that postincisal plane was quite vertical, with some swelling of bone more inferiorly. Area of genial pits missing. Anterior margin of ramus bears well-defined, s/i tall , a/p shallow preangular notch. Surface of gonial angle only modestly rugose; slight vertical swelling below sigmoid notch. Gonial angle damaged; was evidently inwardly inflected, not sharply cornered. Internally, angle bears series of small swellings that culminate in vertically elongate but only modestly developed medial pterygoid tubercle. Quite sharply pointed coronoid process rises fairly well above level of condyle. Sigmoid notch moderately deep, mildly asymmetrical, deepest just behind midpoint of its arc. Sigmoid notch crest runs to midpoint of condyle. Moderately sized, ovoid mandibular foramen points up and back; is roofed by moderately long but not very projecting lingula that is defined on its posteroinferior side by narrow mylohyoid notch. Thin, distinct pillar runs down midline of internal surface of coronoid process to fade out behind region of M1. Faint mylohyoid line rises below M2/3 septum; small (submandibular) fossa lies anteriorly beneath.

Alveolus for upper I1 large, round; I^2 alveolus more compressed. C^1 alveolus very long b/l but very compressed m/d. P^1 had started erupting (part of a septum visible). Three-rooted dm^1 very worn; paracone mesiobuccally positioned with buccally swollen enamel at base (cusp lies over mesiobuccal root); no metacone over distobuccal root; protocone lies over lingual root and quite distal to paracone; apparently thick postprotocrista had been present. dm^2 very worn, molariform with three well-splayed roots. Protocone was quite centrally placed; moderate preprotocrista ran to mesially placed paracone; possibly a somewhat stouter postprotocrista ran to metacone. Hypocone moderately sized; thick postcingulum curves from it gently around to metacone, enclosing a small, relatively deep basin. M^1 fully erupted; more bulbous lingually than buccally; quite long. Protocone quite centrally positioned; distinct preprotocrista runs around front of paracone; base of metacone extends quite far lingually but does not form complete postprotocrista. Hypocone large, swells out distolingual corner of tooth; from hypocone, thick postcingulum runs to base of metacone, enclosing well-defined basin. Trigon basin truncated lingually. Roots bifurcate fairly close to neck.

Lower C root was quite big; posterior wall of alveolus bears faint vertical pillar. P_1 root was quite wide b/l and compressed m/d; alveolus bears faint pillars on both sides. Beneath the two b/l wide alveoli for now missing dm_2 is largely hidden crown of P_2. M_{1-2} both relatively small, generally ovoid teeth, with slightly more bulbous buccal than lingual sides. Buccal cusps slightly centrally shifted; lingual cusps peripherally placed. On both, stout paracristid runs between bases of metaconid and protoconid (which lie opposite one another), encloses small, well-defined basin. Pillar runs directly lingually down internal face of protoconid; "deflecting wrinkle" runs buccodistally to contact mesiolingually distended base of hypoconid, which extends lingually beyond midline of crown. Moderately developed hypoconulid lies just to buccal side of midline in both teeth; at least as preserved on M_2, small posterior fovea lies just lingual to hypoconulid. As seen on M_2, and M_3 still in crypt, enamel thickly crenulated. In both M_{1-2}, talonid basins long and buccolingually truncated.

REFERENCES

Bastin, B. et al. 1986. Fluctuations climatiques enregistrées depuis 125,000 ans dans les couches de remplissage de la grotte Scladina (Province de Namur, Belgique). *Bull. Assoc. Fr. Etude Quat.* 1/2: 168–177.

Bonjean, D. et al. 1997. Grotte Scladina (Sclayn, Belgique): Bilan des découvertes néandertaliennes et analyse du contexte. *Cinq. J. Arch. Namuroise* 5: 19–27.

Otte, M. (ed.) 1992. *Recherches aux grottes de Sclayn*, Vol. 1. *Le Contexte.* Liege, ERAUL.

Otte, M. et al. 1993. Découverte de restes humains immatures dans les niveaux Moustériens de la grotte Scladina à Andenne (Belgique). *Bull. et Mém. Soc. Anthropol. Paris*, n.s. 5: 327–332.

Repository

Direction d'Archéologie, 62 ave des Tilleuls, 4000 Liège, Belgium.

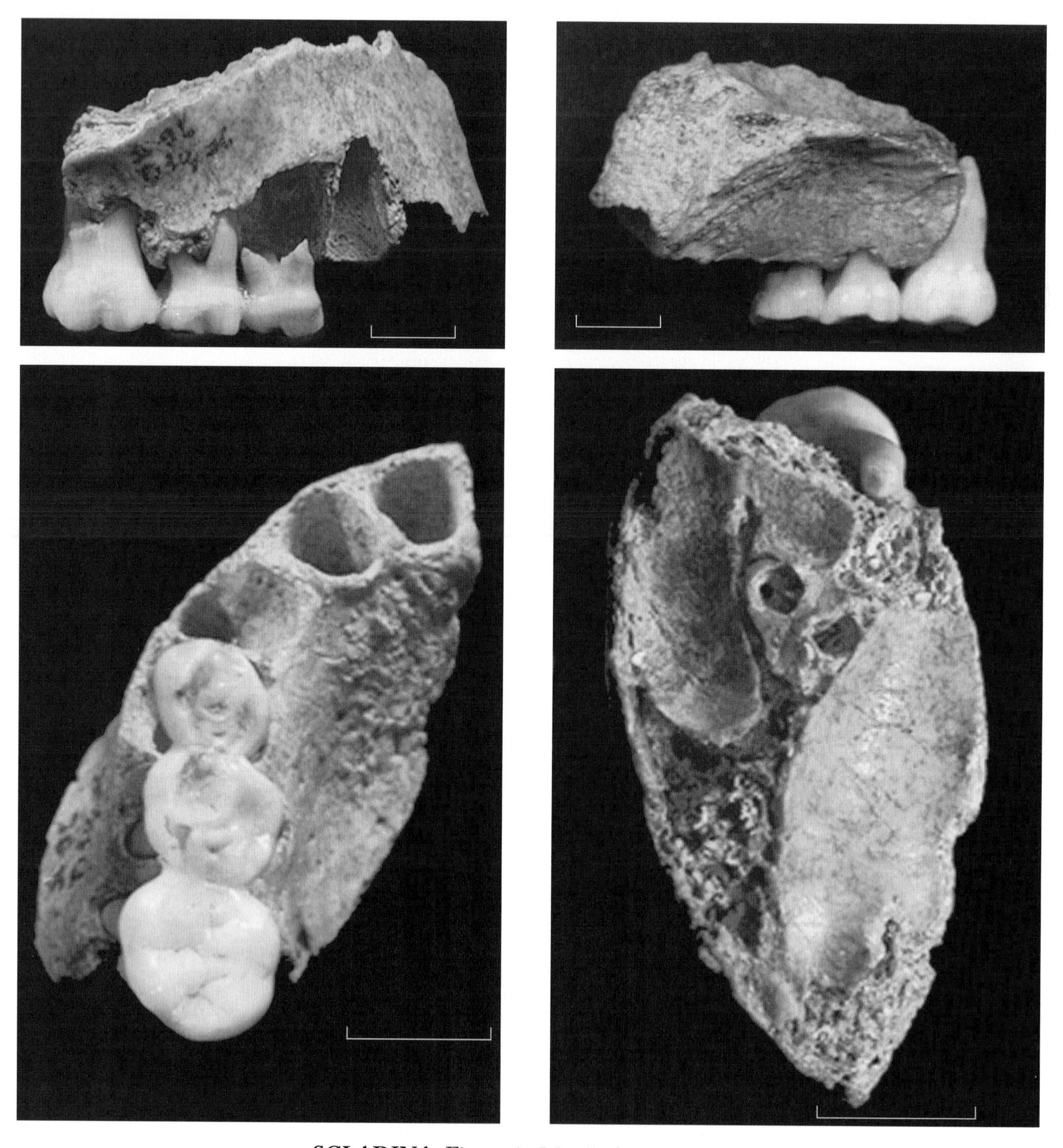

SCLADINA Figure 1. Maxilla (scale = 1 cm).

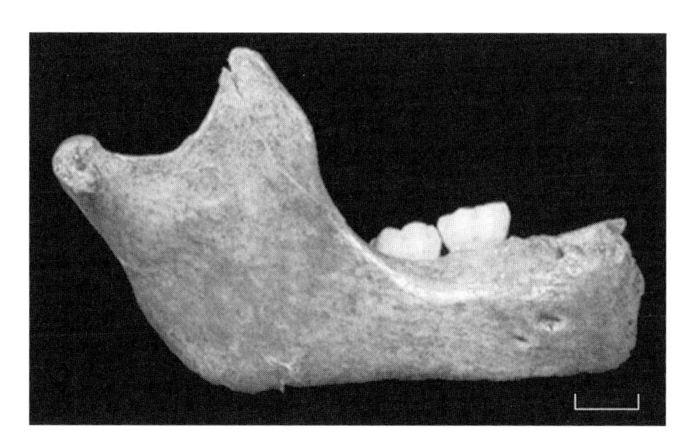

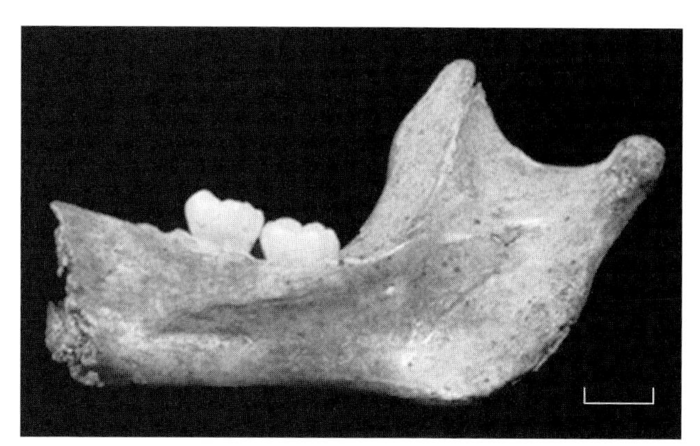

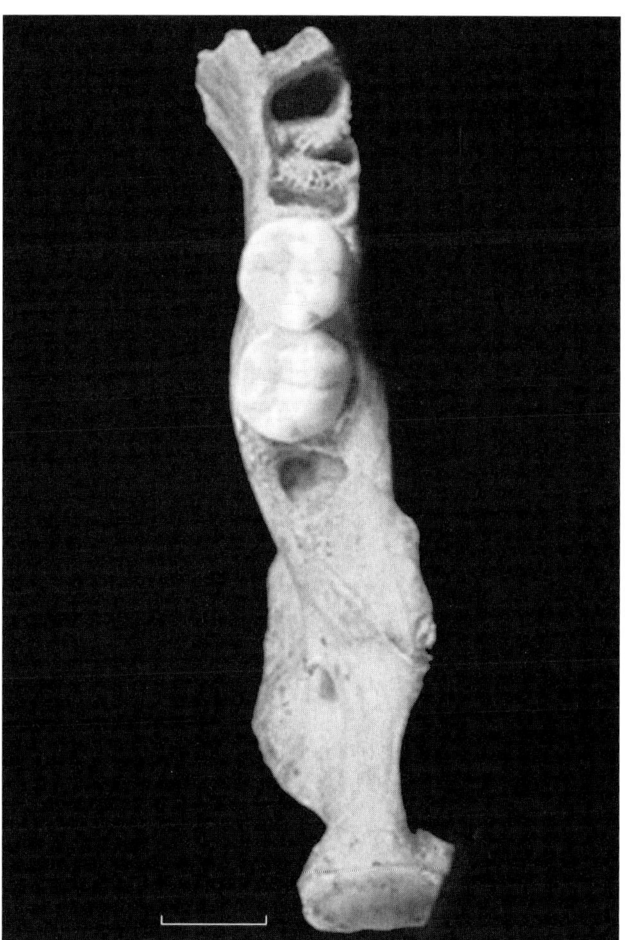

SCLADINA Figure 2. Mandible (scale = 1 cm).

Sipka

Location

Cave site on the N slope of Kotouc Hill, near the village of Stramberk, northern Moravia, Czech Republic.

Discovery

K. Maska, August 1880 (mandible); E. Grepl, September 1966 (isolated premolar).

Material

Juvenile mandible fragment with anterior teeth; isolated LP_1.

Dating and Stratigraphic Context

Stratified cave site, quite well dug by the standards of the time. The hominid was said by Maska (1885) to be part of a burial contemporaneous with the occupation level, containing hearths, in which it was found. The associated fauna contained extinct mammals (Maska, 1885) and was later interpreted as early Würm (Absolon et al., 1933). The isolated premolar apparently came from an equivalent level (Grepl, 1972/3).

Archaeological Context

The fairly abundant associated lithic industry is clearly Mousterian in character. Absolon (1933) gave it its own name, "Sipkian," as a distinct facies of the eastern Mousterian.

Previous Descriptions and Analyses

Fragmentary though it is, this fossil is of considerable historical importance as the first Neanderthal fossil to be widely debated after the discovery of the Feldhofer individual. Schaaffhausen (1880) hailed the Sipka jaw as a representative of *Homo primigenius* (i.e., *Homo neanderthalensis*) but almost immediately met with objections from Virchow (e.g., 1882) that the specimen was an abnormal modern adult. It was not long, however, before the finds at Spy in 1886 confirmed the distinctiveness (and normality) of Neanderthal morphology, and well before the end of the nineteenth century the Sipka fragment was universally accepted as Neanderthal, an allocation that has remained uncontested since. Grepl (1972/3, p. 171) described the isolated premolar as that of a "forerunner of *Homo sapiens*."

Morphology

Juvenile mandible fragment (Cast, K-350, 1880) with R–L I1–2, plus C–P2 in varying stages of eruption. Symphyseal region unembellished but for subalveolar depression; inferior margin seems to have angled upward. Viewed from below, symphyseal region uniformly thick from front to back. Digastric fossae quite large, broad. I1–2 crowns angle back on relatively short but stout roots. C crown tall. P1 large relative to P2; had some lingual development.

References

Absolon, K. 1933. Ograve podstate palaeolitichych industrii ze Sipky a Certovy diri na Morave. *Anthropologie (Praha)* 11: 253–272.

Absolon, K. et al. 1933. *Bericht der czechoslovakischen Subkommission der "Internationalen Geologischkongressen." Brunn 1933*: 1–31.

Grepl, E. 1972/3. Die archäologische Forschung der Höhle Sipka in den Jahren 1966 und 1967. *Acta Archaeol. Carpathica* 13: 161–172.

Maska, K. 1885. Celist predpotapniho cloveka nalezena v Sipce u Stiamberka. *Cas. Vlast. Muz. Spolku Olomuck.* 2: 27–35.

Schaaffhausen, H. 1880. Funde in der Sipkahöhle in Mähren. *Sber. Niederrein Ges. Nat.-u. Heilk.* 260–264.

Virchow, R. 1882. Der Kiefer aus der Schipka-Höhle und der Kiefer von La Naulette. *Z. Ethnol.* 14: 277–310.

Repository

The lower jaw fragment was destroyed in the fire at Mikulov Castle in 1945. A cast (K-350, 1880) remains in the Anthropos Institute, Moravske Muzej, Brno, Czech Republic. Whereabouts of premolar unknown.

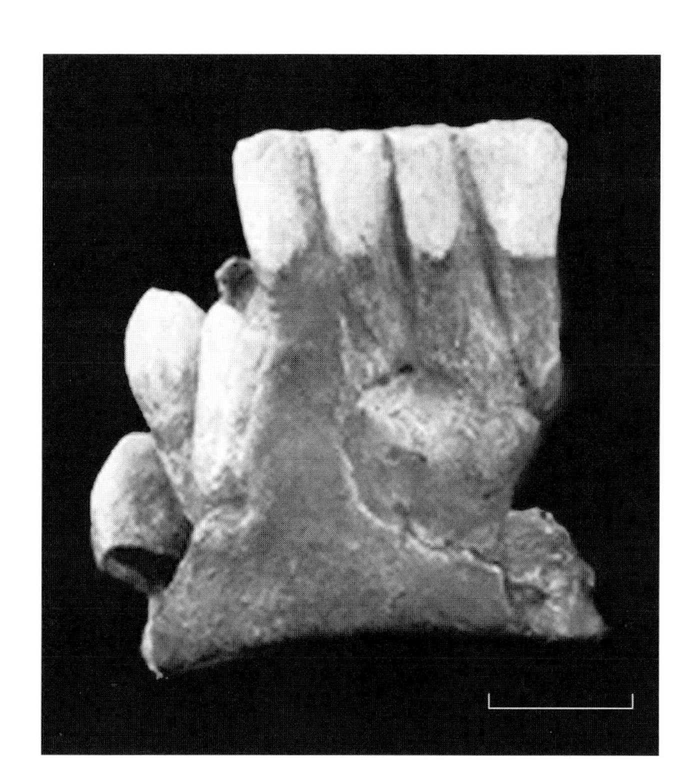

SIPKA Figure 1. Scale = 1 cm.

Spy

Location
Cave in the limestone bluff of Betche-aux-Roches, near Spy village, ca. 15 km ENE of Namur, Belgium.

Discovery
M. Lohest and M. de Puydt, July 1886.

Material
Two partial adult skeletons including calottes and some facial elements (Spy 1 and 2); two teeth and tibia of a juvenile (Spy 3).

Dating and Stratigraphic Context
Stratified terrace deposit in front of cave entrance. The skeletons (possibly burials) were found low in the sequence, in brown clay deposits containing a late Pleistocene fauna indicative of cool conditions. Dating is problematic; Zeuner (1940) estimates the early part of the last glacial.

Archaeological Context
Mousterian.

Previous Descriptions and Analyses
Described by Fraipont and Lohest (1886), who immediately recognized the remains as Neanderthal and documented them more completely the following year (Fraipont and Lohest, 1887). These authors' analysis confirmed that Neanderthal morphology was normal, not pathological. Unfortunately, however, it also suggested to them that their hominid had walked bent-kneed. Krause (1909) placed the Spy hominids in the new species *Homo spyensis*, but their Neanderthal affinities have never subsequently been in doubt, and their fortunes have fluctuated along with those of the Neanderthals as a whole. Estimated cranial capacity is 1553 cc for Spy 1 and 1305 cc for Spy 2 (Holloway, 1985). Vlcek (1993) reports that Fraipont and Lohest calculated the cranial capacity of Spy 2 at 1723 cc, but later E. Patte estimated it at 1425 cc.

Morphology

Spy 1
Adult, possibly male, said to be ca. 35 years old. Calotte with partial maxillae, missing most of facial skeleton and basicranium. Most of L parietal and squamous portion of temporal reconstructed. R maxillary fragment preserves alveolae for I1–2, plus C–M3; smaller L maxilla fragment has C–P2. Also parts of R and L mandibular copora and rami; all lower teeth present. Some isolated teeth. Some potential confusion of postcranial elements between Spy 1 and 2.

Frontal rises relatively steeply from posttoral plane; almost immediately slopes shallowly back. This profile maintained to about midpoint of sagittal suture; then slopes more steeply down toward lambda. At lambda, occipital bulges slightly; bulge truncated by undercutting of nuchal plane. Skull narrowest posttorally; broadest at region of parietomastoid suture. Viewed from rear, cranium is "en bombe."

Smoothly rolled, not very tall s/i supraorbital tori confluent across glabella; taper slightly toward sides. Seen from above, tori retreat gently from glabella. Posttoral plane moderately long, almost flat; dips somewhat above glabellar region. Interorbital region very broad; glabella moderately prominent but flattish. Frontal sinuses large; extend far up into frontal, and laterally beyond orbital midline; were partially subdivided by septa. Frontal lobes did not extend forward as far as region of posttoral plane. Temporal lines most crisply developed as they emerge from zygomatic processes of frontal; quickly turn backward to fade into low muscle scars that arc gently posteriorly, turning in more sharply above posterior end of parietomastoid suture, where they curve forward toward parietal notch.

On the R, part of gracile frontal process of zygoma preserved. Zygoma was probably oriented obliquely laterally and somewhat medially, such that side wall of orbit slanted inward from top to bottom; huge maxillary sinus extends far laterally its body.

In R maxilla, inferolateral corner of nasal aperture preserved; lies high up (suggesting that there was downward "step" to nasal cavity floor). Preserved fragment of nasal margin quite sharp. Maxillary sinus continues well forward along nasal cavity wall, extending up into frontal process of maxilla. Anterior root of zygomatic arch broken off; could have originated relatively close to alveolar margin; was distinctly inclined laterally. Nasoalveolar clivus quite vertical. L maxillary fragment contains C–P2 but no notable morphology.

Posterior root of zygomatic arch originates well anterior to external auditory meatus. Fairly prominent, upwardly sweeping supramastoid crest runs forward into thin, low suprameatal crest. Crest confluent with more laterally swollen posterior root of zygomatic arch. Arch runs straight forward. Small crest runs inferior and parallel to suprameatal crest, terminates inferiorly in small anterior mastoid tubercle level with, and opposite to, external auditory meatus. Very deep mandibular fossa a/p long. Articular eminence steep, wall-like; is restricted both anteriorly and posteriorly; is bounded medially by short, thick medial articular tubercle that continues wall-like morphology of articular eminence medially. Posteriorly, mandibular fossa limited by peaklike elevation of vaginal process, which wraps around tiny, thin styloid process. Vaginal process runs along long axis of ectotympanic tube; fades out well before reaching its lateral end; does not contact mastoid process. Styloid process located very close to carotid foramen and directly in line with mastoid notch behind. Notch broad, shallow anteriorly; deepens posteriorly where it terminates in well-marked U-shaped scar. Large stylomastoid foramen somewhat separated laterally, and slightly posteriorly, from base of styloid process. Ectotympanic tubes of fair diameter but incompletely ossified laterally on both sides, thus short. Carotid foramen large, medially positioned. On the R, much of long but faint Waldeyer's crest preserved.

Mastoid processes thick at base; point directly downward; do not extend far inferiorly (judging from L side, although slightly broken inferiorly, process did not extend inferiorly as far as region of occipitomastoid suture). Parietomastoid suture moderately long, horizontal; flows into moderately long, horizontal anterior lambdoid suture. Occipital wide but low (= very short from lambda to "torus"). Lambdoid suture rises almost at right angles to arc flatly across region of lambda. Occipital "torus" horizontal, well delineated inferiorly by broad infratoral depression; is defined above by horizontally partitioned and pitted suprainiac depression.

As seen on R maxilla, palate quite deep, sloping at front, vertical at sides.

Lambdoid and sagittal sutures partly visible; neither segmented or deeply denticulated.

Internally, frontal crest, thick, low, not very long. What is preserved of petrosals is very broad laterally; region of arcuate eminence low, laterally broad; superior petrous sinus not excavated; no definable subarcuate fossa, although on both sides is sub-subarcuate crease. Jugular fossa relatively deep. Cochlear canaliculus very large, deep. L petrosal broken, exposing two large air spaces. Break on R petrosal, just lateral to jugular fossa, reveals cavernous sinus. Superior sagittal sinus preserved only in the occipital, where it is relatively deep, narrow, and delineated on each side by a crest. L transverse sinus more prominent than R; apparently stayed entirely within occipital as it ran to become a deep, short sigmoid sinus.

Mandible lacks both condyles and region below including mandibular foramina, portions of coronoid processes, angles (including region of medial pterygoid tubercles); contains all teeth. Digastric fossae large, well marked, wide m/l; face obliquely back. Mylohyoid line well developed, very oblique; distinct submandibular fossa below. Symphyseal region broad, arcuate from side to side; external surface bears subalveolar depression that is emphasized by outwardly curving anterior tooth roots, otherwise devoid of morphology. On the L, well-defined inferior marginal tubercle lies below M1; inferior margin of symphyseal region in front of tubercles elevated. Viewed from below, bone of symphyseal region essentially uniformly thick from front to back and not appreciably thicker than corpora behind. Mental foramen (both sides) lies below M1; is not notably large. Internally on symphyseal region, postincisal plane tall s/i, almost vertical. Genial pits confluent above vertical median divider. Retromolar space huge, emphasized by what would have been s/i tall, but shallow, preangular notch. Stout crest runs down center of preserved bases of coronoid processes. On the R, this crest becomes confluent with internal alveolar crest posteriorly ascending mylohyoid line.

Upper dentition represented by RC–M3 and LC–P2. Cs very worn; fairly slender crowns angled in on roots. P1–2 crowns more rectangular than lingually tapering; mesially placed paracone and protocone lie opposite one another. P1s very worn; crowns angled in on roots; crowns somewhat broader b/l and a little longer m/d across paracone. In general, upper molars subequal in width b/l; do not diminish much in size from M1 to M3. M1 highly worn; protocone was probably quite internally placed; metacone was much smaller than paracone and quite close to it; large hypocone swells out distolingual corner of crown, giving skewed outline. M2 also quite worn; seems protocone was somewhat internally placed; metacone was not much smaller than paracone; hypocone only moderately swells out d/l corner of crown. Appears to have been thick postcingulum on both M1 and M2. Wear obscures cusps on M3; as in M2, protocone somewhat inwardly placed, paracone smaller than protocone, and hypocone quite large but does not swell out d/l corner of crown.

All lower anterior teeth very worn, C to C; roots of all curve outward, crowns may have angled back. I1s were quite compressed m/l; somewhat larger I2s probably flared laterally. Lower Cs were probably originally concave lingually, with central pillar. P1s very worn; protoconid was mesially placed; stout crest ran lingually and down toward neck, leaving distinct notch between it and lingual swelling distal to it; distal moiety of tooth was much larger than the mesial. P2s very worn; metaconid and protoconid mesially placed and opposite one another (thus distal part of tooth much larger than mesial part). Molars all ovoid in outline; hypoconulids lie buccal to midline; lingual cusps slightly centrally placed; buccal ones more peripheral (thus buccal sides of crowns look more swollen than straighter lingual sides). M1–2 subequal in size; M3 slightly smaller. Especially as judged by RM1–2, root divergence begins relatively close to crown (= teeth not taurodont).

Spy 2

Partial skeleton said to be of a male ca. 25 years old. Craniodental remains consist of calotte missing facial skeleton, most of basicranium, parts of R and L greater sphenoid wings of sphenoid, small fragments of R (with P2–M2) and L (with M3) maxillary alveolar regions; parts of R (with P1–M3) and L (with M1–3) mandibular corpora and rami, lacking gonial angles, condyles, and coronoid processes. Upper LM1–2 isolated; roots of M2 fit perfectly into alveolus of L maxillary fragment. Also isolated RI^1, LI^2, and LP^{1-2}, as well as RC_1 and LP_2. All teeth very worn. In general, bone of cranial vault not very thick.

Viewed from the side, rise of frontal short vertically, with long sloping plane up to coronal suture. This plane continues to midpoint of parietals, at which point profile slopes back and down occipital to an almost vertical suprainiac plane in which lies pitted depression. Suprainiac plane then sharply undercut by nuchal plane. Viewed from above, skull narrow at posttoral region; there is very little postorbital constriction. Greatest width lies far back on skull, above region of parietomastoid suture. Viewed from behind, braincase is quite “en bombe.”

Supraorbital tori smoothly rolled although not very tall s/i (even medially); taper a bit laterally; retreat in superior view from where glabella would have been; bone vermiculate. Supratoral sulcus flattish but strongly curved posteriorly because of sharp frontal rise; sulcus narrower than in Spy 1. Missing glabellar region reveals moderate frontal sinuses that extend laterally to orbital midline but superiorly only to point at which frontal rises; portions of multiple septa present.

Low, distinct scar of temporal lines arcs sharply up from edge of zygomatic process of frontal, then curves gently up and back, following arc of squamosal suture; continues as far back as region of parietomastoid suture, then curves in steeply toward parietal notch. Squamosal appears to have been relatively short a/p, coronally arced; reconstruction obscures configuration of anterior squamosal suture. Appears that temporal fossa would not have been very deep postorbitally. Part of sphenoid preserved on the R; shows sphenoid sinus had expanded laterally as far as foramen ovale. On the temporal, at the parietal notch, is strong, bulging supramastoid crest that arcs up and back, is continuous anteriorly with lower suprameatal crest that flows into more laterally expansive posterior root of zygomatic arch.

Mandibular fossae very deep (although deepest anteriorly), relatively long a/p, laterally oriented; narrow from side to side although open laterally; bounded anteriorly by flat, wall-like "articular eminence" with sharp edge in front (as in Spy 1). Heel-like medial articular tubercle very prominent; forms as a medial extension of the articular eminence. Vaginal process weakly developed; is most prominent around base of slender styloid process (both sides); may not have reached edge of ectotympanic tube; did not contact mastoid process. Styloid process very medially placed (more than Spy 1); on the L, lies moderately close to very medially placed, fairly large carotid foramen. Both stylomastoid foramina large; on R, foramen is separated noticeably from styloid process; on the L, foramen lies at its base. Styloid process lies in line with broad, moderately deep mastoid notch. Anterior part of notch more filled in than posterior part, which broadens behind into shallow depression (= digastric fossa?).

Moderately sized posterior root of zygomatic arch (seen on L) makes right angle and runs straight forward. Ectotympanic tube rather long but not completely ossified laterally. As in Spy 1, a crest runs up and back from behind external auditory meatus, ending in swollen anterior mastoid tubercle below. Suprameatal crest not as well developed as upwardly curving supramastoid crest. Mastoid process relatively short a/p; apparently did not project much; points downward. Wavy parietomastoid suture is relatively long. Nothing certainly preserved of either paramastoid or occipitomastoid crests. On the R, is most of bulky, anteriorly directed Waldeyer's crest. On the L, jugular fossa broken but still visible; cochlear canaliculus huge.

Occipital plane short, arcuate. Lambdoid suture forms continuous curve, going flatly across the region of lambda. Occipital "torus" was broad, horizontal; upper border of "torus" confined to lower part of suprainiac depression; inferior border well delineated by very broad infratoral depression (which lies above where short anterior lambdoid suture turns to run upward). Suprainiac depression quite long, roughened, not very tall s/i; not so much pitted as dotted with little local elevations.

As seen in R maxillary fragment, palate deep; maxillary sinus penetrated quite far toward alveolar region.

Coronal, sagittal, and lambdoid sutures weakly denticulate, unsegmented.

Internally, frontal lobes lay right over orbital cones. On the L, posterior part of petrosal broken away, revealing interior; huge sinus (vacuous, not pneumaticized) extends from mastoid process along inferior part of body of petrosal (other sinus cavities encased by petrosal may also lie more anteriorly within petromastoid complex, because tomogram reveals this whole, very broad area is greatly sinused). On the L, superior surface of petrosal flat, quite wide m/l; lacks arcuate eminence. There is long, faint groove for superior petrous sinus. There is very small subarcuate fossa on the L, apparently also on the R; also subsubarcuate crease on the R. Superior sagittal sinus most distinct in region of parietals; becomes crestlike on the occipital, thickening to region of internal occipital protuberance. Segments preserved of subequal but poorly preserved transverse sinuses. Sigmoid sinus moderately impressed on both sides. Sigmoid and transverse sinuses confined to occipital and petromastoid regions; do not overlap onto parietals. Cerebellar impressions quite deep.

On R and L mandibular fragments, retromolar space relatively large, with what appears to have been well-defined preangular sulcus above. Small mental foramen lies below M1 on each side. Internally, mandibular foramen long, very compressed, oriented up and somewhat back; very heavy, oblique mylohyoid groove. Lingula missing; apparently ran length of foramen. Low, thick crest descends down center of base of coronoid process; becomes confluent with internal alveolar margin and mylohyoid crest.

RI^1 very worn; crown angles in from neck. LP^{1-2} subequal in b/l dimension; P^1 slightly longer m/d, especially buccally; paracone and protocone, with deep crease between bases, opposite each other and mesially placed (also seen on RP^2). Especially on P^1, crown angles in on root. Root tips bifid, slightly more separated on P^1; deep groove extends more than halfway up root on lingual side. All upper molars very worn;

protocone somewhat inwardly placed; hypocone least distolingually enlarged on M^3, most on M^1. M^2 larger than M^1; M^1 larger than M^3. M^{1-2} metacone much smaller than paracone (more so on M^2); both cusps quite peripheral. M^3 metacone relatively larger. All molars have three roots; as seen especially on M^{1-2}, roots diverge quite close to neck.

RC_1 and RP_1 crowns angle in on root. RC_1 not very large crowned; apparently had quite distinct lingual tubercle. P_1 has notch between mesial crest and lingual swelling; just distal and internal to this is small fovea. P_1 slightly longer m/d than P_2, especially buccally. P_2 seems slightly longer m/d on its lingual than buccal side. P_1 has slightly bifid root tips, two longitudinal grooves buccally; roots bifurcate quite close to the neck. P_1 protoconid somewhat mesially placed. P_2 protoconid and metaconid lie opposite each other, are quite mesially placed. On LP_2 is thick postcingulid distal to metaconid; roots stout. Lower molars very worn, ovoid in plan; buccal cusps more centrally and lingual cusps more peripherally placed; hypoconulids lie just to buccal side of midline of crown. As seen on both M_1s, root bifurcation quite close to neck. Isolated LI^2 barrel shaped. Isolated RI^1 was probably shoveled. Isolated LP^{1-2} with posteriorly shifted buccal cusps, bifid root tips.

REFERENCES

Fraipont, J. and M. Lohest. 1886. La race humaine de Néanderthal ou de Cannstadt, en Belgique. *Bull. Acad. R. Belg., C. Sci.* 12: 741–784.

Fraipont, J. and M. Lohest. 1887. La race humaine de Néanderthal ou de Cannstadt, en Belgique. Recherches ethnographiques sur les ossements humains, découverts dans des dépots quaternaires d'une grotte à Spy et détermination de leur âge géologique. *Arch. Biol. Paris* 7: 587–757.

Holloway, R. L. 1985. The poor brain of *Homo sapiens neanderthalensis*; see what you please. In: E. Delson (ed), *Ancestors; The Hard Evidence*. New York, Alan R. Liss, pp. 319–324.

Krause, W. 1909. Anatomie der Menschenrassen. In: R. Bardeleben (ed), *Handbuch der Anatomie des Menschen*, vol. 1, p. 3.

Vlcek, E. 1993. *Fossile menschenfunde von Weimer-Ehringsdorf*. Stuttgart, Konrad Theiss Verlag.

Zeuner, F. 1940. The age of Neanderthal Man. *Occ. Paps Inst. Archaeol. Univ. Lond.* 3: 3–20.

Repository

Institut Royal des Sciences Naturelles de Belgique, Brussels, Belgium.

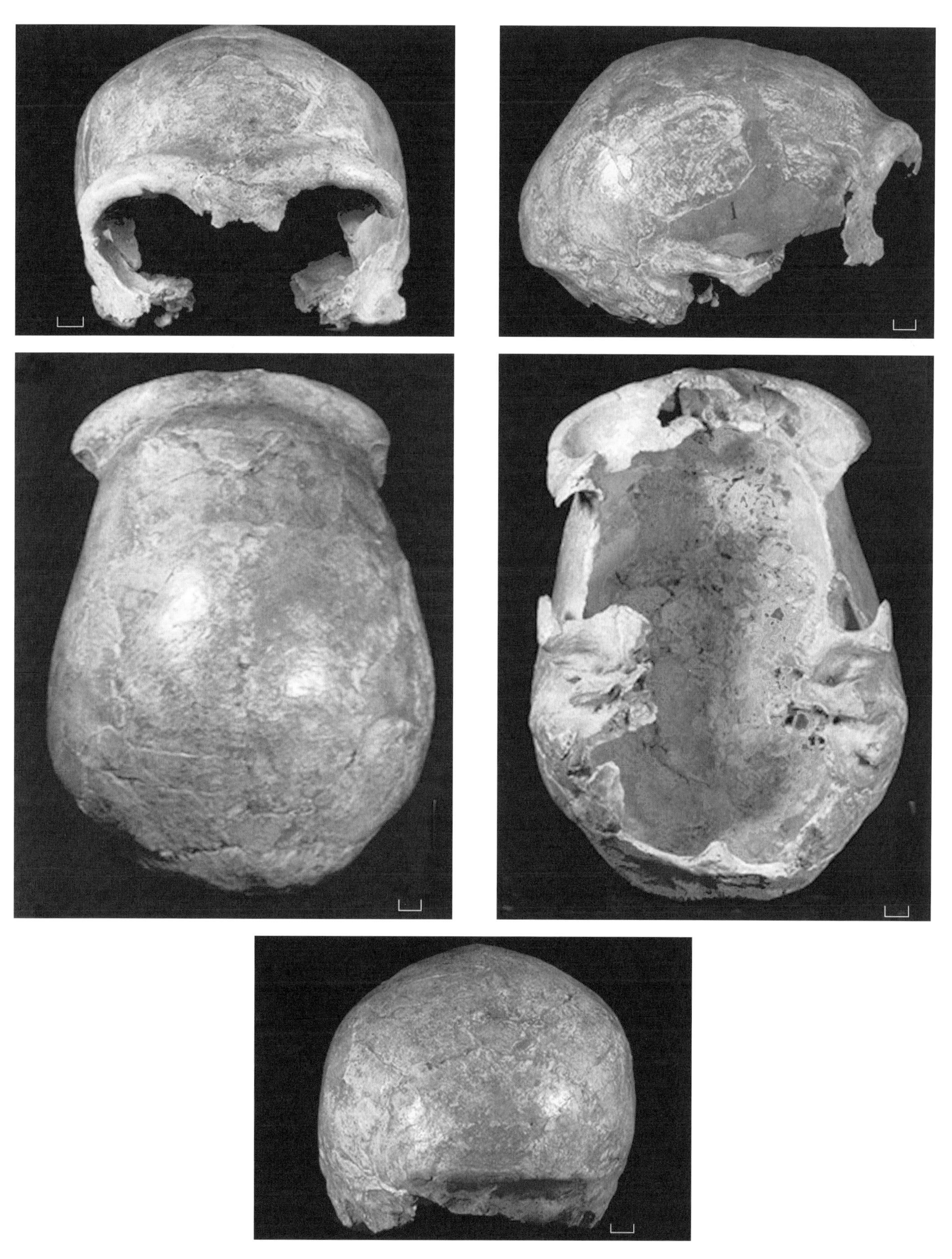

SPY Figure 1. Spy 1 (scale = 1 cm).

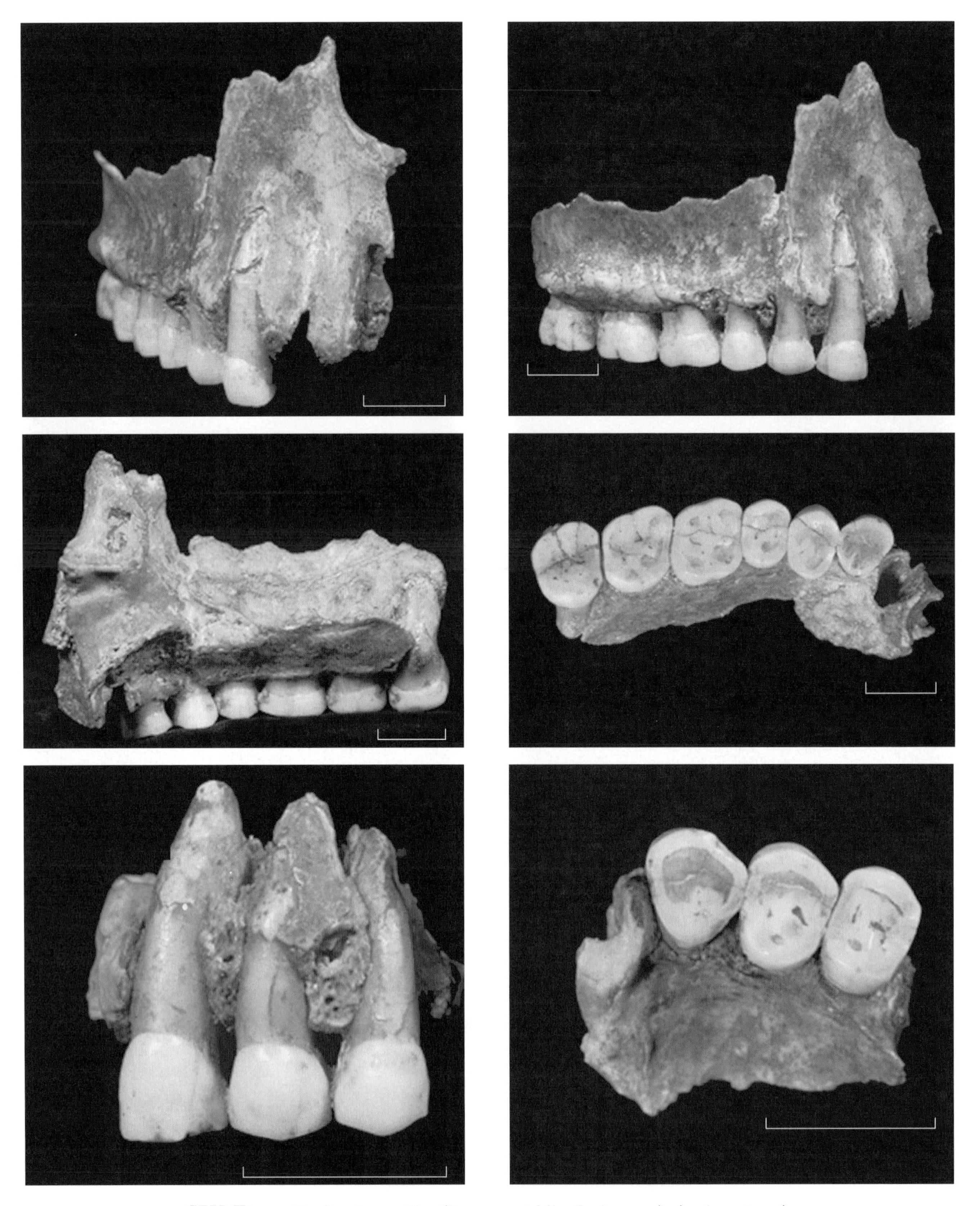

SPY Figure 2. Spy 1 maxillae (R: top, middle; L: bottom), (scale = 1 cm).

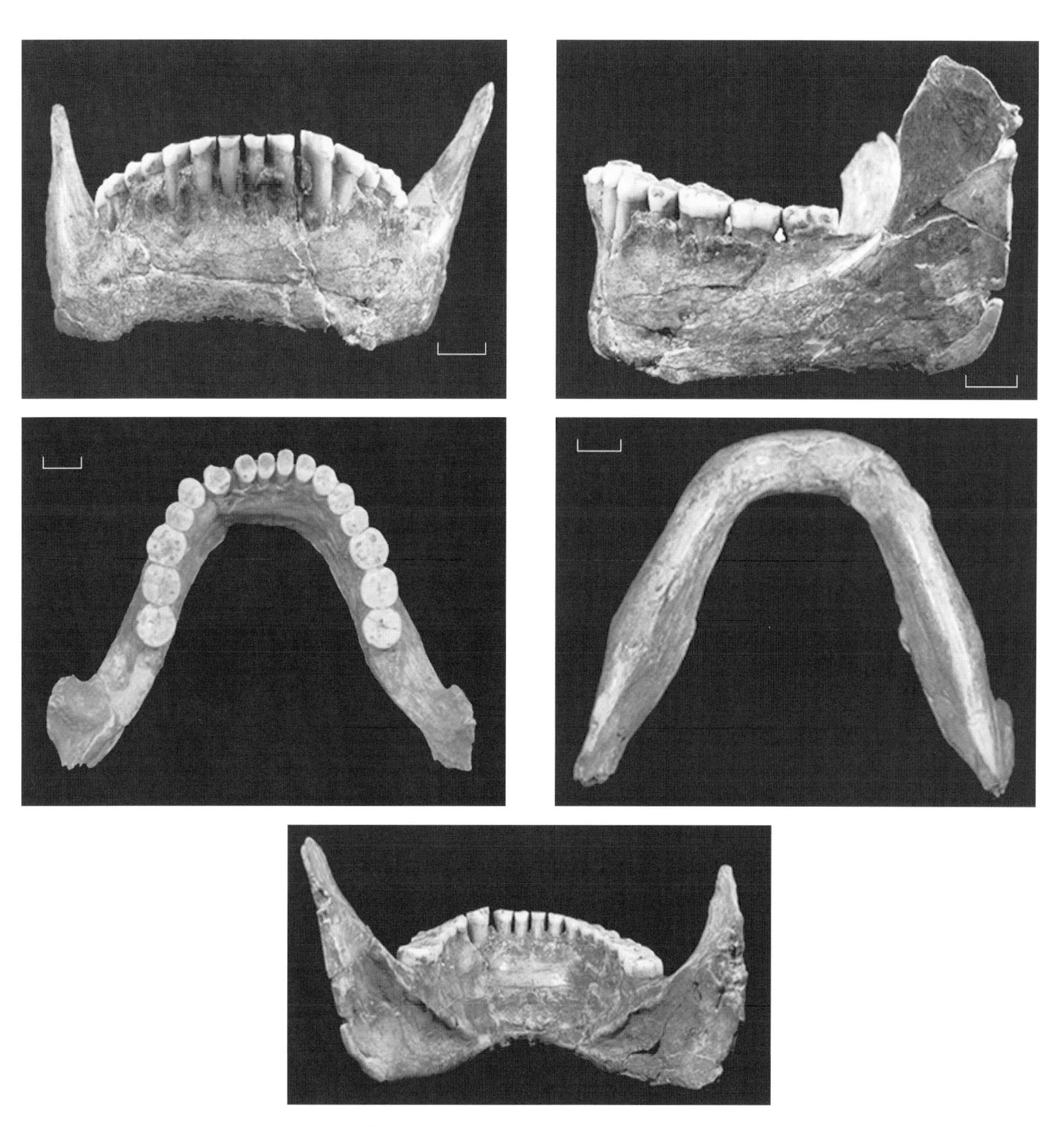

SPY Figure 3. Spy 1 (scale = 1 cm).

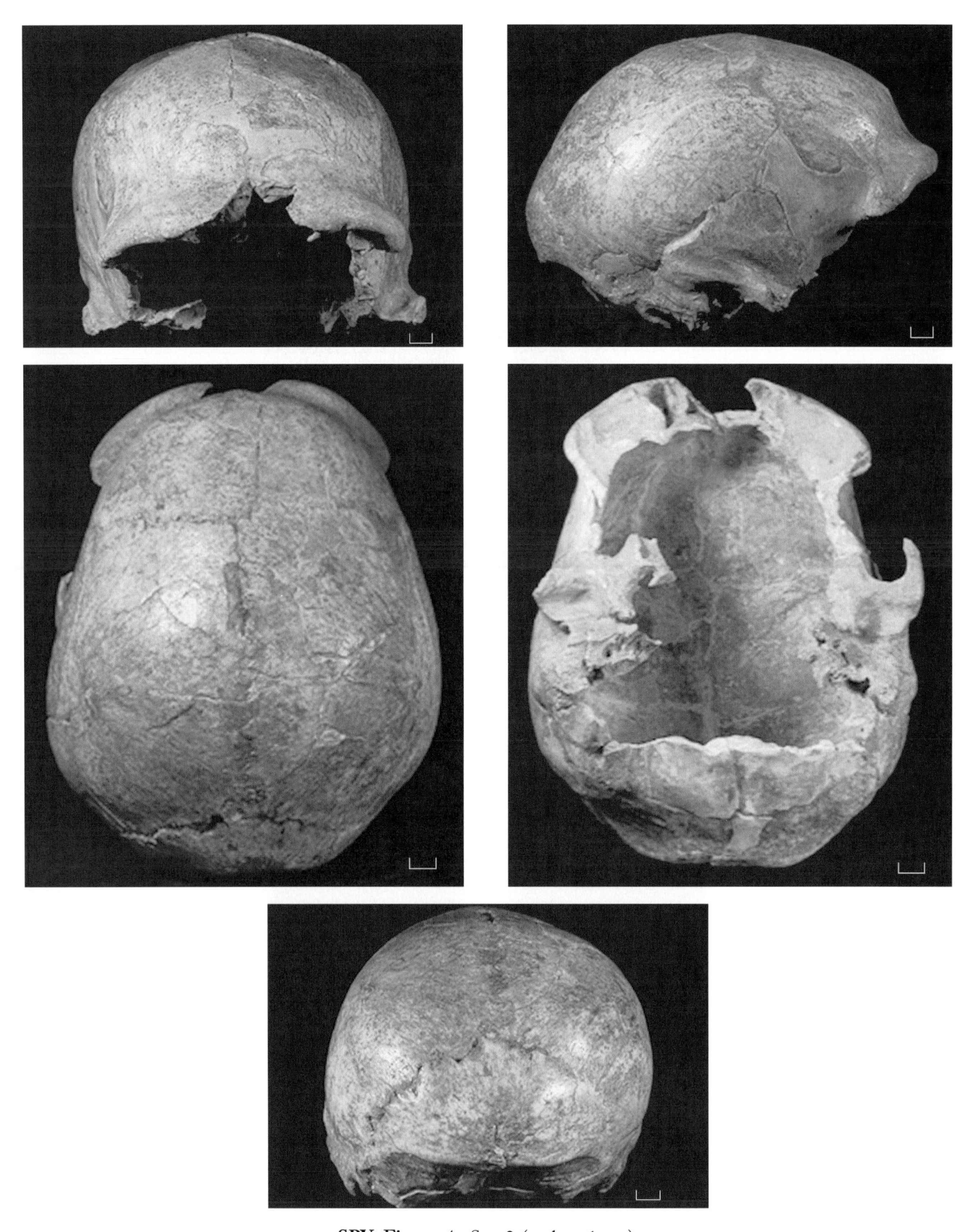

SPY Figure 4. Spy 2 (scale = 1 cm).

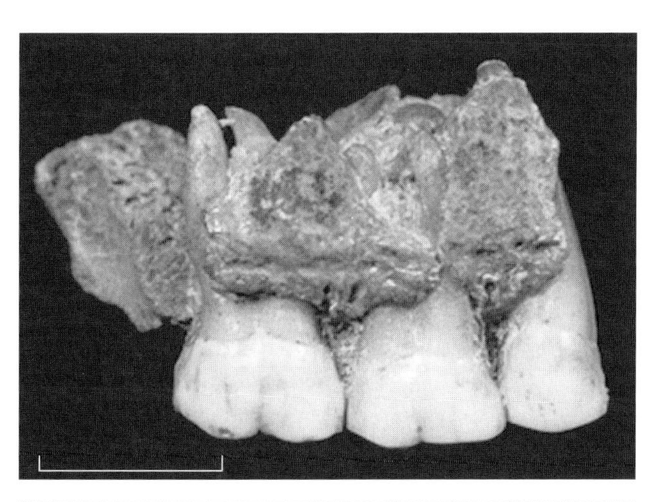

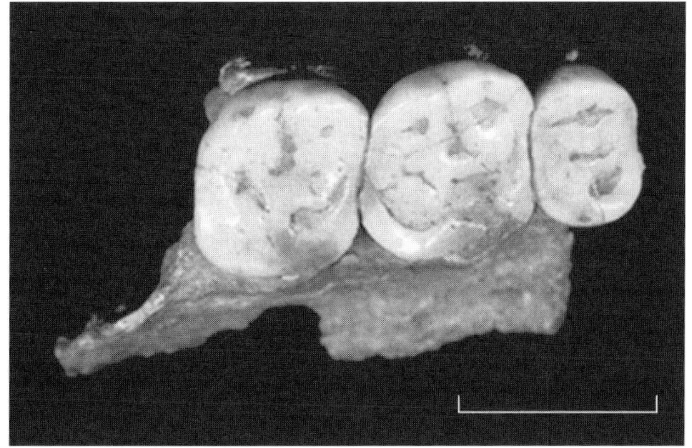

SPY Figure 5. Spy 2 (scale = 1 cm).

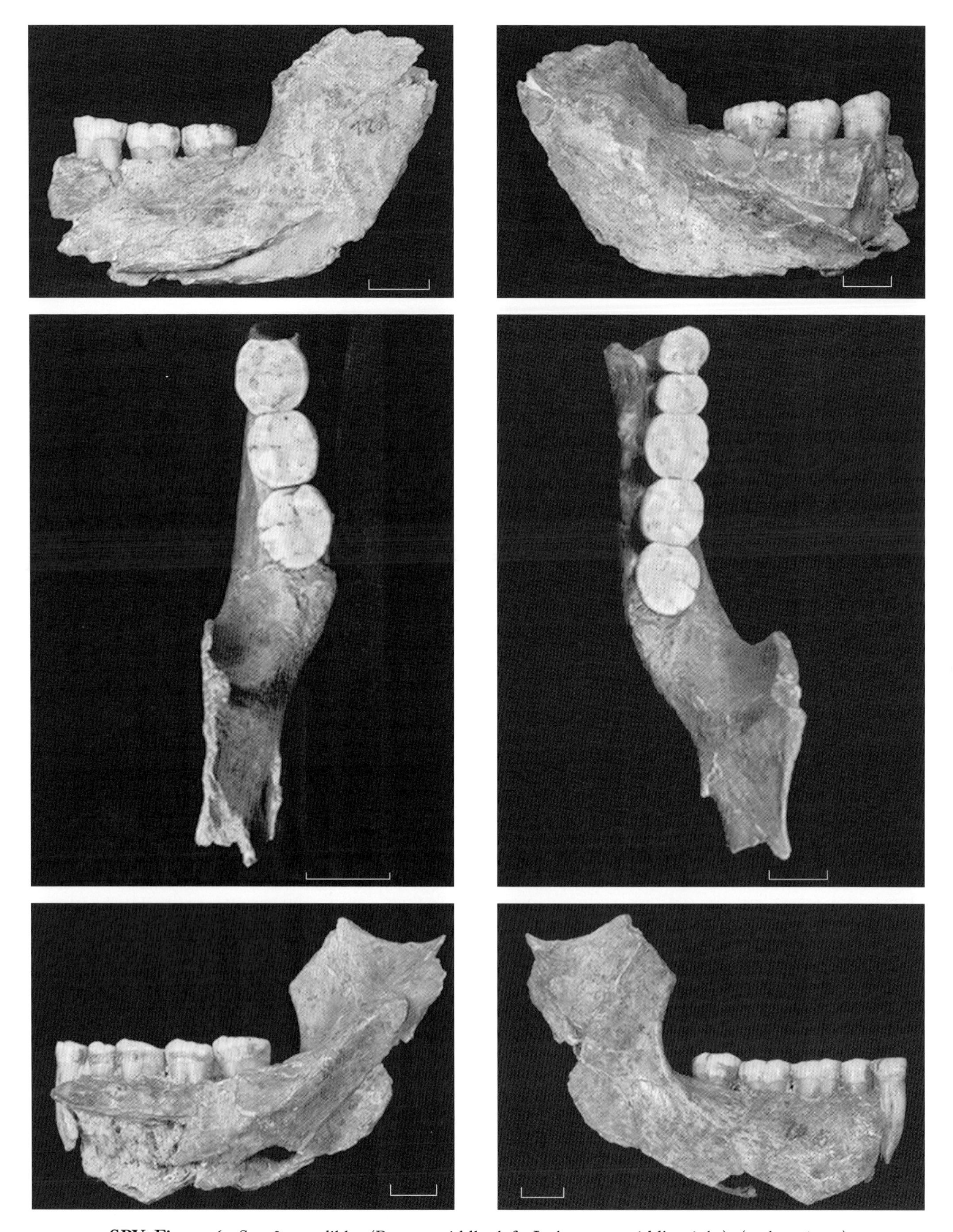

SPY Figure 6. Spy 2 mandibles (R: top, middle, left; L: bottom, middle, right). (scale = 1 cm)

Steinheim

Location
Gravel quarry at the northern limit of Steinheim an der Murr, some 20 km N of Stuttgart, Germany.

Discovery
K. Sigrist, July 1933.

Material
Somewhat crushed adult cranium, possibly female.

Dating and Stratigraphic Context
The Sigrist gravel pit has several layers of sands and gravels, with four distinguishable faunas apparently represented (Adam, 1954a,b). The Steinheim cranium was found in the layer containing the second oldest of these faunas, which is indicative of interglacial conditions. Most recent authorities agree that the interglacial in question is the Penultimate (Holsteinian) and that the fauna thus dates from somewhere in isotope stage 7. If so, the cranium is around 225 ka old.

Archaeological Context
There is no direct association with artifacts.

Previous Descriptions and Analyses
For all its importance, this cranium has yet to be monographed. It was announced by Berckhemer (1933), who allocated it to the new species *Homo steinheimensis* in 1936 and to *Homo (Protanthropus) steinheimensis* in 1937. Weinert (1936) provided a rather longer description in which he pointed to a combination of Neanderthal and modern features that for long continued to puzzle others (e.g., Boule and Vallois, 1957). Morant (1938), followed by several subsequent authors, noted similarities to the Swanscombe specimen. Stringer et al. (1984) preferred to sidestep the matter of Steinheim's affinities, whereas Day (1986), trying to give everyone something, took the contorted view that here is a female (hence lightly built) "example of a *Homo erectus/Homo sapiens* transitional form that is at the root of the European Neanderthal side-branch" (p. 82). Altogether this specimen has proven so intractable that it has not received the detailed attention it deserves. Olivier and Tissier (1975) estimated cranial capacity at 1100 cc.

Morphology
Adult. Extremely weathered, small cranium, crushed extensively on L; missing part of L frontal, parietal, sphenoid, orbital region and maxilla, and anterior part of palate; also missing most of nuchal region anterior to infratoral sulcus, including lateral parts and basiocciput, and R zygomatic arch. Teeth preserved are upper RP2–M3 and LM1–3, little worn; alveoli for RC–P1 also present. R side of skull looks somewhat distorted; angle of face to braincase may be close to natural position, although originally face may have been more forwardly placed.

Face appears somewhat large relative to neurocranium. In profile, cranium relatively long, low. Short frontal gently curves back from a/p moderately long posttoral plane. Top of braincase fairly flat until strong curve (beginning posterior to level of mastoid process) down onto rather vertical occipital plane. Nuchal plane steeply angled forward below infratoral undercutting. Viewed from behind, maximum cranial width occurs low, at relatively small, but distinct, supramastoid protuberances; were it not for these protuberances, maximum width would be higher, above high point of squamosal suture, where hints of parietal eminences give braincase outline an angled, "roofed" appearance. Above protuberances, sidewalls of braincase fairly straight/very gently bulging to curve in strongly at region of "eminences" to relatively long and straight, slightly angled up upper parietal surfaces. Viewed from above, and preserved on the R, braincase bulges broadly not quite two-thirds back; short posterior profile angles in gently to broad, shallow arc of occipital; longer anterior profile angles in gently to extraordinarily shallow postorbital constriction.

Preserved R supraorbital torus smoothly rolled, fairly s/i tall, laterally tapering, apparently confluent with (probably) prominent glabella; together appear quite anteriorly protrusive. Torus appears not to have retreated from glabella (when viewed from above). Continuous posttoral and postglabellar planes relatively long a/p. Break in L supraorbital region exposes laterally very expansive frontal sinus; sinus does not extend very far up into frontal. Large R orbit seems undistorted; inferior border slopes upward to medioinferior, rounded corner (producing "aviator glasses" shape); lateroinferior "corner" lies well below level of top of nasal aperture. Infraorbital groove (in orbital floor) long relative to infraorbital canal, but both quite short, laterally placed. Interorbital region would have been broad.

Nasal aperture quite large relative to skull size. Nasal bones were probably flexed quite sharply; relationship of nasion to this flexure uncertain. Nasal bones were narrow, keeled in midline (somewhat projecting medially, not broad and flat). Preserved portion (on the R) of lateral crest of aperture margin quite sharp; probably curved forward superiorly, giving very long nasomaxillary suture and long frontal process; sutures in area obliterated. Floor of nasal cavity depressed below inferior rim. Low, peaked, anteriorly constricted medial projection (penetrated by thin, vertical creaselike foramen) within nasal cavity; medial projection confluent with superior nasal crest (of unknown height) above. Matrix obscures other nasal cavity morphology. Frontal processes of maxilla not noticeably inflated superiorly. Region of maxilla from margin of nasal aperture to infraorbital foramen somewhat inflated; area below infraorbital foramen modestly concave. Infraorbital foramen not very large; opens downward, with long, vertical groove below it continuing down to alveolar crest. Infraorbital region s/i tall, relatively flat, vertical, forwardly facing.

Relatively s/i tall anterior root of zygomatic arch (as seen on the R) arises close to alveolar margin, above level of M1; angles steeply up and out (viewed from front). Anterior squamosal raised and alisphenoid medially concave, producing distinct posterior temporal fossa. Distinct angulation at anteroinferior corner of squamosal delineates infratemporal fossa below from temporal fossa above. Squamous portion of temporal apparently short a/p, not very tall s/i. Parietomastoid suture relatively long and horizontal.

Posterior root of zygomatic arch originates anterior to external auditory meatus. Suprameatal crest well developed; confluent supramastoid crest that lies right above V-shaped, quite downwardly extended parietal notch; notch lies above middle of mastoid process (both sides). Mandibular fossae long a/p, moderately to quite deep; fronted by thin articular wall. Articular surface extends anteriorly onto temporal; medially bounded by moderately sized medial articular tubercle (seen on the R). Poorly developed vaginal process runs along midline of ectotympanic tube, with low but marked peak in its middle around depression for styloid process. Styloid pit present, and quite medially placed, in line with mastoid notch. Very large stylomastoid foramen (as seen on the L) medially placed; lies lateral to, and somewhat apart from, styloid pit. Both ectotympanic tubes broken; were probably short. Somewhat posteriorly oriented carotid foramen quite large and medially placed, right by small, m/l thin, a/p oriented foramen lacerum against basiocciput. Mastoid process small, thin a/p at base; slightly broken distally but could have been somewhat projecting, with slight anterior orientation. Mastoid notch quite narrow, constricted, long, and straight. On the L especially, occipitomastoid suture preserved; no evidence of any crest lying along it.

Anterior lambdoid suture long and horizontal on both sides (especially well visible on the L); small ossicle on the L where lambdoid suture flexes upward. Occipital plane rather vertical, as high as broad. Lambdoid suture peaks at lambda (thus creating equilateral triangle with occipital "torus" as base). Occipital "torus" low, horizontal, modestly wide, not posteriorly projecting; shallow infratoral depression defines it from below. No external occipital protuberance. Faint external occipital crest on relatively upright preserved infratoral nuchal surface. Broad area above occipital "torus" slightly depressed bilaterally, giving two very shallow "suprainiac depressions," one on either side of midline, and extensively defining "torus" superiorly. Basiocciput relatively long and narrow with slight central bump surrounded by depressions; pharyngeal fossae tiny.

Palate (on the R) undistorted, deep, with steep sides and an anterior slope. Posterior nasal spine very long, protruding. Medial and lateral pterygoid processes somewhat distorted on both sides (extremely so on the L) but convergent both superiorly and inferiorly. Foramen ovale (as better seen on the R) lies lateral and quite posterior to superior root of lateral pterygoid plate.

Internally, petrosal (as seen on the L) broad, flat superiorly; no arcuate eminence; bone was pulled away from squamosal during fossilization, revealing extensive sinus formation interiorly. Large cochlear canaliculus preserved on the R and large anterior condylar canal on the L. No superior petrous sinus. Sigmoid sinus very short. Transverse sinuses relatively deeply impressed; the R may have been larger than the L.

Upper RC–P1 alveolae, RP2–M3, and LM1–M3 preserved. All upper teeth relatively little worn. RC alveolus very big (especially relative to small skull and cheek teeth); comes in obliquely to inferior border of nasal aperture. RP1 alveolus quite compressed mesiodistally. RP2 short m/d and narrower b/l than M1; buccal and especially lingual sides of crown somewhat bulbous. Paracone taller, more compressed b/l than broader-based protocone; m/d groove between para- and protocone truncated mesially and more so distally by para- and metaconules, respectively.

Enamel surfaces of Ms somewhat wrinkled, with tiny accessory conules. M1 longer m/d and narrower b/l than M2; M3s small, triangular. M1–3 paracone dominant cusp; on M1–2, cuspules on buccal side of paracone. M1–3 protocones internally situated, creating truncated trigon basins. On M2 and especially M3, metacone smaller than paracone. M1 hypocone very swollen distally; more swollen buccally on M2. M1–2 postprotocrista more distinct than preprotocrista. M1–2 lingual surfaces bear Carabelli's pits and/or grooves around lingual surface of protocone.

References

Adam, K. D. 1954a. Die zeitliche Stellung des Urmenschen Fundschicht von Steinheim an der Murr Innerhalb des Pleistozans. *Eiszeitalter v. Gegenwart* 4/5: 18–21.

Adam, K. D. 1954b. Die Mittelpleistozanen Faunen von Steinheim an der Murr (Wurttemberg). *Quarternaria* 1: 131–144.

Berckhemer, F. 1933. Ein Menschen-Schädel aus den diluvialen Schottern von Steinheim a. d. Murr. *Anthropol. Anz.* 10: 318–321.

Berckhemer, F. 1936. Der Urmenschenschädel aus den zwischeneiszeitlichen Floss-Schottern von Steinheim a. d. Murr. *Forschn. Fortschr.* 12: 349–350.

Berckhemer, 1937. Bemerkungen zu H. Weinert's Abhandlung "Der Urmenschenschädel von Steinheim." *Verh. Ges. Phys. Anthropol.* 2: 49–58.

Boule, M. and H. Vallois. 1957. *Fossil Men.* London, Thames and Hudson.

Day, M. 1986. *A Guide to Fossil Man*, 4th ed. Chicago, University of Chicago Press.

Morant, G. 1938. The form of the Swanscombe skull. *J. R. Anthropol. Inst.* 68: 67–97.

Olivier G. and H. Tissier. 1975. Determination of cranial capacity in fossil men. *Am. J. Phys. Anthropol.* 43: 353–362.

Stringer, C. et al. 1984. The origin of anatomically modern humans in western Europe. In: F. Smith and F. Spencer (eds), *The Origins of Modern Humans.* New York, Alan R. Liss, pp. 51–135.

Weinert, H. 1936. Der Urmenschenschädel von Steinheim. *Z. Morph. Anthropol.* 35: 463–518.

Repository

Staatliches Museum für Naturkunde, Rosenstein 1, 7091 Stuttgart, Germany.

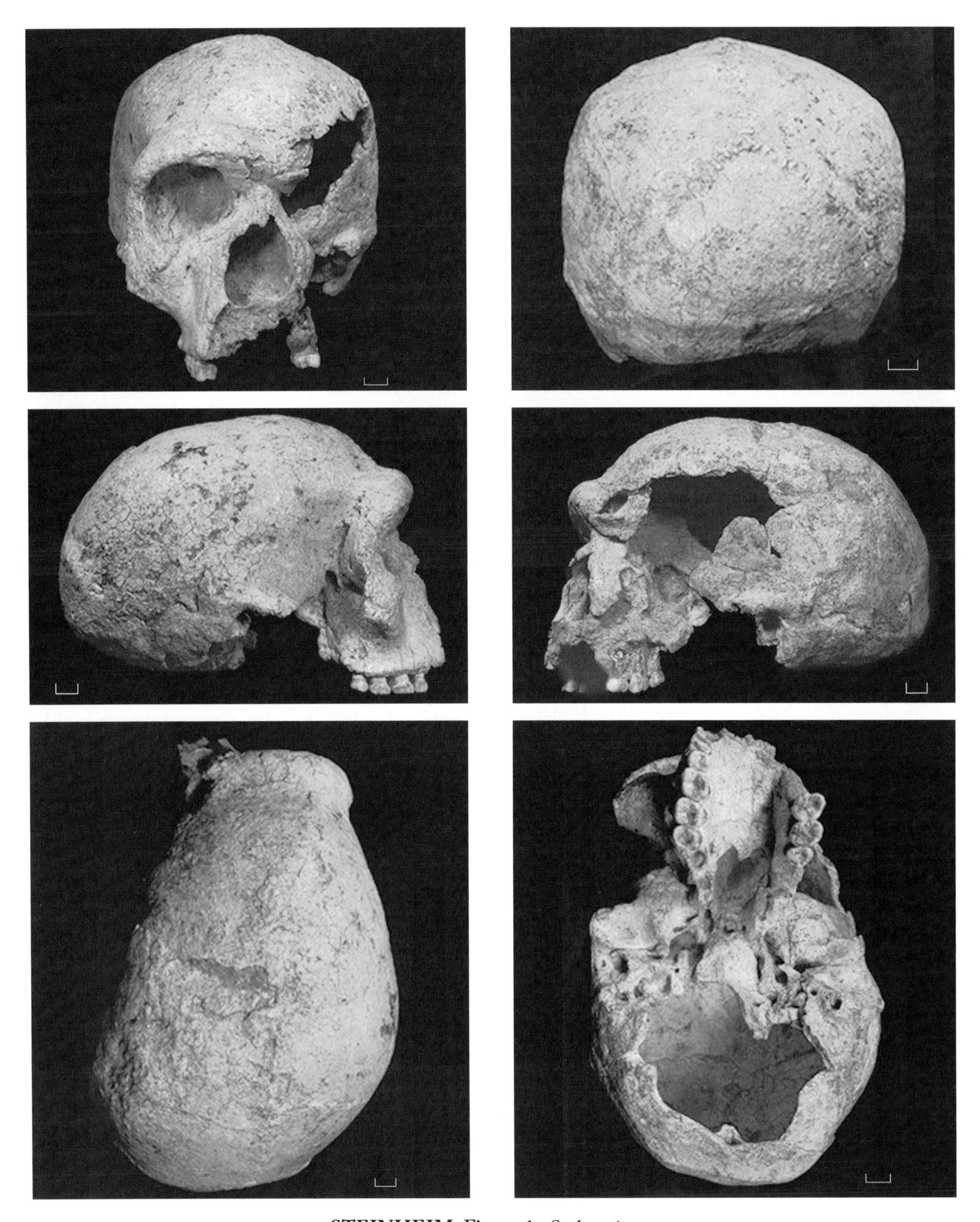

STEINHEIM Figure 1. Scale = 1 cm.

STEINHEIM Figure 2. Including close-up of nasal cavity crest with small medial projection, (not to scale).

Subalyuk

Location

Near the village of Cserepfalu, Borsod county, northern Hungary.

Discovery

Excavations of J. Dancza, 1932.

Material

Adult mandible and associated postcranial elements (Subalyuk 1); juvenile braincase with associated maxilla, some unerupted tooth crowns and some postcranial fragments (Subalyuk 2).

Dating and Stratigraphic Context

Regrettably, this site was dug very rapidly and not as carefully as it might have been. According to reports in Bartucz et al. (1938), the human remains were found fairly deep in the sediment pile at Subalyuk and scattered over a wide area. In his report on the remains, Bartucz (in Bartucz et al., 1938) notes that the adult bones were darker colored than the juvenile ones, leading him to suggest that the two sets might have come from different strata or at least might have fossilized differently. However, both are now believed to come from Layer 11 (L. Kordos, personal communication). Mottl (in Bartucz et al., 1938) concluded that the associated fauna is of early Würm aspect; according to L. Kordos (personal communication) the microfauna is strongly indicative of a continental steppe climate, with correlation to the *Lagurus* biozone suggesting an age of 70–60 ka.

Archaeological Context

According to Kadic (1940) the associated lithics are of Eastern Mousterian type; traces of fire, although no hearths, were found in the sediments.

Previous Descriptions and Analyses

The Subalyuk 1 mandible was first described by Szabo (1935), who compared it to eastern European Neanderthals. This specimen was later said by Bartucz (in Bartucz, 1938) to belong to a female Neanderthal of about 40–45 years. In the same contribution Bartucz described the Subalyuk 2 juvenile remains as those of a 6- to 7-year-old Neanderthal. Numerous subsequent authors (see review by Tillier et al., 1996) have referred to these fossils, invariably in the context of Neanderthal evolution in eastern Europe. Thoma (1963) reduced the age of the juvenile to 3–4 years, whereas Tillier et al. (1996) reduced that of the adult to perhaps 25–35 years. Tillier et al. also produced the most recent of several reconstructions of the juvenile cranium.

Morphology

Collection includes a fragmentary and reconstructed child's braincase with associated maxilla and some unerupted tooth crowns and fragments of a second isolated cranium; also an adult partial mandible.

Subalyuk 1

Adult mandible with symphyseal region, RP1–LC, LP2–M3, and part of L ramus; partially reconstructed inferiorly. Corpus not very thick m/l or tall s/i. Enough inferior margin preserved to show that corpus was shallow. Symphyseal region relatively broad, gently arced from side to side. In profile, symphysis tilted somewhat forward; roots of Is curve out gently over bone below, which is essentially straight down to inferior margin. Inferior margin was probably quite flat posteriorly, with faint inferior marginal tubercle below P2; in front of tubercle, inferior margin rises somewhat toward symphysis. Slight subalveolar depression, but otherwise no symphyseal morphology. Viewed from below, as seen on the L, symphyseal region uniformly thick from front to back. Moderate to large mental foramen under M1. Fairly expansive digastric fossa preserved on the R. Retromolar space was large. Anterior edge of L ramus strongly tilted back; bone posteriorly quite thin m/l. Only posteriormost part of mylohyoid line preserved; quite pronounced. Mandibular foramen compressed, oriented up and back; margin s/i tall; no lingula. In front of this foramen, stout, ridgelike pillar courses down from below region of missing coronoid process to become confluent with pronounced internal alveolar crest. Bone posterior to this pillar quite hollowed out.

Well-worn, forwardly placed anterior teeth rise higher than posterior teeth. Although worn, I1–2 still tall crowned. Roots of I1–2 very long; stout and thick b/l, although compressed m/d. I1–2 lack lingual tubercles or margocristids. I1s parallel sided; I2s flare a bit distally. Cs very stout rooted; crowns bear distinct margocristids, especially distally, and low central lingual keel but no lingual swelling. RP1 very large, stout rooted. Protoconid lingually placed, producing long buccal slope. Mesial crest comes down from tip of protoconid, turning inward to terminate at base of low lingual swelling; crest defines small but deep fovea. Distal crest links paraconid with lingual swelling, enclosing slightly larger fovea. LP2 worn; had substantial metaconid and small but distinct trigonid basin; crest running down from protoconid arcs across to base of metaconid, enclosing slightly larger fovea. M1–3 somewhat rounded in outline; fairly heavily worn; had substantial hypoconulids and quite large talonid basins. M3 has, and M1–2 probably had, very deep trigonid basins. Radiographs show large pulp cavities that increase in size from M1 to M3 (crowns do not decrease in size distally). Root cleft low on M1, lower on M2; M3 roots divide only at their very tips.

Subalyuk 2

Calvaria. Some problems with reconstruction: orbital roof and frontal contribution to frontonasal suture angled backward too far; also questionable orientation of alisphenoid and foramen ovale, which should be further lateral and down.

Cranium was broad posteriorly; maximum breadth would have been low on vault. Supraorbital margins thin but thicken very slightly medially. Orbital rim quite sharp; no sign of supraciliary development. Frontonasal suture better preserved on R frontal, where sinus development has only just begun. No sign of separation of infratemporal and temporal fossae by shelf on alisphenoid. L squamous portion of temporal better preserved than the R; was probably low. Parietomastoid suture long, fairly horizontal.

Mastoid processes have begun to differentiate; better-preserved L mastoid quite heavily roughened by muscle markings. Mastoid notch very small, narrow. In the R petrosal, floor of ectotympanic tube not ossified; is partial foramen of Huschke. Stylomastoid foramina fairly large; lie medial to mastoid process. On the R, base of stout styloid process preserved, lying about as far medial to stylomastoid foramen as latter is from mastoid process. On the R, very stout vaginal process rises abruptly from lateral margin of very medially placed carotid foramen; peaks where had wrapped around styloid process, which had left large impression; fades out laterally without contacting mastoid process.

Judging from reconstruction, occipital was broad but not very tall. Lambdoid suture was apparently patent; may have had ossicles near lambda. Small suprainiac depression present on occiput; no sign of torus or infratoral depression. Muscle scarring indicating nuchal plane begins fairly low, continues forward to quite stout incipient Waldeyer's crest. This latter fades out just medial to what would have been a substantial occipitomastoid crest. Part of foramen magnum preserved; would have been very large and long. Only superior arm of contribution to occipital condylar canal has begun to fuse with basisphenoid; canal itself is closed over. Foramen ovale bounded by thick margin of bone on medial side; foramen spinosum quite large, with fully ossified medial border; these two foramina separated by broad band of bone.

Sagittal and medial coronoid sutures were apparently closed, suggesting some sort of developmental abnormality (synostosis?) that may be associated with presence of postbregmatic depression. Metopic suture visible (unless feature is simply a break).

Internally, region of arcuate eminence very broad, prominent; trace of superior petrous sinus laterally; cicatrix of subarcuate fossa almost closed over, and area rather flat. Superior sagittal sinus visible on the R just where it deviates laterally.

Maxilla. Partially reconstructed; contains alveoli for Ldi1–2 and preserves rest of deciduous teeth and crowns of both M1s. On the R, part of lateral nasal aperture preserved; inferior margin of aperture is complete.

Anterior nasal spines almost intact; are bifid, large, forward projecting. Spinal crest runs back and lateralward; inferior margin of nasal aperture runs up toward crest, creating depression (prenasal fossa) between the two. Nasal cavity floor somewhat depressed. Nasoalveolar clivus long and vertical. Nasal aperture was moderately large. Preserved on inner wall of frontal process is broken but huge medial projection; upper part of projection merges into low crest that continues on up inner wall of frontal process; projection is very broad from front to back and may have been pneumaticized. Lacrimal groove was very broad, deep, and apparently unroofed. Maxillary sinus expanded forward to just behind dc root.

Anterior teeth all curve lingually; roots rather long for deciduous teeth. Rdi1 tall, with good margocristae, especially medially; lingual surface not excavated or shoveled, and lacks any tubercle. Rdi2 more flared laterally, with more rounded mesial edge; lingual surface embellished only with low margocristae. Upper Ldc looks as if tooth germ had begun to divide before crown formation. Nonpathological Rdc crown rather slender; lingually, slight central keel lingually and foveae bounded by margocristids on either side; distal edge swells more than mesial edge, so apex is mesially shifted.

dm1s must have had three roots, the two buccal roots more splayed on the R than on the L, but in both cases, cleft between roots lies well above neck. Paracones, protocones, and hypocones distinct; compressed, so rather ridgelike, producing continuous ridge down each side of crown, enclosing narrow basins. Well-defined preprotocristae and postprotocristae enclose constricted but deep trigon basins. Distolingually distended hypocones have long postcingula that meet the postprotocristae, thus demarcating long talon basins. Above paracone is buccal swelling near neck of tooth.

dm2s had three distinct roots, well splayed and dividing high above neck. Cusps more crestlike than cusplike. Protocones and hypocones quite centrally placed, producing long lingual slope. Carabelli's pits present, quite deep. Strong preprotocrista runs to apex of paracone; stout postprotocrista courses up metacone. Trigon basin very constricted but deep. Protocone fold runs down to base of hypocone. Hypocone large and distolingually swollen; well-developed postcingulum runs to metacone and encircles a large and deep talon basin.

M1s have fully developed crowns, with roots just beginning to form. Protocone and hypocone both centrally shifted, producing long lingual slope. Strong preprotocristae arc out and back anteriorly on both sides to meet paracones, thus forming very constricted but deep trigon basins. Large hypocones lingually distended; are met at their bases by long protocone folds. Large, well-defined talonid basins produced by prominent postcingula that arc out from hypocones and swing in to meet metacones. Enamel within these basins moderately wrinkled.

Nasal bone. L; immature individual. Long, narrow, strongly flexed ca. 1 cm below nasion. Nasal crest huge.

Upper Incisors. Crowns of upper RI1–2, not quite fully formed. I1 crown as m/d wide as s/i tall; somewhat excavated lingually, with moderate margocristae; not greatly shoveled; lacks lingual tubercle; mesial margin straight; distal margin flares a little. I2 crown would not have been tall; has large margocristae that converge toward neck, producing marked barrel shape.

Also miscellaneous cranial fragments with patent sutures.

REFERENCES

Bartucz, L. et al. 1938. A Cserepfalui Mussolini-Barlang (Subalyuk). *Geol. Hung. (ser. Paleontol.)* 14: 1–320.

Kadic, O. 1940. Topographische, Morphologische und Stratigraphische Verhaltnisse der Höhle. *Geol. Hung. (Palaeontol.)* 14: 29–46.

Szabo, J. 1935. L'homme moustérien de la grotte Mussolini (Hongrie). Etude de la mandibule. *Bull. Mém. Soc. Anthropol. Paris* 6: 22–30.

Thoma, A. 1963. The dentition of the Subalyuk child. *Z. Morph. Anthropol.* 54: 127–150.

Tillier, A.-M. et al. 1996. The Subalyuk Neanderthal remains (Hungary): a re-examination. *Ann. Hist.-Nat. Mus. Nat. Hung.* 88: 233–270.

Repository

Hungarian Natural History Museum, Ludovika tér.2, H-1083 Budapest, Hungary.

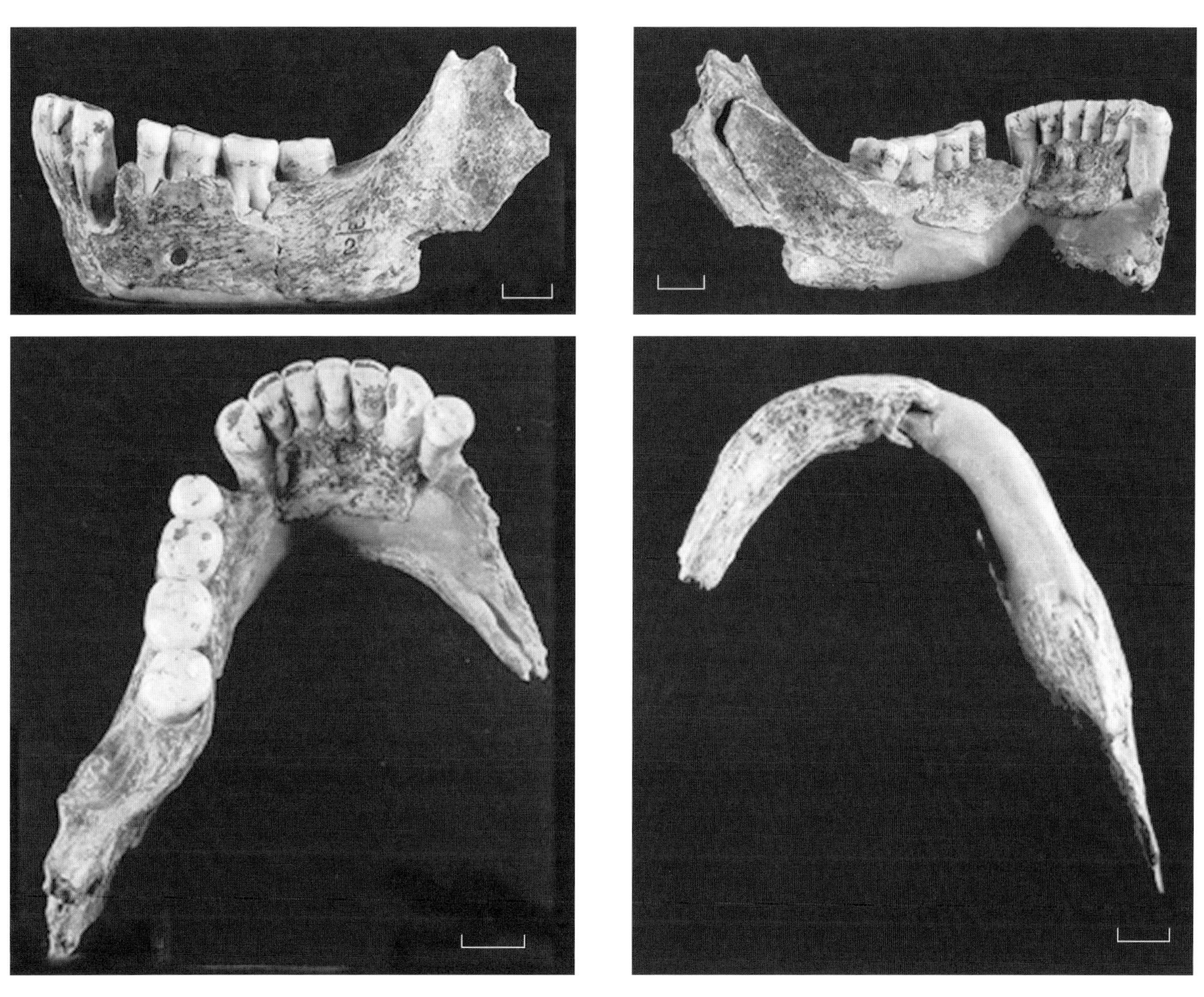

SUBALYUK Figure 1. Subalyuk 1 (scale = 1 cm).

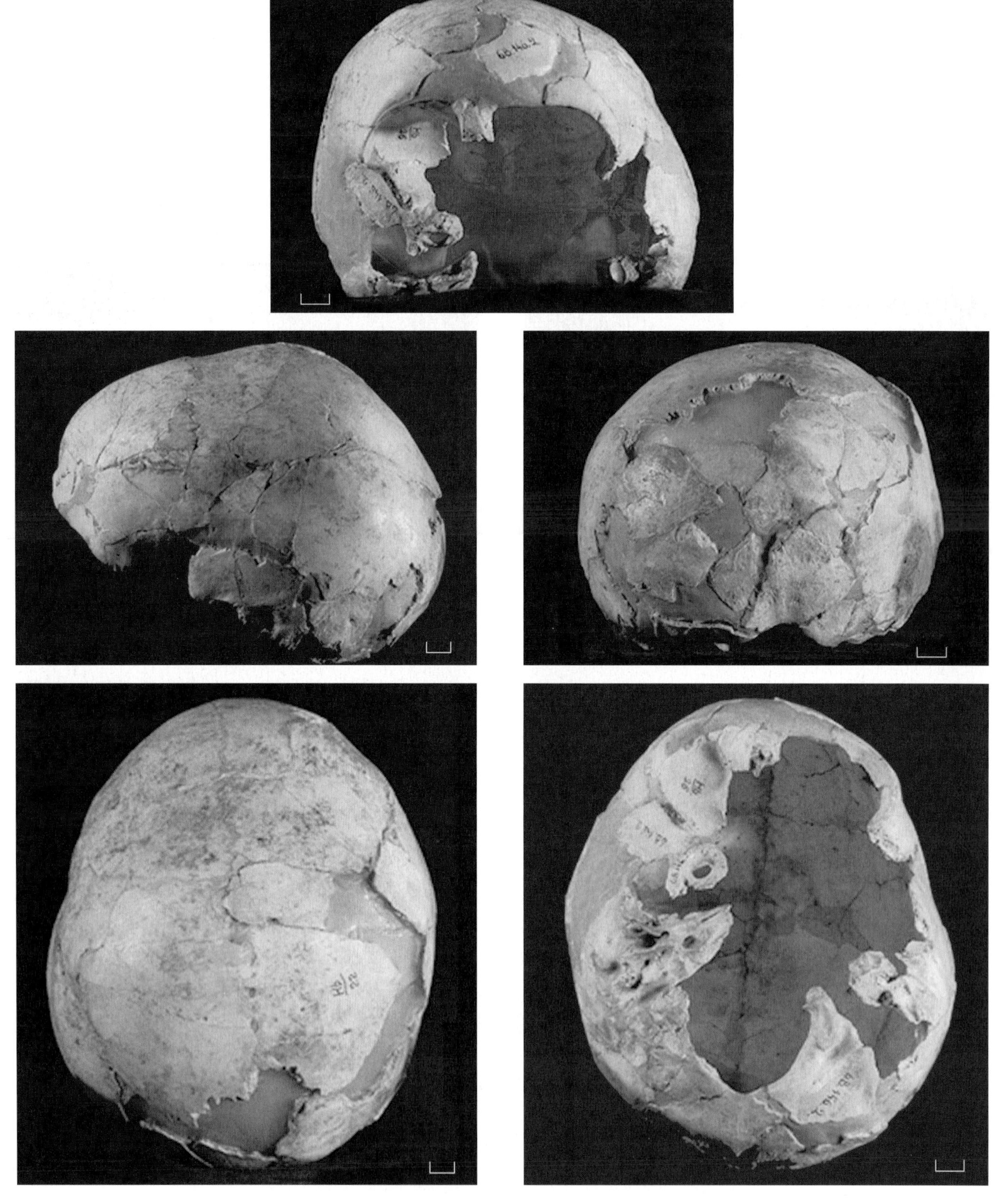

SUBALYUK Figure 2. Subalyuk 2 (scale = 1 cm).

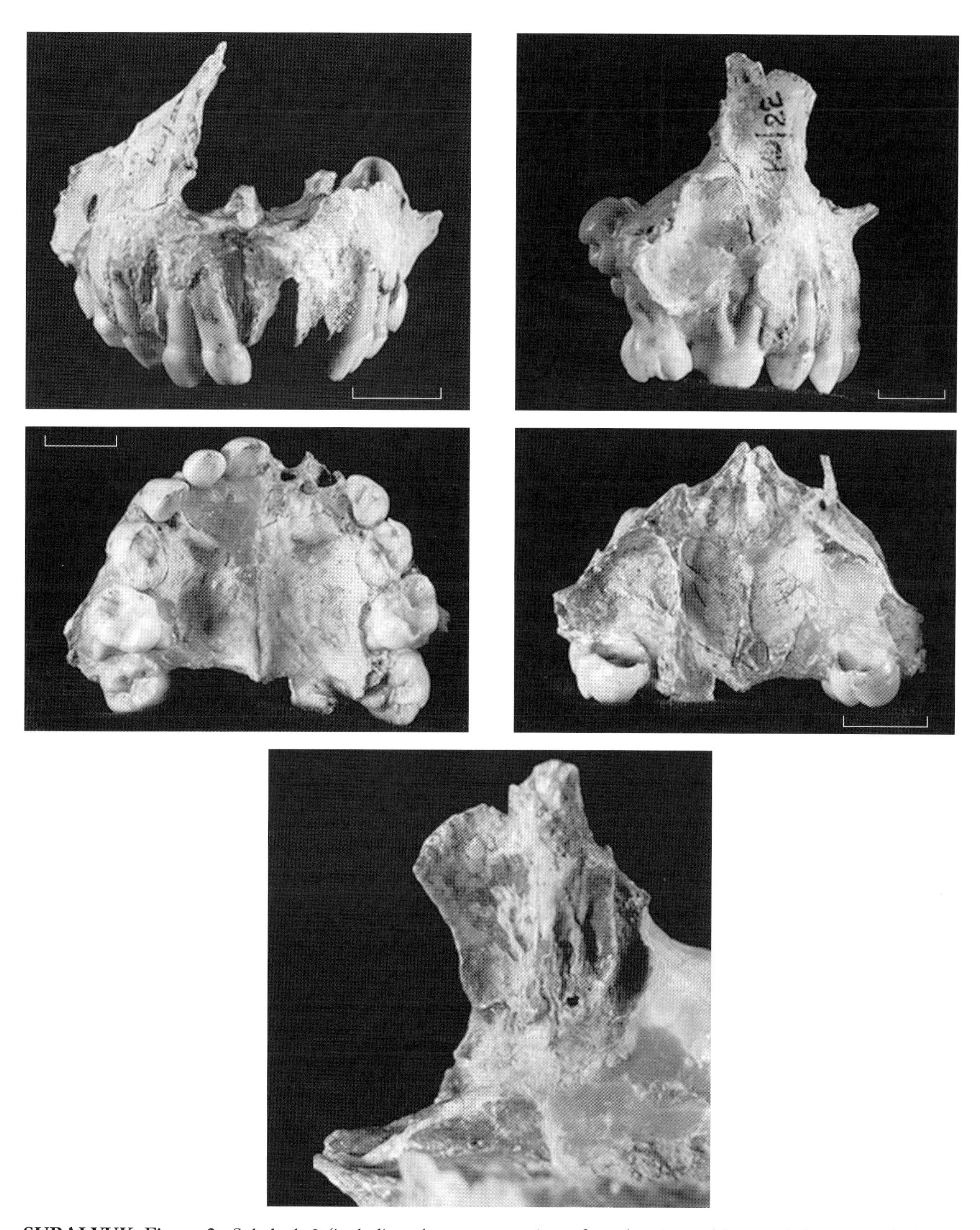

SUBALYUK Figure 3. Subalyuk 2 (including close-up, rear view of nasal cavity and low medial projection), (scale = 1 cm).

Svitavka

Location

Brickyard just E of Svitavka village, near Boskovice, some 50 km N of Brno, Moravia, Czech Republic.

Discovery

L. Smolikova, May 1962.

Material

Highly fragmentary adolescent cranium with some postcranials.

Dating and Stratigraphic Context

Stratified loess with paleosols (Smolikova, 1962). Vlcek (1967) suggests an age of ca. 26–25 ka for the hominid, but in view of its archaeological association others believe it may be older, perhaps 30 ka (V. Kuzelka, personal communication).

Archaeological Context

Aurignacian (B. Klima, quoted in Oakley et al., 1971).

Previous Descriptions and Analyses

The Svitavka hominid was described by Vlcek (1967) as a modern *Homo sapiens* with particular resemblances to Brno 3 and Dolni Vestonice 3.

Morphology

Highly fragmentary, possibly male (based on greater sciatic notch), adolescent cranium. Sutures patent; dental wear is minimal. Face reconstructed; mandible distorted but complete except for coronoid processes. L and R temporals lack petrosals. M3s impacted above and below.

Supraorbital region flows vertically into frontal. Superciliary area clearly bipartite, with glabellar "butterfly" and lateral plates distinguishable but not prominent. Frontal sinuses confined to region of the "butterfly," separated by median septum. Nasal bones relatively short, slope down from nasion. Frontal processes of maxilla thin, not protrusive. Canine fossae well developed. Lower margin of nasal aperture rounded; nasal floor flat. Nasoalveolar clivus short and sloping. Maxillary sinus large but does not intrude into nasal cavity. Body of zygoma faces straight forward; zygomatic arches flare slightly. Posterior root of zygomatic arch originates over auditory meatus, with faint crests behind. Mastoid process incompletely developed. Broken vaginal process was very low, did run to edge of ectotympanic tube. Mastoid notch narrow; paramastoid crest probably adjacent. Parietomastoid suture was very short. All sutures differentiated.

Mandibular symphyseal region bears well-defined mental trigon with slight symphyseal keel. Digastric fossa small. Genial tubercle single, very low. Moderately sized mental foramen lies below P2; accessory foramen on the L. No mylohyoid line, but submandibular fossae present. Mandibular foramina incomplete. Sigmoid notch crest courses to lateral aspect of condyle. Tooth roots small, slender.

All upper tooth roots short. P^1 larger than P^2. M^1 with very small talon basin; M^{2-3} malformed. P_2 slightly larger than P_1. Ms decrease in size from M_1

to M_3; lack distinct trigonid basin; only M_1 has hypoconulid (small).

REFERENCES

Oakley, K. et al. 1971. *Catalogue of Fossil Hominids. Part II: Europe*. London, British Museum (Natural History).

Smolikova, L. 1962. Nalez kosternich pazustatku pleistocenniho cloveka ve Svitavce u Boskovic. *Cas. Mineral. Geol.* 7: 361–363.

Vlcek, E. 1967. Der jungpleistozäne Menschenfund aus Svitavka in Mähren. *Anthropos* 19: 262–270.

Repository

Narodni Muzeum, Vaclavske nam. 68, 11000 Praha, Czech Republic.

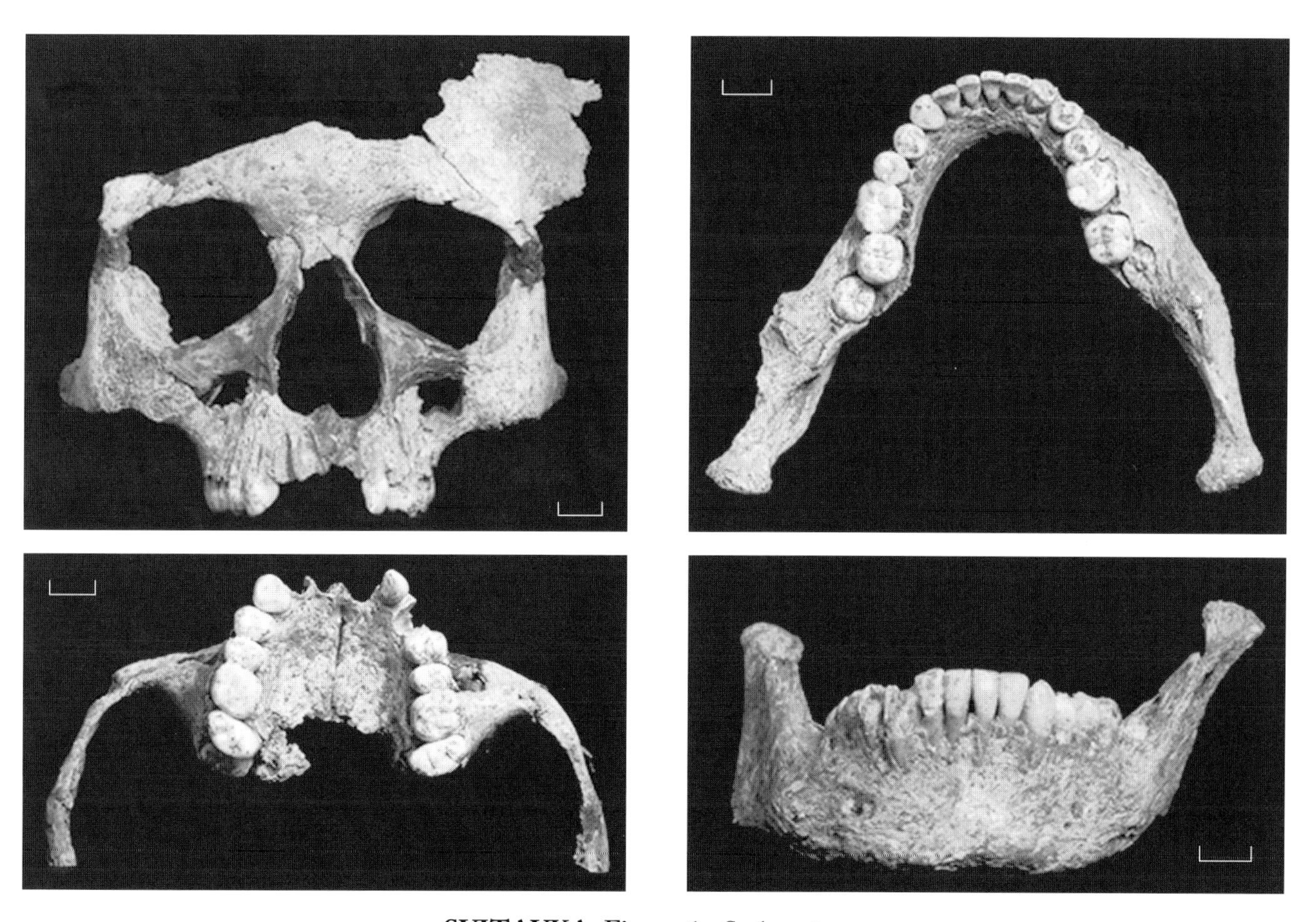

SVITAVKA Figure 1. Scale = 1 cm.

Swanscombe

Location
Barnfield gravel pit, Swanscombe, Kent, England.

Discovery
A. Marston, June 1935 (occipital) and March 1936 (left parietal); J. Wymer and A, Gibson, July 1955 (right parietal).

Material
Separated left and right parietal and occipital of the same adult individual, possibly a female.

Dating and Stratigraphic Context
The gravel pit reveals a stratified sequence of loams and gravels of late middle Pleistocene age, most likely spanning the period between about 350 and 250 ka. On the basis of a quite abundant fauna, the Upper Middle Gravels that yielded the hominid are correlated to the Hoxnian (i.e., Holsteinian or Mindel-Riss) interglacial (Oakley, 1952, 1957). The whole Swanscombe sequence is of post-Anglian age (i.e., more recent than isotope stage 12, hence under 400 ka), but whether one or more major climatic cycles is represented at the site remains unclear.

Archaeological Context
The Upper Middle Gravels have yielded handaxes and flake implements characterized as middle Acheulean (Hawkes et al., 1938; Wymer, 1964).

Previous Descriptions and Analyses
Morant (in Clark et al., 1938) emphasized metric similarities between the two cranial bones then known and those of *Homo sapiens* and suggested affinities with the Steinheim hominid. Various subsequent authors (e.g., Breitinger, 1955; Howell, 1960, Stringer et al., 1984, Wolpoff, 1996) have supported this affinity, whereas Boule and Vallois (1957) saw Swanscombe as a "Presapiens" form distinct from a Steinheim-Neanderthal lineage. Weiner and Campbell (1964) regarded Swanscombe as representing yet another "variety" of *Homo sapiens*, paralleling Ngandong, Kabwe, and the Neanderthals. Wolpoff (1980) found that both Swanscombe and Vérteszöllös lay in an intermediate position between *Homo erectus* and the Neanderthals, the former being more evolved. Santa Luca (1978) stressed similarities to Neanderthals, and indeed there is a growing feeling that Swanscombe represents a Neanderthal forerunner (see Stringer and Gamble, 1993). Montandon (1943) made the Swanscombe hominid the holotype of *Homo sapiens protosapiens.* Estimated cranial capacity is 1325 cc (Olivier and Tissier, 1975).

Morphology
Occipital and both parietals; R parietal slightly damaged along sagittal suture anteriorly and along squamous suture.

Bone of parietals very thick, especially from lambdoid suture to prominent parietal eminence. Braincase would have been ovoid in cross section; broadest at or below top of the squamous suture (preserved on the

L). On the better preserved L parietal, two subcutaneous lesions, one near the lambdoid, one near the coronal suture; on the R parietal one lesion, near lambda. Parietomastoid suture (as preserved on the L) was very long, horizontal. Parietal contribution to parietal notch long, wedge shaped.

Occipital does not bulge; is angled well below lambda. Occipital plane broader than tall s/i. Anterior lambdoid sutures short; behind, undifferentiated lambdoid suture curves up to arc around lambda. Occipital "torus" goes from side to side; is lowest in midline; infratoral sulcus fairly shallow. Suprainiac fossa broad but shallow; character of bone makes it impossible to say whether it was pitted. Faint Waldeyer's crest on the L; even fainter on the R. Foramen magnum long, narrow; condyles far forward. Basiocciput broad; thin by foramen magnum; thickens markedly toward synchondrosis. Sphenoid sinus penetrated far back, impressing itself into basiocciput. Transverse sinus (on the L) leaves superior sagittal sinus well above internal occipital protuberance.

Coronal, sagittal, and lambdoid sutures undifferentiated.

REFERENCES

Boule, M., and H. Vallois. 1957. *Fossil Men.* London, Thames and Hudson.

Breitinger, E. 1955. Das Schädelfragmentes von Swanscombe. *Mitt. Anthropol. Ges. Wien* 84/85: 1–45.

Clark, W. le Gros et al. 1938. Report of the Swanscombe Committee. *J. R. Anthropol. Soc.* 68: 17–98.

Hawkes, C. et al., 1938. The industries of the Barnfield Pit. *J. R. Anthropol. Inst.* 68: 30–47.

Howell, F. 1960. European and northwest African middle Pleistocene hominids. *Curr. Anthropol.* 1: 195–232.

Montandon, G. 1943. *L'Homme Préhistorique et les Préhumains.* Paris, Payot.

Oakley, K. 1954. Swanscombe Man. *Proc. Geol. Assn.* 63: 271–300.

Oakley, K. 1957. Stratigraphical age of the Swanscombe skull. *Am. J. Phys. Anthropol.* 15: 253–260.

Olivier G. and H. Tissier. 1975. Determination of cranial capacity in fossil men. *Am. J. Phys. Anthropol.* 43: 353–362.

Santa Luca, A. 1978. A re-examination of presumed Neandertal-like fossils. J. Hum. Evol. 7: 619–636.

Stringer, C. and C. Gamble. 1993. *In Search of the Neanderthals.* London, Thames and Hudson.

Stringer, C. et al. 1984. The origin of anatomically modern humans in western Europe. In: F. Smith and F. Spencer (eds), *The Origins of Modern Humans.* New York, Alan R. Liss, pp. 51–135.

Weiner, J. and B. Campbell. 1964. The taxonomic status of the Swanscombe skull. In: C. Ovey (ed), *The Swanscombe Skull.* London, Royal Anthropological Institute, pp. 175–209.

Wolpoff, M. 1980. Cranial remains of European middle Pleistocene hominids. *J. Hum. Evol.* 9: 339–358.

Wolpoff, M. 1996. *Human Evolution.* New York, McGraw-Hill.

Wymer, J. 1964. Excavations at Barnfield Pit, 1955–1960. In: C. Ovey (ed), *The Swanscombe Skull.* London, Royal Anthropological Institute, pp. 19–61.

Repository

The Natural History Museum, Cromwell Road, London SW7 5BD, England.

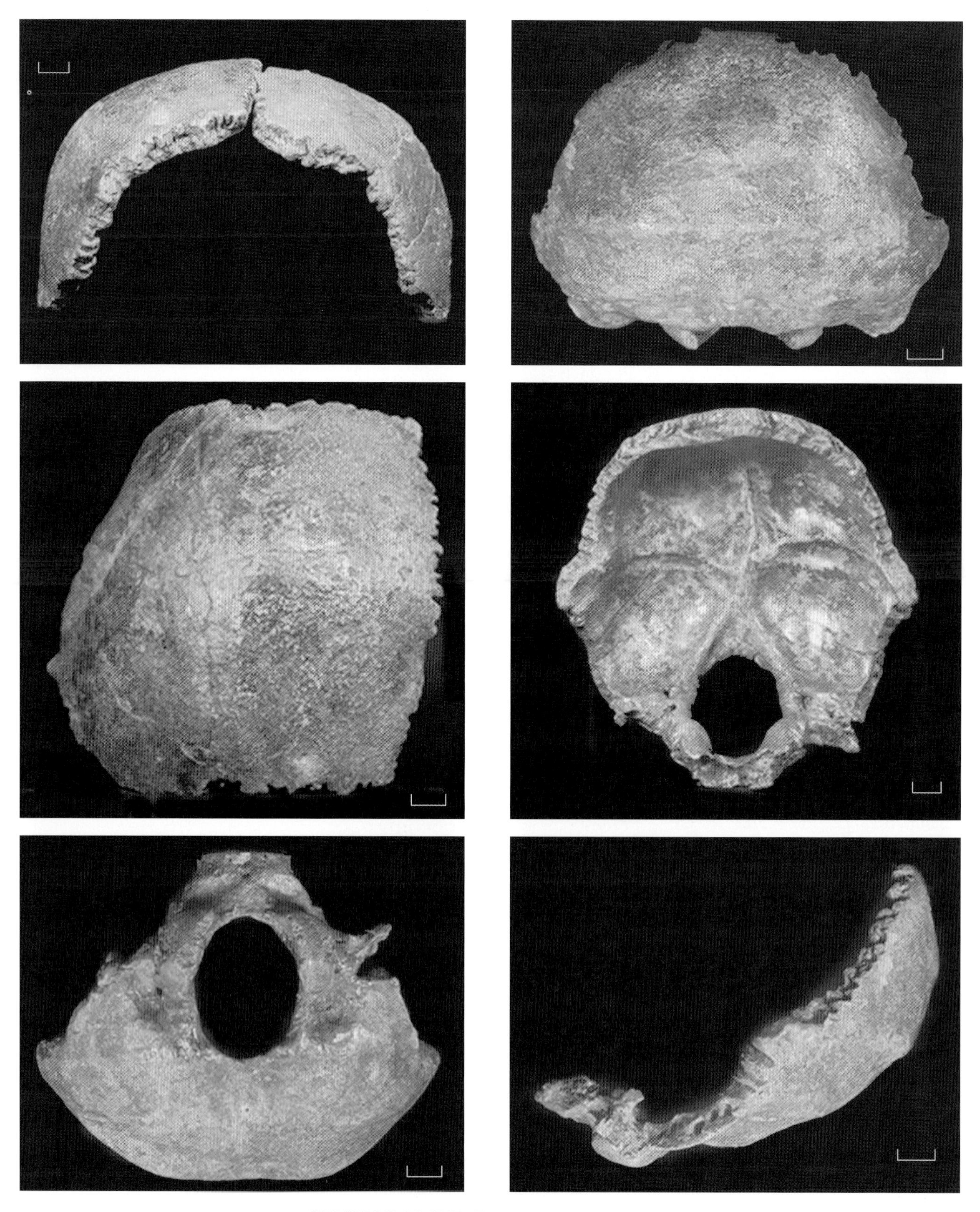

SWANSCOMBE Figure 1. Scale = 1 cm.

Velika Pecina

Location

Cave site near Goranec, 12 km NW of Ivanec, Varazdin district, NW Croatia.

Discovery

Excavations of M. Malez, August 1961.

Material

Partial frontal bone of adult.

Dating and Stratigraphic Context

The hominid fragment came from Level J, near the top of the cave sequence. Associated fauna suggested the Würm I/II interstadial to Malez (1963), but charcoal from the overlying (top) Level I gave a Gröningen radiocarbon date of 34 ka. However, a direct AMS date of just over 5 ka has recently been reported by Smith et al. (1999), who conclude that the specimen is thereby "remove[d] from the list of chronometrically dated early modern humans in Europe" (1999, p. 12284).

Archaeological Context

The industries represented in Level J were said by Malez (1963) to span the Mousterian to "Proto-Aurignacian."

Previous Descriptions and Analyses

Malez (1963, 1965) assigned the frontal fragment to *Homo* aff. *neanderthalensis*, but Smith (1984, p. 176) noted that it "falls clearly within the early modern *H. sapiens* group."

Morphology

R frontal extending from medial part of L orbit through R orbit and including frontonasal suture. Very thick bone. Bipartite supraorbital torus; medial "butterfly" portion taller s/i but continuous across moderately swollen glabella; glabella with slight indentation. Lateral part of torus thin, bulges slightly at zygomaticofrontal suture. Frontal sinus confined to glabella region, intrudes only very slightly above orbit. Internally, frontal crest distinct but quite low.

References

Malez, M. 1963. Istrazivanje Pleistocenske stratigrafije I faune u 1962 godini. *Ljet. Jugosl. Akad. Znan. Umjetn.* 69: 305–313.

Malez, M. 1965. Nalazista fosilnih hominida u Hrvatskoj. *Geoloski Vjesn.* 18: 309–324.

Smith, F. 1984. Fossil hominids from the Upper Pleistocene of Central Europe and the origin of modern Europeans. In: F. Smith and F. Spencer (eds), *The Origins of Modern Humans.* New York, Alan R. Liss. pp. 137–209.

Smith, F. et al. 1999. Direct radiocarbon dates for Vindija G1 and Velika Pecina Late Pleistocene hominid remains. *Proc. Natl. Acad. Sci. USA* 96: 12281–12286.

Repository

Institute of Paleontology and Quaternary Geology, Zagreb, Croatia.

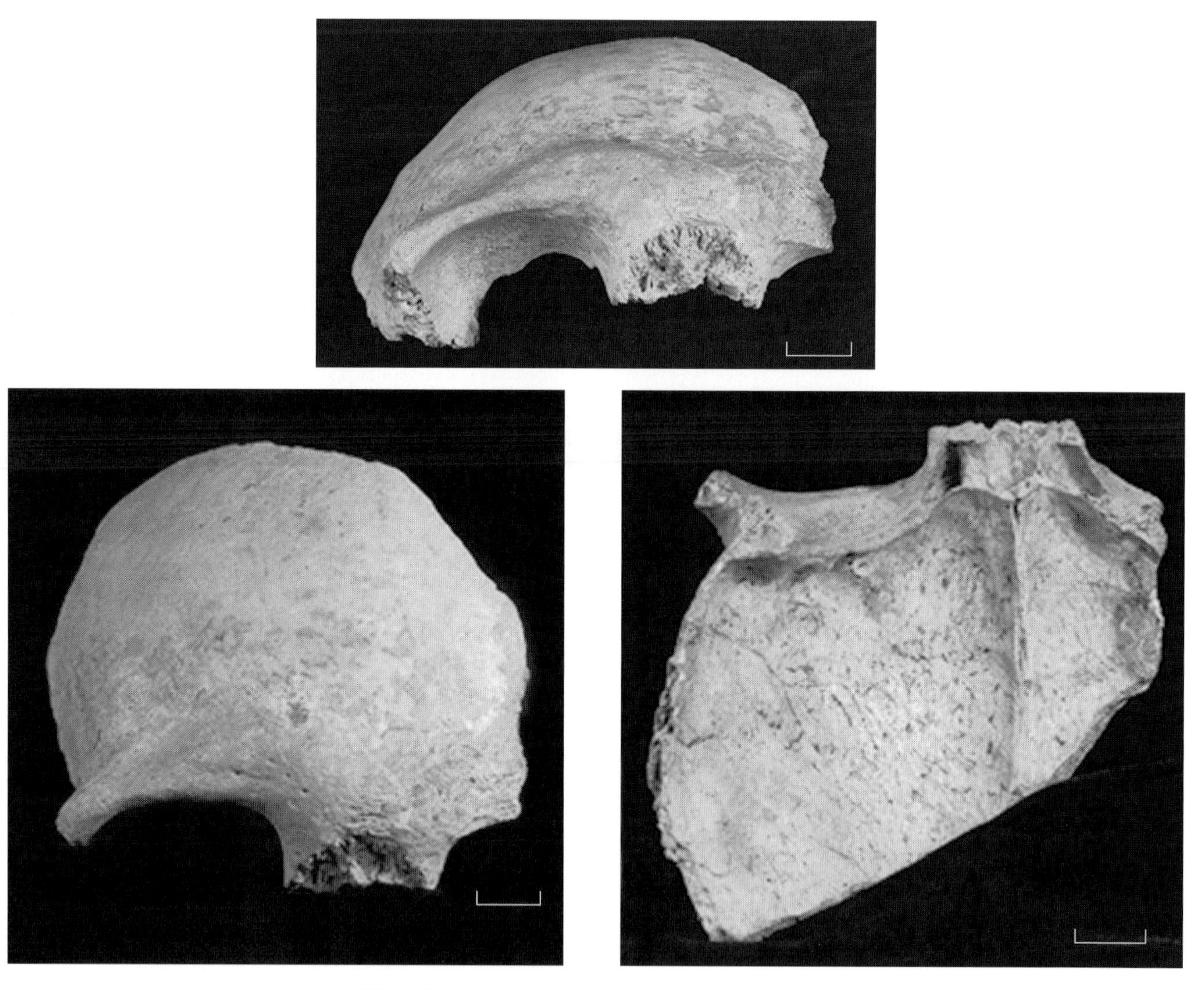

VELIKA PECINA Figure 1. Scale = 1 cm.

VÉRTESZÖLLÖS

LOCATION

Travertine quarry north of the village of Vértesszöllös, some 15 km W of Budapest, Hungary; now a museum.

DISCOVERY

Excavations directed by L. Vértes, February 1965 (when Vértesszöllös 1 was recognized in material excavated the previous year) and August 1965 (Vértesszöllös 2).

MATERIAL

Several lower deciduous teeth (Vértesszöllös 1) and a partial adult occipital bone (Vértesszöllös 2).

DATING AND STRATIGRAPHIC CONTEXT

Four occupation levels have been described in the Vértesszöllös travertine deposits (Kretzoi and Vértes, 1965), which lie on a terrace of the Atalér River that was correlated with the "Mindel" terraces 4–5 of the Danube Valley (Pécsi, 1990). The human remains come from the lowest occupation layer. The associated fauna suggests relatively temperate conditions. Thus, if the assemblage dates from a temperate interlude within the Mindel (Elster) Glacial, a date of ca. 400 ka would be appropriate for the hominids. This sits fairly well with U-series dates of from 250 to 350 ka obtained by Cherdyntsev (1971) for travertines younger and approximately contemporary with the hominids. Schwarcz and Latham (1990), however, have since analyzed what they consider to be more reliable (less contaminated) samples, obtaining ages of 185–210 ka for the lower occupation layer. These later dates place the hominids in isotope stage 7 or possibly in early stage 6.

ARCHAEOLOGICAL CONTEXT

The hominids are associated with a Lower Paleolithic stone industry dubbed the Buda by Kretzoi and Vértes (1965). This industry is characterized by small "choppers" and flake tools, made mostly from small quartzite pebbles and was compared most closely by its describers to the Clactonian. Shallow depressions in the archaeological deposits contain concentrations of bone fragments, some of which show signs of burning.

PREVIOUS DESCRIPTIONS AND ANALYSES

In the first announcement of the Vértesszöllös hominids, Vértes (1965) referred to them as *Homo erectus*. Thoma (1966) more fancifully made the occipital the holotype of the new taxon *Homo (erectus* seu *sapiens) palaeohungaricus* (a subspecies at the boundary between *Homo erectus* and *Homo sapiens*), thereby setting the stage for an ongoing debate over whether the Vértesszöllös hominid is *Homo erectus* (e.g., Wolpoff, 1977) or "archaic *Homo sapiens*" (e.g., Stringer, Howell, and Melentis, 1979). The latter pointed specifically to resemblances to the Petralona cranium. Another notion, expressed by Steslicka (1968), is that of Neanderthal affinity. Estimated cranial capacity is 1325 cc (Thoma, 1981).

MORPHOLOGY

Vérteszöllös 1

Lower, probably R, dm2 with metaconid and enough else to show huge trigonid basin and thickened crest-like structure descending from metaconid to talonid basin. Talonid basin seems to have been quite broad, with cusps well spaced. Enamel seems moderately thick; pulp cavity was quite small. Roots missing.

Rdc[1] crown, broken just below neck. Tip worn flat. Would have been quite asymmetrical. Root appears to have been angled relative to crown, which bulges buccally. Lacks lingual tubercle. Margocristae broad but faint.

Fragment probably of a lower P (not a molar as thought by Thoma), with two small cusps separated by narrow cleft. A groove runs at right angles to this cleft, from which the enamel rises to the break.

Vérteszöllös 2

Greater part of occipital bone, with lambdoid sutures intact but missing area of foramen magnum, although occipitomastoid suture present on the R.

Occipital tall; not V shaped in profile. Sharp angle between nuchal and occipital planes; with nuchal plane held horizontal, occipital plane rises sharply and appears to be relatively short up to lambda. Part of lambdoid sutural area missing on the L side; is clear the suture was peaked rather than continuously curved across lambda. Anterior lambdoid suture virtually undistinguished; occipitomastoid suture descends almost straight down. Faint elevation along occipitomastoid suture; small Waldeyer's crests present bilaterally. Muscle markings on occipital base faint at best. More or less continuous, quite deep sulcus essentially along region of superior nuchal line creates from below pronounced, but only moderately wide, occipital "torus" that inflects downward at lateral margins of sulcus. No suprainiac depression above "torus." "Torus" thickest (because sulcus deepest) laterally; "torus" more extensive laterally on the R than on the L.

Bone thick laterally, especially over the region of transverse sinuses; thins markedly toward lambda (but never very thin). Superior sagittal sinus represented only by faint depression just superior to interior occipital protuberance, where it veers off to flow into the R transverse sinus.

REFERENCES

Cherdyntsev, V. 1971. *Uranium-234.* Tel Aviv, Israel Program for Scientific Translations.

Kretzoi, M. and L. Vértes. 1965. Upper Biharian (Intermindel) Pebble-industry occupation site in western Hungary. *Curr. Anthropol.* 6: 74–87.

Pécsi, M. 1990. Geomorphological position and absolute age of the Vérteszöllös Lower Palaeolithic site. In: M. Kretzoi and V. Dobosi (eds), *Vérteszöllös: Site, Man and Culture.* Budapest, Akadémiai Kiado, pp. 27–62.

Thoma, A. 1966. L'occipital de l'Homme mindélien de Vérteszöllös. *L'Anthropologie* 70: 495–534.

Thoma, A. 1981. The position of the Vérteszöllös find in relation to *Homo erectus.* In: B.A. Sigmon and J.S. Cybulski (eds), *Papers in Honor of Davidson Black.* Toronto, University of Toronto, pp. 105–114.

Schwarcz, H. and A. Latham. 1990. Absolute age determination of travertines from Vérteszöllös. In: M. Kretzoi and V. Dobosi (eds), *Vérteszöllös: Site, Man and Culture.* Budapest, Akadémiai Kiado, pp. 549–555.

Stringer, C., C. Howell and J. Melentis. 1979. The significance of the fossil hominid skull from Petralona, Greece. *J. Archaeol. Sci.* 6: 235–253.

Steslicka, W. 1968. W sprawie Stanowiska systematcznego dolnoplejstcenskiej kosci potylicznej z Vérteszöllös. *Przeg. Antrop.* 2: 267–274.

Vértes, L. 1965. Discovery of *Homo erectus* in Hungary. *Antiquity* 39: 303.

Wolpoff, M. 1977. Some notes on the Vérteszöllös occipital. *Am. J. Phys. Anthropol.* 357–364.

Repository

Magyar Nemzeti Muzeum, Budapest, Hungary.

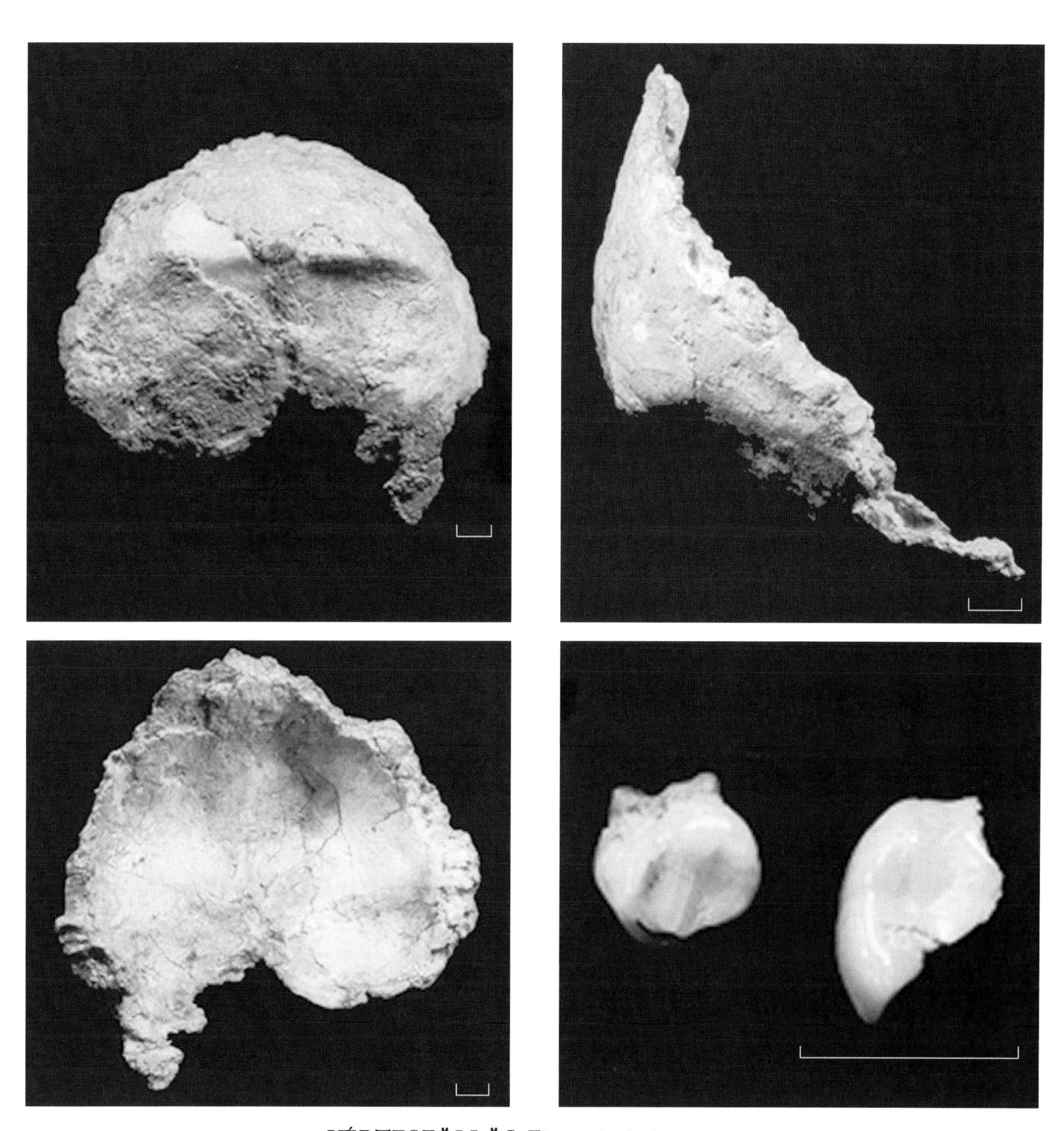

VÉRTESZÖLLÖS Figure 1. Scale = 1 cm.

VINDIJA

LOCATION
Karst cave close to the regional center of Ivanec, 55 km NNE of Zagreb, Croatia.

DISCOVERY
Excavations directed by M. Malez, 1974–1981.

MATERIAL
There are three chronologically distinct sets of hominid remains from this site (Wolpoff et al., 1981). The most recent of them are some 34 fragmentary specimens, mostly cranial, from level D. A little older are a few pieces from levels F and G_1. Oldest is the collection of fragments from level G_3.

DATING AND STRATIGRAPHIC CONTEXT
The deposits at Vindija have been divided into 13 layers, of which layers D–L are of Pleistocene age (Wolpoff et al., 1981). The uppermost Pleistocene layer, D, is a fine sandy deposit that has yielded hominid fossils; it dates from less than 27 ka, which is the ^{14}C age of one subunit of the underlying layer F, which has also yielded a few hominid fragments. The bulk of the Vindija hominid sample comes from layer G_3, a sandy subunit dated by amino acid racemization on bone to 42 ka (Protsch, quoted in Smith et al., 1985), a fairly plausible maximum date, if open to some technical question, especially because a cave bear long bone from the overlying level G_1 has recently given an AMS radiocarbon date of 33 ka (Karavanic, 1995) and Smith et al. (1999) have recently reported direct AMS dates on G_1 Neanderthal bone of 29–28 ka (making these specimens the youngest Neanderthals yet reported from Central Europe and in the same time range as the latest Neanderthals from Iberia).

ARCHAEOLOGICAL CONTEXT
The deposits at Vindija are relatively poor in tools, suggesting only sporadic occupation of the site. Artifacts from layer D have been described by Karavanic (1995) as late Gravettian, whereas toward the bottom of layer F Aurignacian implements begin to be encountered. The earliest Aurignacian tools come from level G1. Below G_1, and down through level K, industries are described as Mousterian (Wolpoff et al., 1981).

PREVIOUS DESCRIPTIONS AND ANALYSES
Wolpoff et al. (1981) described the Gravettian level D hominids as fully modern *Homo sapiens* and also placed the sparse Aurignacians from levels F and G_1 in this category. However, they found considerable "overlap" between these later forms and the earlier Mousterian fossils of level G_1, which they characterized as Neanderthals (in their view a variant of "archaic *Homo sapiens*"). They thus perceived continuity between the "progressive" Neanderthals and moderns at this site, the former evolving into the latter in situ, a scenario that had already been proposed by Smith and Ranyard (1980). But in reporting further finds, mostly from G_3, Smith et al. (1985) hinted that the

Vindija G_3 Neanderthals were less distinct from those of Krapina than Wolpoff et al. had claimed, thus by implication weakening the argument for continuity.

MORPHOLOGY

Mouterian Group (Levels G1 and G3)

Cranial Elements. There are nine frontal fragments from Level G3. As preserved, supraorbital tori quite thin, continuous over orbit, smoothly rolled, but truncated and sharply angled at superior orbital margin and quite aggressively protrusive. Post- or supratoral sulcus short with relatively tall frontals behind. Frontal sinuses sometimes small but often capacious, extending up into frontal and laterally to midpoint of orbits. Occasional supraorbital notch, but it does demarcate supraorbital features.

Number of other braincase fragments in collection, none of them very large. Sagittal and coronal sutures uniformly interdigitated, with shallow denticulations (e.g., Vi-74/208). None of the specimens very thick boned. Most remarkable are the following. Vi-75/205: small part of occipital, from midline, with some evidence of horizontal occipital torus and suprainiac depression. Vi-80/301 (G3): central part of occiput; long, horizontal torus divided into two halves by shallow ridge, with well-developed sulcus beneath and very large, pitted suprainiac fossa above; internally, R transverse sinus veers off sharply from superior sagittal sinus, with accessory transverse sinus running to meet it from the internal occipital protuberance. Vi-77/307 (G1): virtually complete L zygoma. Three large surface foramina, one below inferior orbital margin; inferior portion of well excavated for maxillary sinus; superior portion deep.

Mandibles and Lower Teeth. There are five specimens from G3. They are characterized by the following features. Thin corpora. Possibly subalveolar depression but essentially featureless, broadly arcing or straight symphysial region that may rise slightly vertical. No inferior marginal tubercles. Moderate to large mental foramen under P2 or M1; possibly accessory foramen. Gonial region truncated, smooth externally, somewhat inflected, and with slight vertical elevation. Small to large, distinct retromolar space. Digastric fossae variably developed. Genial pits (can be pronounced) or genial tubercles (can be faint). Low, straight, sloping, possibly marked mylohyoid line. Submandibular fossa not distinguishable or very distinct. Low but extensive medial pterygoid tubercle. Mandibular foramen compressed, horizontally or obliquely facing. Lingula possibly lacking. Strong crest from region of lingula descends to meet strong internal alveolar crest. Broad, shallow sigmoid notch, deepest at midpoint. Sigmoid notch crest runs to lateral aspect of condyle.

Alveoli in general large and deep; those for Is may be slightly procumbent. C somewhat concave lingually, with margocristids; buccally, crown curves back from root. P1 dominated by protoconid; thick paracristid runs to basal lingual; distal to swelling is a tiny talonid delineated by a crest. M1 possibly smaller than M2. Where visible, roots separate low on M1. M2 roots slightly separated to columnar. M3 roots columnar. All Ms ovoid/rounded occlusally with well-defined trigonid basins, small, buccally placed hypoconulids, well-separated hypoconids and entoconids, accessory cusps between metaconid and entoconid, large, broad talonid basins, some enamel wrinkling; lack centroconids. Accessory cusp lies between metaconid and entoconid.

Aurignacian Group (Upper Levels D and E)

Cranial Elements. There are three fragments of parietal and one of parietal with almost complete occipital attached. Judging by latter, skull was tall, with peaked lambdoid suture and narrow occipital; occipital bulges slightly in side view and devoid of torus and depression. Parietals thick with Pacchionian depressions. Sagittal suture deeply interdigitated, with four distinct segments. Anterior lambdoid suture short. There is a L greater wing of sphenoid, lacking distinction between temporal and infratemporal fossae.

Mandible and Lower Teeth. There is a R mandibular fragment with alveoli for anterior teeth, Rdm1–2 and M1, and crypt for M2. Distinct mental trigon present. Mental foramen under posterior root of dm1. No genial pit; area of genial tubercles roughened. Digastric fossae indistinct. dm1 with two splayed roots that bifurcate near neck; region of protoconid quite bulbous buccally, distended medially; paracristid completely encloses minuscule fovea. Cusps (now worn) surround a small talonid basin; were probably not ever tall or distinct. dm2 with two distinct roots that diverge at neck; no evidence of trigonid basin; talonid basin indistinct; hypoconulid centrally placed. M1 enamel slightly wrinkled, with small anterior fovea; no trigonid basin; talonid basin small; hypoconulid medium sized, centrally placed.

REFERENCES

Karavanic, I. 1995. Upper Paleolithic occupation levels and late-occurring Neandertal at Vindija Cave (Croatia) in the context of Central Europe and the Balkans. *J. Anthropol. Res.* 51: 10–35.

Smith, F. and G. Ranyard. 1980. Evolution of the supra-orbital region in Upper Pleistocene fossil hominids from south-central Europe. *Am. J. Phys. Anthropol.* 53: 589–610.

Smith, F. et al. 1985. Additional Upper Paleolithic human remains from Vindija Cave, Croatia, Yugoslavia. *Am. J. Phys. Anthropol.* 68: 375–383.

Smith, F. et al. 1999. Direct radiocarbon dates for Vindija G1 and Velika Pecina Late Pleistocene hominid remains. *Proc. Natl. Acad. Sci. USA* 96: 12281–12286.

Wolpoff, M. et al. 1981. Upper Pleistocene human remains from Vindija Cave, Croatia, Yugoslavia. *Am. J. Phys. Anthropol.* 54: 499–545.

Repository

Institute of Paleontology and Quaternary Geology, Croatian Academy of Sciences and Arts, 41000 Zagreb, Croatia.

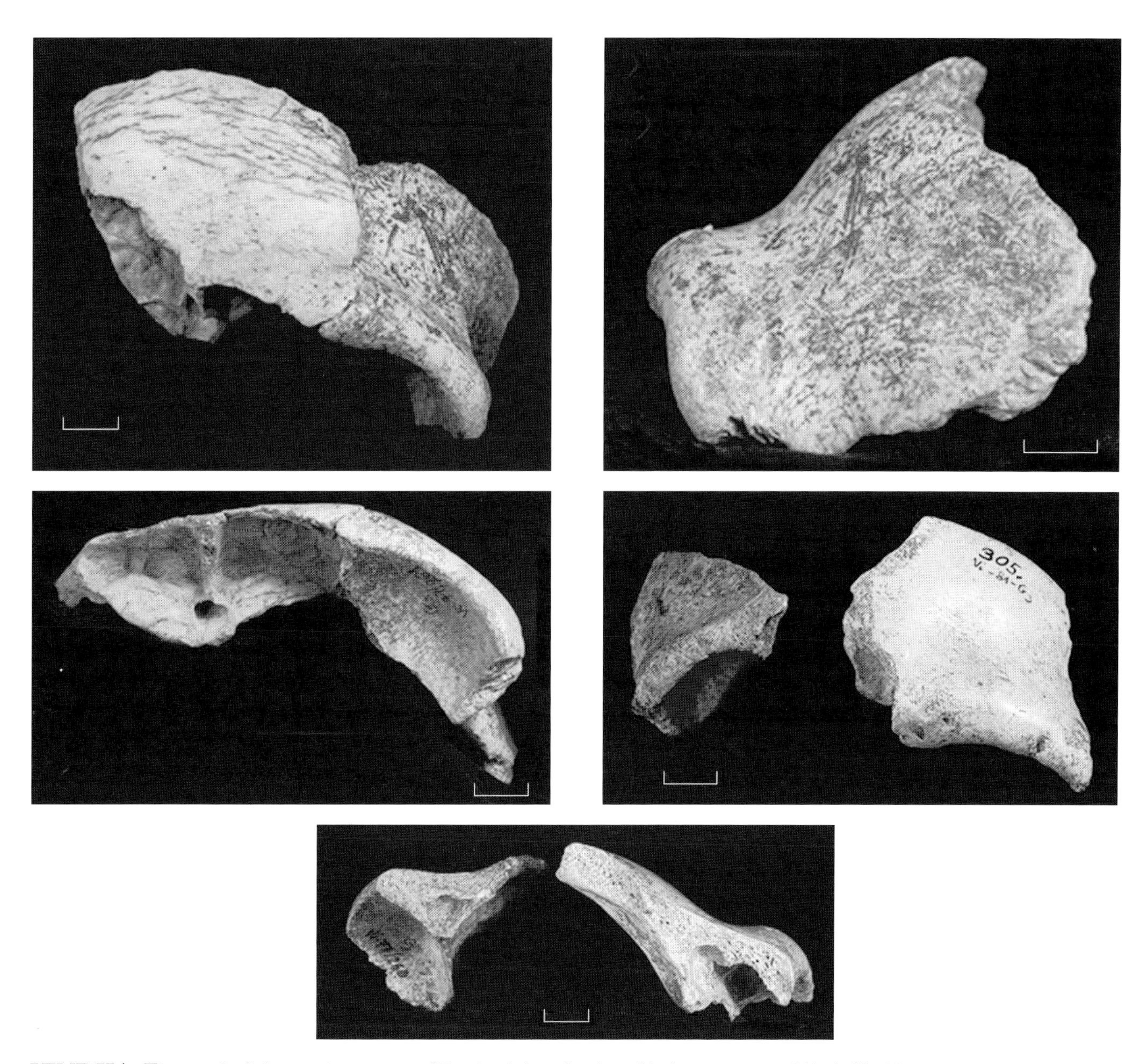

VINDIJA Figure 1. Mousterian group: Vi-93-275 and 25-34(?) (top row, middle left); Vi-77-260g3 and Vi-81-G3 (middle right, bottom), (scale = 1 cm).

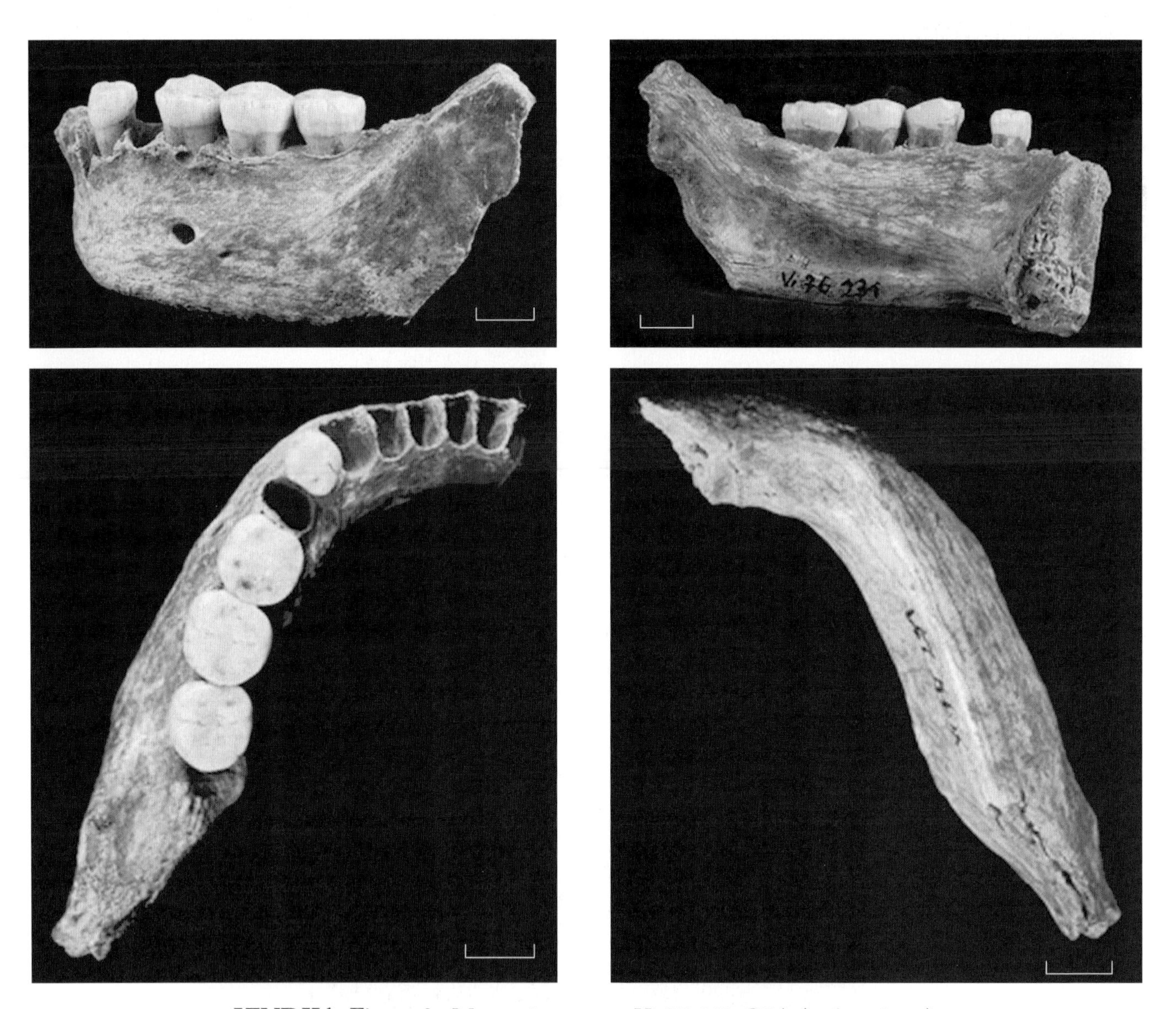

VINDIJA Figure 2. Mousterian group: Vi-76-213-G3A (scale = 1 cm).

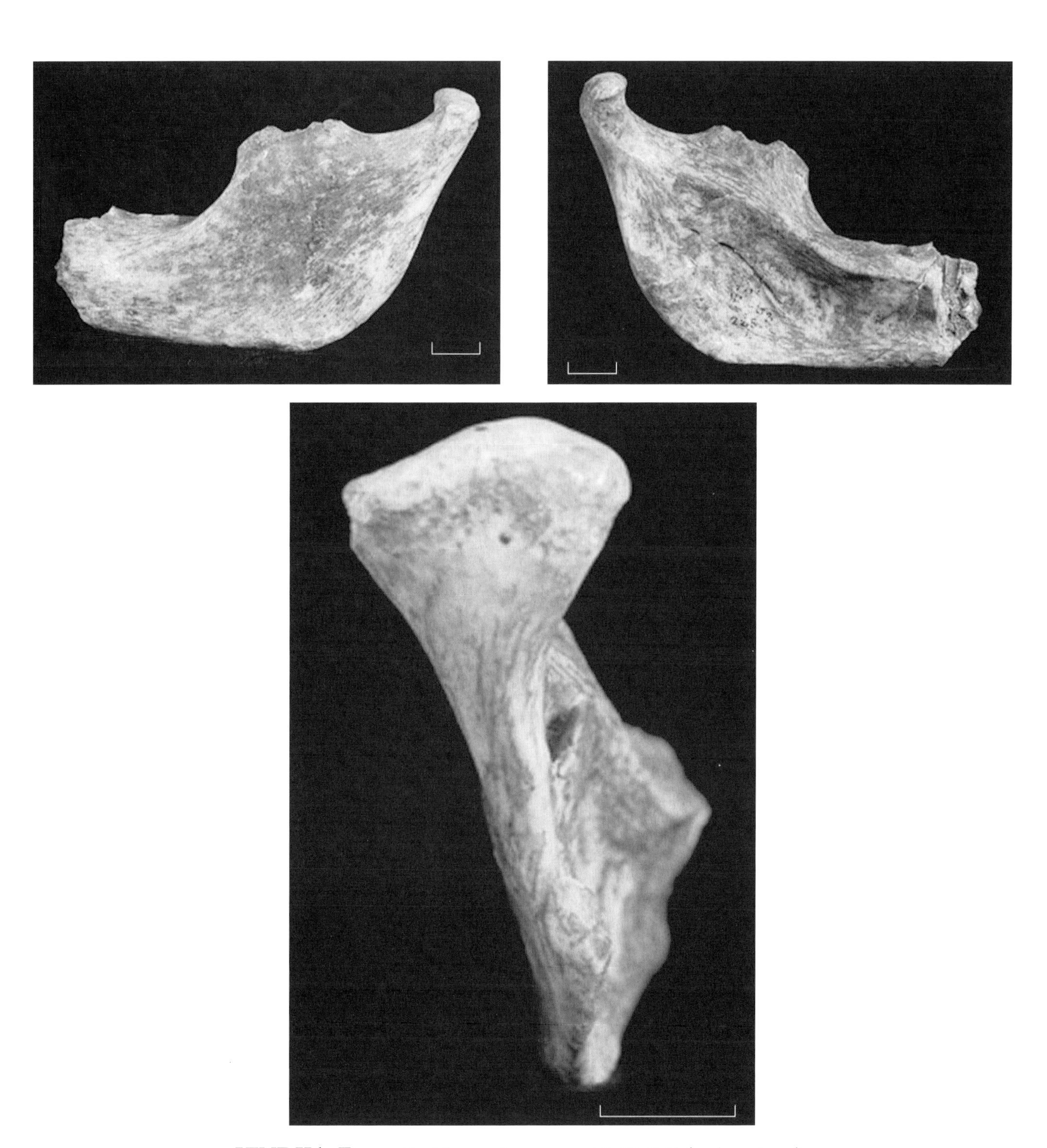

VINDIJA Figure 3. Mousterian group: Vi-76-265 (scale = 1 cm).

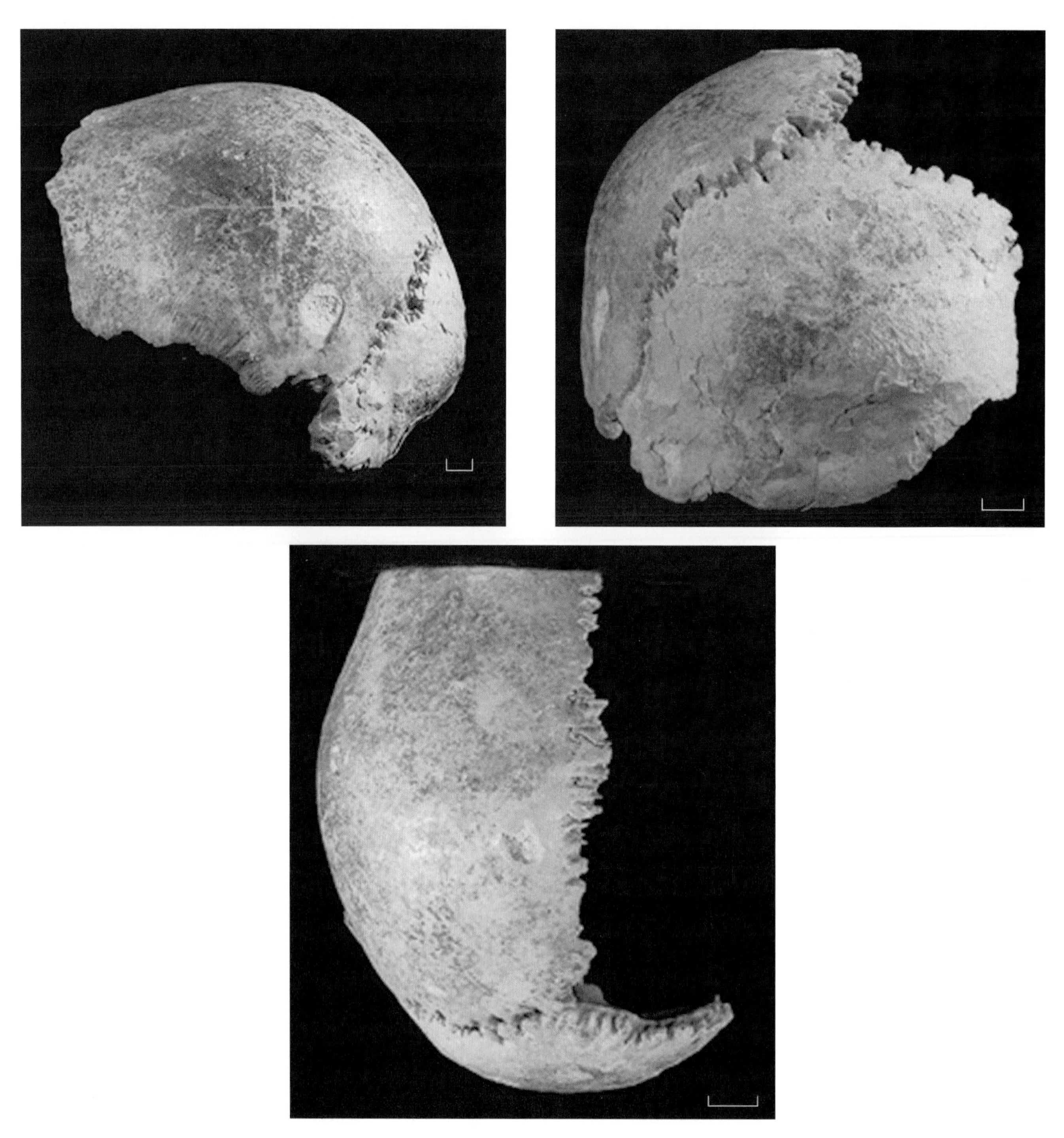

VINDIJA Figure 4. Aurignacian group: Vi-(19741)-210 (scale = 1 cm).

Vogelherd (Stetten)

Location

Limestone cave in the valley of the Lone river, 1 km N of Stetten village, near Oberholzingen, Baden-Württemberg, Germany.

Discovery

G. Riek, July 1931.

Material

Almost complete calvaria, with partial mandible (Vogelherd 1); partial cranium (Vogelherd 2); most of a right humerus (Vogelherd 3).

Dating and Stratigraphic Context

The SW entrance of the small cave of Vogelherd contained about 2 m of sediment running from the earliest Aurignacian through the Gravettian to the Magdalenian (directly overlain by a thin layer of Medieval occupation). The Vogelherd 1 and 3 partial skull and humerus (plus two vertebrae) were found at the bottom of the Aurignacian layer, just above bedrock (Riek, 1932, 1934); this level has been radiocarbon dated at about 32 ka (Hahn, 1977, 1981). The associated fauna confirms cold climatic conditions (Riek, 1934). The Vogelherd 2 partial cranium was also found low in the deposits but evidently in a pit; Riek (1934) considered it intrusive from the Upper Paleolithic, and it is not considered further here.

Archaeological Context

The Vogelherd 1 and 3 specimens are associated with an Aurignacian industry characterized by end scrapers made on thick blades and flakes and blades with flat retouch as well as split-base bone points (Gamble, 1986). This site is particularly notable for the spectacular series of small animal figurines carved in mammoth ivory.

Previous Descriptions and Analyses

The Vogelherd Aurignacians were most fully described by Gieseler (1937), who recognized that the Vogelherd 1 cranium was fully modern *Homo sapiens* but suggested (mistakenly) that the Vogelherd 3 humerus might be that of a Neanderthal. Wolpoff (1996) considers the Vogelherd 1 cranium comparable to the early modern Mladec population.

Morphology

Vogelherd 1

Almost complete calvaria, lacking face and part of petrosal. Partial mandible also present. Consists of complete R ramus and corpus extending to the L well past symphysis to P2 alveolus; RM1–3 quite worn but in place; other teeth are represented by alveoli. Bone not very permineralized. Calvaria fairly thin boned, moderately long, vault quite rounded; only slight flattening in front of, and slight bulge behind, lambda. In posterior profile, sides lean in gently upward, curve smoothly over top of cranium.

Medial portion of supraorbital region and glabella lacking; on the R is preserved lateral terminus of glabella "butterfly" and lateral to it a thin, platelike, posteriorly oriented supraorbital rim. Exposed frontal sinuses extend quite far up into frontal, to a level at least equal to highest point on squamosal suture; laterally, sinuses extend on the R up to, and on the L slightly beyond, midpoint of supraorbital margin; sinuses contain traces of septa (indicating multifocular conformation). Orbital roofs angle bluntly back onto supraorbital margins. Frontal lobes extended fully forward over orbital cones. Laterally, relatively short oblique plane lies behind supraorbital margins, behind which frontal rises sharply. Centrally, frontal bone was more domed; would have risen straight up behind glabellar region. Distinct frontal eminences lacking. Postorbital constriction not pronounced. Very thick temporal ridges rise almost vertically from zygomaticofrontal suture. Behind coronal suture, on both sides, temporal lines gradually fade posteriorly, becoming almost imperceptible by their point of recurvature. Parietals lack distinct eminences; form flat plane between apex of cranium and lambda.

Anterior portion of temporal fossa was not deeply excavated. Especially on the R, a sharp angulation along a muscle scar on the alisphenoid delineates an infratemporal fossa. Anterior portion of the relatively long, not very tall squamosal flows smoothly into curvature of alisphenoid. Squamosal suture not very strongly arced.

M/l wide mandibular fossae angle forward and out; are basically quite narrow but have been lengthened a/p by remodeling of articular eminence in front and of ectotympanic tube behind. Medial side of mandibular fossae not closed off by medial articular tubercle. Postglenoid plate m/l wide but low; extends laterally beyond margin of auditory meatus. Rather modest vaginal process most salient at point where it wraps around the quite centrally placed, relatively thin styloid process; fades out laterally along relatively short ectotympanic tube. Ectotympanic tube closely appressed to front of mastoid process. As seen on the L, modestly sized stylomastoid foramen lies lateral and posterior to base of styloid process.

Posterior root of zygomatic arch takes origin in front of somewhat ovoid, slightly anteriorly tilted, moderately sized auditory meatus. This root does not project out strongly from side of skull; posteriorly grades into low and somewhat sharp suprameatal crest that is confluent posteriorly with thicker and upwardly curving supramastoid crest. Parietomastoid sutures long, fairly horizontal. Vertical parietal notches lie somewhat in middle of basally quite bulky, long mastoid processes. Mastoid processes broken at their tips, exposing variety of medium to large air cells; orientation of processes not determinable. Occipitomastoid sutures preserved on both sides; run in shallow grooves; are bounded at least medially by slight elevation of bone. Quite well-developed, sharp Waldeyer's crests lie somewhat medial to these sutures, bilaterally.

Occipital plane tall, broad; bulges smoothly from top to bottom and from side to side. Lambdoid suture rises sharply from asterion to peak bluntly at lambda. At a point more or less level with asterion, inward sagittal curve of occipital bulge interrupted by wide, shallow depression that lies just above superior nuchal line. In midline, just below this shallow depression, lies broad roughening in thickened central area of laterally and downwardly curving superior nuchal line. Below superior nuchal line, in midline, lies depressed muscle attachment area; anterior to it, nuchal plane curves quite sharply forward to foramen magnum. Foramen magnum relatively small, ovoid, broadest posteriorly. Relatively long occipital condyles more convex anteriorly than posteriorly; lie quite anteriorly along margin of foramen magnum. Basiocciput does not angle upward greatly anterior to foramen magnum; basicranium thus essentially flat.

Breakage exposes very large, multiple sphenoidal sinuses; they pervade superior roots of damaged medial and lateral pterygoid plates. These plates were not confluent superiorly. Superior root of medial pterygoid plates lie laterally, well beyond margin of basisphenoid. Foramina ovales lie in line with, and at roots of, lateral pterygoid plates; also lie well within the sphenoid. Relatively large foramina spinosa (behind ovales) not enclosed medially by bone (i.e., open medially). Sphenooccipital synchondrosis obliterated. Basiocciput relatively wide, long, thick, its surface highly textured. Partially preserved on the R, appears that foramen lacerum was small and anteromedial portion of petrosal quite tapered. Fairly large carotid foramina (L larger than R) face slightly backward. Jugular foramina (R much larger than L) point slightly forward.

The three major sutures segmented; sagittal and lambdoid deeply denticulated.

Internally, preserved portion of frontal crest quite prominent, especially inferiorly. No signs of sutures interiorly. Numerous Pacchionian depressions. On both sides, anterior branch of middle meningeal artery very large. Moderately wide petrosals bear broad superior petrous sinuses and large, domed arcuate eminences. Subarcuate fossae indented, not fully closed off.

Mandibular corpus narrow from side to side; deepens anteriorly below Ms to reach its deepest point below the Ps. Inferior border of corpus curves somewhat upward from this point forward to symphysis. Externally, symphyseal region flattish across outwardly inflected I1–2 roots, with even a slight depression in this region. Central keel broad, low; takes origin below region between I1s and expands inferiorly, becoming confluent with very protrusive inferior margin that is almost straight across from side to side and that corners bluntly bilaterally below the Cs. Central keel bounded bilaterally by upper and lower pair of depressions; upper ones shallower. Moderately sized mental foramen lies under P2. Viewed from below, rather small digastric fossae quite posteriorly oriented; symphyseal region only minimally thicker at midline than elsewhere along thickened inferior margin. Interior surface of symphysis quite broadly curved, vertical; distinct twin genial tubercles lie inferiorly along surface. Mylohyoid line quite rugose; submandibular fossa below only moderately excavated. M3 lies well in front of slightly backwardly tilted anterior border of ramus, which is inferiorly slightly concave in profile, but does not have well-defined preangular notch. Gonial region thick, slightly outwardly reflected; bears strong muscle scars inferiorly on both internal and external surfaces. Gonial angle obtuse but not cut off. Coronoid process slender, somewhat peaked; lies markedly above level of damaged and highly remodeled condyle. Sigmoid notch deepest right at base of condyle. Sigmoid notch crest rather laterally displaced. Large, compressed lingula-lacking mandibular foramen oriented almost straight upward.

The three preserved lower Ms subequal in size; roots bifurcate some distance below cervix, are too worn to say much else.

Vogelherd 2

Plagiocephalic partial cranium, lacking facial skeleton and most of cranial base. Slightly thicker boned than Vogelherd 1, also wider across frontals and parietals, giving more rounded outline from above.

Lateral portions of supraorbital region preserved, along with lateralmost extents of glabella "butterfly" clearly distinct from lateral plates. "Butterfly" was evidently quite strongly swollen anteriorly; lateral plates quite vertically oriented. Orbital roofs angle quite acutely into supraorbital margins. Sulcus would have defined glabellar "butterfly" from above and anteriorly swelling frontal dome from below. On the L, frontal eminence rather prominent; on both sides, fairly rugose but not very long temporal ridge rises quite vertically from behind rather short zygomatic process of frontal. Temporal ridges fade quickly, well in front of coronal suture, to become faint temporal lines that arc quite high along parietals.

Frontal sinuses expand full width and height of glabellar "butterfly"; also pervade orbital roof, over which frontal lobes extended fully. Postorbital constriction minimal. Appears that squamosal plane would have flowed smoothly into alisphenoid. Judging by preserved squamosal sutures on the parietal, appears that squamosals were long and not very tall s/i and squamosal suture only shallowly curved. L parietal notch vertical, a/p long; lies over midline of broken mastoid process.

Mandibular fossa relatively wide m/l, deep, somewhat long a/p; is oriented obliquely anteriorly. Its articular surface defined anteriorly by almost vertical articular eminence; posteriorly, extends over anterior border of ectotympanic tube. Sheetlike vaginal process damaged; certainly extended full length of ectotympanic tube; contacted mastoid process or came very close to it; probably peaked around very massive preserved base of styloid process. At very base of styloid process, posteriorly, lies medium-sized stylomastoid foramen. Paramastoid crest low but distinct; lies right at base of mastoid process, well lateral of occipitomastoid suture, which appears not to have borne a crest. On the L, very weak Waldeyer's crest preserved quite medial and posterior to occipitomastoid suture.

Mastoid process was long a/p, swollen laterally at its base, pervaded internally by small to medium air cells. As seen on L, posterior portion of zygomatic arch takes root anterior to relatively large, ovoid, slightly anteriorly inclined auditory meatus; angled moderately out from side of skull, then curved forward, probably running straight to zygoma, thereby enclosing narrow temporal fossa; flows behind into low, broad suprameatal crest that is in turn confluent with equally developed supramastoid crest that turns almost vertically upward.

Posterior margin of foramen magnum preserved; is broadly curved. Posterior to it, and separated by thin median crest, lie two deep but confined muscular depressions. Median crest runs back to tiny, laterally compressed, crestlike external occipital protuberance that lies at inferiormost point of bow-shaped superior nuchal line. Large, scallop-shaped muscular impressions below superior nuchal line, bilaterally. Sinuous highest nuchal line visible well above superior nuchal line, and parallels its contour.

Lambdoid suture rises steeply upward from asterion, arcs across region of lambda. Missing from skull is large Inca bone that would have occupied the upper one-third of the broad and tall occipital plane. Parietomastoid suture long, sinuous; runs back from somewhat vertically oriented parietal notch on the L.

Major cranial sutures variably closed internally but almost fully patent externally; are strongly denticulated, and less strongly, though distinctly, segmented.

REFERENCES

Gamble, C. 1986. *The Palaeolithic Settlement of Europe.* Cambridge, Cambridge University Press.

Gieseler, W. 1937. Bericht über die jungpaläolithischen Skelettreste von Stetten ob Lontal. *Verh. Ges. Phys. Anthropol.* 8: 41–48.

Hahn, J. 1977. *Aurignacien: Das ältere Jungpaläolithikum in Mittel- und Osteuropa.* Köln, Fundamenta Reihe.

Hahn, J. 1981. Abfolge und Umwelt der jungeren Altsteinzeit in Südwestdeutschland. *Fundberichte aus Baden-Württemberg* 6: 1–27.

Riek, G. 1932. Paläolithische Station mit Tierplastiken und Menschlichen Skelettresten bei Stetten ob Lontal. *Germania* 16: 1–8.

Riek, G. 1934. *Die Eiszeit Jägerstation am Vogelherd im Lonetal.* Tübingen, pp. 1–338.

Wolpoff, M. 1996. *Human Evolution* (1996/7 ed.). New York, McGraw-Hill.

Repository

Osteologische Sammlung, Eberhard-Karls-Universität, Tubingen, Germany.

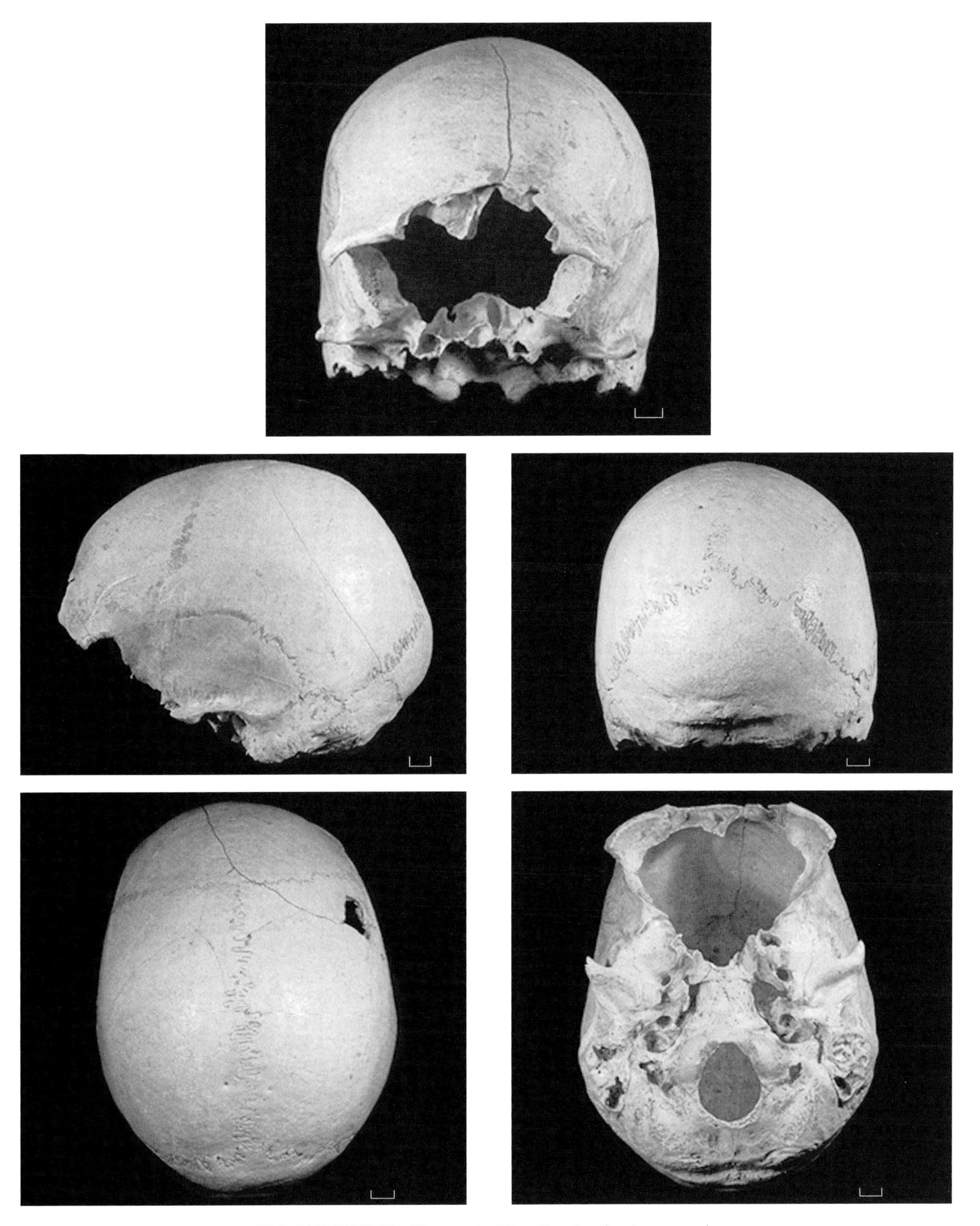

VOGELHERD Figure 1. Vogelherd 1 (scale = 1 cm).

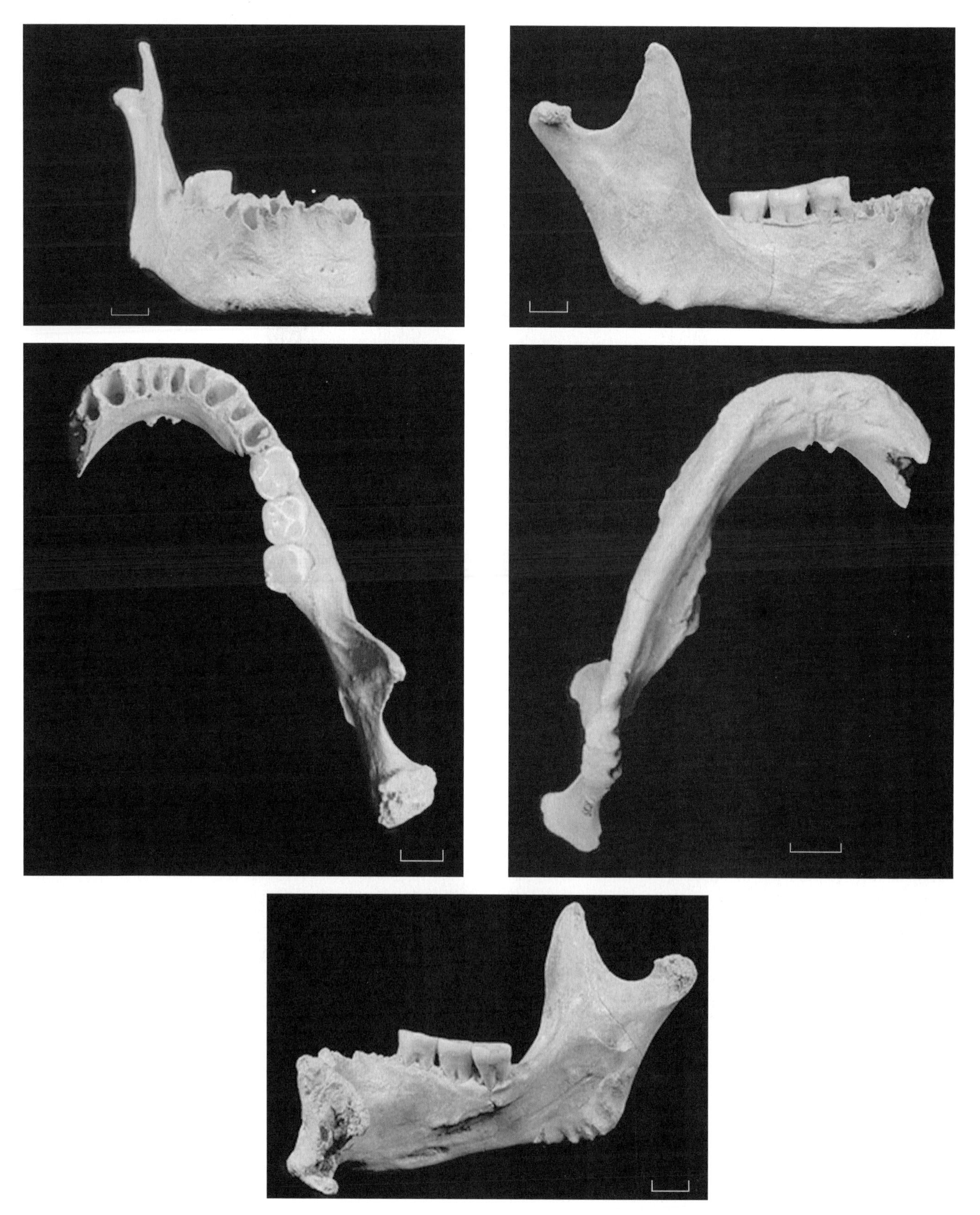

VOGELHERD Figure 2. Vogelherd 1 (scale = 1 cm).

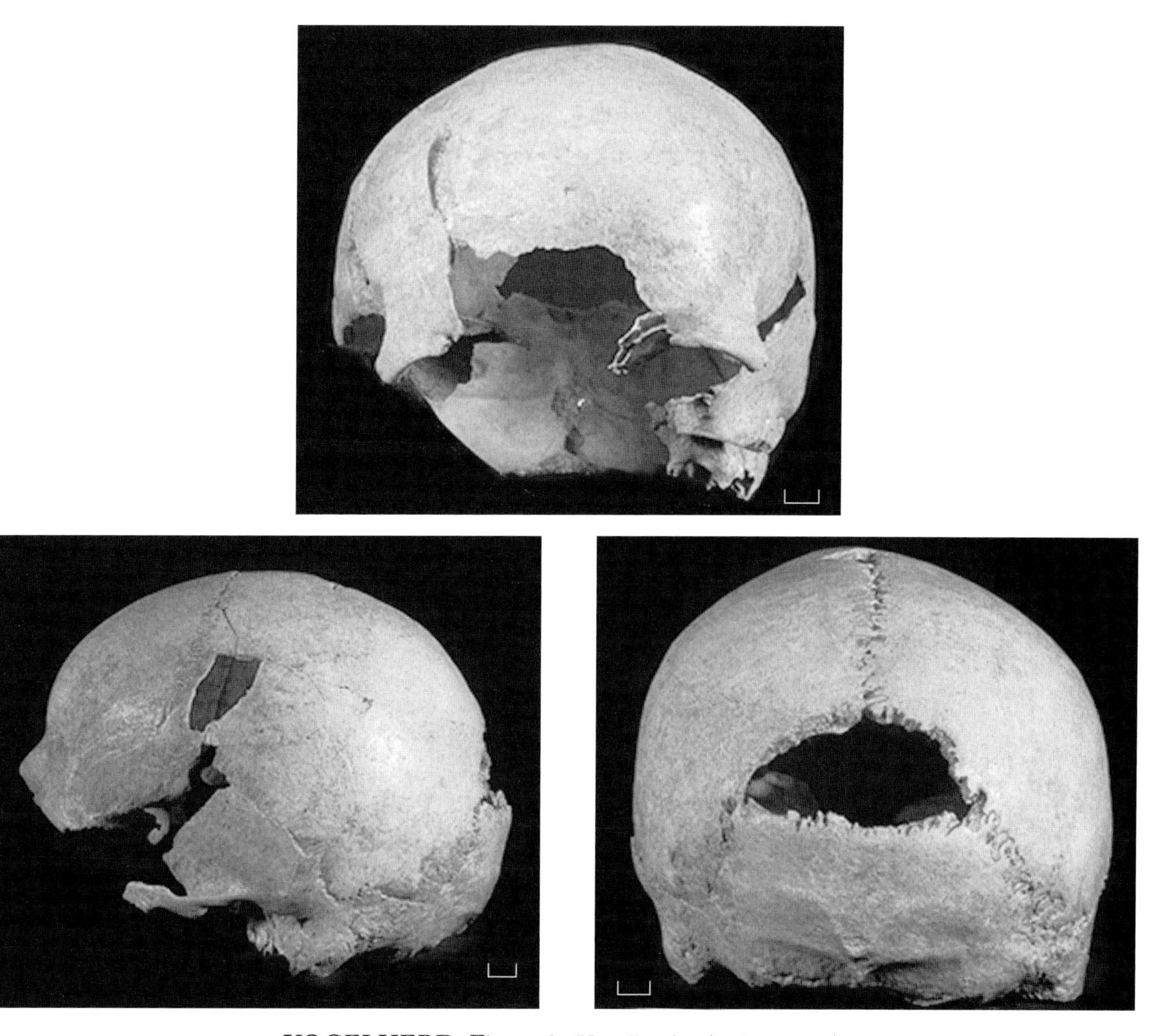

VOGELHERD Figure 3. Vogelherd 2 (scale = 1 cm).

Zafarraya

Location

Karst cave in the northeastern extremity of the province of Malagá, Spain, at the municipal boundary of the town of Alcaucin.

Discovery

Excavations of C. Barroso-Ruiz and P. Medina Lara, 1980–1983; of C. Barroso-Ruiz and J.-J. Hublin, 1992.

Material

One almost complete mandible and some postcranial fragments.

Dating and Stratigraphic Context

Five Mousterian layers, A–E, were initially identified in a secondary fissure just within the entrance to the cave (Hublin et al., 1995). Later excavations further inside the cave and in harder sediments used arbitrary 5-cm levels. Human remains were discovered low down in the 5-m-thick sequence, in layers D and E. Hublin et al. (1995) reported radiocarbon and U-series dates on bone and teeth, respectively, taken from the Mousterian deposits. These included one U-series date of 33.4 ka on a sample found directly in association with hominid remains. Samples found slightly higher in the Mousterian horizon averaged 29.8 ka by U-series. Radiocarbon dates for stratigraphically equivalent specimens came in much younger, at 22–24 ka. Overall, Hublin et al. concluded that Zafarraya provided evidence for the persistence of Neanderthals until at least 35 ka and of the Mousterian until under 30 ka. Because the undated uppermost Mousterian levels at Zafarraya have been greatly disturbed, Barroso-Ruiz (personal communication) suspects that the Mousterian lasted at the site until 27 ka or even less.

Archaeological Context

Throughout the excavated section the lithics recovered are of typical Mousterian type, showing Levallois technique (Barroso-Ruiz et al., 1983, 1984). There was apparently little stoneworking activity in the cave itself, but the partial hominid mandible was found within a distinct hearth, along with other burned bone and tooth fragments (Hublin et al., 1995).

Previous Descriptions and Analyses

From the time of their initial description by Garcia Sanchez (1986), the Zafarraya hominids have universally been regarded as Neanderthal.

Morphology

No. 8698 is largely complete mandible, missing part of R coronoid, RM3, and LI1–2. Broken and glued at midline. Corpora of moderate size, not notably thick m/l but quite tall s/i. Rami quite long a/p. All teeth very worn, chipped.

Viewed from the front, symphyseal region moderately broad, with relatively strong curve across from side to side. Bilaterally, are quite pronounced subalveolar fossa, with possibly another centrally; all lie in region between the canines. Area below these fossae smoothly curved. In profile, incisor roots slightly overhang symphyseal region, emphasizing subalveolar depressions below. Also in profile, bone slightly curves out inferiorly, down and around. Two mental foramina on each side, larger on the L. On the L, both foramina lie below M1; on the R, one lies below M1 the other a bit more anteriorly. Just slightly posterior to these foramina, inferior margins of corpora slightly thickened and distended; above these distensions are swellings out of the bone that extend posteriorly from the foramina. Right across front of jaw, inferior margin raised anterior to low, blunt inferior marginal tubercles.

Anterior root of ramus rises from below lower M3. Short retromolar space accentuated by shallow preangular notch. Gonial region relatively thin, truncated posteriorly into straight margin. Externally some muscle scars run obliquely down and back to slightly everted inferior gonial margin. Internally, medial pterygoid tubercle large, rugose, broad but low; below is smaller but still distinct secondary tubercle.

Internally, somewhat short postincisal plane slightly concave in region just behind incisors, then descends more vertically toward damaged genial region. Digastric fossae quite deep, long a/p; are not very broad m/l; may not have been separated at midline. On both sides, mylohyoid line long, oblique; scarring most prominent below M3s. Long submandibular fossae quite excavated. As seen most completely on the L, low, broad pillar runs up and back from relatively broad, flat internal alveolar crest and up midline of broad-based coronoid process to its thin peak. Just anterior to large, compressed, somewhat superiorly pointing mandibular foramina, another low, broad crest comes off internal alveolar crest and fades out below neck of condyle. Mandibular foramina lie just above level of tooth rows; are bounded by short, vertical, posteriorly expanded lingulae that show angular edge superiorly. Small but distinct mylohyoid grooves originate well below mandibular foramina.

Both condyles extended medially; lateral margins distended downward into tall, vertical, a/p compressed tubercles; posterior surfaces flat. R condyle bulkier than L; superior surface is flat, straight across. As seen on the L, coronoid process slightly taller than superiorly flat condyle. Thin sigmoid notch crest runs obliquely down toward neck of condyle, reaches lowest point just anterior to condyle. As seen more clearly on the R, sigmoid notch crest terminates lateral to midpoint of condyle.

With exception of Cs, teeth relatively small. Buccal faces of Cs may have been tilted back a bit. Cs had some development of lingual tubercle that was delineated distally by shallow depression. More noticeably on the L than on the R, P1 was larger than P2. As seen on the L, P1 has somewhat mesially placed lingual groove between small anterior fovea and crest that runs down protoconid and turns back to enclose slightly larger posterior fovea. Distal corners of P2s swollen out; buccal slopes of P1s apparently bore at least one lingual groove. M2 was apparently larger than M1; M1 may have been a bit larger than M3. As seen on the M1–2s, roots bifurcate well below neck; were apparently not separated on M3s.

REFERENCES

Barroso-Ruiz, C. et al. 1983. Avance al estudio cultural antropologico y paleontologico de la cueva del "Boquete de Zafarraya" (Alcaucin, Malagá). *Antropol. Paleoecol. Hum.* 3: 3–14.

Barroso-Ruiz, C. et al. 1984. Le gisement mousterien de la grotte du Boquete de Zafarraya (Alcaucin-Andalousie). *L'Anthropologie* 88: 133–134.

Garcia Sanchez, M. 1986. Estudio preliminar de los restos neandertalenses del Boquete de Zafarraya (Alcaucin, Malagá). In: *Homenaje a L. Siret* (1934–1984). Seville, Junta de Andalucia.

Hublin, J.-J. et al. 1995. The Mousterian site of Zafarraya (Andalucía, Spain): dating and implications on the Palaeolithic peopling process of Western Europe. *C. R. Acad. Sci. Paris* 321: 931–937.

Repository

Muséo Provincial de Malagá, Malagá, Spain.

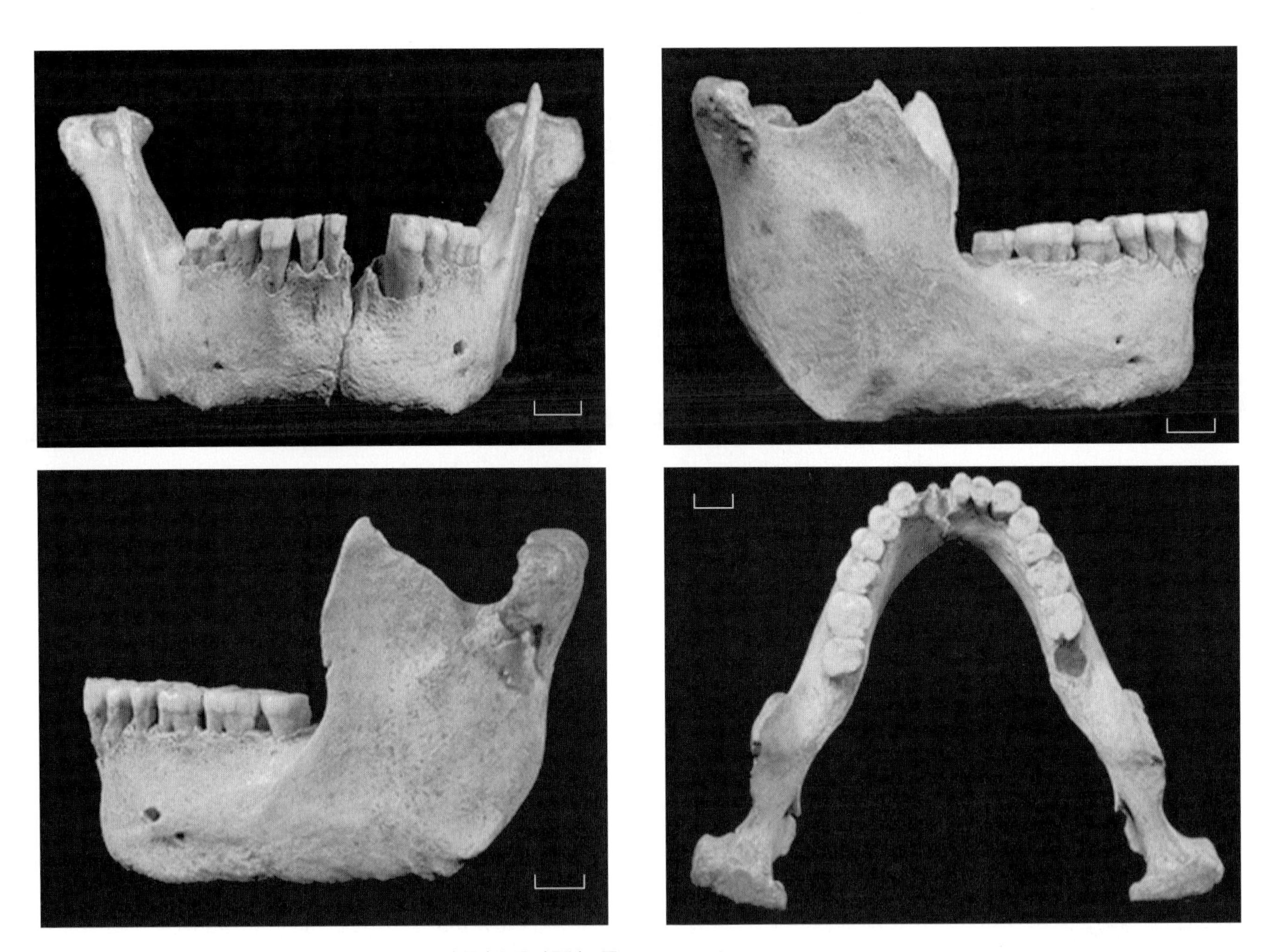

ZAFARRAYA Figure 1. Scale = 1 cm.

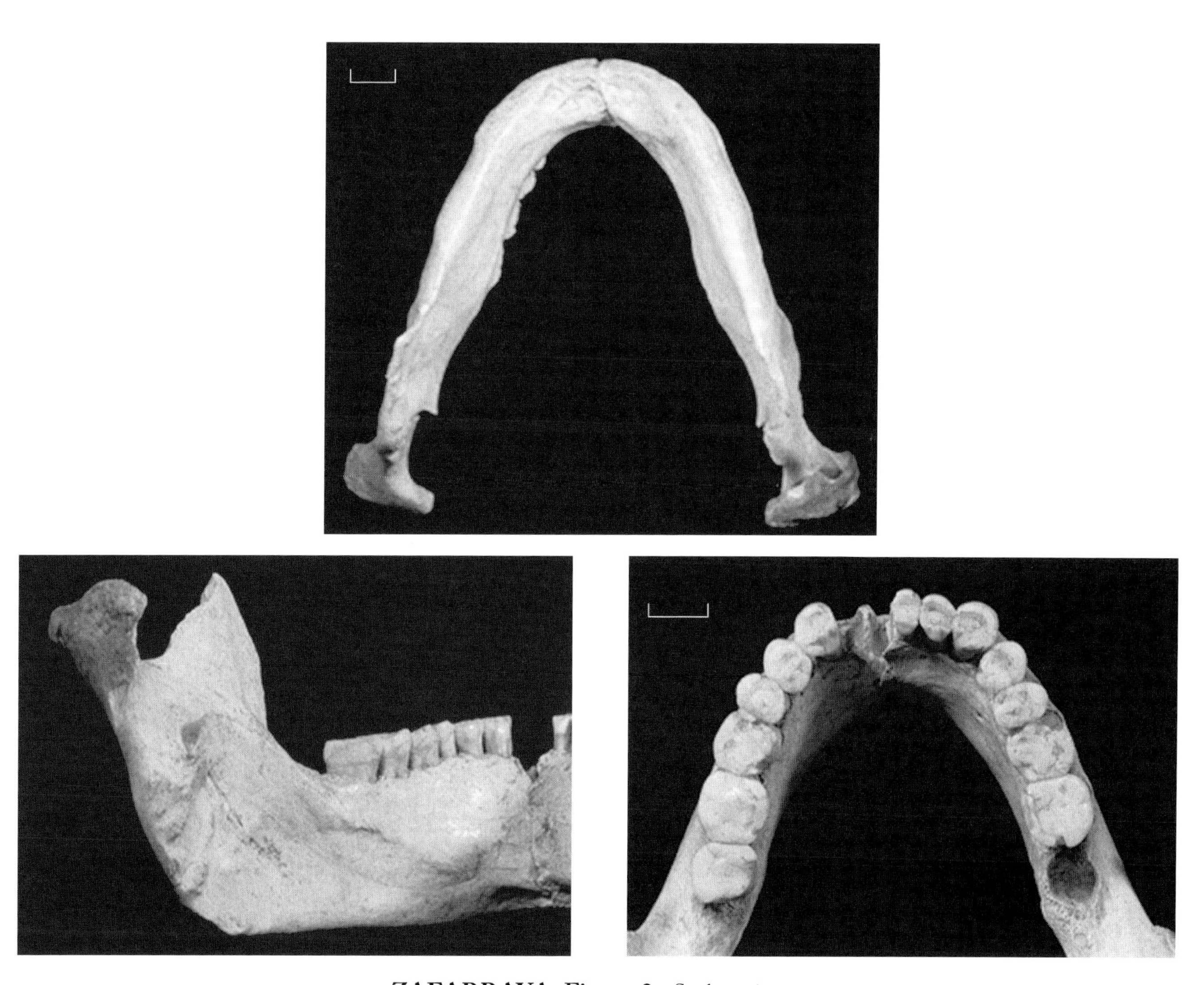

ZAFARRAYA Figure 2. Scale = 1 cm.

Zlaty Kun

Location

Cave site 500 m from Koneprusy village, near Beroun, Central Bohemia, Czech Republic.

Discovery

Laborers, November 1950 (ZK 1); excavations of F. Prosek, 1951–2 (ZK 2).

Material

Adult calvaria plus zygomatic bone (ZK 1); adult frontal, mandible and some postcranial bones (ZK 2). Now considered to represent one individual.

Dating and Stratigraphic Context

Said to date to 32–27 ka (L. Kuzelka, personal communication). Fossils found in debris that had fallen from a karst chimney above, at a depth of some 14 m (Jelinek and Orvanova, 1999).

Archaeological Context

cf. Szeletian (Jelinek and Orvanova, 1999). Sparse industry contains rounded Middle Paleolithic flakes plus part of a blade and some crude bone implements.

Previous Descriptions and Analyses

Mostly described in a series of papers by Vlcek (e.g., 1955, 1957); anatomically modern.

Morphology

Adult. Single large partial skull, formerly thought to belong to two individuals. Catalog number of NMP AP 40/70 2-15 includes some postcranials. Lacks most of R face, most of L side of skull, and basicranium; L and partial R zygomas present. RC^1–M^2, LC_1–M_2, and RI_2–M_2 present; all highly worn and uninformative. Mandible present; as reconstructed does not articulate well with mandibular fossae and is misaligned with maxilla.

Skull long, narrow, fairly gracile looking, but fairly thick boned. Parietal eminences well marked; maximum cranial width falls quite high. Frontal low but quite vertical, with distinct eminences. Glabellar "butterfly" shallow but broad and tall; lateral portion of superciliary arch moderately large, flat. L and R parts of frontal sinus very small; confined to area just behind nasion. Inferior lateral corner of orbit squared off. On the L, two zygomaticofacial foramina lie below inferior orbital margin. Nasal aperture narrow, relatively tall, with low, rough and horizontal conchal crest. Inferior margin of aperture rolled, with no distinct edge. Floor of nasal fossa was probably not depressed. Maxillary sinus extended forward to region of the C, very slightly invading frontal processes; also partially penetrated zygoma. Posterior root of zygomatic arch originates anterior to auditory meatus. Suprameatal crest low but crisp; is separated from stout and upwardly curving supramastoid crest. Mastoid process not greatly protruding; points forward and down. Mastoid notch broad and shallow; lacks digastric fossa behind. Paramastoid crest low, posteriorly placed; is separated by narrow groove (running from stylomastoid foramen) from large occipitomastoid crest. Bone missing medial to this.

Vaginal process originates quite anteriorly (carotid foramen impossible to identify); runs obliquely back and out to peak around what would have been a medium-sized styloid process; lateral to this it was broken but evidently did not reach edge of ectotympanic tube. Stylomastoid foramen lies just at base of where styloid process would have been. Parietomastoid suture very short. Lambdoid suture slopes up from asterion, peaks at lambda. Occiput broad, triangular; bulges below lambda. Occipital plane defined below by bow-shaped superior nuchal line with large, crescentic bulge in its center. Bone appears eroded above this, creating false "depression." Bilateral depressions below superior nuchal line quite deep; nuchal plane itself rather smooth although convex from side to side. No clear ridge at inferior margin of superior nuchal line. Highest nuchal line appears to be double.

Internally, some Pacchionian depressions lie anteriorly. Petrosal very broad, flat superiorly; region of arcuate eminence very flat but peaks slightly in middle and lateral to edge. Subarcuate fossa not completely closed over; bone of region slightly puffed out. No groove for superior petrous sinus. Cranial sutures segmented. Roots of all teeth short and slender.

Mandible deep, with tall mental trigon with symphysial keel. Digastric fossae moderate in size. Genial tubercles present. Mental foramen lies below P2. Gonial angle greater than 90 degrees; is not "cut off." External surface of ramus has modest masseteric markings; some small pterygoid roughenings internally, low on the gonial margin. Mylohyoid lines faint; downwardly oblique. Mandibular foramina incomplete. Sigmoid notch crest terminates laterally on condyle.

REFERENCES

Jelinek, J. and E. Orvanova. 1999. *Hominid Remains: An Update, Vol. 9: Czech and Slovak Republics.* Brussels, Royal Belgian Institute of Natural Sciences, pp. 1–118.

Vlcek, E. 1955. New finds of the Diluvial Man in the Bohemian Karst, Czechoslovakia. *Actes Cong. Panaf. De Prehist.* Paris: 783–789.

Vlcek, E. 1957. Pleistocenni clovek z jeskyne na Zlatem Koni u Koneprus. *Anthropozoikum* 6: 283–311.

Repository

Department of Anthropology, National Museum of Natural History, Vaclavske Nam, 11579 Praha, Czech Republic.

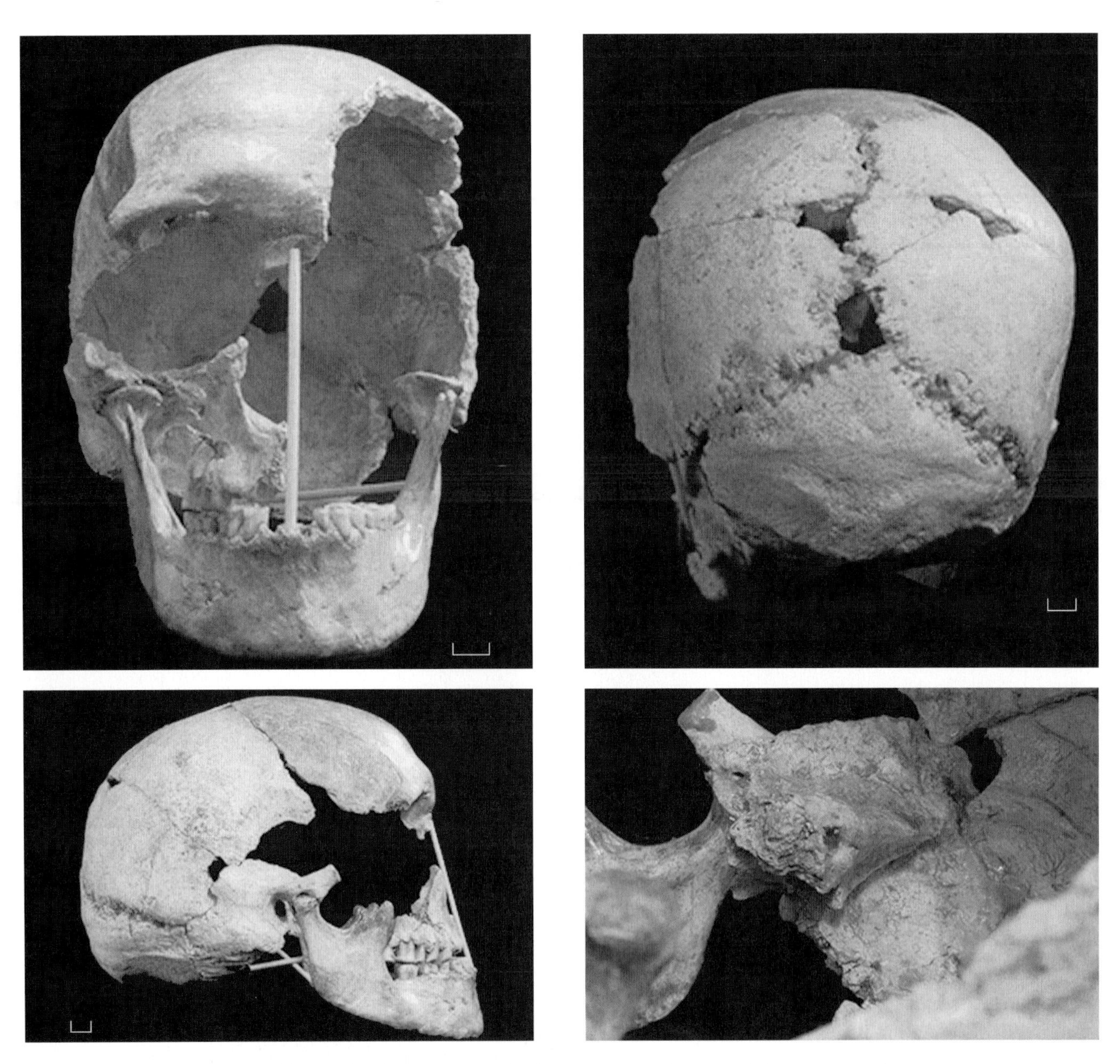

ZLATY KUN Figure 1. Including close-up of petrosal (scale = 1 cm).